Sensors VI

Technology, Systems and Applications

Sensors Series

Series Editor: **B E Jones**

Other books in the series

Current Advances in Sensors
Edited by B E Jones

Solid State Gas Sensors
Edited by P T Moseley and B C Tofield

Techniques and Mechanisms in Gas Sensing
Edited by P T Moseley, J O W Norris and D E Williams

Hall Effect Devices
R S Popović

Sensors: Technology, Systems and Applications
Edited by K T V Grattan

Thin Film Resistive Sensors
Edited by P Ciureanu and S Middlehoek

Biosensors: Microelectrochemical Devices
M Lambrechts and W Sansen

Sensors VI

Technology, Systems and Applications

Edited by

K T V Grattan
City University, London

and

A T Augousti
Kingston University

Institute of Physics Publishing
Bristol and Philadelphia

British Library Cataloguing-in-Publication Data

A catalogue record for this book is available from the British Library

ISBN: 0 7503 0316 6

Library of Congress Cataloging-in-Publication Data are available

Series Editor: **Professor B E Jones**, Brunel University

Published by Institute of Physics Publishing, wholly owned by The Institute of Physics, London

Institute of Physics Publishing, Techno House, Redcliffe Way, Bristol BS1 6NX, UK

US Editorial Office: Institute of Physics Publishing, The Public Ledger Building, Suite 1035, Independence Square, Philadelphia, PA 19106, USA

Printed in the UK by Galliard (Printers) Ltd, Great Yarmouth, Norfolk

Preface

In spite of the fact that *measurement* has been a preoccupation of civilized mankind since earliest times, both for trade and barter, it is only in the last few decades that the demand for new sensor technologies, to enable those measurements to be made, has been met with the rapid development of *new sensors and systems.*

Today, measurement systems are of immense importance in a wide range of industrial, domestic and similar environments, together with the sensors and transducers which underpin the use of the technology. They are key elements in the rapidly evolving field of *measurement and instrumentation* and particularly so with the incorporation of intelligence and microprocessors into sensor systems through VLSI technology. For example, solid state technologies combining micromachining open up new possibilities in terms of performance, size and power consumption. In addition, optical technologies continue to develop with fibre optics, and traditional sensing methods such as ultrasonics, electrochemistry and mechanical techniques are also flourishing. This, coupled with the use of sophisticated software packages to model and evaluate both individual transducers and sensor systems, has led to their incorporation in an ever increasing variety of applications. This requires that sensors must operate well in the field, and often under some of the most extreme conditions of temperature, pressure and humidity that are necessary for industrial processes and, for example, in the aerospace environment.

The need for accurate and reliable sensors to monitor the natural environment continues to be driven by legislation in the European Community, and on a wider international front, and there is still much work to be done to meet the demands of the users to offer new and better sensor methods.

This work illustrates some of the new developments in *sensors: technology, systems and applications* over recent years. It builds upon the first volume published two years ago, and documents the efforts of a wide variety of specialist authors, many international experts in their disciplines, from the UK, the EC and further afield. The volume records the Proceedings of the Sixth Conference on Sensors and their Applications which was held in Manchester, England, on 12–15th September, 1993. The Conference had returned to Manchester, the venue in 1983 for the first Conference in the series, and during the intervening time some of the most exciting aspects of sensor science, technology and their applications have been presented, including, in 1987, the first of the highly successful Eurosensors Conferences. The text is topical, and the material has been carefully refereed to ensure a consistent quality of this record of the Conference.

The editors are very grateful to the members of the Technical Programme Committee and the Conference Organizing Committee for their efforts in ensuring that the high standards set by previous Conferences have been maintained in the selection of material for this volume, and for their support for *Sensors and their Applications VI.*

K T V Grattan
City University, London

A T Augousti
Kingston University

Contents

Section A

GAS SENSORS

Advanced sensors for the gas industry

R D Pride
British Gas, Gas Research Centre,
Ashby Road, Loughborough, Leicestershire, LE11 3QU

ABSTRACT: British Gas supplies gas and provides equipment and service to the commercial and industrial sectors and over 17.5 million domestic customers. A wide variety of sensors is required in an industry that monitors and controls the flow of gas from exploration well to final customer and in many cases specialised sensors have been or are being developed to meet the specific demands of the industry. The paper provides an overview of some of these applications and the sensors, particularly for gas detection, that have been developed to meet the industries' needs.

1. INTRODUCTION

In order to carry out its business of finding, distributing and ensuring the safe, efficient use of natural gas by its customers, whilst ensuring minimal environmental impact, British Gas utilizes a wide range of information supplied to it from a diverse variety of sensors. This paper provides an overview of some of these sensors and their associated technologies and considers emerging new developments with the emphasis on gas detection. The paper is structured to review sensor applications from well-head to customer.

In an industry where safety is paramount, emphasis is always placed on well proven technology. New developments that often look interesting on the laboratory bench require exhaustive testing to ensure reliable long-term operation before being accepted for general use. It is often this aspect that results, firstly in very long lead times before real benefits can accrue to the industry from new sensor techniques, and secondly, in the high development costs and long delays in the realisation of profits for the sensor manufacturer. However, the size of the industry provides many opportunities for large scale adoption of any particular proven sensor technology, and further, a successful development for British Gas can now be exploited on a world wide basis through its Technology Transfer or Venture Capital programmes within the Global Gas business unit. There is no shortage of original ideas that warrant investigation, and to this end British Gas supports many university sensor research groups and centres of expertise throughout the country to explore novel sensing concepts.

2.0 THE GAS PRODUCTION PLATFORM

The traditional method for ensuring that there are no leaks in major gas installations is to monitor a large number of points using catalytic pellistor sensors. These well establised devices rely on low temperature ($\tilde{}500^{\circ}C$) catalytic oxidation of methane using palladium and thorium salts, are installed in a bridge circuit with compensating elements and provide a resistance change of typically 0.25% for a 1% change in the lower explosive limit or LEL. (For methane in air 1%LEL~500ppm). Unfortunately, these sensors are prone to long term drift in stability and also may be poisoned or inhibited by a number of gases and vapours. Particularly important in the gas industry is H_2S, which is to be found in significant quantities in some of the newer fields that are currently being developed, for example the Karachaganak field in Kazakhstan. In order to avoid false alarms these sensors are normally arranged so that a voting system identifies a dangerous situation. It is therefore normal policy to recalibrate sensors at 3 monthly intervals, which is an expensive operation since there may be several hundreds of units on any one platform. Intelligent sensors or sensors that could be tested without recourse to visiting each site would therefore offer financial benefit.

Open path IR sensors offer one possible route forward for both gases. The attenuation of a specific frequency IR beam that is preferentialy absorbed compared with a reference beam is a well established technique in closed systems and commercially available units for open path lengths of 200m and more are currently being evaluated. The method is not affected by chemical poisoning and removes the cost penalty of point installation which therefore potentially offers a lower cost testing procedure, but of course responds to total gas content in the path and not concentration, eg 0.1LELm=10%LEL over 1m. A novel development[1,2], originally by British Gas, Plessey and British Coal demonstrated the potential of methane detection using a single broadband tungsten halogen source and a low finesse scanning Fabry Perot etalon filter. The filter acts as a comb, having the same separation of transmission peaks as the methane absorption lines, resulting in a reduction of signal when methane is detected compared with the reference signal when the filter lines are not coincident; an approximate form of correlation spectroscopy. The receiving diode output is proportional to methane concentration. The development is currently being taken to a prototype stage by a commercial manufacturer using a recently devloped, potentially low cost, Fabry-Perot filter. More into the future British Gas is supporting developments such as OTIM (Optical Transform Image Modulation), a passive technique which uses an interferometer to detect a change in interference pattern produced by a gas in ambient light. It has already been shown to detect NO_2 and offers the further potential of volume detection, as opposed to just line of sight.

Another approach to provide distributed detection is to use optical fibre detectors, with their intrinsic safety advantages. In a LINK project, British Gas, in collaboration with Strathclyde University, British Telecom Labs, AEA Technology and Sieger LTD, have demonstrated that a 'D' shaped fibre provides an enhanced evanescent wave that is absorbed in the presence of methane[3]. Optical fibres transmit light through a core by total internal reflection at the boundary with a cladding layer; some light, referred to as the evanescent wave, travels outside the core/cladding interface and can be absorbed. If the cladding is removed and

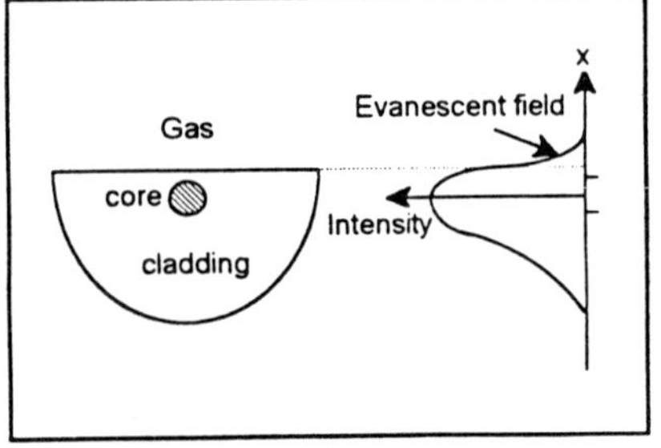

Figure 1
'D' Shaped Optical Fibre

replaced by air, part of the evanescent field energy will be lost if the D-fibre becomes surrounded by an absorbing gas, methane for example. In a first demonstration of the effect in a distributed optical fibre sensor, a 50% concentration of methane in air has been detected at a wavelength of 1.6μm, where methane has an IR absorbtion band. Modelling indicates that by adding a high refractive index cladding overlayer to the 'D' fibre, (see figure 1) the evanescent wave can be made to extend further beyond the fibre cladding into the surrounding gas medium, thus increasing the sensitivity by a factor of at least ten.

3.0 THE RECEPTION TERMINAL & PROCESSING

When gas arrives at the terminal it requires processing to bring it into specification and odourising before distribution in the national transmission network. Different fields provide gas of different composition (table 1) and some mixing of the various gases in the distribution system ensures that gas of an acceptable calorific value and Wobbe Number is finally sent out. Natural gas is a complex mix of gases and any new developments in gas sensing must consider cross specificity between gas species. Gas chromatographs are used to sample the gas composition and gas calorimeters that burn the gas to determine calorific value have now been superseded by interpolation from equations of state, composition and density measurements. Only limited numbers of very expensive instruments have been required in the past but with the move to common carriage there is an increasing need for more detailed monitoring and the opportunity exists for the development of new lower cost measurement techniques. For example, the concept of using mass production micromachining techniques to develop a gas chromatograph on a silicon chip has been reported in the past[4]. However, such developments are likely to be more appropriate to field monitoring in the distribution system, discussed in section 5. The statutory limit concentration for H_2S is 3.3ppm and is traditionally monitored using an optical measurement of lead acetate paper. The formulation of the odourant composition, shown in table 2, ensures that gas is easily recognisable, but there is also a need for lower cost methods of measuring sulphur based components in the transmission system.

A further addition to the gas is the inclusion of monoethyleneglycol or MEG in concentrations of up to 300mg/m^3 via large 'Foggers'. This is required in order to maintain the swelling of packing material used in pipe joints on the older cast iron mains system. British Gas has developed what is believed to be the first industrial application of a biosensor, to detect the concentration of MEG in natural gas. A biosensor requires a suitable enzyme that will react with the MEG and produce a chemical that can be easily measured. One of the novel features of this system was the identification and isolation of a suitable enzyme 'dehydrogenase' from soil samples taken from areas where MEG was known to have been

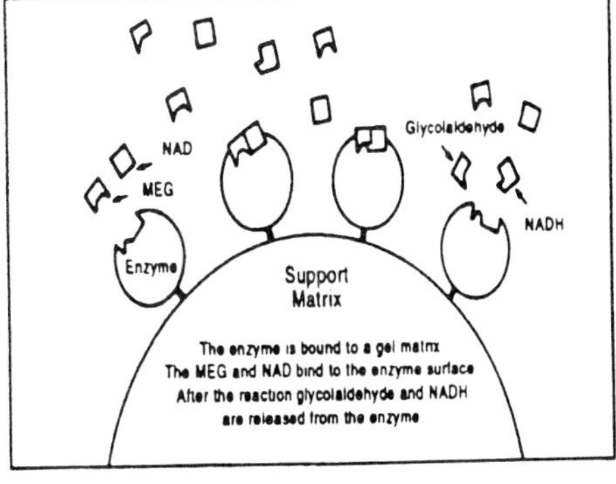

Figure 2
The Immobilised Enzyme Reaction

spilled. In order for the enzymes to catalyse the MEG, a cofactor called Nicotinamide Adenine Dinucleotide (NAD) is also required. The reaction releases hydrogen which bonds to the NAD producing NADH (figure 2) and an electrochemical cell is then used to re-oxidise the NADH and provide an electrical signal proportional to its concentration. This sensor system is now deployed in the British Gas regions to ensure acceptable levels of MEG are maintained.

The flow of gas into the transmission system is generally monitored by orifice plate and differential pressure cell but there are also some 4 multi-path ultrasonic gas meters undergoing evaluation. These measurements form part of the fiscal accounting system.

Component	Analysis/volume %		
	Leman	Frigg	Sleipner
Methane	94.81	95.72	80.77
Nitrogen	1.22	.53	.50
Carbon Dioxide	.04	.31	6.12
Ethane	3.00	3.37	7.34
Propane	.55	.03	3.17
n-Butane	.10	.005	.74
Isobutane	.09	.008	.42
Pentanes	.06	.0008	.43
Hexanes	.03	.005	.24
Heptanes	.02	.004	.18
Octanes	.01	.005	.07
Nonanes	.00	.01	.02
Benzene	.03	-	-
Toluene	.01	-	-
Helium	.03	.003	-
CV/MJ per cubic meter	38.84	38.58	-

Table 1
Gas Compositions of 3 Fields

GAS ODORANT (BE)		
COMPOSITION:	3	SULPHUR BASED COMPONENTS
C_2H_5SH	6%	ETHYL MERCAPTAN
$(CH_3)_3CSH$	22%	TERTIARY BUTYL MERCAPTAN
$(C_2H_5)_2S$	72%	DIETHYL SULPHIDE
CONCENTRATION:		$16mg/M^3$ OR 4.5ppm

Table 2
Gas Odourant Composition

MAJOR		MINOR	
FLUE GASES	TYPICAL VALUES %	FLUE GASES	TYPICAL VALUESppm
N_2	70	NOx	100
H_2O	15	CO	50
CO2	7	H_2	25
O_2	7	N_2O	2
		SOX	1
		CH_2O	<1

TABLE 3
Flue Gas Constituents*

* Non-condensing boiler, 200°C, Natural Gas, 25% excess air.

4.0 TRANSMISSION & STORAGE

Gas is transmitted from the terminals to the regions through a high pressure (~75bar) National Transmission System in the UK by compressors, powered by gas turbines such as the RB211 and Avon. The status of these compressors is continuously monitored at each local site, employing a substantial number of sensors, possibly upwards of 300 sensors per compressor, to monitor parameters such as temperature and vibration. There are some 180 stations located around the country which monitor information on the gas pressure which is transmitted back to the main grid co-ordination centre at Hinckley. Here some 500 pressures (mainly strain gauge), 150 temperatures (mainly Pt. resistance), and flow measurements from over 100 sites are continuously monitored in order to make decisions on the throughput of the grid. Flow and pressure measurements of gas both into and out of the transmission system provide a measure of any possible leakage.

Gas is stored by several methods in the grid, including changes in the transmission system pressure, storage in exhausted gas fields and storage of liquified natural gas (LNG). There are some twenty LNG tanks in the system with a combined output of 90 million cubic meters. A major concern for liquified storage is the detection of any leakage and this has been traditionally ensured by arrays of several hundreds of thermocouples placed around the base of the storage tanks. Two British Gas developments providing fully distributed, intrinsically safe, sensing are now successfully operating at storage sites. The first relies on the property of optical glass fibres where the ratio of core to cladding refractive index changes significantly when the temperature of the fibre drops due to liquid gas impinging on the surface of the fibre. Below -55°C this results in loss of light, supplied by an LED, from the core into the cladding and hence loss of signal at the fibre end which is used to trigger an alarm output when signal attenuation exceeds 1.5dB[1]. This system, which has been in use since the mid-80s, will support several kilometers of fibre, encased in stainless steel for protection, and has been shown to have a MTBF of 14 years. In an alternative prototype system, limited to less than 50m at present, a polyethylene tube is fed with a low flow rate of CO_2. Exposure of the tube to LNG causes the CO_2 to solidify in the tube and the resulting pressure drop is detected by a simple pressure switch (figure 3).

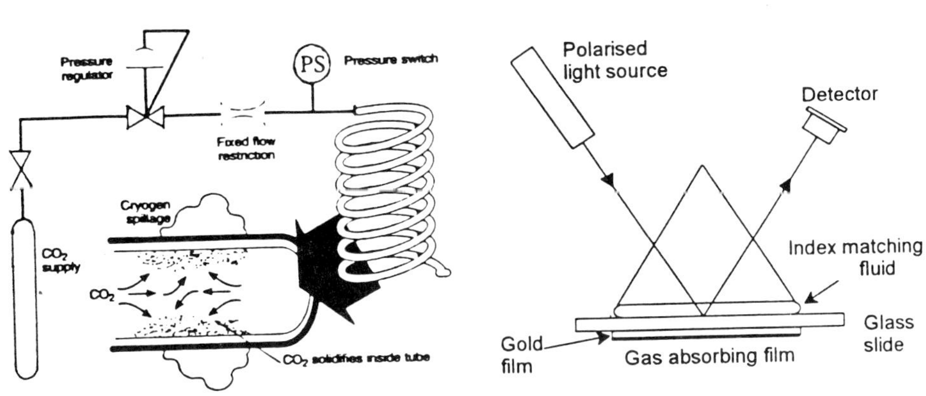

Figure 3
Cryogenic Leak Detector

Figure 4
SPR Configuration

5.0 DISTRIBUTION

The regional distribution system operates around 10bar and again the main measurement is pressure, with some 1500 points being monitored by telemetry systems back to the 12 regional headquarters. There are some 30,000 governor sub-stations where the pressure is further reduced to the 20-30 millibar level, but many of these are monitored by local systems only.

The main tool for monitoring gas leakage on the distribution system is the Gascoseeker. This instrument was developed by British Gas and employs pellistors in a bridge network to measure gas concentrations up to 5% by volume (100%LEL) and further thermal conductivity (katharometer) sensors providing measurement to 100% gas in air. This is a well established instrument and in the Mark 2 version there is the added advantage of a fully computerised calibration system.

A new optical technique, employing the principle of Surface Plasmon Resonance (SPR), offering the potential of a low cost hand held instrument for the measurement of MEG 'on the district', has also been developed by British Gas[5]. SPR involves the interaction of the electric component of light with surface plasmons (quantised charge oscillations) in a thin metal film. If the frequency and wave vector of the light match those of the surface plasmons then light is absorbed. In order for SPR to occur light is coupled into the thin film using a prism, see figure 4, and a plot of angle of incidence against reflectance shows a minimum at the position of resonance. This minimum is dependent on the dielectric constant of the material in contact with the exposed face of the metal film, and if this is affected by the absorption of a gas then a change in the position of the minimum may be observed. The search for suitable materials has shown that polypyrrole is suitable for MEG in the range of interest and the optical bench-top configuration required for measurement has now been reduced in size into a prototype hand held instrument.

6.0 UTILISATION

6.1 Domestic

Most modern gas appliances incorporate safety devices such as flame failure or vitiation sensors to prevent the flow of unburnt gas, but the leakage of gas is of concern to all, and the inclusion of the odourant coupled with the efficient biological sensor with which we are all equipped, provides one of the best methods of leak detection down to below 2%LEL. However there are instances where the human sense of smell can fail and to address this situation in particular, many different gas detectors have been evaluated by British Gas over the years and this research has contributed to the formulation of a new British Standard 7348 for domestic gas detectors. For domestic applications where long term reliability is essential, manufacturers have been continuously improving their products, particularly with regard to semi-conducting oxide sensor technology, and today there are two products approved to the standard. The response of a semi-conductor sensor is non-linear, but changes in resistance are more than an order of magnitude greater than for pellistor sensors below 20%LEL. Long term stability has been assessed in special test rigs, and from field trials of samples in staff houses figure 5 shows typical results. Unfortunately, to date, semi-conductor gas sensors require mains power operation and this places two constraints on domestic gas alarms. The first is the basic unit cost and the second is the cost of installation.

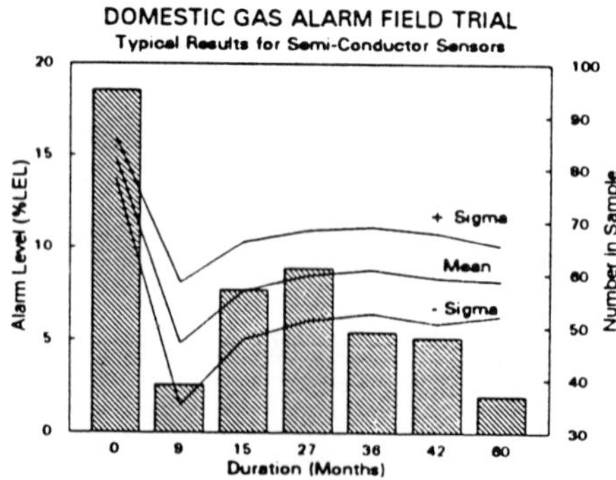

<u>Figure 5</u>

Domestic Gas Alarm Field Trial

The BS recommends the use of an unswitched fused outlet - this then avoids the possibility of an activated alarm being switched off and causing a potential explosion hazard from a mains generated spark. In order to overcome these problems low power consumption sensor technologies need to be explored with typical contenders being piezo electric, Surface Acoustic Wave, SAW, and various optical techniques such as SPR and optical reflectance. However, the overriding problem is one of material selection in order to achieve a unique methane response.

99% of all uncontrolled gas leaks reported to British Gas are attended to within one hour. The traditional approach is to use soap solution sprayed or painted onto the suspect pipework, but an alternative method is to use a leak tracer incorporating a gas sensor. The Gascoseeker, referred to earlier, utilizes a pellistor sensor and is an accurate, regularly recalibrated instrument. However, it is used regularly in domestic environments where a range of poisons that affect pellistors are known to exist. A review of available devices that use semi-conductor sensors has been undertaken and they were generally found to suffer from a range of deficiencies - not least being high cost, compared with soap solution. Certain features were considered essential, such as the inclusion of a sample pump, and a full specification for an instrument resulted in the field testing of some 60 prototype instruments (figure 6) over a period of several months. This work demonstrated the reliability of the sensors and also areas where further improvements could be made.

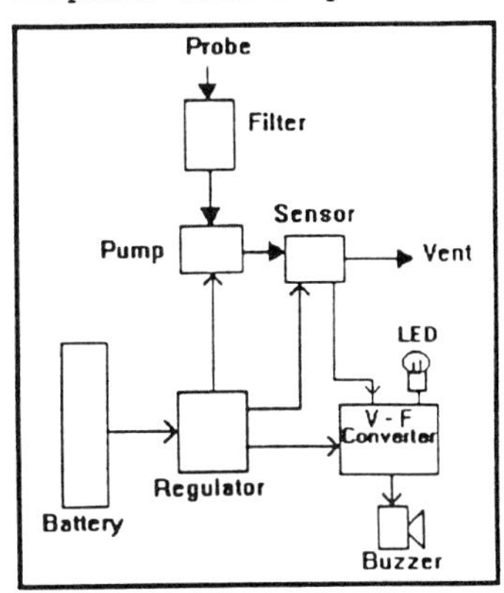

Figure 6
Prototype Gas Leak Tracer

Another major concern is the satisfactory combustion performance of gas appliances. Following substantial field measurements, the technique of monitoring the ratio of CO to CO_2 in the combustion products of natural gas has been shown to provide a suitable measure of whether a gas boiler requires dismantling and cleaning[6]. The sensors employed to do this are electro-chemical cells for CO and O_2, and the instrument computes the CO_2 and the CO/CO_2 ratio. There are currently 10,000 of these instruments performing satisfactorily in the field, but there is room for future improvements. A detailed study of cell performance has shown that the CO cell suffers from a gradual loss in sensitivity over the initial 6 months of the two year cell life (figure 7) and therefore recalibration at intervals is required. Additionally, the cost of cells is also a major factor, with the O_2 cell currently employed only having a one year life. With new fabrication methods, smaller, lower cost CO cells, cells with reduced H_2 cross-sensitivity, and O_2 cells with extended lives of up to 5 years are now becoming available. The need to monitor further combustion products such as NO_2 is an area that may be influenced by future legislation.

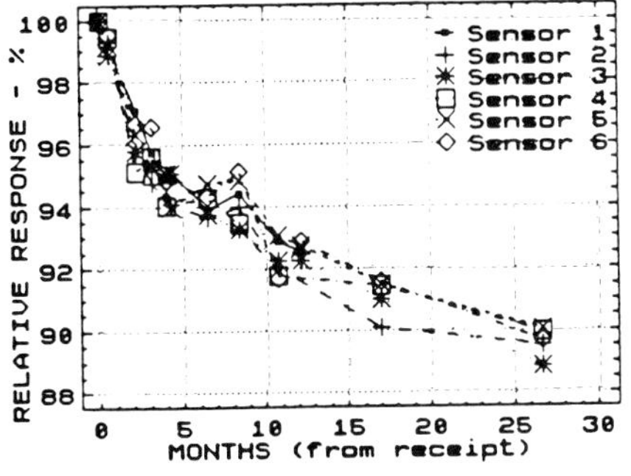

Figure 7
CO Cell Sensitivity Loss

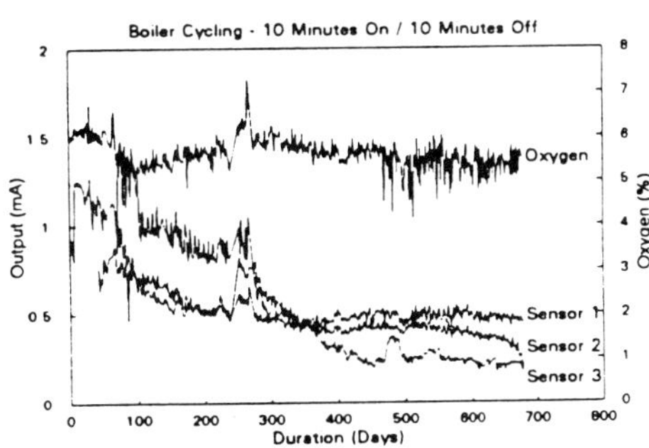

Figure 8
O_2 Sensor Long Term Test Results

Areas of appliance control that are currently under development for domestic equipment require very low cost reliable sensors for the detection of combustion products, either in the flue way or as safety devices mounted within an appliance casing. Further, work on diagnostic systems for domestic appliances could well include in the future, additional sensors, for example for NOx, H_2, CO or formaldehyde. These requirements may be met in the long term by the development of zirconia based amperometric sensors for flue gas O_2[7], or semi-conducting oxide based sensors for O_2 and other flue gases. Flue gas is an aggressive, wet environment, see table 3, and cross-sensitivity of sensors between gases and long term integrity of sensors are important aspects. Results shown in figure 8 from long term tests on O_2 sensors, developed at Middlesex University under a British Gas contract, suggest that at least five years reliable operation is already possible.

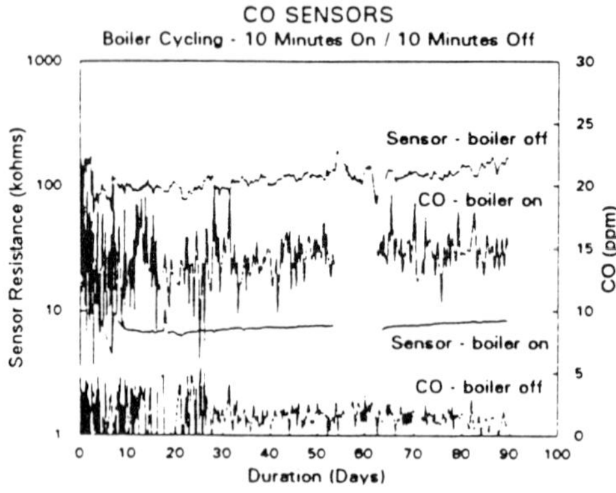

Figure 9
CO Sensor Life Test Results

British Gas is collaborating in several programmes on semi-conducting sensors. With CERAM Research, funded by the DTI/LINK scheme, an investigation into the response of a wide range of ceramics to toxic gases is establishing material responses when fabricated as simple pellets. A DTI/LINK scheme with Swansea University and Morgan Materials Technology Ltd is developing pre-production type sensors for CO and H_2, based on screen printing techniques and materials originally developed at Swansea. Some of these sensors are already undergoing long term evaluation on boiler flue test rigs similar to those previously reported for O_2 sensors. A preliminary typical plot of response to an on/off cycling boiler is shown in figure 9.

A further programme on semi-conducting sensors is funded under BRITE/EURAM and is aimed at developing a wide range of both flammable and toxic gas sensors produced by both thick and thin film techniques. The main British Gas role is to provide application specifications, measure basic responses to gases - particularly wet gases, in an automated bench-top test facility shown in figure 10, and to undertake long term environmental testing.

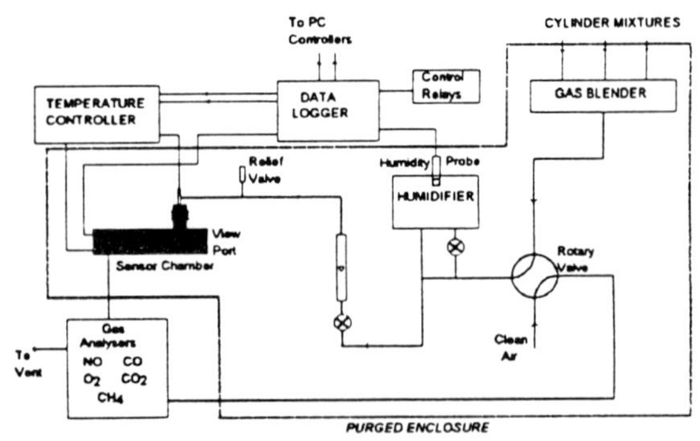

Figure 10
Automated Testing Of Sensors

6.2 Industrial/Commercial

The control of air/gas ratio in industrial or commercial burners is important in maximising efficiency, minimising flue gas emissions and maintaining product quality. The most advanced systems monitor and control the ratio through measurement of air and gas flows. One product, manufactured under licence from British Gas, uses low cost thermistors as a form of hot wire anenometer. Other industrial gas flow sensors have been developed for control, commissioning and load monitoring. Examples include a miniature averaging pitot 'Microprobe', now manufactured under license, that utilises a micro-engineered flow sensor, and a novel flow meter based on a fluidic oscillator[8].

Oxygen trim systems have been available for large boilers for several years. These are designed to maintain a boiler at maximum efficiency while compensating for any change in atmospheric conditions or gas quality through measurement and control of the oxygen in the flue gases. Furnaces may also incorporate oxygen measurement for the control of efficiency, but in addition to monitor and control the combustion atmosphere, which can often determine the quality of the product. Advances in oxygen sensors, especially in the automotive market where precise air/fuel ratio is required for the correct operation of a catalytic converter, has lead to low cost and robust sensors being made available. British Gas has developed several instruments based on these new sensors, and hopes through their use, to extend the range of plant for which such controls are economically viable.

In some specialised industrial processes combustion is deliberately maintained at or below stoichiometric in order to subject the product to a reducing atmosphere. Under these conditions CO is a suitable control parameter and the development of low cost, high temperature sensors suitable for furnace conditions, is a priority. To meet this need British Gas is collaborating with Keele University on a DTI/ETIS funded project. Another industrial gas fired process that would benefit from sensor development is drying. Pottery, paper and wood drying all take place at temperatures beyond the range of most commercially available humidity sensors. The development of a low cost, high temperature humidity sensor would open the way for the control of humidity and an improvement in the efficiency of these processes.

Natural Gas Vehicles (NGVs) are being developed by British Gas, initially for fleet use and possibly in the longer term for the private car user. NGVs are already in use in some countries, such as Canada, Italy and New Zealand but opportunities for sensors exist for control systems and safety monitoring both on the vehicle and at filling stations.

7.0 ENVIRONMENTAL MONITORING

Instrumentation for environmental measurements is available to detect either a wide range of gases, such as gas chromatographs, or may be developed to detect a discrete range of gases of particular interest and at the right price. On large sites and for the broader environmental and safety applications British Gas has employed two techniques. The first is referred to as an Area Survey Vehicle (ASV) and includes various gas analysis systems for total hydrocarbon and methane analysis. These coupled with a global satellite positioning system, and sophisticated computer models for predicting plume formation, are used to ensure that British Gas activities have minimal environmental effect. The second technique is the use of its own mobile LIDAR system[9] which it has used for many years as a research tool in the study of the atmospheric dispersion of gas releases. High energy pulses of UV from a xenon chloride excimer laser are scattered with a frequency shift characteristic of the molecules in the path. The intensity of the Raman scattered light is proportional to concentration, and range is calculated from the timing of the propagation pulse, enabling 1-20% methane over 50-1000m to be detected.

However, for routine monitoring, and where a discrete set of gases is expected to be monitored, lower cost solutions have been sought. One example was the development of a demonstration 4 sensor instrument by Kent University under a British Gas contract. This comprised a tin dioxide single crystal sensor for CO, a Pb phythalocyanine semi-conducting polymer for NO_2, an electro-chemical cell for O_2 and a standard Taguchi sensor for CH_4. The novelty of this development was the calculation of concentration by determining the rate of change of response of the polymer to NO_2. Kent has also been exploring the SPR technique discussed earlier for NO_2 detection.

Sensors that operate at or near room temperature offer the opportunity of long life battery operation and hence reduced cost portable equipment. Several other universities have been supported by scholarships in the development of such novel sensors, and of particular note is the work at Durham University on resistance measurements of conducting polymers deposited using Langmuir Blodgett, LB, evaporation and spin coating techniques for NO_2, H_2S and SO_2. Their work has shown that doping of polyaniline, although giving low resistance films, results in long term instability due to the loss of dopant. The use of de-protonated polyaniline, although resulting in high resistance materials, leads to more stable sensors.

Also the work at Coventry University on optical reflectance measurements at specific wavelengths of LB deposited organic films for NO_2 is yielding interesting results. At the last Sensors & their Applications conference[10] they reported good sensitivity down to 1ppm for phthalocyanine films, but also a slow response. With new materials and techniques they are now able to improve response times to tens of seconds.

The concept of the electronic nose, developed at Warwick University, for gas and vapour detection, is now well known and the application of statistical techniques and neural networks to provide sophisticated pattern recognition of responses from many sensors is under investigation at a number of centres, UMIST for example. However, for the generally small gas molecules that British Gas is mainly interested in, a reasonable response of the sensor to the gas is an essential starting point, thus emphasising the underlying materials chemistry problem.

8.0 CONCLUSION

The paper has tried to demonstrate that a diverse range of sensor techniques is employed in gas detection to meet the wide variety and exacting needs of the gas industry. In practice sensors do not have simple individual gases to detect but complex mixtures from which specific species have to be identified. Emphasis has been placed on the reliability of sensor systems and the continuing search for improved techniques that offer features such as lower cost, improved specificity, intrinsic safety and distributed sensing.

9.0 REFERENCES

1. D Pinchbeck, Optical fibres in the gas industry. Gas Engineering & Management, Nov.91.

2. J Dakin, C Wade, D Pinchbeck, J Wykes. A novel optical fibre methane sensor. Int. J. of Opt.Sensors, Vol.2, Aug.87

3. B Culshaw, F Muhammad, G Stewart, D Pinchbeck, et al. Evanescent wave methane detection using optical fibres. Electron. Lett.,28,2232. 1992

4. S Terry et al. A gas chromatograph air analyser fabricated on a silicon wafer. IEEE Trans. on Electron Devices. Vol.ED26-pp204, 79

5. S Peacock, H Jaggers. Advanced sensors based on surface plasmon resonance. 1992 Int. Gas Research Conf., USA.

6. J Newcombe, M Price, A Sussex. Servicing-right first time. 126th AGM of the IGE, June 89.

7. K Hargreaves, N Ovenden, R Sauba & J Cotton. The application and performance of combustion sensors in domestic gas boiler control systems. Sensors, Technology, Systems and Applications. P113, Adam Hilger, 91.

8. D Churchill. Developing a flow sensor for industrial applications. Relay, Feb.91. BG Publications Midlands Research Station, Solihull.

9. R Bilbe,S Bullman & F Swaffield. An improved Raman LIDAR system for the remote measurement of natural gas releases into the atmosphere. Meas. Sci. Technol.1 495-499 1990.

10. L Miller, A Newton, C Sykesud & D Walton. Optical gas sensing using Langmuir-Blodgett films. Sensors, Technology, Systems and Applications. P139, Adam Hilger, 91.

ACKNOWLEDGEMENT

The author would like to thank British Gas for permission to publish the paper and to all his colleagues both within British Gas and at Universities who have provided helpful assistance in its preparation.

A surface plasmon resonance gas sensor using crown-ether substituted phthalocyanine films

S. J. Peacock, V. Rivalle and J. D. Wright
Centre for Materials Research, Chemical Laboratory, University of Kent, Canterbury, Kent. CT2 7NH
and
H. C. Jaggers
British Gas plc, Gas Research Centre, Ashby Road, Loughborough, Leicestershire. LE11 3QU

ABSTRACT: A small hand held sensor for nitrogen dioxide based on the technique of surface plasmon resonance (SPR) has been developed. A range of phthalocyanine compounds based on crown ethers were used as the sensing films. These compounds responded to nitrogen dioxide at ambient temperatures and sensitivities of several orders of magnitude greater than that hitherto obtained in SPR experiments are reported. The application and operation of these materials is discussed.

1. INTRODUCTION

Surface plasmon resonance (SPR) is potentially a very sensitive technique for gas sensing and several applications have recently been reported (1,2). The technique has an advantage over other gas sensing methods as it can operate at room temperature and optical fibres can be used for the light input and output. This means that both low power consumption and intrinsic safety can be readily achieved. For SPR to be successful, sensing films that can be laid down easily and uniformly need to be developed. In addition such films need to be active at or near ambient temperatures.

The availability of suitable films has been severely limited and this has slowed the progress of SPR as a sensing technique. However, recently it has been shown that films of crown-ether-substituted phthalocyanines will respond to nitrogen dioxide at room temperatures in conductometric experiments (3). Also these films can be spin coated onto thin metal surfaces making them ideal substrates for SPR studies (4).

In this paper we report the development of a hand held sensor and demonstrate that SPR is capable of sensing nitrogen dioxide at sub ppm levels using crown ether phthalocyanine films.

2. EXPERIMENTAL

An apparatus based on the Kretschmann prism configuration described previously (5) was used in all of the laboratory experiments.

For optimum response of the SPR effect it is essential that reproducible metal films are deposited. Both gold and silver films produce the SPR effect in the visible part of the spectrum with silver giving the sharper and hence more sensitive response. However, untreated silver will tarnish in the presence of nitrogen dioxide and although a thin coating (about 1 nm) of nickel will prevent this (6) we used gold films only in this work. These films showed no change in SPR even when exposed to high concentrations of nitrogen dioxide. Theoretical calculations show that an optimum thickness for gold to exhibit SPR is 40nm and this thickness was used in all our experiments.

Phthalocyanines tetra substituted with 18-crown-6, 15-crown-5 and 21-crown-7 groups were used as sensing films. Work was also carried out with a 15-crown-5 phthalocyanine in which one oxygen in each crown was replaced with N-$COCH_3$. This compound is subsequently referred to as a B_2 phthalocyanine. All films were spun from a chloroform solution using the techniques described by Sambles and Wright (4) onto gold deposited on microscope slides.

3. DEVELOPMENT OF A PROTOTYPE SENSOR

That a convergent beam approach can be used to detect the position of SPR is well documented with the first experimental set up reported by Kretschmann (7). This method has the advantage that no moving parts are employed and miniaturisation is possible, thus making it ideal as the basis of a small hand held sensor. The prototype sensor is shown in figure 1 and utilises a visible diode laser module as the light source which is both collimated and polarised.

Cylindrical lenses first expand and then contract the beam so that a sheet of light is focussed onto the gold. The SPR is seen as the dark absorption line in the fan of light being reflected from the gold film. Gas or vapour interaction with the film is seen as a shift in the position of absorption and shallowing of the SPR is seen as the absorption line becoming less distinct. For all the SPR information to be recorded the whole of the SPR curve needs to be monitored with a detector such as a linear CCD array but by using a single photodiode coupled to a pinhole or as in this instrument a drawn out optical fibre a simple and inexpensive instrument can be designed.

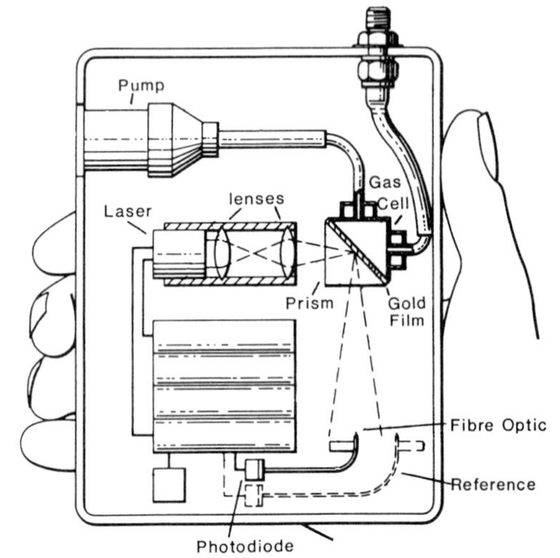

Fig. 1. *The prototype sensor*

The success of the sensor depends on selecting the optimum sized pinhole such that only a small portion of the SPR output on or close to the minimum is sampled. For clean air at this point little or no light impinges on the pinhole but if a shift is observed on exposure to gas or vapour a corresponding increase in light intensity is seen as the position of minimum moves away from the sampling point. Previous work (8) has shown the validity of this approach for sensing monoethylene glycol with polypyrrole films. However, provided that the SPR is well characterised over the range of conditions being investigated then the method should be applicable to a wide range of sensing applications.

4. SPR SENSING OF NITROGEN DIOXIDE

Initially all four phthalocyanines were exposed to 100ppm of nitrogen dioxide over a time period of 180 minutes. All the resultant SPR curves showed marked changes compared to the curve for air alone. A typical set of curves (for the 18-6 crown) is shown in figure 2.

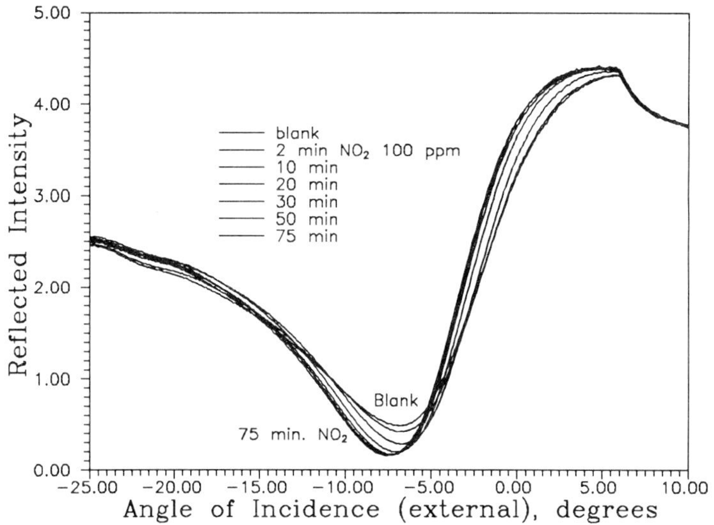

Fig. 2. *SPR plots for the 18-6 crown exposed to nitrogen dioxide over a 100 minute period in order of exposure time.*

An interesting feature is the deepening of the resonance on exposure to nitrogen dioxide. This was similar for all the phthalocyanines. In the very early stages a shift towards the critical angle is seen but as the reaction proceeds a shift away and towards higher angles occurs. Experiments were then carried out at lower concentrations to determine the limit of detection. Figure 3 shows the results for 1ppm absorption of nitrogen dioxide over a 60 minute period.

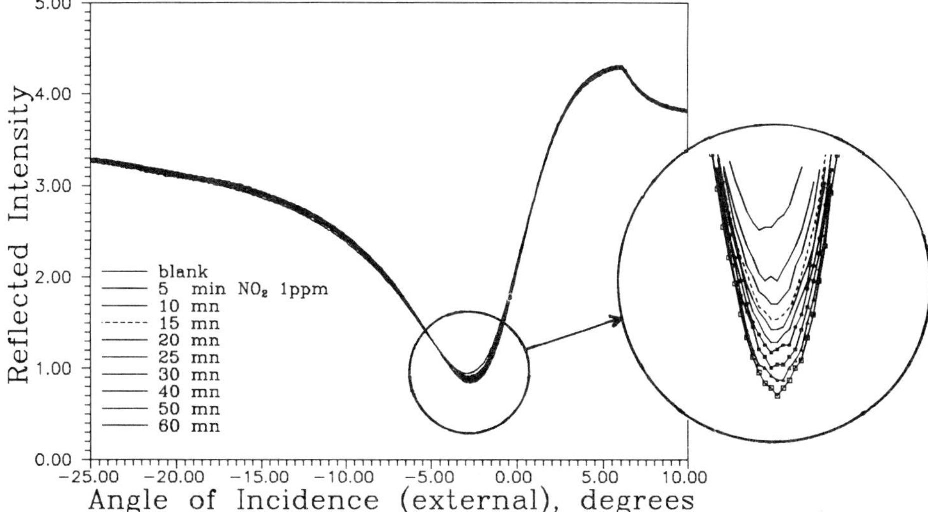

Fig. 3. *SPR curves for 1ppm nitrogen dioxide absorption on the 18-6 crown in order of exposure time.*

Subsequent experiments showed the limit of detection to be 0.1 ppm but these results were a little noisy. This result is, however, several orders of magnitude better than that reported for any other phthalocyanine used in a SPR experiment and is sufficiently large to be detected by the prototype instrument.

A potential problem with all these films is the time take for the system to reach equilibrium. This is common with most phthalocyanines and is observed in conductivity experiments (9). One way to overcome the problem is to predict where equilibrium would occur electronically by carrying out the experiment under very well defined conditions. Whilst such an approach works well for conductivity experiments (9) more work is needed before it could be applied here. An alternative approach is to pretreat the films in some way. Lead phthalocyanine, for example, can be heat treated to increase reversibility (10). Treatment of films of the crowned phthalocyanines with aqueous potassium chloride has been found to give a much faster response and reversal (3) but this is impractical for the thin films used here.

The effect of alternating nitrogen dioxide and air on the depth of the SPR curve for the 21-7 crown is shown in figure 4. Although the change from air to nitrogen dioxide is constant with time there is a gradual decline in response for the change in depth of the SPR. This suggests that equilibrium is not being reached in the timescale of the experiment. However, if the films were saturated overnight in an atmosphere of nitrogen dioxide then they were found to reach equilibrium more rapidly and figure 4 shows the effect of saturation with nitrogen dioxide on the response of the 15-5 and B2 crowns to 30 ppm of nitrogen dioxide.

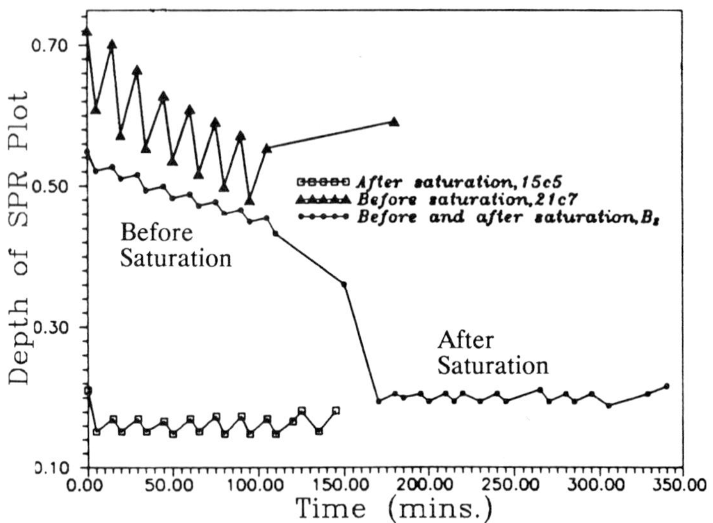

Fig. 4. *Comparison of three of the crowns showing the effect of reversibility on the 15-5, 21-7 and B2 crowns and the effect of saturation on the 15-5 and B2 compounds. Alternate points represent the switching between air and nitrogen dioxide (30 ppm) on the depth of the SPR curve.*

5. DISCUSSION

All crowns show similar trends which are somewhat unusual as they show a deepening of the SPR curve on exposure to gas rather than the more usual shallowing. From these experiments the order of sensitivity towards nitrogen oxide appears to be 21-7, 18-6, 15-5 and B2 with the 21-7 being the most sensitive. However, from figure 4 it can be seen that the reversibility of the 15-5 and B2 crowns is better than the 21-7 particularly after these two films have been saturated with nitrogen dioxide overnight although saturation experiments need to be carried out with the 21-7 before this observation can be verified. Saturation also helps to overcome the problem of the length of time to reach equilibrium with the saturated films readily reaching equilibrium within a few minutes.

Usually the shift of the SPR curve is towards higher angles coupled with a shallowing. This can be attributed to either an increase in the film thickness or an increase in the refractive index of the film (or both). In our experiments we see an increase in the depth of the SPR curve suggesting a decrease in the imaginary component in the permittivity. If this is the case then a reduction in the optical absorption on exposure to nitrogen dioxide should be seen at the measuring wavelength. Figure 5 shows the spectrum for the 18-6 crown before and after exposure to nitrogen dioxide and confirms at least qualitatively that such change is occurring.

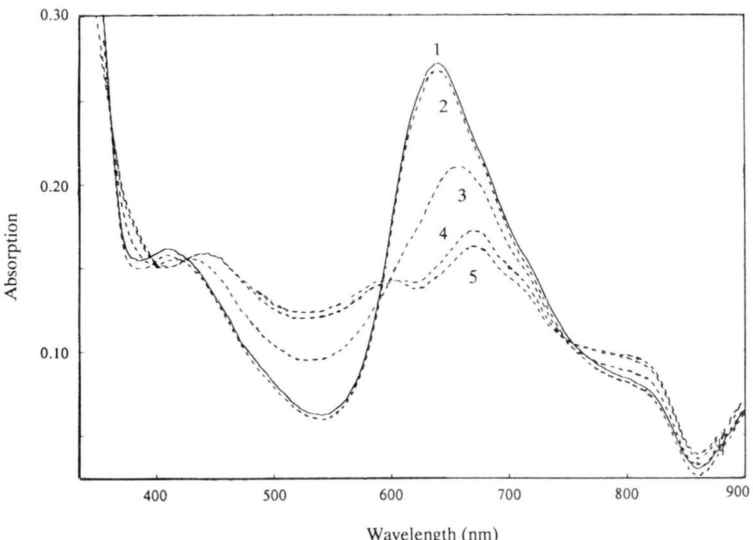

Fig. 5. *Absorption spectrum for the 18-6 crowned phthalocyanine (1) and after increasing exposure to nitrogen dioxide (2-5).*

For a conventional copper phthalocyanine Petty and his group (6) showed that although they saw a similar change in the absorption spectrum, the SPR showed a shallowing. They concluded that mechanism was made up of at least two stages but could only speculate on their nature. However, in their work they were using nitrogen dioxide concentrations of 300ppm and such high concentrations are probably influencing the reaction mechanism.

There are several features in this work that suggest that at least two effects are taking place. During the initial stages of absorption the SPR moves to the right before it moves to the left. Also when reversal of the absorption takes place the SPR does not return to its original position, although saturation can produce a constant base line. Physically the mechanism of the SPR changes can be modelled but more work is needed before a detailed mechanism for the chemical and absorption processes can be proposed.

6. CONCLUSIONS

SPR is a sensitive technique for gas sensing and a simple hand held prototype can be built. This prototype was initially tested with monoethylene glycol vapour but has application to other gases and vapours as well including nitrogen dioxide. For this gas a group of crown ether substituted phthalocyanines have been identified as suitable sensing substrates with a 21-7 crown apparently giving the greatest sensitivity. However, after pretreatment by saturation with nitrogen dioxide overnight the 15-5 and B2 crown gave repeatable results and may be more suitable as sensing films. More work is necessary before a mechanism for the process can be determined.

ACKNOWLEDGEMENT

We thank Professor R. J. M. Nolte and Professor Ö. Bekâroğlu for supplying the substituted phthalocyanines.

REFERENCES

1. P. S. Vukusic, G. P. Bryan-Brown and J. R. Sambles, *Sensors and Actuators B*, **8**, (1992), 155.
2. H. C. Jaggers, S. J. Peacock and J. S. Shaw, *SENSORS, Technology, Systems and Applications ed K T V Grattan,* Adam Hilger Series on Sensors, (1991), 133.
3. P. Roisin, J. D. Wright, R. J. M. Nolte, O. E. Sieken and S. C. Thorpe, *J. Mater. Chem.*, **2**,(1992), 131.
4. P. S. Vukusic, J. R. Sambles and J. D. Wright, *J. Mater. Chem.*, **2**, (1992), 1105.
5. H. C. Jaggers, S. J. Peacock and J. S. Shaw, *Anal. Proceedings*, **28**, (1991), 341.
6. D. G. Zhu, M. C. Petty and M. Harris, *Sensors and Actuators B*, **2**, (1990), 265.
7. E. Kretschmann, *Opt. Commun.*, **26**, (1978), 41.
8. H. C. Jaggers and S. J. Peacock, *Proceedings Int. Gas Research Conf.*, (1992), 784, Orlando, Florida.
9. A. V. Chadwick, A. Wilson and J. D. Wright, *Sensors and Actuators B*, **4**, (1992), 499.
10. T. A. Jones, B. Bott and S. C. Thorpe, *Sensors and Actuators B*, **17**, (1989), 476.

A novel miniature zirconia gas sensor with pseudo-reference. Part 2: combined amperometric and potentiometric operation for monitoring combustion systems

M Benammar & WC Maskell

Energy Technology Centre, Middlesex University, Bounds Green Road, London N11 2NQ, UK.

Abstract

Single chamber amperometric zirconia oxygen sensors require simple electronics but do not enable the lean and rich regions to be distinguished when operated in the flue of a combustion system. A solution previously proposed entails the addition of a second chamber containing a piped reference gas such as air. In this paper the solution presented involves the generation of a pseudo-reference gas, which always contains excess oxygen but the composition is not fixed, by continuous electrochemical pumping of oxygen into the second chamber. The oxygen concentration in the sample gas was determined amperometrically. Lean and rich regions were distinguished potentiometrically. A double chamber device was constructed and successfully operated in the flue of a gas-burning system over a wide range of air-to-fuel ratios from rich to lean.

1. Introduction

The determination of the air-to-fuel ratio (A/F) in combustion systems is a major application of solid-state oxygen sensors. Boiler systems are generally operated in the lean region in order to optimise efficiency and minimise emissions; this can be achieved by using electronic management systems incorporating oxygen gas sensors. Amperometric sensors may be assembled to enclose an internal volume and incorporate a pore or porous material providing a diffusion path between the internal volume and external atmosphere. When operated in excess oxygen, amperometric sensors (Fig.1) operating in the limiting current mode (current plateau region) provide an output current (I_{lim}) proportional to the oxygen concentration in the sample gas [1],

$$I_{lim}=4F\sigma_{O2}\,P_{O2} \tag{1}$$

where F and P_{O2} are the Faraday constant and the oxygen partial pressure in the sample gas respectively. σ_{O2} is the leak conductance with respect to oxygen, $\sigma_{O2}=D_{O2}S/RTL$ where D_{O2} is the oxygen diffusion coefficient, S and L are respectively the cross-sectional area and length of the leak, and R and T are respectively the gas constant and operating temperature.

Amperometric sensors can be made small in size at relatively low cost and require simple electronics; this makes them suitable candidates for small-size combustion systems operating with excess oxygen. However these sensors do not distinguish the two sides of stoichiometry; the reason for this has been explained previously [2]. This can be a problem in combustion applications where there exists a risk of crossing into the fuel-rich region. The classic potentiometric (Lambda) sensor with reference oxygen partial pressure (Fig.2) obtained from a stable reference gas (eg. air) or from a metal-metal oxide redox couple [3,4] enables the two sides of stoichiometry to be distinguished (Fig.3) but provides a logarithmic EMF output, E,

$$E=(RT/4F)\ \ln(P_{O2}/P_{ref}) \qquad (2)$$

where P_{ref} is the oxygen partial pressure of the reference gas.

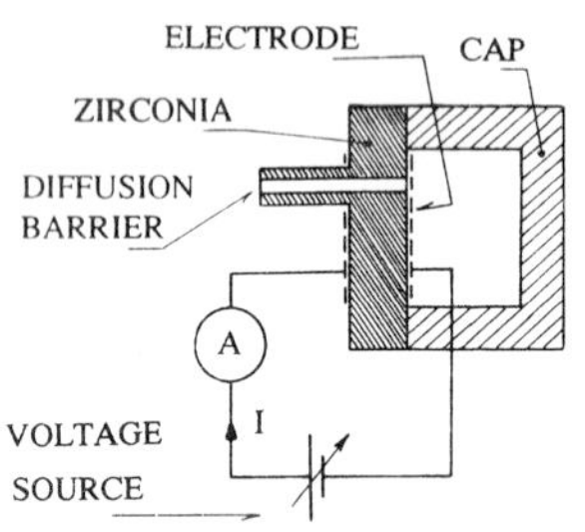

Fig.1: Schematic diagram of a single-cell amperometric device operating in the limiting-current mode.

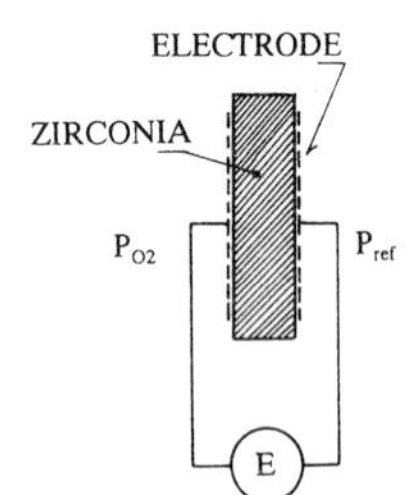

Fig.2: Schematic diagram of a potentiometric sensor with stable reference.

Dietz [1] pointed out the advantage of combining the information provided by amperometric and potentiometric sensors: the amperometric device provides a measure of the air-to-fuel ratio in the air-excess region and the potentiometric sensor enables the two sides of stoichiometry to be distinguished. Subsequent authors have translated this idea into practical devices [5-8] using air as reference gas for the potentiometric cell. The arrangement is shown in Fig.(4).

The provision of a reference gas with fixed composition to the sensor adds complexity and cost. However, since the reference gas is only used to enable the two sides of stoichiometry to be distinguished, its precise composition is not critical provided that it is maintained with excess oxygen (Fig.3). By using a double chamber device as shown in Fig.(5) a pseudo-reference oxygen partial pressure may be generated continuously by pumping oxygen electrochemically into the internal volume of chamber (B). By ensuring that the reference is maintained with excess-oxygen at all times the sensor enables the two sides of stoichiometry to be distinguished. Sensors based upon this principle have been described [2,9,10]. In this communication a sensor comprising a zirconia double chamber arrengement and operating in a simple mode is presented.

2. Principle of operation

A schematic diagram of the sensor is shown in Fig.(5) It consists of two chambers, chamber (A) acting to sense the A/F and chamber (B) providing a pseudo-reference gas against which to refer all electrode potentials. Each chamber incorporates a diffusion path between its internal volume and external atmosphere (i.e. sample gas). A constant current, I_B, pumps oxygen electrochemically into chamber (B). If I_B is of a sufficient magnitude then the pseudo-reference electrode (PRE) within chamber (B) is maintained with excess oxygen at all times but the oxygen partial pressure is not precisely defined. The sensor cell (A) is operated in the limiting-current mode: a constant voltage is applied to the pump of the cell (A), and the resulting current is measured. The Nernst EMF between the two internal volumes of the device is also monitored. The latter enables the two sides of stoichiometry to be distinguished as explained above.

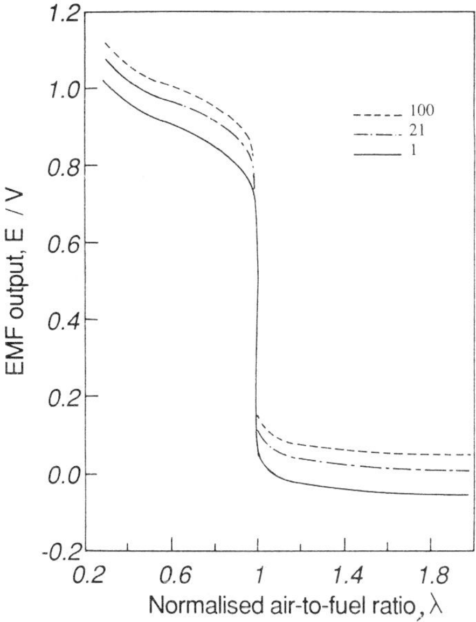

Fig.3: Theoretical output of a potentiometric cell operated at 750°C vs normalised air-to-fuel ratio (λ) for combustion of methane in air (wet analysis). Values for P_{ref} (kPa) are indicated on the figure. (λ=actual air-to-fuel ratio/stoichiometric air-to-fuel ratio).

3. Experimental

The double chamber device was similar in construction (Fig.6) to the single chamber device reported previously [11,12]. The diffusion holes in cell (A) and (B) were laser-drilled in the zirconia [13]; these were each approximately 700μm long with average diameters of 100μm and 50μm for cells (A) and (B) respectively. Thick film platinum heaters [12,14,15] were affixed with a glass on either side of the device enabling operation in the range 500-800°C. The schematic diagram of the electronic circuit used for operating the sensor is shown in Fig.(5).

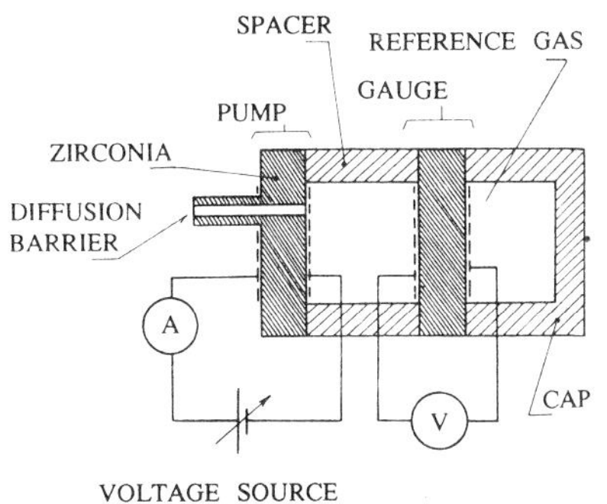

Fig.4: Schematic diagram of an amperometric sensor with stable reference.

The device was first tested in a mixture of air and nitrogen, precisely controlled using commercial mass-flow meters (from Brooks Instruments). Tests were then conducted in the flue of a premixed gas-burner where the air-to-fuel ratio was set by adjusting the flows of air and natural gas (92–96% methane, the remaining 4-8% divided approximately equally between nitrogen and propane) at the inlet to the burner. All tests were carried out at a sensor temperature of 700°C.

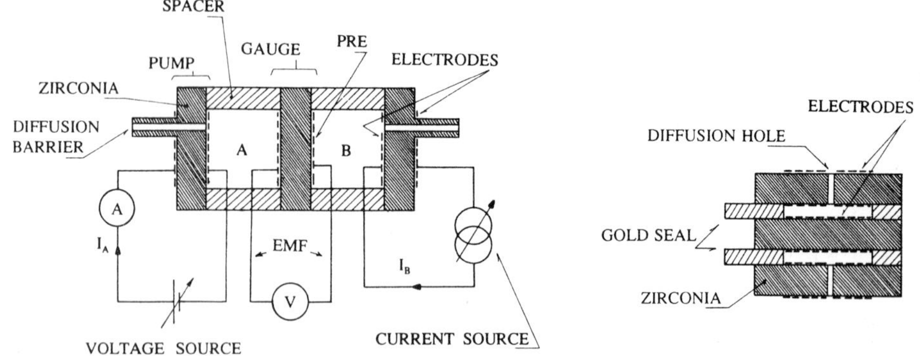

Fig.5: Schematic diagram of the amperometric sensor with pseudo-reference. Excess oxygen is maintained in cell (B) by the continuous electrochemical pumping of oxygen into it.

Fig.6: Cross-section of the double-chamber device.

4. Results and discussion

Cells (A) and (B) were separately tested in the amperometric mode in air-nitrogen mixtures using the arrangement of Fig.(1). Theory indicates the pump current to be given by eqn.(1). Figure (7) demonstrates the predicted linearity between the limiting current (located by the dashed line) and P_{O2} for both cells. Using the approximate value of 700μm for the length of the holes and a value of 160 mm^2 s^{-1} for the diffusion coefficient, D_{O2}, of oxygen in nitrogen at 700°C [16], the mean effective diameters of the holes in cell (A) and cell (B) were estimated using eqn.(1); these were 130μm and 60μm respectively.

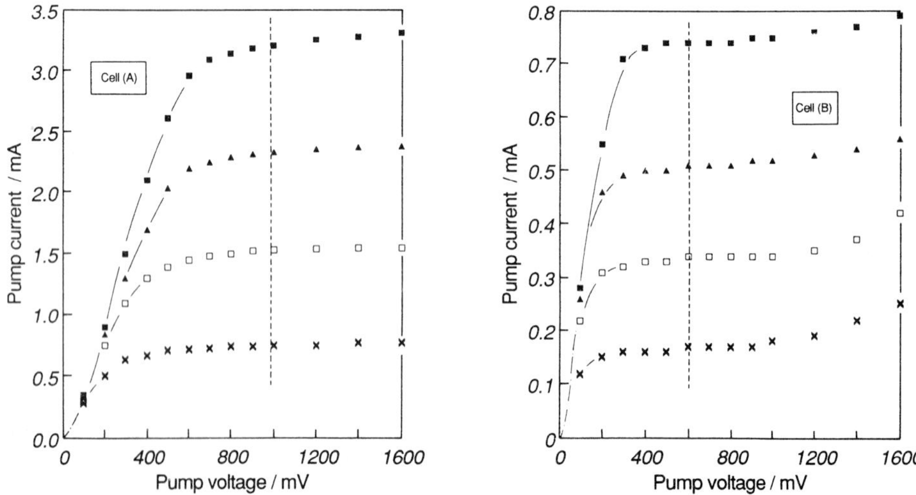

Fig.7: Chambers (A) and (B) operated separately in the amperometric mode (Fig.1). Results of operation at 700°C in a mixture of air and nitrogen with variable oxygen concentration: ×, 5%; □, 10%; ▲, 15%; ■, 21%. The limiting current, I_{lim} is located by the dashed line.

The sensor was then operated in the flue gas: cell (A) was operated in the limiting-current mode while cell (B) was used to generate the pseudo-reference. Figure (8) shows the output current and gauge EMF in the normalised air-to-fuel ratio, λ, range (0.7<λ<1.6) for various magnitudes of the pumping current I_B. The current I_A behaved similarly to that reported previously by Copcutt and Maskell [17] with the current always of the same sign but tending to zero at stoichiometry. The gauge showed the expected potentiometric characteristic (Fig.3) with a large (~ 700 mV) potential step at stoichiometry. With small pumping currents (0.5, 1.0 mA) the gauge voltage returned to a low value (~ 200 mV) with progressively reducing air-to-fuel ratio. The reason for this was that the pumping current, I_B, was insufficient to maintain excess oxygen in the pseudo-reference under these conditions. However, provided that the current I_B was sufficient (3 mA with the sensor used) to maintain excess-oxygen at the gauge electrode in chamber (B), the two sides of stoichiometry were unambiguously distinguished.

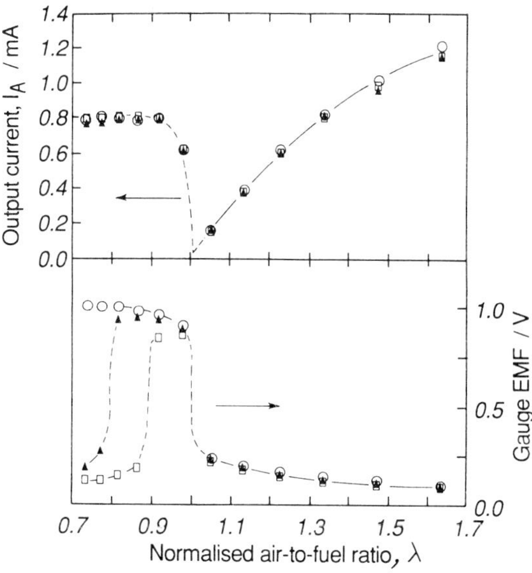

Fig.8: Characteristics of the double chamber device operated in the combustion products of a gas burning system. Cell (A) was operated in the limiting-current mode with a pump voltage of 500mV. Cell (B) generated the pseudo-reference using various values for the pumping current I_B (mA): □ , *0.5;* ▲ , *1;* ○ , *3. T=973K.*

5. Conclusion

A novel double chamber zirconia gas sensor has been described not requiring the provision of a piped reference gas. Chamber (A) acted as a conventional amperometric sensor allowing determination of the air-to-fuel ratio in the excess-air region. A pseudo-reference was generated by pumping oxygen continuously into the second chamber (B). Provided that the pumping current was sufficient, then chamber (B) contained excess oxygen and provided a sufficiently stable reference gas enabling the two sides of stoichiometry to be distinguished potentiometrically. The sensor was successfully operated in flue gases covering a wide range of air-to-fuel values on both sides of stoichiometry.

6. Acknowledgement

This work was partly supported by a grant from the Council for Funding Polytechnics and Colleges (PCFC).

7. References

[1] H Dietz, Solid State Ionics 6 (1982) 175-183.

[2] M Benammar and WC Maskell, A novel miniature zirconia gas sensor with pseudo-reference. Part 1: Amperometric operation providing unambiguous determination of air-to-fuel ratio, Applied Physics A, in press.

[3] C Déportes, M Hénault, F Tasset and G Vitter, US Patent 4045319 (1977).

[4] WC Maskell and BCH Steele, J Appl Electrochem 16 (1986) 475-489.

[5] WC Vassel, EM Logothetis and RE Hetrick, SAE paper 841250 (1984).

[6] S Soejima and S Mase, SAE paper 850378 (1985).

[7] EM Logothetis, WC Vassell, RE Hetrick and WJ Kaiser (1985), Transducers '85, 3rd International Conference on Solid-State Sensors and Actuators, IEEE Catalog number 85CH2127-9.

[8] S Ueno, N Ichikawa, S Suzuki and K Terakado, SAE paper 860409 (1987).

[9] S Suzuki, T Sasayama, M Miki, H Yokono, S Iwanaga and S Ueno, SAE paper 850379 (1985).

[10] WC Maskell (1991), UK Patent GB 2 208 007.

[11] H Kaneko, WC Maskell and BCH Steele, Solid State Ionics 22 (1987) 161-172.

[12] M Benammar and WC Maskell, "Miniaturised solid-state pump-gauge oxygen sensors: practical aspects", Proc Int School of Materials Science and Technology (Solid State Ionics for Sensors and Electrochromics), W Weppner (Ed), 1-12 July 1992, Erice-Sicily, Italy, in press.

[13] WC Maskell and BCH Steele, Solid State Ionics 28-30 (1988) 1677-1681.

[14] WC Maskell (1988), UK Registered Design 1053348.

[15] M Benammar and WC Maskell, J Phys E: Sci Instrum 22 (1989) 933-936.

[16] RH Perry (late editor), DW Green (editor), "Perry's Chemical Engineer's Handbook", 6th edition, McGraw-Hill, 1984, p.3.285.

[17] RC Copcutt and WC Maskell, Solid State Ionics 53-56 (1992) 119-125.

An integrated approach to an artificial nose based on ASICs and conducting polymers

J V Hatfield, P J Hicks, P Neaves, K C Persaud[†] and P Travers[†]

Departments of Electrical Engineering & Electronics and [†]Instrumentation & Analytical Science, UMIST, PO Box 88, Manchester M60 1QD

ABSTRACT: Electrically conducting organic polymers based on heterocyclic molecules display reversible changes in conductivity when exposed to polar volatile chemicals. This paper describes progress that has been made towards realising an artificial nose based on arrays of these polymers. The polymers are interrogated for resistance changes by means of an Application Specific Integrated Circuit (ASIC) realised in BiCMOS technology. The ASIC and the polymer array are housed on a single thick-film ceramic substrate.

1. INTRODUCTION

A number of gas sensor technologies have been reported which suffer from drawbacks of various kinds. These include high power consumption for heating elements of tin oxide and platinum pellistor type elements, poor stability and sensitivity of piezoelectric devices, susceptibility to poisoning by sulphur containing compounds in metal oxide semiconductors, and slow response times of electrochemical fuel cells. One useful set of materials that may be utilised as sensors in an electronic nose is that of electrically conducting organic polymers based on heterocyclic molecules. These polymers display reversible changes in conductivity when exposed to polar volatile chemicals. Unlike many commercially available gas sensors, rapid adsorption and desorption kinetics are observed at ambient temperatures.

Pelosi and Persaud have investigated the gas sensing properties of a large number of conducting polymers, [1]. The concentration-response profiles are almost linear over a wide concentration range. This is advantageous as simple computational methods may be used for information processing. The materials do not display high specificity to individual gases but they can be chemically tailored to enhance differences in response to particular classes of polar molecules.

Identification of a specific chemical thus makes use of an array of sensors of different specificities. The compound of interest is identified when the entire response pattern of such an array is identical with that stored in a computer memory. In this respect these sensor arrays behave very similarly to olfactory sensor arrays in the biological system. Different polymers made from modified monomer units show broad overlapping response profiles to different volatile compounds. Hence, arrays of these sensors should behave very similarly to olfactory sensor arrays in the biological system. This paper outlines the development at UMIST of an integrated artificial nose exploiting Application Specific Integrated Circuit (ASIC) techniques in conjunction with arrays of conducting polymers.

2. THE UMIST CONDUCTING POLYMER SENSOR ARRAY

Miniature arrays consisting of up to twenty different conducting polymer materials have now been realised. A microprocessor driven circuit measures changes in resistance of individual sensor elements. It interrogates the sensor array at user defined intervals and data are stored in memory. Each sensor element changes in resistance when exposed to a volatile compound. However, the degree of response to a given substance depends on the type of polymer element used so that a pattern of resistance changes can be recorded and processed to produce a set of descriptors for that particular substance. The sensor responses are normalised to represent relative changes in resistance and thus concentration-independent patterns can be produced.

The response of these conducting polymer sensor arrays to ammonia, methylamine, dimethylamine and trimethylamine has been investigated. Controlled concentrations of the four amines were generated using Kin-Tek permeation tubes (Air Products). An aqueous solution of the amine is contained within a polymer membrane which allows the vapour to permeate through at a controlled rate. The polymers have the characteristics of p-type semiconductors and show a high sensitivity to amines. It is believed that this is the result of the bound amine molecules acting as electron donors and thereby reducing the number of holes, [2] . The concentration independent patterns produced by a twenty sensor array are shown in Figure 1.

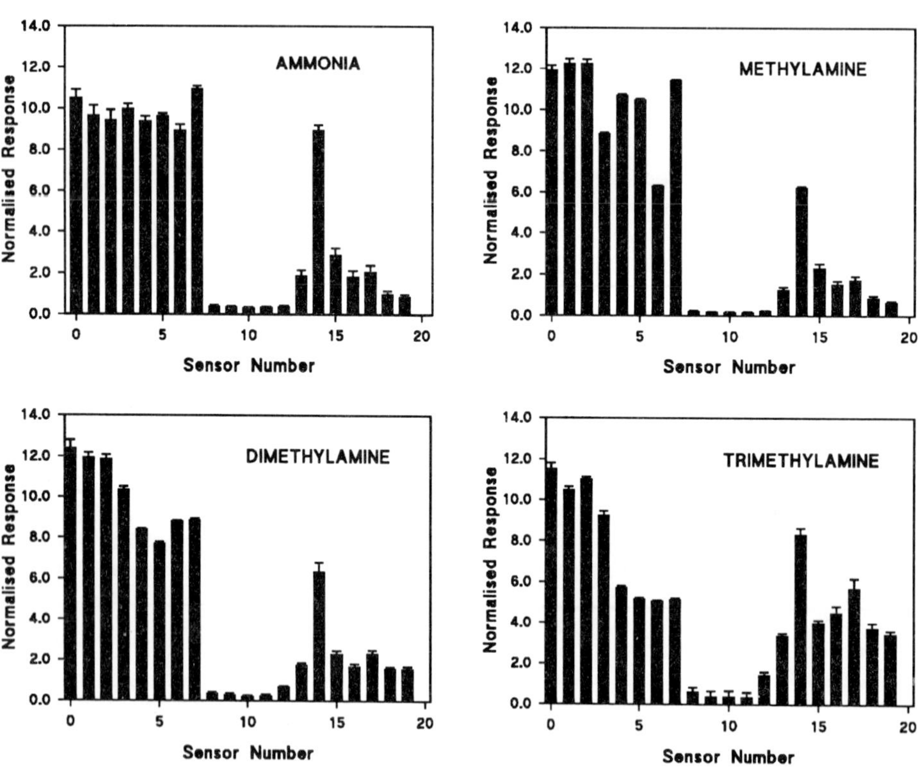

Fig. 1. Average patterns generated by a 2-D element array in response to the four amines. Error bars indicate the standard deviations in the relative response of each sensor

For ammonia, the concentration-response profile of the same array has been plotted (Figure 2) using the average resistance change across all twenty sensor elements. Response and recovery times are material dependent but typical values are 10 seconds (100% response) and 8 minutes (90% recovery). At concentrations below 1ppm the linear relationship between change in resistance and concentration breaks down. When exposed to amines at these low levels there is a steady increase in resistance (Figure 3) indicating accumulation of amine molecules at the sensor surface. The initial rate at which the resistance increases is found to be concentration dependent, and can be used to construct a calibration curve (Figure 4).

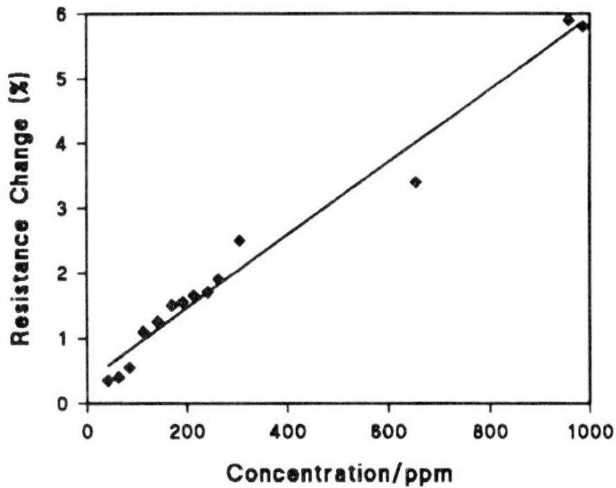

Fig. 2. Average resistance change plotted against concentration for exposure to ammonia. A line of least squares fit has been drawn through the data points.

3. THE INTEGRATED NOSE

The integrated nose is a hybrid constructed on a 90mmX40mm multi-layer, thick-film ceramic substrate, (Figure 5). At one end, an array of 32 pairs of exposed gold electrodes surround 3 sides of a ground plane. All are routed back to an 84-pin Application Specific Integrated Circuit (ASIC), designed in-house, and fabricated by a commercial 2μm BiCMOS process under the EUROCHIP initiative. A 16-bit self calibrating Analogue to Digital converter (A to D) with a serial output is also mounted on the substrate. The ASIC, performs two distinct functions. In deposition mode it controls the deposition of the conducting polymers onto the gold electrodes. Part of the polymer deposition also involves a lift-off technique well known in IC manufacture. In measurement mode the ASIC can perform several different functions. It can measure changes in resistance of individual sensor elements when exposed to a volatile compound. The ASIC can also perform correlation measurements between elements.

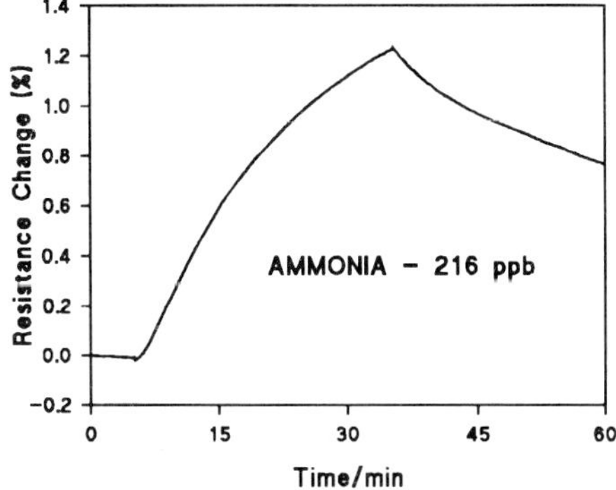

Fig. 3. Average resistance change for a 20 sensor array when exposed to 216ppb ammonia.

A discrete component, microprocessor controlled system has been in operation for some time. In this system a constant current is forced across the required sensor and the resulting voltage is offset by a D to A converter and amplified by a precision instrumentation amplifier. The resulting signal is then filtered and digitised.

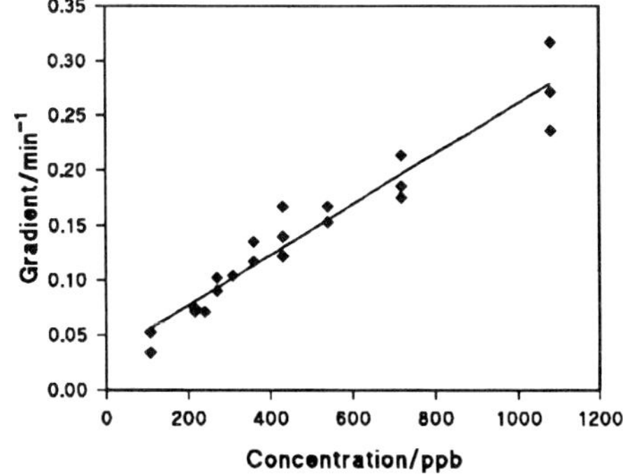

Fig. 4. Initial gradient on exposure to ammonia plotted against concentration. A line of least squares fit has been drawn through the data points.

4. A SIGNAL PROCESSING ASIC FOR AN ELECTRONIC NOSE

It was felt for various reasons that the discrete component configuration was not an appropriate one to transfer to a single chip solution. The design philosophy is firmly rooted in the voltage domain. From the point of view of analogue IC design the basic building blocks available, MOS and bipolar transistors, are current output devices.

The current mode approach [3], therefore, offers many advantages:
Increased dynamic range: currents from hundreds of nano-amps to milli-amps can be accommodated. This is not restricted by the trend to lower supply voltages.
Increased bandwidth: voltage swings are small (≈hundreds of milli-volts). Hence signals can change at a greater rate.
Reduced power consumption: because voltage swings are generally small, power consumption, even with bipolar devices, can be greatly reduced.
Greater flexibility: common functions such as addition, subtraction and multiplication are both easier to implement and more accurate.

Of great importance, and a deciding factor, in the present application is that the multiplexing of signals can be achieved more easily in less physical space than the voltage-mode counterpart. This enables poor quality and, therefore, physically small analogue switches to be employed.

The ASIC can be divided into three parts: an analogue "front-end", a serial interface and control circuitry. The front end, where possible, utilises current-mode building blocks. A simplified block-diagram is shown in Figure 6 with control signals omitted for clarity. The front-end allows *changes* in current to be measured or it can perform cross-correlation between two sensors. The use of correlation as a measurement technique is well known and can provide dramatic noise improvements and measurement accuracy, [4]. Control of the ASIC is via an I^2C serial bus, which is an industry standard 3-wire bus. It is envisaged that this will ultimately be carried out by an on-board micro-controller.

Fig. 5. The integrated UMIST nose. Substrate measures 4x9cm. The ASIC package is an 84pin leadless chip carrier. The 16-bit A to D is not mounted.

4.1 Analogue front end

The measurement of current is as follows. A sensor is selected and one of four voltages is applied across the sensor. This is necessary since the basal resistance of the UMIST sensors vary between 1kΩ and 100kΩ. The output of the multiplexer is a current. This current is then offset and amplified by the analogue multiplier and converted to a voltage. Note (Figure 6), that the multiplexer has two outputs. During simple current measurements, one of the outputs is used to set the gain of the analogue multiplier.If a cross-correlation is being performed then these two outputs represent the current from the two selected sensors.

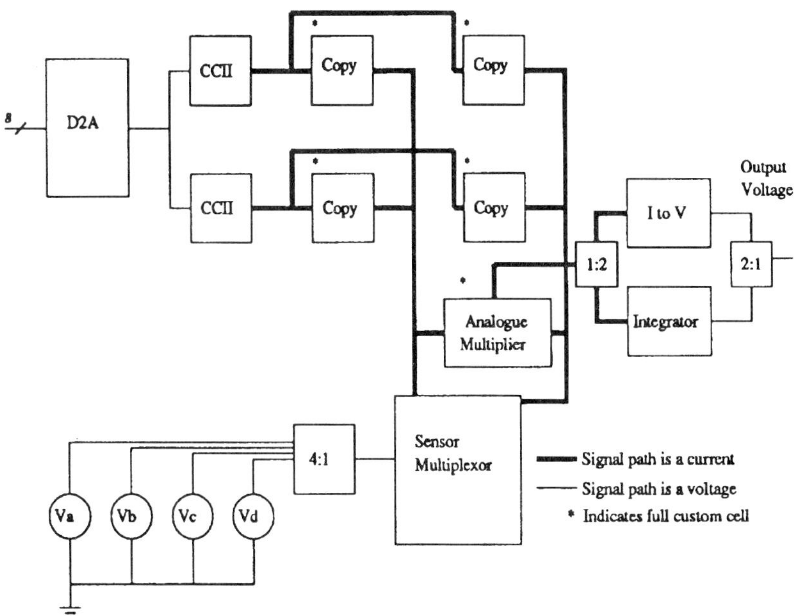

Fig. 6. Block diagram of ASIC analogue "front-end" circuitry.

The current offset is provided by an 8-bit D to A, two current conveyors and four current copiers. The current conveyors provide accurate voltage to current conversion. The current copiers behave in much the same way as a voltage sample and hold. They allow a current to be accurately "memorised", [5]. Thus a coarse current can be set and memorised with one current copier and the remaining smaller offset can be set by the other current copier. Addition of these two currents is trivial. This particular technique allows a low resolution D to A, with a consequent saving in silicon area, to set accurate currents. The analogue multiplier provides current gains of 1, 10, or 100 as required, [6].

4.2 Digital interface

A block diagram of the digital interface is shown in Figure 7. The digital interface allows the user to control the front end and interface to the external A to D converter. Control of this hybrid circuit/sensor array is via the I^2C bus. Note that the hybrid constitutes a *bus slave*, all data transfers are initiated from a bus master.

CONCLUSIONS

We have developed deposition techniques that allow arrays of different conducting polymers to be deposited on a thick film substrate. Application Specific Integrated Circuits employing novel signal processing techniques for monitoring resistance changes have been fabricated by SGS-Thomson under the EUROCHIP initiative. The basic cells have been tested and found to be functional.

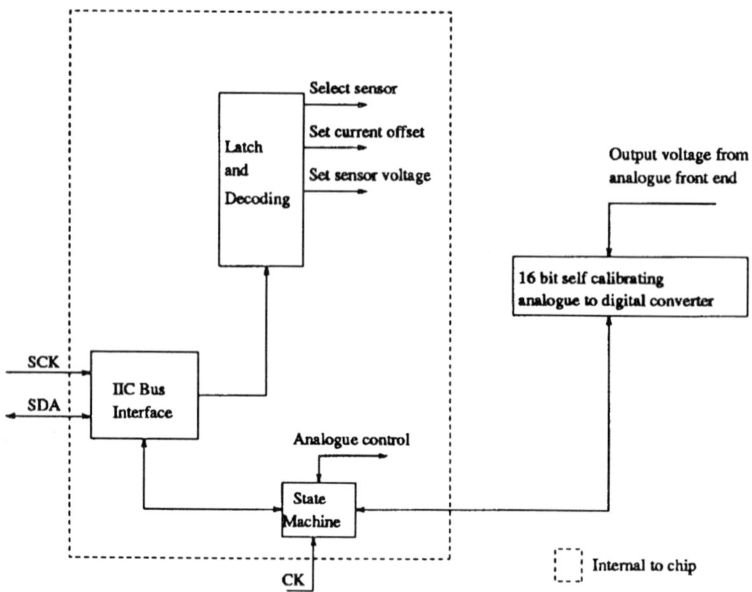

Fig. 7. Block diagram of ASIC digital interface circuitry.

REFERENCES

[1] P Pelosi and K C Persaud, 'Gas sensors: towards an artificial nose', *Sensors and Sensory Systems for Advanced Robots.* (Ed P Dario) NATO ASI Seroes F: Computer and Systems Science, Springer-Verlag, Berlin, 1988, pp361-382.

[2] J J Miasic, A Hooper and B C Tofield, 'Conducting Polymer Gas Sensors', *J. Chem. Soc., Faraday Trans. 1*, 1986, **82**, pp1117-1126.

[3] 'Special Issue on Current-Mode Analogue Signal Processing', *IEE Proceedings*, 1990, **137**, Part G, pp61-184.

[4] J Jordan, P Bishop and B Kiani, 'Correlation-Based Measurement Systems', *Ellis Horwood Ltd*, ISBN 0-7458-0627-9, 1989.

[5] G Wegmann and E A Vittoz, 'Very accurate dynamic current mirrors', *Electronics Letters*, 1989, **25**, pp644-646.

[6] B Gilbert, 'A new wide-band amplifier technique', *IEEE Journal of Solid-State Circuits*, 1968, **SC-3**, No. 4, pp353-365.

ACKNOWLEDGEMENTS

The authors acknowledge with gratitude the Science and Engineering Research Council of the UK for providing funding for this project, and EUROCHIP for making commercial micro fabrication facilities available.

Oxygen sensing with perovskite thin films

Michael L. Post and Brian W. Sanders

Institute for Environmental Chemistry, National Research Council of Canada, Montreal Road, Ottawa, Ontario, K1A 0R6, Canada.

ABSTRACT: The solid state chemistry of the non-stoichiometric perovskite $SrFeO_{2.5+x}$ has been exploited to provide a sensing material which is potentially specific to gaseous oxygen. The principle of the sensor is based upon bulk structural changes which occur as the oxygen composition of the perovskite varies within the range of the phase limits, $0 \leq x \leq 0.5$. The technique of pulsed laser ablation has been used to deposit $SrFeO_{2.5}$ and $SrFeO_3$ as thin films upon sapphire substrates, and their reactivity with oxygen has been investigated. The optical and crystallographic properties of the films are reported, and a possible mechanism of sensor transduction for the films is identified.

1. INTRODUCTION

A diverse range of metal oxides have been used to provide the active sensing material used for chemical sensing devices. These oxides have been shown (1-3) to undergo changes in physical properties upon exposure to a large selection of gaseous species. In the many studies of sensor materials which have been reported, there has been an overwhelming emphasis upon the change in properties which accompany surface reactivity. The mechanisms which depend upon the surface chemistry of the metal oxides, such as those of the redox type, result in an intrinsic lack of specificity when used for sensing gases within a close structural family or for those with similar oxidative chemistry.

The present work is directed toward oxygen sensing with a material which, if an appropriate transduction principle is selected, provides a potentially specific response to oxygen. The material for which data is presented is the non-stoichiometric perovskite $SrFeO_{2.5+x}$ where the oxygen composition lies in the range $0 \leq x \leq 0.5$, and within which four phases have been identified (4). For these investigations, the perovskite has been deposited as a thin film by pulsed laser ablation. This technique provides a reproducible method of preparing samples which facilitates the subsequent measurements which have been made. It is also likely that an effective method of integration of a material of this type into a prototype sensing device will also utilise thin film form. Some thermodynamic properties of the system $SrFeO_{2.5+x} + O_2$ have already been reported, (5). Given here is an outline of the structural transformations which have been exploited to provide a sensor transduction mechanism which is based on the changing optical properties of $SrFeO_{2.5+x}$ as the oxygen stoichiometry varies. A comparison of some of the properties of the bulk powder state with the thin films are made.

2. EXPERIMENTAL

Bulk powder samples of $SrFeO_{2.5+x}$ were prepared by sintering stoichiometrically mixed proportions of high purity, (better than 99.99%), $SrCO_3$ and Fe_2O_3 at temperature T=1350K. Phase and elemental purity were confirmed by X-ray diffraction and ICP-mass spectrometry. Materials with compositions at each of the phase limits were prepared from the parent batch of $SrFeO_{2.5+x}$ by treatment at moderate temperature either under vacuum, (for x=0), or in oxygen, (for $x\approx0.5$), in a stainless steel reactor (5). The technique of pulsed laser ablation was employed to deposit thin films from a compressed and sintered target of either $SrFeO_{2.5}$ or $SrFeO_3$ upon optically transparent sapphire wafers which were single crystal with $(1\bar{1}02)$ orientation. The laser was of the excimer type operating with KrF, which produces radiation of wavelength λ=248nm. Films were deposited upon a heated substrate ($500 < T < 1120$K) in a chamber which was either under vacuum or at low oxygen pressure, $p(O_2) \approx 30$Pa. Using these conditions, films of approximate thickness 300nm were obtained in 20 minutes. Other details of the laser ablation process are reported elsewhere (6). X-ray diffraction data were collected for powder samples and for thin films using a Scintag P2000 diffractometer, operating in θ-θ mode with monochromated Cu-K_α radiation. The thin films were exposed to a series of changing oxygen pressures $\{10^{-2} < p(O_2) < 110\text{x}10^3\text{Pa}\}$ in a stainless steel reactor which was attached to a manifold to allow gas manipulation. Products of the as-deposited and oxygen treated films were studied by X-ray diffraction and by uv-vis spectroscopy. Spectroscopic measurements were made in transmission, with a reference of uncoated sapphire $(1\bar{1}02)$, over the wavelength range $200 \leq \lambda \leq 900$nm, using either a Varian Cary 1E or a HP8450A spectrophotometer.

3. RESULTS AND DISCUSSION

The X-ray diffraction spectra of powdered $SrFeO_{2.5+x}$ at the upper and lower phase limits are shown in Fig.1. The oxygen rich composition has the cubic perovskite structure; the oxygen deplete phase is orthorhombic brownmillerite (4,7), which is a structure type related to an orthorhombic distortion of the perovskite lattice (8). The thin films, prepared from a target of either composition under the conditions described above, exhibited a strong preferential orientation, Fig.2. Favoured deposition planes were (110) and (100), and the relative proportion of each of these components was found to be dependant upon the substrate temperature during deposition (6). The changes in the X-ray spectrum, as a result of changing x, close to 2θ=32.5° where the (110)

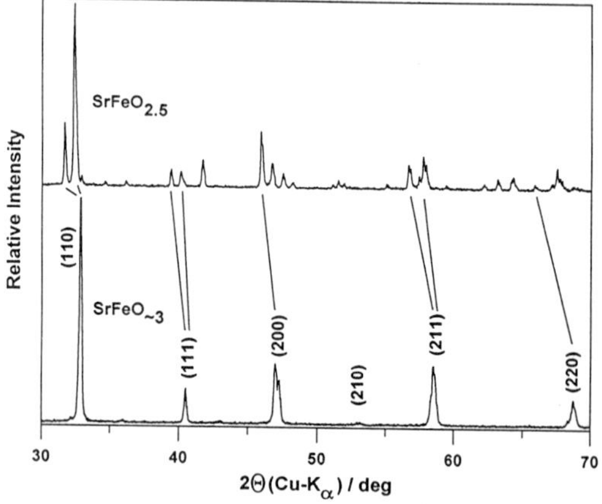

Figure1. X-ray diffraction spectra for bulk powder samples of $SrFeO_{2.5+x}$ taken with Cu-K_α radiation. The lower curve shows the structure close to the cubic form, (x = 0.88); the upper curve shows the orthorhombic-brownmillerite form, (x = 0). Lines indicate the related (hkl) of each symmetry.

reflection of the cubic form occurs, are shown in more detail in Fig.3.

The oxygen stoichiometry of the thin films can be changed after deposition by subjecting them to thermal and oxygen treatments. In this manner, the composition of the films are shifted between the oxygen deplete and oxygen rich condition, and *vice versa*. The result of these treatments upon the films are shown by the X-ray spectra, (Fig.4), which change in consequence to the structural changes driven by a variation in x. Vacuum treatment at $T \approx 600$K produces the oxygen deplete structure, (x=0), while exposure to oxygen $\{p(O_2) \approx 50\text{kPa}\}$ at T≈420K, followed by cooling, results in the cubic, oxygen rich phase ($x \approx 0.5$). The films at each phase limit

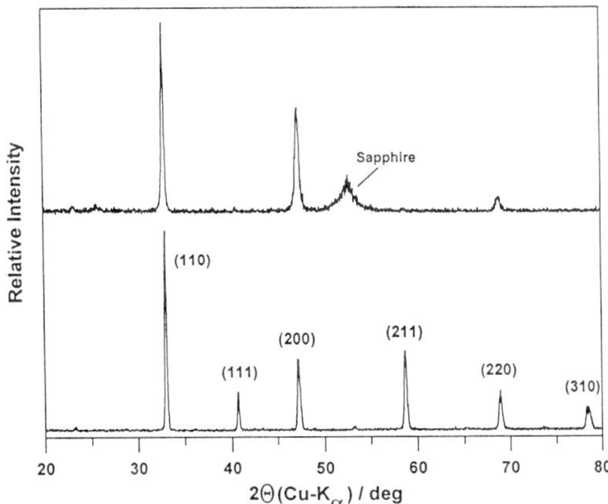

Figure 2. X-ray diffraction spectra taken with Cu-K_α radiation for $SrFeO_{\sim 3}$ as a bulk powder, (lower curve), and as an oriented thin film on sapphire, (upper curve).

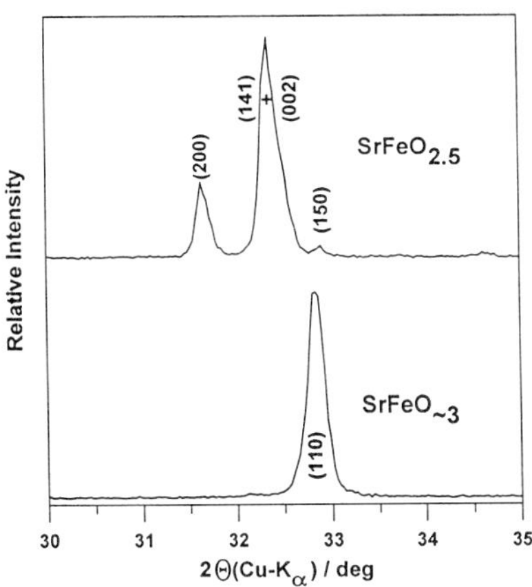

Figure 3. X-ray diffraction spectra taken with Cu-K_α radiation for $SrFeO_{2.5+x}$ as a bulk powder with $x \approx 0.5$, cubic, (lower curve); and $x = 0$, orthorhombic, (upper curve). Miller indices for the reflections are referred to each respective lattice.

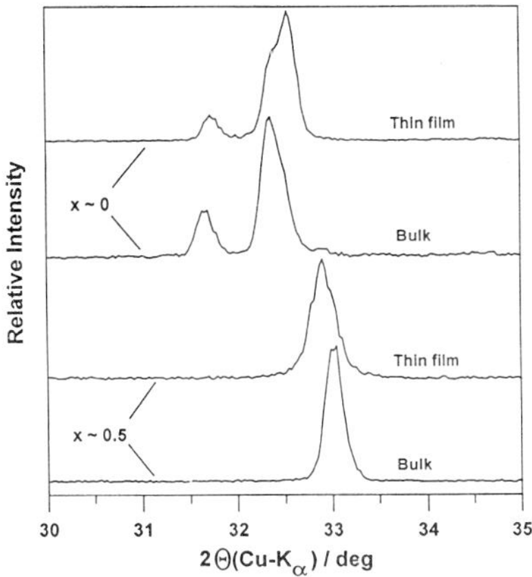

Figure 4. X-ray diffraction spectra for $SrFeO_{2.5+x}$ in the region of the cubic-(110) reflection taken with Cu-K_α radiation. Bulk powder and thin film spectra are compared; the lower two curves are for $x \approx 0.5$; the upper two curves are for $x = 0$

also have a distinctly different optical appearance. For x=0, the film is yellow and transparent; for $x \approx 0.5$ the film has a metallic sheen and is reflective. These features are shown quantitatively in Fig.5 by the respective uv-vis spectra of films obtained at each phase limit. The transmittance for a film of $SrFeO_{\sim 3}$ is lower than that for a film of $SrFeO_{2.5}$ at all wavelengths, λ>340nm. The maximum difference in transmittance between these two forms lies at $\lambda \approx 400$nm, and is about two orders of magnitude. Optical properties such as those which are dependant upon changes in oxygen composition, offer a possible mechanism of transduction for $SrFeO_{2.5+x}$ when used as an oxygen sensor. Although the data presented here are for thin films which are reacting with oxygen in a batch process, equivalent information is now becoming available (9) for *in situ* reactivity. The pressure-composition isotherms for the $SrFeO_{2.5+x}$ + O_2 reaction (5), show that, at $T \approx 620$K for instance, a range in oxygen pressure of 0 to ~10^5Pa is required to move the composition between x=0 and $x \approx 0.5$. This range identifies the oxygen pressures over which an accessible sensor response would be provided by conversion between the structural states. The linearity of the spectroscopic changes with x, over this pressure range, are currently being determined.

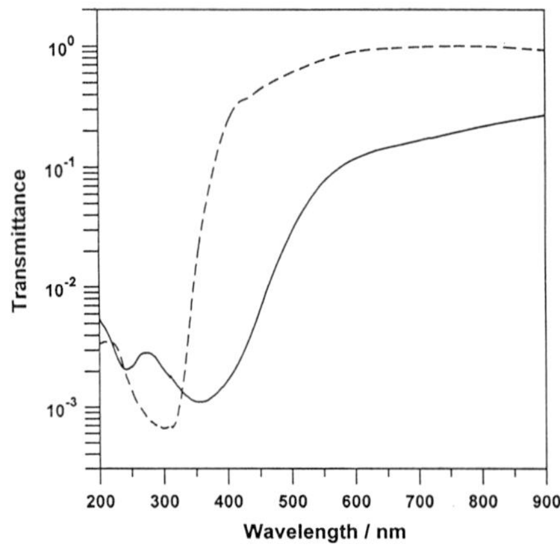

Figure 5. Uv-vis spectra for $SrFeO_{2.5+x}$ shown in transmittance. The full curve is for the oxygen rich, ($x \approx 0.5$), composition; the broken curve for the oxygen deplete, (x = 0), form.

4. CONCLUSIONS

The non-stoichiometric perovskite $SrFeO_{2.5+x}$ exists over a composition range $0 \leq x \leq 0.5$ and can be prepared in the form of thin films by using pulsed laser ablation of the parent material. The films are deposited as oriented products, and provide X-ray spectra from which oxygen stoichiometry can be determined. At compositions close to the phase limits, the bulk reactivity of the films with oxygen parallels that observed for powder samples. The optical properties of the films, in both transmission and reflection, vary as the structural lattice changes through the oxygen stoichiometry range. The thermodynamics of the $SrFeO_{2.5+x}$ + O_2 system indicate that, at elevated temperature, the films can be applied to sensing the presence of oxygen over a dynamic range approaching five orders of magnitude. The change in optical properties which accompany the bulk structural changes of $SrFeO_{2.5+x}$, provides the basis for a transduction mechanism for use of this material as a thin film oxygen sensor.

5. REFERENCES

1) S. Yamauchi (ed.), *Chemical Sensor Technology*, Vol. 4, Kodansha, Tokyo; Elsevier, Amsterdam, 1992, and articles therein.

2) P. Ciureanu and S. Middelhoek (eds.), *Thin Film Resistive Sensors*, IOP Publishing, Bristol, 1992, and articles therein.

3) T. Grandke and W. H. Ko (eds.), *Sensors: A Comprehensive Survey*, Vol 1, VCH Publishers, New york, 1989, and articles therein.
4) Y. Takeda, K. Kanno, T. Takada, O. Yamamoto, M. Takano, N. Nakayama and Y Bando, *J. Solid State Chem.*, **63**, 237, 1986.
5) M. L. Post, B. W. Sanders and P. Kennepohl, *Sensors and Actuators **B***, 1993, In Press.
6) B. W. Sanders and M. L. Post, *Proceedings of the Materials Research Society Meeting*, Boston, USA., Nov. 1992, In Press.
7) J. B. MacChesney, R. C. Sherwood and J. F. Potter, *J. Chem. Phys.*, **43**, 1907, 1965.
8) J.-C. Grenier, N. Ea, M. Pouchard and P. Hagenmuller, *J. Solid State Chem.*, **58**, 243, 1985.
9) M. L. Post, B. W. Sanders and J. Yao; Manuscript in Preparation.

Electrochemical zirconia sensor model for reducing gas mixtures

A.D.Brailsford, M.Yussouff and E.M.Logothetis

Physics Department, Ford Research Laboratory, Dearborn, MI-48121-2053, USA.

ABSTRACT: A model describing the response of an electrochemical gas sensor to oxygen and a binary mixture of reducing gases is described. The step-like change in output voltage occurs at an inverse redox ratio which, in general, is a complicated function of gas transport parameters and reaction rate constants for the individual physical processes on the porous electrode surfaces. However, for catalytic electrodes, a rule of mixtures is shown to hold for the location of the output step. Numerical examples indicate that the model is capable of reproducing several aspects of the behavior of sensors of this type.

1. INTRODUCTION

In earlier publications a first principles general model [1] for electrical type of metal oxide gas sensors was described and applied to the zirconia based sensor [2]. The latter is an electrochemical cell having one electrode exposed to a reference atmosphere (e.g. air) and the other electrode exposed to the measurement gas. In thermodynamic equilibrium [3], it is well known that the emf of the cell is determined only by the equilibrium oxygen partial pressures in the reference and measurement gases. Away from thermodynamic equilibrium [4], however, the response of the sensor depends upon the specific composition of the measurement gas. In the previous treatment of the zirconia sensor [2], its response was calculated for the case of a single reducing gas such as CO or H_2 mixed with oxygen and nitrogen. The case of a gas mixture containing two reducing gases, e.g. CO and H_2, together with O_2 and N_2, is considered here in order to expose new features that can arise in more complex measurement gas environments.

2. THE MODEL

It is presumed that there is a significant concentration of mobile positively charged oxygen vacancies inside the zirconia. Adsorbed oxygen atoms on the air-side electrode (e.g. porous Pt) recombine with such vacancies at the air/electrode/zirconia-interface, thereby depleting the vacancy concentration near this electrode. (It is assumed that electrons are easily transferred from and to the electrodes as and when required for charge conservation.) Similarly, reducing gas molecules (e.g. carbon monoxide, hydrogen etc) adsorbed on the measurement electrode (e.g. porous Pt) consume oxygen adatoms, thereby creating additional oxygen vacancies near this electrode. Variation of the oxygen vacancy concentration inside the zirconia near the latter electrode that results from changes in the measurement gas composition produces an electrostatic field, and thus an emf across the material, causing the sensor to act as a Nernst cell. In steady state, the open circuit sensor emf at the absolute temperature T is given by [2]

$$V_0 = (k_B T / 2e_0) \ln [\theta_{0AIR} / \theta_{0EG}] \quad (1)$$

where k_B is the Boltzmann constant, e_0 is the electronic charge and θ_0 is the fractional occupancy of electrode surface sites by adsorbed single oxygen atoms. The essential challenge of the model is thus the determination of θ_0 for different measurement gas compositions.

3. TWO REDUCING GASES: THE RULE OF MIXTURES

A specific gas mixture consisting of O_2 , CO , CO_2 , H_2 and H_2O (denoted in general as species α , with α = 1, 2, 3, 4 and 5 respectively) in an inactive carrier gas, nitrogen, is considered. The active gases adsorb and desorb on the measurement electrode. The adsorbed O_2 molecules then dissociate to adatoms which combine with adsorbed CO to produce CO_2 and with adsorbed H_2 to produce H_2O. Finally, in the so called water-gas reaction, adsorbed CO reacts with adsorbed H_2O to produce CO_2 and H_2 , and, conversely, CO_2 reacts with H_2 to produce CO and H_2O.
Using the above reaction scheme, it is possible to write down the steady state equations that govern the various fractional occupancies of the surface lattice sites, θ_α . For example, for adsorbed oxygen molecules (O_{2ads}), one gets

$$\dot{\theta}_1 = 0 = \dot{n}_1 /\omega - k_D \theta_1 + (1/2) k_R \theta_0^2 \quad (2)$$

where, $\dot{n}_\alpha$ is the mean rate of addition of particles of species α to the surface, k_D is the rate constant for the surface dissociation of adsorbed oxygen molecule and k_R is its formation rate from oxygen adatoms. The area density of adsorption sites, ω, is taken the same for all chemical components. Similar detailed rate equations can be formulated for other species. However, for present purposes it suffices to note the set of constraints that must hold for any reaction scheme :

$$\dot{n}_2 + \dot{n}_3 = 0 \quad , \; \dot{n}_4 + \dot{n}_5 = 0 \quad \text{and} \quad 2\,\dot{n}_1 - \dot{n}_2 - \dot{n}_4 = 0 \; . \tag{3}$$

Such relations are independent of the details of the surface chemistry and express the conservation of carbon, hydrogen and oxygen nuclei respectively.

Insertion of the sensor into the gas system modifies the local concentrations of all reactive species and induces diffusive mass transfer. Consider a cylindrical sensor with outer (measurement) electrode at $r = r_s$. Let the partial pressures of the gases at $r = L$ ($> r_s$), far away from the measurement electrode, be denoted by p_α^∞ . The partial pressures of the same gases at the sensor surface at $r = r_s$, denoted by p_α^0, differ from the prescribed values at $r = L$ due to surface interactions. Transfer processes between the surface and gas are given by

$$\dot{n}_\alpha / \omega = k_\alpha^a \, p_\alpha^0 - k_\alpha^d \, \theta_\alpha \quad , \tag{4}$$

where k_α^a and k_α^d are adsorption and desorption rates. Further, assuming that the gas phase diffusion is solely responsible for mass transport of the various species from $r = L$ to the sensor surface, one can write

$$\dot{n}_\alpha / \omega = D_\alpha \, (p_\alpha^\infty - p_\alpha^0)(\kappa / D_1) \; . \tag{5}$$

Here, D_α is the diffusion coefficient of the species α in the gas phase and κ is the mass transfer coefficient for oxygen.

Eqs.(4) and (5) allow the θ_α to be expressed in terms of the changes in partial pressures induced near the measurement electrode of the sensor. Additionally, the conservation conditions enable three of these local partial pressures to be expressed in terms of their asymptotic values and the pressure drops of the remaining two (here chosen to be O_2 and CO). Thereby one finds for adsorbates of the two reducing species in particular that:

$$\theta_2 = K_2 p_1^{\infty} [R_2 - \beta_2 y] , \theta_4 = K_4 p_1^{\infty} [R_4 - \beta_4 (2x - y)] , \quad (6)$$

where x and y are defined by the equations

$$p_1^{\infty} x = (p_1^{\infty} - p_1^{0}) , \quad p_1^{\infty} y = D_2 (p_2^{\infty} - p_2^{0}) / D_1 ; \quad (7)$$

and $\beta_\alpha = (\kappa / k_\alpha^a + D_1 /D_\alpha)$, $K_\alpha = (k_\alpha^a / k_\alpha^d)$ and $R_\alpha = (p_\alpha^{\infty} /p_1^{\infty})$.

A similar procedure, augmented by the use of Eq.(2), yields the complementary oxygen adatoms fractional occupancy

$$\theta_0 = [(2 p_1^{\infty}/k_R) (k_D k_1^a/k_1^d) [1 - (1+b) x]]^{1/2}, \quad (8a)$$

where $b = (1 + k_1^d /k_D) \kappa /k_1^a$. (8b)

From these relations, a rule of mixtures may be derived quite simply provided the electrode behaves as a catalytic agent for the reactions of oxygen with the reducing gases (This entails the rapid reaction of adsorbed reducing gas molecules with the oxygen adatoms before they are able to desorb from the electrode surface). Two distinct regimes for the adatom population are expected [1]. In the first regime, the reducing gas adsorbates will be in excess, with $\theta_0 \simeq 0$, while in the second regime, the oxygen adatoms will be dominant, with adsorbed CO and H_2 vanishingly small ($\theta_2 \simeq 0$ and $\theta_4 \simeq 0$). It is clear from Eq.(1) that the cell emf is expected to be large in the first regime and (relatively) small in the second regime. Therefore, qualitatively a step in the sensor voltage is expected at that point where the two regimes coincide [1]. This, as may be confirmed from Eqs.(6) - (8) by eliminating y and equating the expressions for x , occurs at a critical value λ_M of the inverse redox ratio [5] $\lambda = 2/(R_2 + R_4)$ given by

$$\lambda_M = (p_2^{\infty} \lambda_c + p_4^{\infty} \lambda_H)/(p_2^{\infty} + p_4^{\infty}) \quad (9)$$

where $\lambda_c = (1+b)/\beta_2$ represents the position of the voltage step [1,2] for the extreme case ($p_4^{\infty} = 0$) of only one reducing gas, CO, in the gas mixture. Similarly, $\lambda_H = (1+b)/\beta_4$ represents the position of the voltage step for the extreme case ($p_2^{\infty} = 0$) of only one reducing gas, H_2 , in the gas mixture.

4. DISCUSSION

In order to obtain more specific indications of the variation of the sensor output with varying gas composition, it is necessary to work with a detailed chemical rate theory prescription for all atomistic processes occurring on the electrode surface. Such a process has been carried out for the reaction scheme enumerated earlier using extensions of the method for one reducing gas that is described in our earlier investigation [2]. While it appears inappropriate to enter into the full detail here, some examples of the outcome of such detailed computations afford a clearer insight into the nature of the predictions of the model now considered. As an example, a selection of results for reducing gas mixtures is presented in Figs.1 to 3, corresponding to some specific values of the set of parameters. The cases of only CO and only H_2 in the mixture are shown as broken curves in Fig.1, while the solid curve describes the emf when both CO and H_2 are present in the mixture. These curves were obtained for the partial pressure values shown in the figure. The shift in λ_M due to changes in the diffusion and adsorption of the reducing species relative to those of oxygen is shown in Fig.2. The parameter γ_4 is the ratio of the adsorption rate of oxygen to that of hydrogen. In Fig.3, we depict in more detail the changes in the sensor emf as the adsorption rate of H_2 is varied with respect to that of O_2. The rule of mixtures, Eq.(9), holds

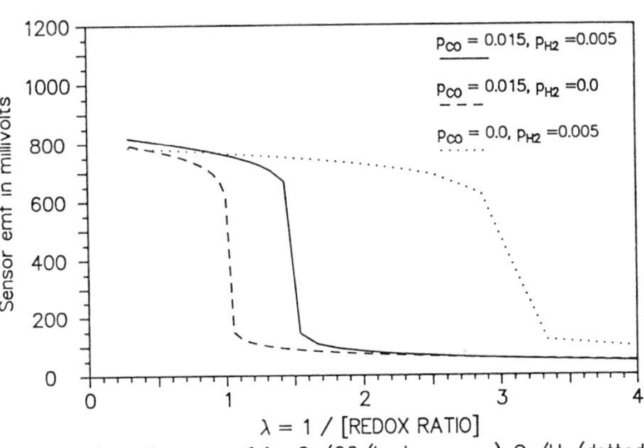

Fig.1 : Sensor emf for O_2/CO (broken curve), O_2/H_2 (dotted curve) and the mixture O_2/CO/H_2 (solid curve).

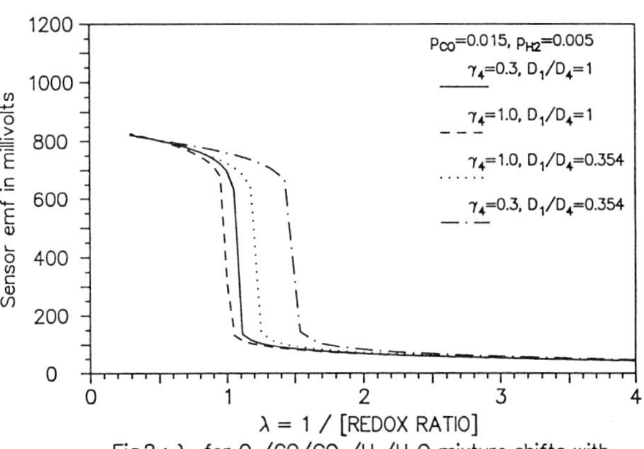

Fig.2 : λ_M for O_2/CO/CO_2/H_2/H_2O mixture shifts with different diffusion and adsorption ratios.

for the transition point in each case. Finally, Fig.4 depicts the change in the value of λ_M with the ratio of partial pressures of reducing gases. Using the known results [5] for the ratio of CO to H_2 in the combustion products of selected fuels, one derives the different values of λ_M marked on Fig.4 for benzene, gasoline, isooctane, ethane and methane. The water-gas reaction does not change the general conservation conditions expressed by Eq.(3) and hence does not affect λ_M . In summary, the formalism appears to be able to describe correctly the observed trends in the sensor behavior. However, much more work remains to be done before the modeling of this technically important sensor can be claimed to have reached full maturity.

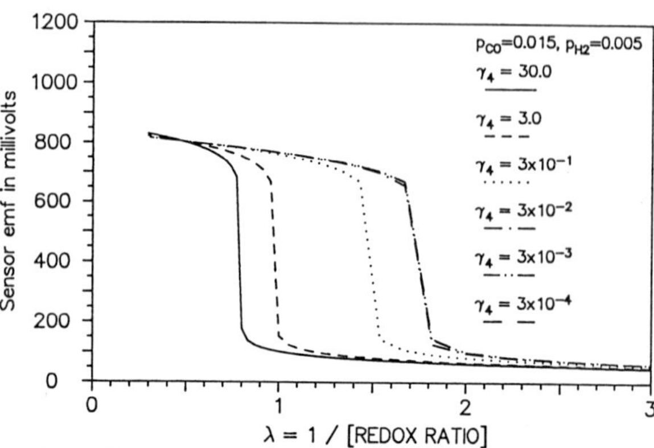

Fig.3 : Changes in sensor emf due to incease in H_2 adsorption rate (= const/ γ_4) in $O_2/CO/CO_2/H_2/H_2O$ mixture.

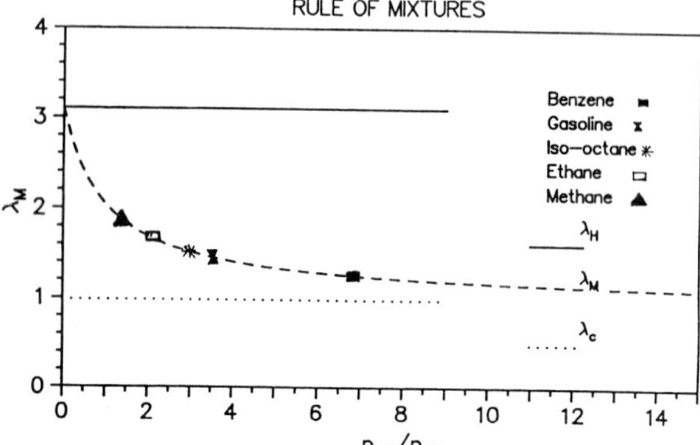

Fig.4 : Transition point λ_M as a function of p_{CO}/p_{H2} in the $O_2/CO/H_2$ mixture (assumed values: λ_c =1.0 and λ_H =3.1). Special points denote λ_M for different fuels.

5. REFERENCES

1. A.D.Brailsford and E.M.Logothetis, *Sensors and Actuators*, 7(1985) 39-67.

2. A.D.Brailsford, M.Yussouff and E.M.Logothetis, 4th International Meeting on Chemical Sensors, Tokyo, Sept. 13-17 (1992).

3. E.M.Logothetis, Advances in Ceramics, 3, p. 388 (A.H.Heuer and L.W. Hobbs, ed. American Ceramic Soc.) (1981).

4. T.Wang et al, Paper # 930352, SAE Automotive Engineering Congress, Detroit, March 1993.

5. J. F. Hepburn , Paper # 920799, SAE Automotive Engineering Congress, Detroit, February 1992.

An optical fibre based vapour sensor for spoilage in meat

V. P. Shiers, C. L. Honeybourne
Faculty of Applied Sciences
The University of the West of England, Bristol, BS16 1QY

ABSTRACT: A sensor is described to detect diacetyl, one of the chemicals that evolves from red meats at the onset of spoilage. This method employs optical fibres to transmit light to a remote detection point. The transmission spectra of several dyes coated onto filter paper are modified in the near Ultra Violet region on exposure to diacetyl vapour. The use of quantum mechanics and molecular modelling to 'shift' the transmission change to higher wavelengths i.e. to produce a visible colour change, is discussed.

1. INTRODUCTION

Sensor research has reached the forefront of technology in the past decade. Sensors are available to measure a wide range of physical[1-3], chemical[4-7], biochemical[8,9] and biomedical[10] parameters. Of interest here are optical chemical sensors. At present there are sensors of this type that monitor the presence of toxic gases[11-15] as well as oxygen[16,17], anaesthetic halothane,[18] etc. Here a sensor is proposed that is designed for use in the food industry. An optical fibre based chemical sensor is described that will detect spoilage in meat.

It is documented in the literature[19,20] that several chemicals evolve from meat during the spoilage process and that specific chemicals evolve during certain stages of this process. Some of the vapours that evolve from bad meat are ethanol, 2-methyl propanol, 3-methyl-1-butanol, 2,hydroxybutanone, ethyl acetate, toluene, 1-pentanol, dimethyl sulphide and methanethiol. If one of these compounds were present from the onset of spoilage and if a sensor could detect it then in turn the meat spoilage would be recognised. The chemical chosen that fulfils these criteria was butane 2,3 dione or diacetyl as it will be referred to here. The most common reaction of diketones, such as diacetyl is with diamine compounds (Figure 1). Therefore the method used to detect the diacetyl vapour was by measuring the near Ultra Violet / visible spectra of diamine dyes before and after exposure to diacetyl vapour. This was done using optical fibres to transmit the light so that remote sensing was possible. The equipment used for these experiments is shown in Figure 2.

2: EXPERIMENTAL

In total 15 dyes have been used which react with diacetyl vapour, due to limitations on space only two will be discussed here. These are 2,3 diaminonaphthalene both the acidified and non acidified forms are used and a chemical referred to as Dye 3. This is a dye synthesised specifically for the purpose of reaction with diacetyl. It is a metal complex of a diamine dye and it is investigated to see if the metal catalyses the vapour reaction in any way.

Figure1: Dye - Diacetyl reaction scheme

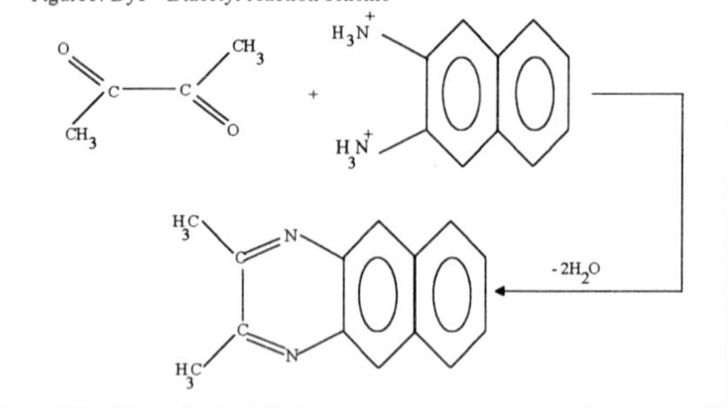

A method of delivering the vapour of liquids of low volatility using diffusion tubes was employed[21]. This involved quantitatively determining the amount of vapour that evolved up a narrow diameter (4mm) tube. The diacetyl was kept in an enclosed flask and the tube dipped into it. The amount of diacetyl diffusing out of the top of the tube was measured gravimetrically and spectrophotometrically. A supply of blended air flowed over the top of the tube and took with it a known amount of the diacetyl vapour. Thus a low concentration of vapour can be delivered into the optical flow chamber. The concentration of vapour in the flow is dependent upon the carrier gas (air) flow, temperature of the vapour and the dimensions of the diffusion tube.

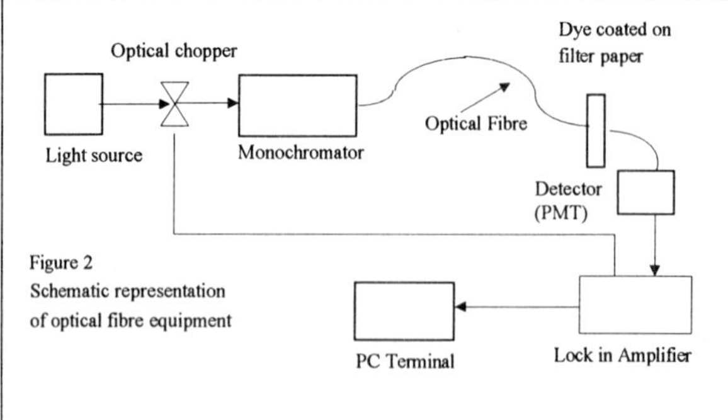

Figure 2
Schematic representation
of optical fibre equipment

A dye solution of known concentration was made up in ethanol and then acidified by adding an excess of concentrated hydrochloric acid. A small amount of this solution (10μl) was then added using a micro syringe to a piece of filter paper. The solvent was allowed to air dry, resulting in a small amount of dye being deposited onto the filter paper. This was then placed in-between two plastic optical fibres as shown in figure 2. The first fibre takes the light from the monochromator and directs it onto the paper; the second fibre collects the transmitted light and guides it to the photomultiplier tube for detection. The filter paper and the incoming and outgoing optical fibres were enclosed in a sealed vapour flow rig, directly linked to the delivery apparatus. The UV / visible transmission spectrum of the dye was then measured. The vapour was allowed into the chamber for five minutes then the transmission spectrum was recorded again. If the data from the first (blank dye) run is L, and the data from the second run (exposed dye) is M, then,

$$\frac{M}{L} \times 100 = \text{Normalised \% transmittance} \qquad \text{(Equation 1)}$$

For all the dyes examined, several experiments were carried out; the effect of the initial dye concentration on the extent of transmittance change; the effect of decreasing concentrations of diacetyl vapour; observation of reaction, if any, when the dye solution is not acidified; the relationship between number of 10μl additions or 'loadings' deposited onto filter paper and transmittance change. These experiments enabled the limits of detection to be evaluated for all the dyes and from these results the 'best' available dye was selected for analysis of real meat samples. The extent of selectivity and reversibility was also studied.

3: RESULTS

The dyes selected were initially exposed to a high concentration of diacetyl (Spectra 1, 2 & 3). The purpose of this was to discover the major characteristics of the transmittance spectra i.e. where peaks occur in the near UV and visible regions. The dye solutions tested were both acidified and non acidified with concentrated hydrochloric acid. These spectra are calculated using Equation 1 above.

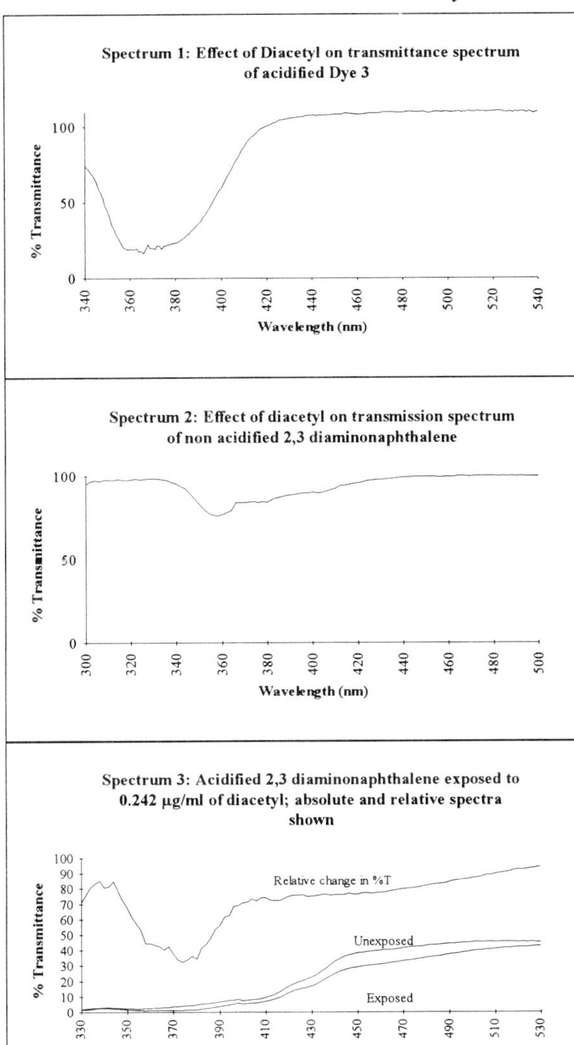

No reaction is observed for the non acidified dye 3, however the acidified equivalent (Spectrum 1) exhibits a peak at 355 - 385nm. Non acidified 2,3 diaminonaphthalene (Spectrum 2) has a peak at 355 - 365nm. The acidified form of this dye (Spectrum 3) has a large peak at approximately 360 - 385nm and a broad shoulder well into the visible region. This shoulder is responsible for a faint visible colour change that appears when concentrated dye solutions are used. In addition, Spectrum 3 shows the absolute transmittance spectra of unexposed and exposed acidified 2,3 diaminonaphthalene. The absolute, unexposed spectrum has very low transmittance up to 400nm and so any modification due to the vapour reaction appears as a large percentage change. However between 430 and 530nm there is a reduction in transmittance for the exposed dye, but because the absolute value is considerably higher, the percentage change appears a lot smaller in magnitude.

The effect of decreasing diacetyl concentration was examined for all the dyes. Vapour concentrations of 0.242, 0.11, 0.0367, 0.025 and 0.013µg/ml of diacetyl in air were delivered to the flow chamber. It was found, as expected that as diacetyl concentration increases the change in % transmittance increases. However for each dye the extent of change was different. It can be seen from spectrum 4 that the acidified 2,3 diaminonaphthalene is more sensitive (i.e. higher transmittance change is observed) over the range of all vapour concentrations used. In addition, for all dyes tested, the acidified form of the dye is always more sensitive than the non acidified. In fact often the non acidified dye showed no evidence of reaction at all. It was found that the acidified form of 2,3 diaminonaphthalene was the most sensitive to the lowest concentration of diacetyl and this sensitivity was increased by employing the following methods.

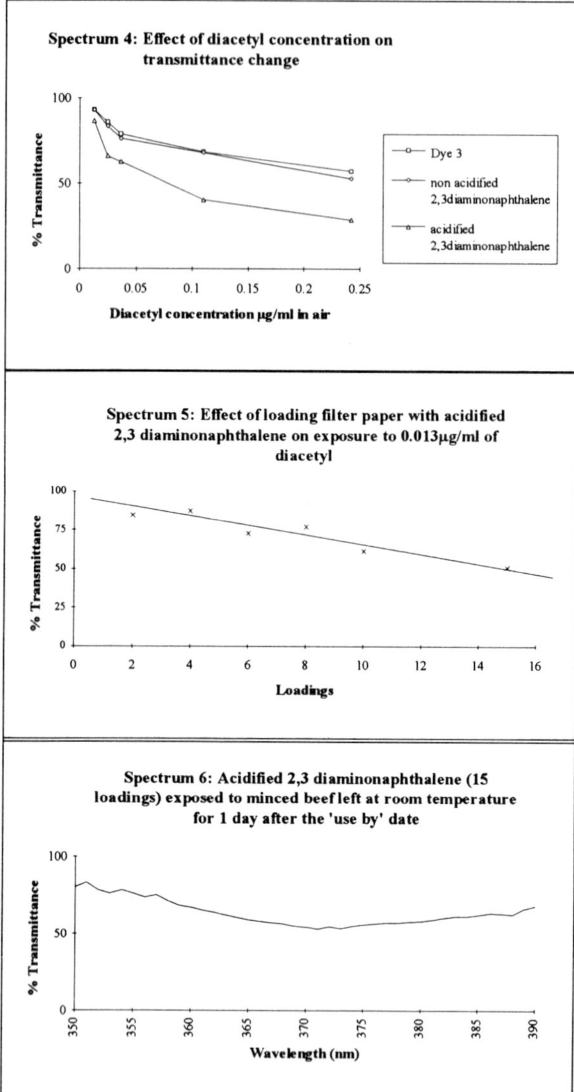

It has been shown by experiment that sensitivity increases by increasing the concentration of the initial dye solution up to a limit at which point the sensitivity actually begins to decrease. Therefore the optimum concentration of acidified 2,3 naphthalene (0.02mol/litre) was used for the 'loading' process. In this case it was found that change in % transmittance increased with loadings and it was decided that 15 loadings would be taken as the means for preparing the best dye for vapour reaction. As can be seen from Spectrum 5 the percentage transmittance improves by approximately 30% with 15 loadings. This result was expected since, with increasing loading of the dye onto the paper, the effective thickness of the paper is increased and so the absolute transmittance will inevitably be reduced. This in turn means that any spectral change on exposure to the diacetyl will result in a larger % transmittance change.

Using this dye set-up a sample of real meat was analysed. A portion of minced beef was placed inside the optical flow chamber along with the dye on the filter paper. The meat had reached its 'use by' date. Analysis of the dye after keeping the meat at room temperature for 24 hours revealed a decrease in transmittance of about 50% (Spectrum 6).

The reaction time was followed by a fixed (peak) wavelength run against time. The dye (15 loadings) was placed in the optical flow cell and the experiment was started. After several minutes the diacetyl flow was turned on, and when reaction was complete (4 - 5 minutes), i.e. no further change in transmittance was observed, the vapour was turned off and the flow cell was purged with pure air. There was no change in transmittance and so the dye - diacetyl reaction was irreversible.

Although diacetyl was used here as a measure of meat spoilage, other compounds are also known to evolve from spoiling meat; 2 hydroxybutanone (acetoin), ethylacetate and toluene are substances that fall into this category and these were tested for any response to the 'best' dye. A small reaction is observed for high concentrations of acetoin but an appreciable diacetyl reaction still occurs to a dye previously exposed to acetoin. No reaction was recorded with the other compounds.

4. QUANTUM MECHANICAL MODELLING

It has been shown that the major transmittance changes occur in the near Ultra Violet region usually between 350 and 390nm. It is desired that this transmittance peak is shifted into the visible. To do this a quantum mechanical modelling program was used. The molecular structure of a dye was fed into the computer and the program calculates the effect that diacetyl will have upon the electronic spectrum. In this way it was possible to add extra functional groups or ring systems to a molecule in order to 'synthesise' theoretically a molecule that will produce a visible colour change. It is believed from results so far that addition of further benzene rings to the molecule will increase the wavelength of the spectral change.

5. APPLICATIONS

If a dye were available that exhibited a visible colour change, it is proposed that it could be printed onto the surface of a gas permeable meat pack. In this way the diacetyl vapour could evolve through the packaging and react with the dye. This would be a more useful indication than the present method of 'use by' dates, which always err on the side of caution.

A network of optical fibres could be incorporated into the quality control system of a meat refrigeration warehouse. The dye could be coated onto the ends of a multiplexed fibre system, and air surrounding the individual consignments could be wafted onto the dye. If the diacetyl is detected, the reflection spectrum of the dye would be modified. This could be constantly monitored remotely, in the quality control room.

The production of a hand held device could improve other aspects of meat testing. Presently in some countries the freshness of large quantities of meat is detected by sensory methods (i.e. the human nose). The equipment in figure 2 could be adapted using an LED as the light source, filters instead of the monochromator and a photodiode as the detector. Because the dye - diacetyl reaction is irreversible a reel of dye coated filter paper or nylon strip could be employed so that the after each sampling the reel could be automatically spooled on.

6. CONCLUSIONS

It has been shown that several diamine dyes react with diacetyl vapour as an indicator for spoilage in meat. Acidification of the dyes either causes reaction to occur or enhances the extent of transmittance change. The extent of spectral modification is found to be dependent upon the concentration of initial dye solution, concentration of vapour exposed to the dye and the thickness of the dye on the filter paper (i.e. loadings). The quantum mechanical modelling has been used to develop the best structure of a dye that exhibits a visible colour change. The dye was found to be irreversible, but that is not a problem, since there must be no ambiguity concerning the freshness of meat; once the meat has spoiled it does not get any better. Optical dye sensors could be developed using the methods described above to detect spoilage in fish, fruit, dairy produce and even beer.

7. REFERENCES

1.C. Mariller, M. Lequine; SPIE **798** *Fiber Optic Sensors II* 1987 121-130
2. A. Hartog, G Gamble; *Physics World* 1991 March 45-49
3. J. Turan, S. Petrik; *Chemike Listy* 1989 **83**(5) 521-530
4. A.L. Harmer, R. Narayanaswamy; *Chemical Sensors*; 1987 ed. T.E. Edmonds Chapter 13
5.W. Rudolf Seitz; *Analytical Chemistry* January 1984 **56**(1) 16A-34A
6. W.Rudolf Seitz; *CRC Critical Reviews in Analytical Chemistry* 1988 **19**(2) 135-173
7. John O.W. Norris; *Analyst* November 1989 **114** 1359-1372
8. C.A. Villarruel, D.D. Dominguez, A. Dandridge; SPIE **798** *Fiber Optic Sensors II* 1987 225-229
9. R. Narayanaswamy; *Biosensors & Bioelectronics* 1991 **6** 467-475
10. P.Bechi, F.Pucciani, R.Falciai, F.Baldini, F.Cosi, A.Bini, F.Milanesi; *Sensors & Actuators B* 1992 **7** 775-779
11. R. Narayanaswamy, F. Sevilla,III; *Analyst* April 1988 **113** 661-663
12. S.A. Momin, R. Narayanaswamy; *Analytica Chimica Acta* 1991 **244** 71-79
13. M. Archenault, H. Gagnaire, J.P. Goure, N. Jaffrezic-Renault; *Sensors & Actuators B* 1992 **8** 161-166
14. J.M. Charlesworth, C.A. McDonald; *Sensors & Actuators B* 1992 **8** 137-142
15. Shinzo Muto, Akitoshi Ando, Tatsuo Ochiai, Hiroshi Ito, Hitoshi Sawada, Akira Tanaka; *Japanese J. Applied Physics* 1989 **28**(1) 125-127
16. R. Narayanaswamy; SPIE **798** *Fiber Optic Sensors II* 1987 249-252
17. F. Baldini, M. Bacci, F. Cosi, A. Del Bianco; *Sensors & Actuators B* 1992 **7** 752-757
18. J.A. Barnard Howie, P. Hawkins; *Analyst* 1993 **118** 35-40
19. R.H. Dainty, R.A. Edwards, C.M. Hibbard; *J. App. Bacteriol*; 1985 **59** 303-309
20. G. Stanley, K.J. Shaw, A.F. Egan; *Appl. & Env. Microbiology* 1981 816-818
21. A.P. Altshuller, I.R. Cohen; *Anal Chem* 1960 **32** 802-810

8. ACKNOWLEDGEMENTS

V.P.S. wishes to thank the AFRC for a research assistantship and C.L.H. wishes to thank the AFRC for the receipt of Grant FG44/514.

Gas sensing comparative properties of $LuPc_2$ and CuPc

A. Pauly, J.-P. Blanc, S. Dogo, J.-P. Germain, C. Maleysson, M. Passard
Laboratoire d' Electronique (U.R.A. 830 C.N.R.S.), Université Blaise Pascal,
63177 AUBIERE Cedex - FRANCE

ABSTRACT : $LuPc_2$ and CuPc thin films are doped at different temperatures with various concentrations of NO_2 gas diluted in N_2, and their conductivities are recorded in-situ during doping and dedoping. A model is developped to explain the conductivity variations with time and with NO_2 concentration. Results show the influence of oxidation potential value of the phthalocyanine in its ability to be used as oxidizing-gas sensing material.

1. INTRODUCTION

Phthalocyanines (Pc's) are known to exhibit large conductivity enhancements on exposure to oxidizing gases such as NO_2. This property has initiated a great number of studies on the use of Pc's as sensitive layers in semiconductor-type gas-sensors. Attention has essentially been focused on metallic phthalocyanines such as PbPc, CuPc, ZnPc [1] and little work has been devoted to less "classical" phthalocyanines. We have been interested in the behavior of lutetium bisphthalocyanine ($LuPc_2$) as sensing material for oxidizing gas because of its low oxidation potential compared to other Pc's. In this paper, the sensitivity to NO_2 of thin films of $LuPc_2$ is compared to that of CuPc. A model is proposed to describe the conductivity variations of both Pc's upon NO_2 doping. Application to NO_2 sensing is considered.

2. EXPERIMENTAL DETAILS

$LuPc_2$ and CuPc were synthesized at E.S.P.C.I.# Paris. They were sublimed under vacuum and deposited as thin films (300 nm thick) on alumina substrates at room temperature. The substrates (3 mm x 5 mm) were fitted with interdigital Pt electrodes on one side for measurements of the film conductivities and a heating resistor on the other side. Films were doped by a flow of NO_2 diluted in N_2 (at concentrations of 2 to 300 ppm) and dedoped by pure N_2 flow (60 l/h). Throughout each doping-dedoping experiment, the film temperature was kept constant and conductivity was recorded versus time.

3. RESULTS

3.1 $LuPc_2$

$LuPc_2$ thin films were doped at various temperatures with 10 ppm NO_2. A new film was used for each doping, all films being deposited at the same time.

During doping time σ increases up to a maximum then decreases, followed by a σ increase for dedoping in N_2 flow (fig. 1a). This is observed at T = 25 °C and T = 60 °C. At T = 100 °C the conductivity maximum is wide and on dedoping with N_2 the conductivity increase is weak (fig. 1b). The film behavior at 160 or 200 °C is different : a conductivity maximum is still observed during the doping phase but dedoping in N_2 makes σ decrease (fig. 1c).

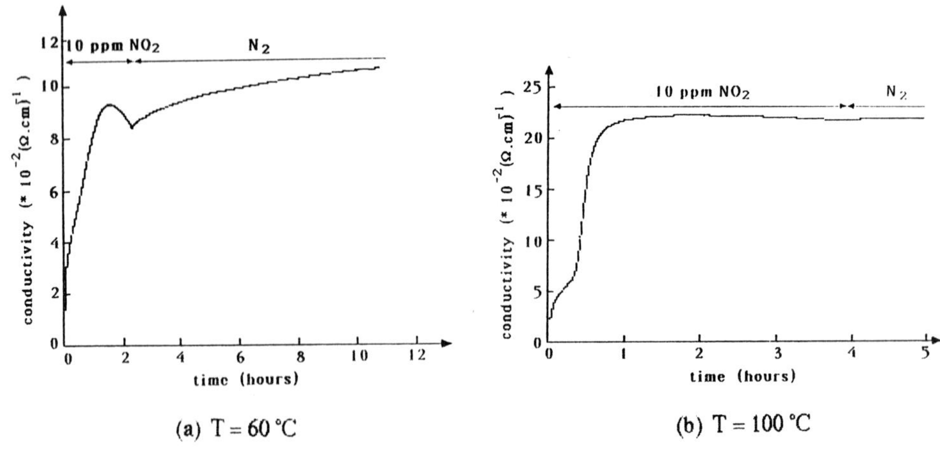

(a) T = 60 °C

(b) T = 100 °C

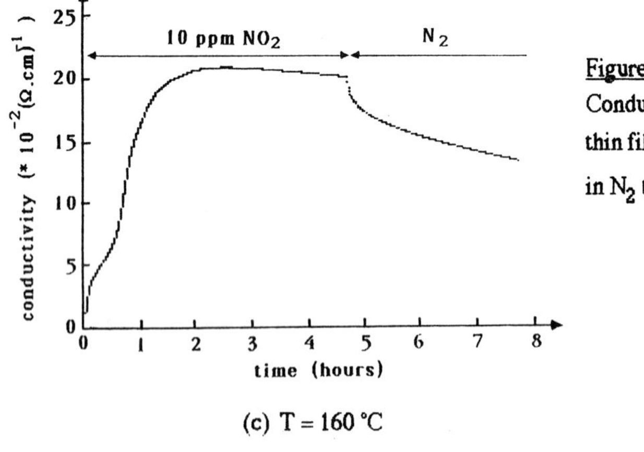

(c) T = 160 °C

Figure 1 :
Conductivity versus time of $LuPc_2$ thin films exposed to 10 ppm NO_2 in N_2 then to pure N_2 at:
a) 60 °C
b) 100 °C
c) 160 °C

3.2 CuPc

For each experiment temperature (25, 60, 100, 160 °C) a new CuPc film was used. On each film, heat-treatment at 200 °C for 10 minutes in N_2 flow, NO_2 doping and N_2 dedoping were repeatedly applied for the successive NO_2 concentrations c = 2, 10, 50, 100 and 300 ppm. Depending on the dopant concentration, the conductivity either reaches a more or less stabilized value (for c = 2, 10, 50 or 100 ppm) or goes through a maximum and decreases (for c = 300 ppm), but on N_2 dedoping a decrease of σ is always observed. The conductivity maxima are summed up in table 1 and conductivity changes with time at T = 25 °C are shown on figure 2.

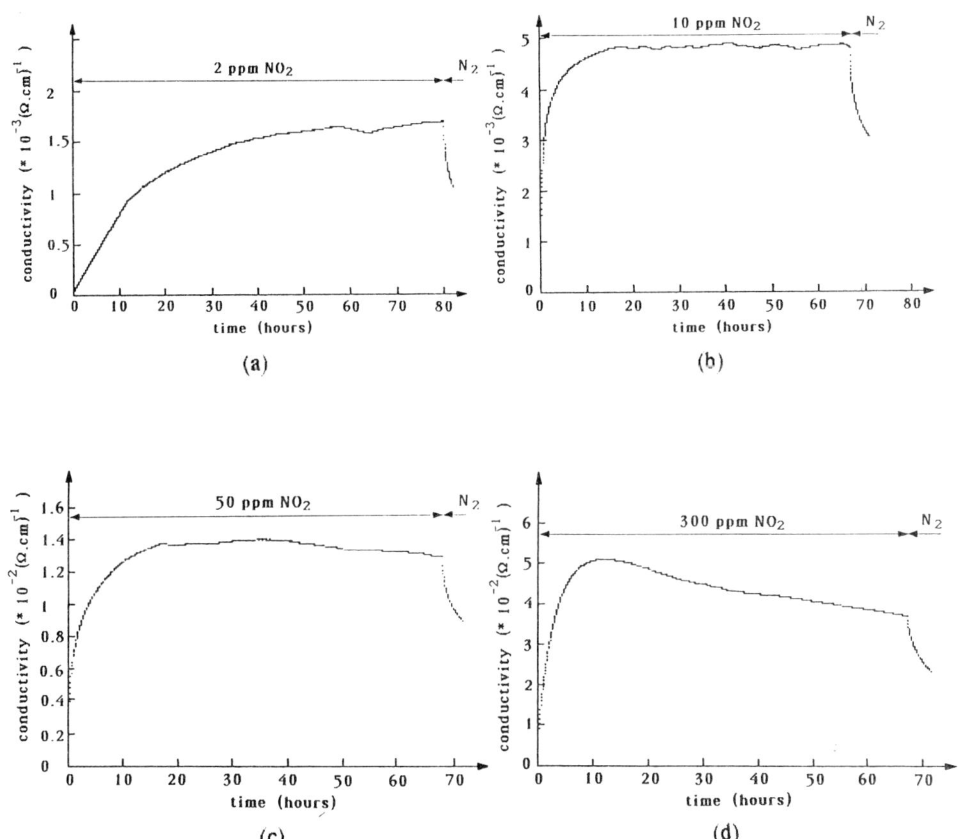

Figure 2 : Conductivity versus time of a CuPc thin film, at 25 °C, during doping with : a) 2 ppm, b) 10 ppm, c) 50 ppm, d) 300 ppm of NO_2 in N_2, and dedoping with pure N_2.

TABLE 1 : conductivity maxima of CuPc films (σ, in $(\Omega.cm)^{-1}$) during doping at various concentrations of NO_2 in N_2 (c, in ppm) and various temperatures (T, in °C).

T	c = 2	c = 10	c = 50	c = 100
25	1.8×10^{-3}	4.7×10^{-3}	1.4×10^{-2}	
60	$\approx 5.3 \times 10^{-4}$	1.8×10^{-3}	3×10^{-3}	
100	9×10^{-4}	1.5×10^{-3}	2.5×10^{-3}	
160	1.2×10^{-3}		2.3×10^{-3}	5.5×10^{-3}

4. DISCUSSION

4.1 Model of gas - phthalocyanine interaction

First. the NO_2 gas is absorbed in the film. Electrons are transfered from MPc to some NO_2 already adsorbed, and the holes left in the phthalocyanine can be delocalized according to the following equilibria :

$$\begin{array}{c} MPc \\ + mNO_2gas \end{array} \underset{\text{absorption}}{\rightleftharpoons} MPc,mNO_2abs \underset{\text{charge transfer (CT)}}{\overset{E_{CT}}{\rightleftharpoons}} MPc^+,NO_2^-,(m-1)NO_2abs \underset{\text{charge delocalization (D)}}{\overset{E_D}{\rightleftharpoons}} \begin{array}{c} MPc,NO_2^-,(m-1)NO_2abs \\ + hole \end{array}$$

m: number of adsorbed molecules per MPc molecule.

Absorption includes physisorption of the gaseous NO_2 molecules on the film surface, their diffusion in the film bulk and the displacement by NO_2 of other gaseous species previously adsorbed on the same sites. From Nernst equations the extrinsic hole concentration in the doped film can be deduced :

$[hole] = \sqrt{[NO_2abs]^m \cdot [MPc]} \exp(-\Delta E/2kT)$ with $\Delta E = (E_{CT} + E_D)$, where :

$[MPc]$ = concentration of molecular units of macrocycles MPc

$[NO_2abs]$ = concentration of molecular units of NO_2 absorbed in the Pc film.

Neglecting the intrinsic carriers, then the film conductivity is : $\sigma = e \cdot \mu \cdot [hole]$ where μ is the mobility of the holes in the phthalocyanine.

At equilibrium, the concentration of absorbed gaseous molecules can be related to the external gaseous partial pressure p (proportional to c) with the help of Langmuir or Freundlich isotherm : $[NO_2abs]$ is proportional to p^n , $0 \le n \le 1$, and σ is proportional to $c^{n.m/2}$ at constant temperature.

4.2 Particular case of $LuPc_2$

$LuPc_2$ is characterized by a low oxidation potential (0.19 V vs SCE) compared to other phthalocyanines [2]: E_{CT} is very low, so it is more easily doped by oxidizing gases.

During NO_2 doping, holes are created by oxidation of $LuPc_2$ into $LuPc_2^+$. At low doping level hole mobility μ can be taken as constant. But when doping level increases, the concentration of ionized species $LuPc_2^+$ becomes so important that hopping of holes from one oxidized molecule to a neutral one is more and more difficult ($LuPc_2^+, LuPc_2^+ \longrightarrow LuPc_2^{++}, LuPc_2$ is very unlikely because it requires 1,2 eV [3]). When $[LuPc_2^+]$ exceeds a value c_0, μ drops; this effect can overcome the increase in hole concentration due to the doping going on, and consequently $\sigma = e.\mu.[hole]$ decreases. On dedoping NO_2 molecules desorb and $[LuPc_2^+]$ decreases, freeing molecules for the remaining holes to hop; μ increases back and so does σ; when $[LuPc_2^+]$ is low enough so that μ can be considered as constant, then σ decreases at the same rate as [hole].

The experimental variations of $LuPc_2$ conductivity observed during NO_2 dopings at T = 25, 60, or 100 °C (§ 3, fig. 1 a and b) can be described by this model : the conductivity maximum is caused by the decrease in the charge carrier mobility when $[LuPc_2^+]$ in the doped part of the film becomes too high [3].

At T = 160 or 200 °C (fig. 1 c), the conductivity maximum cannot be ascribed to the same phenomenon since σ decreases on dedoping; because of the high temperatures, the absorption equilibrium is displaced to the left, so the concentration of gaseous molecules interacting with the film is low and $[LuPc_2^+]$ remains less than c_0; the conductivity maximum is probably due to destruction of the film.

4.3 Case of CuPc

Because the oxidation potential of CuPc (1.0 V vs SCE) is higher than that of $LuPc_2$, the concentration of CuPc molecules oxidized by NO_2 (in the concentration range 2 to 300 ppm) is never high enough to cause the drop of μ. On doping the CuPc film conductivity increases up to a stable value (fig. 2 a and b) and it decreases on dedoping; high concentration of NO_2 (300 ppm) causes the destruction of the film and makes σ decrease during doping (fig. 2 d).

5. APPLICATION TO GAS SENSING

Compared to CuPc, $LuPc_2$ cannot be a good material for strong oxidizing gas sensing because of its particular behavior on doping : the conductivity maximum may be not directly dependent of gas concentration. As previously shown [4], CuPc exhibits interesting properties for gas sensing : exposure to NO_2 makes σ dramatically increase by many orders of magnitude and it is sensitive to very small NO_2 concentration variations. From the experimental results of NO_2 dopings, the conductivity dependence on NO_2 concentration c can be plotted as log σ vs log c for each temperature; straight lines of slope α are obtained showing good agreement with the model since σ has been derived to be proportional to $c^{n.m/2}$, with $n.m/2 = \alpha$.

By definition α is the relative sensitivity of the MPc film to the doping gas: $\alpha = (\Delta\sigma/\sigma)/(\Delta c/c)$

TABLE 2 : relative sensitivity α of CuPc thin films to NO_2 in the concentration range 2 to 50 ppm at various film temperatures (T, in °C).

T	25	60	100	160
α	0.64	0.54	0.32	0.28

Relative sensitivity of CuPc to NO_2 decreases when film temperature increases; this behavior has already been observed with other phthalocyanines [5].

6. CONCLUSION

Our experimental results show that application of phthalocyanine thin films to gas-sensing is no more possible when the material is too much oxidized. With a strong oxidizing gas such as NO_2, the strong concentration of ionised phthalocyanines is reached with very low partial pressures of NO_2. This is the case of $LuPc_2$ and is not found with NO_2 partial pressures up to 300 ppm in the case of CuPc; this implies that $LuPc_2$ is problematic to be used as NO_2 gas sensor but could be a good sensor material for weak oxidizing gases, whereas phthalocyanine with relatively high oxidation potential (eg PcCu) are suitable for sensing strong oxidants.

7. REFERENCES

[1]: J.D. Wright, Progress in surface Science, 31 1989 1.
J.D. Wright, Molecular Crystals, Cambridge, Cambridge University Press, 1989.

[2]: B. Boudjema, Thèse d' Etat, Lyon I, 1987, p. 97

[3]: A. Pauly, J.-P. Blanc, S. Dogo, J.-P. Germain, C. Maleysson, Synth. Met., 57 (1993) 3754.

[4]: S. Dogo, J.-P. Germain, C. Maleysson, A. Pauly, Thin Solid Films, 219 (1992) 244 and 251.

[5]: G. Berthet, J.-P. Blanc, J.-P. Germain, A. Larbi, C. Maleysson, H. Robert, Synth. Met., 18 (1987) 715.

: The authors express their deep gratitude to Professor J. SIMON and M. BOUVET for samples providing.

A laser-based sensor system for trace NO_x ($x = 1, 2$) detection in atmospheric air

A.Clark, R.M.Deas, C.Kosmidis*, K.W.D.Ledingham, A.Marshall, J.Sander and R.P.Singhal,
Department of Physics and Astronomy, University of Glasgow,
Glasgow G12 8QQ, Scotland.

M.Campbell and R.Zheng,
Department of Physical Sciences, Glasgow Caledonian University,
Glasgow G4 OBA, Scotland.

*Permanent address: Dept. of Physics, University of Ioannina, Ioannina, Greece.

ABSTRACT: The existence of NO_x (x=1,2) in the atmosphere has implications from an environmental viewpoint since it plays an important role in the formation of tropospheric O_3 and nitric acid (HNO_3) and also in the degradation of organic pollutants. This paper describes the use of Resonance Enhanced Multiphoton Ionisation (REMPI) to selectively detect trace quantities of NO photolysed from NO_2 in real air samples (at atmospheric pressure) in a miniature ionisation chamber. The present level of NO_2 detection is better than 150ppb and sensitivity enhancements of more than 50 are expected.

1.INTRODUCTION

It has been known for some time that combustion of fossil fuels and biomass burning principally have led to the presence of NO_x (x=1,2) in the atmosphere. Nitric oxide (NO) is one of the key species in the photochemistry of the troposphere especially in the production of tropospheric ozone and nitric acid. In addition NO_x molecules play an important part in the degradation of some organic pollutants. Further, the detection of NO can also be used to identify the presence of nitrocompounds e.g. explosives (1,2,3). Nitrogen oxides are present in the upper atmosphere where typical concentrations vary from ppm to ppb (by volume). The average concentrations of NO_2 in American cities is about 35ppb and in polluted air in both the U.S.A. and Europe these concentrations can exceed 150ppb. NO is present in similar concentrations(4). It is therefore against these concentrations that one must weigh the effectiveness of a detection technique. In this paper no distinction is made between NO and NO_2 since in the present procedure NO^+ is detected from NO directly or after photolysis from NO_2 and therefore the nitrogen oxides will be described generally as NO_x. It however should be pointed out that since a mixture of NO and NO_2 will normally be measured in atmospheric samples NO will be detected with greater sensitivity than NO_2(3). There are a number of different techniques for the detection of NO_x and these will be briefly reviewed. They fall roughly into two categories-chemical based and laser based. Some of the chemical techniques are:

Metal Phthalocyanine Thin films (5,6): These films become conductive when NO_2 is adsorbed on to them. Detection sensitivities are modest with sub ppm analysis difficult but they have the attraction of being cheap and easy to operate.

Ion Chromatography (7): Triethanolamine (TEA)-impregnated glass fibres can be made to collect NO_2 but 24 hour or indeed week long collection of samples is necessary to reach ppb sensitivity and it is subject to chemical interferences.

Chemiluminescence (8): NO reacts with ozone to form excited state NO_2. The excited NO_2 molecules then decay by emitting IR chemiluminescence in the range 600-3000nm, which can be a very sensitive detection procedure approaching ppt levels. There are difficulties with interferences and quantification for this sophisticated technique however.

There are a number of different laser based NO_x detection procedures using laser induced fluorescence (LIF) and resonance enhanced multiphoton ionisation, with and without jet cooling (9-11), in high vacuum time of flight mass spectrometers as well as atmospheric pressure ion chambers. The sensitivities reached by these techniques can be as low as ppt but all measurements have been made in ideal laboratory conditions using high purity gases or at high altitude (12). The purpose of this paper is to present a procedure which is easy to use, with real air samples, that can be used in an urban or industrially polluted area. In situ or real time (recorded) measurements of NO_x can be made so that the large variations of NO_x concentrations which occur with time of day, season, latitude and altitude can be analysed.

2 EXPERIMENTAL.

The experimental system is based on a technique used to measure the concentrations of aromatic compounds in air (13). The principle of detection of the NO molecules is via a resonant two photon ionisation process ($A^2\Sigma(v''=0) \leftarrow X^2\Pi(v''=0)$) at 226nm. The ions are detected in an ionisation chamber filled with ordinary air from a gas bottle or ordinary laboratory air at atmospheric pressure. A schematic diagram of the experimental set-up is shown in Fig. 1.

The laser source consists of a Lumonics EPD 330 dye laser which is pumped by a Lumonics TE 860M-3 excimer laser operated at frequencies of between 10-15Hz. Coumarin 47 and 102 laser dyes are used to span the wavelength range 448-524nm. UV radiation in the range 224-262nm is obtained by frequency doubling the

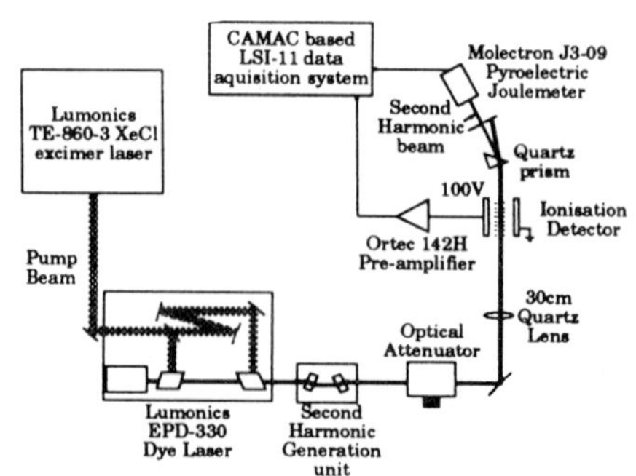

Fig. 1: Experimental Arrangement.

dye output using a BBO 'B' cut crystal. This is mounted on an Inrad Model 5-12 autotracking unit which adjusts the crystal angle for optimum phase matching during a wavelength scan. UV light obtained from this system has a pulse duration of 6ns, a peak flux of 500kWcm^{-2} and a bandwidth of 0.015nm. The beam intensity within the chamber was measured using a Molectron J3-09 joulemeter.

A Newport attenuator was operated manually to ensure that the photon flux remained reasonably constant during the course of a wavelength scan. The miniature ionisation chamber is shown in Fig.2.

It consists of two parallel stainless steel plates of dimensions 2x2.5cm with a separation of 1.8cm which are placed in cast metal box. A variable high voltage supply was connected to the plates via an Ortec 142H preamplifier and the system was normally operated at 100V. The position of the focussed laser beam was approximately 1mm from the positive electrode.

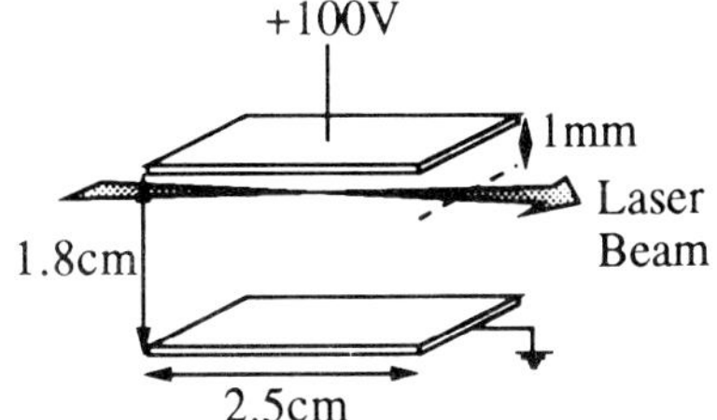

Fig. 2: Detector Parameters.

Gas samples were introduced to the container either by opening the box to laboratory air or by flowing calibrated air + 1ppm NO_2 (BOC special gases) through the counter at atmospheric pressure. The ionisation current is collected through the preamplifier, and if required, further amplified by a Tennelec TC 205A linear amplifier. The amplified signal is then digitised and stored on floppy disc using a CAMAC LSI-11 controlled ADC system.

3.RESULTS

Fig.3 shows a wavelength scan between 224-232nm in the ionisation chamber with air+1ppm NO_2 flowing through the system.

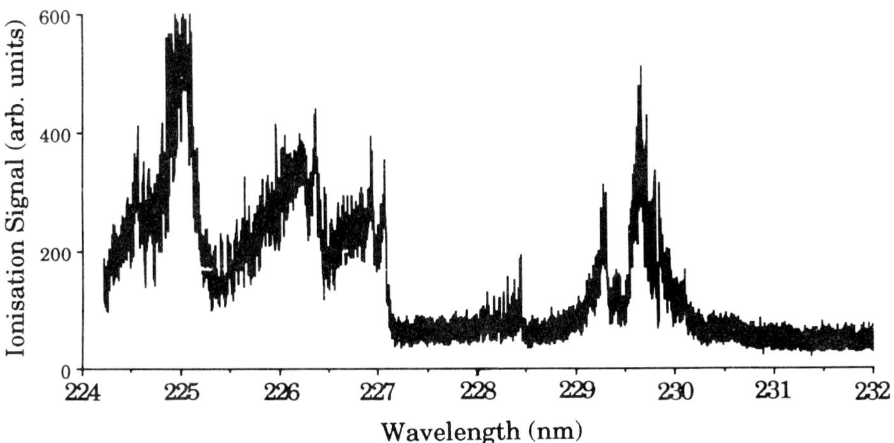

Fig. 3: Ionisation spectrum of 1ppm NO_2 gas in air.

The characteristic double headed two photon ionisation spectrum of NO at 226.4 and 227nm is clearly seen and the other features visible in the spectrum at 225nm, 228.5nm and between 229-230nm belong to multiphoton ionisation processes in molecular oxygen (11). Fig.4 is a background spectrum taken with static laboratory air and again shows the strong oxygen molecular peaks with a prominent oxygen atomic peak at 225.7nm. A typical spectrum of the laser flux recorded simultaneously with the ionisation signal is also shown.

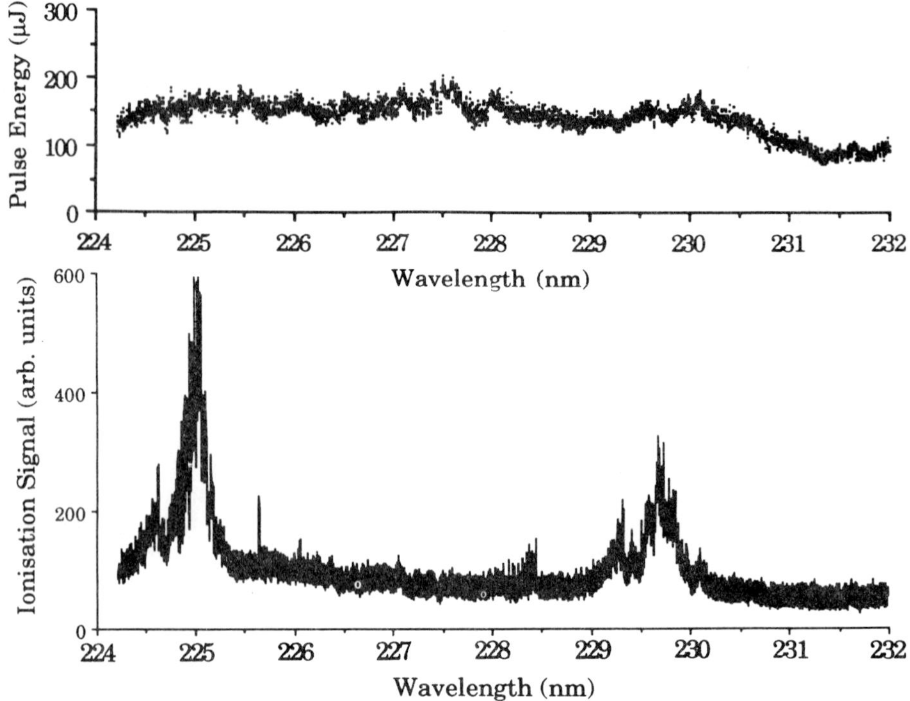

Fig. 4: Ionisation spectrum of laboratory air, with a typical laser pulse energy profile.

Just above the noise can be distinguished the background NO signal corresponding to a concentration level in the laboratory of NO_x at that time of about 150ppb. At the present stage of development of the procedure this represents the detection limit. This type of spectrum could be taken with the stainless steel plates out of the box and the only purpose of the container is to reduce the electrical interference when the laser is pulsing. Each of the points in the spectra shown in Figs. 3 and 4 corresponds to a single laser pulse and by recording and averaging over several thousand laser shots, as well as increasing the laser flux considerably, a sensitivity down to a few ppb is expected.

4. CONCLUSIONS

A sensor for the detection of NO_x has been developed. It is a laser based system operated in conjunction with a simple ionisation chamber. The laser, operating at about 226nm, photolyses the NO_x and then ionises the resulting NO neutral molecules via a two photon process. The ions are detected in the ionisation chamber. The procedure has been tested in open atmosphere conditions as well as with calibrated air samples. At present the sensitivity is about 150ppb but sensitivities down to a few ppb are confidently expected. The ionisation chambers can be made very small - a few ml of air is all that is necessary to make a measurement. The in situ and recorded time sampling of atmospheric conditions in urban areas consists of opening these small volume counters to the air, closing them and then transporting the counters back to the laboratory where they are irradiated by laser light and the ionisation compared with standard samples of NO_x in air. Alternatively if a measurement at a specific site is required, possibly as a function of altitude, the laser system can be housed in a mobile van and the laser directed vertically. The sensor can be elevated to any required height.

5. ACKNOWLEDGEMENTS.

This work has been carried out with the support of the Procurement Executive, Defence Research Agency. R.M.D and J.S. are indebted to SERC for postgraduate funding and R.Z. wishes to thank the Caledonian University for support.

REFERENCES

1) A.Marshall, A.Clark, R.Jennings, K.W.D.Ledingham, J.Sander and R.P.Singhal, Int. J. Mass Spectrom. Ion Proc., 116, 143-156, 1992.
2) A.Clark, K.W.D.Ledingham, A.Marshall, J.Sander and R.P.Singhal, to be published in The Analyst 1993.
3) G.W.Lemire, J.B.Simeonsson and R.C.Sausa, Anal. Chem., 65, 529-533, 1993.
4) J.A.Logan, J. Geophys. Res., 88, 10785-10807, 1983.
5) G.P.Rigby, A.Wilson and J.D.Wright, Sensors: Technology, Systems and Applications, Ed K.T.V.Grattan, Adam Hilger, Bristol 1991, 121-126.
6) Y.Sadaoka, T.A.Jones and W.Gopel, Sensors and Actuators B1, 148-153, 1990.
7) J.E.Sickles, P.M.Grohse, L.L.Hodson, C.A.Salmons, K.W.Cox, A.R.Turner and E.D.Estes, Anal. Chem., 62, 338-346, 1990.
8) O.C. Zafiriou and M.B. True, Environ. Sci. Technol., 20, 594-596, 1986.
9) S.Guizard, D.Chapoulard, M.Horani and D.Gauyacq, Appl.Phys. B., 48, 471-477, 1989.
10) M.Hippler, A.J.Yates and J.Pfab, Optogalvanic Spectroscopy, Inst. Phys. Conf. Ser., 113, 303-306, 1990.
11) J.C.Miller, Anal.Chem., 58, 1702-1705, 1986.
12) J.M.Hoell, G .L.Gregory, D.S.McDougal, A.L.Torres, D.D.Davis, J.Bradshaw, M.O.Rodgers, B.A.Ridley and M.A Carroll, J. Geophys. Res., 92, 1995-2008, 1987.
13) A.Marshall, A.Clark, K.W.D.Ledingham, R.P.Singhal and M.Campbell, Sensors: Technolgy, Systems and Applications, Ed K.T.V.Grattan, Adam Hilger, Bristol 1991,151-155.
14) A.Clark, C.Kosmidis, R.M.Deas, K.W.D.Ledingham, A.Marshall, J.Sander and R.P.Singhal, to be published.

Gas sensitivity of erbium diphthalocyanine

R.A.Collins, A.T.J.Parr and A.Krier
Applied Physics Division, School of Physics and Materials,
Lancaster University, Lancaster LA1 4YB, U.K.

ABSTRACT: The effects of exposure to chlorine, ammonia and nitrogen dioxide on the optical and conductive properties of $ErPc_2$ have been studied. Chlorine has the greatest effect, indicating that $ErPc_2$ may prove of interest in the development of a thin film chlorine sensor. However, a number of questions remain concerning synthesis of the material and the the gas sensing mechanisms.

1.INTRODUCTION

The metal phthalocyanines (MPc's) have assumed an increasingly important role in fundamental studies of gas sensitive semiconductors during the last two docados (Wright 1989). Lead phthalocyanine has received considerable attention with particular reference to the development of a nitrogen dioxide sensor (Bott and Jones 1984) and associated studies have concerned a variety of other MPc's. In recent years the metal diphthalocyanines (Moskalev and Kirin 1965) been the focus of increasing interest in view of their electrical and optical properties. There have been few studies of their gas sensitivity although preliminary work on $LuPc_2$ (Even et al 1992) and $YbPc_2$ and $DyPc_2$ (Jeffery and Collins) have indicated that they may be useful materials for the development of viable thin film gas sensors.

The present work is concerned with erbium diphthalocyanine prepared in-house and exposed to chlorine, ammonia and nitrogen dioxide. The material has been characterised in thin film form using a variety of analytical techniques and has been studied in terms of both its electrical and optical properties.

2.EXPERIMENTAL

$ErPc_2$ was prepared by the method of Moskalev and Kirin by reacting the dried metal acetate with 1,2 dicyanobenzene (phthalonitrile) in the molar ratio of 1:10. This mixture was heated in an air oven at 300°C for 30 minutes.This yielded a dark blue liquid. This was removed from the oven and allowed to cool in air forming a dark blue solid. Material was also prepared by increasing the heating time until the solid formed rather than removing the material from the oven whilst still in liquid form. The hardened material was ground to powder form and then washed sequentially in acetic anhydride, acetone, dimethyl formaldehyde (DMF), acetone and finally chloroform to remove any unreacted material. The remaining material was then dried to constant mass by heating at 150°C. For electrical and some optical measurements pre-cleaned polyborosilicate glass substrates were used. $ErPc_2$ thin films were prepared by thermal evaporation under vacuum. The films were typically of 200 nm thickness and were coloured green. Films were scanned immediately after preparation in a double beam Pye Unicam SP8-100 UV-Vis spectrophotometer between wavelengths 850 and 350 nm. Prepared powder was also dissolved in DMF or chloroform and spectra taken between the same limits to provide comparison with reference spectra for confirmation of material purity. Individual devices were then exposed to flowing chlorine gas for 30 seconds, sealed in phials containing the test gas and transported to the spectrophotometer. $ErPc_2$ was also vacuum sublimed onto dry NaCl discs for use in a Perkin Elmer 1720-X Fourier Transform Infra-red Spectrometer. Scans were performed in the range 1600 - 370 cm^{-1} (6.25 - 27 μm). These samples were also scanned before and after gaseous exposure. Additional measurements were taken by mixing small quantities of $ErPc_2$ with dry KBr powder and then pressing (13 tonnes pressure) into (blue opaque) discs for use in the FTIR spectrometer. Spectra were taken after exposure of the discs to chlorine and also for discs where the powders had been exposed prior to pressing.

Electrical characteristics were taken in a sealed dark cell. Each device array consisted of 8 devices which were tested concurrently. A d.c. voltage of 0.5 V was applied sequentially to each device and the d.c. dark current monitored automatically over a 3 hour test period.

3. OPTICAL RESULTS

Work on diphthalocyanines has been plagued by a number of problems. The most serious has concerned material preparation and specifically the period for which material should be baked at 300°C. Various authors have advocated different preparation conditions and there is some controversy (Moskalev and Kirin 1965, 1970, Clarisse and Riou 1987) concerning whether the solid left after filtration or the filtrate should be regarded as the phthalocyanine. The baking temperature is known to be of importance in the distinction between preparing mono- or diphthalocyanines. The present results relate to material baked for about 80 minutes at 300°C and then removed as a crystalline dark blue solid. Our preliminary studies using alternative conditions indicate the the $ErPc_2$ yield is considerably lower if the material is removed from the oven in the liquid phase.

Figure 1 shows UV-Vis spectra for $ErPc_2$ in DMF and chloroform and are in reasonable agreement with published data (Moskalev and Alimova 1975). Impurities in the DMF lead to a redox reaction which alters the colour of the solution according to the scheme

$$\underset{\text{blue}}{[M^{3+}Pc_2]^-} \Leftrightarrow \underset{\text{green}}{[M^{3+}Pc_2]^o} \Leftrightarrow \underset{\text{red}}{[M^{3+}Pc_2]^+}$$

This is the mechanism whereby diphthalocyanines change colour and in the present case the green solution turned deep blue over a period of hours due to transfer of electrons between the Pc_2 π orbital and impurities in the DMF solvent.

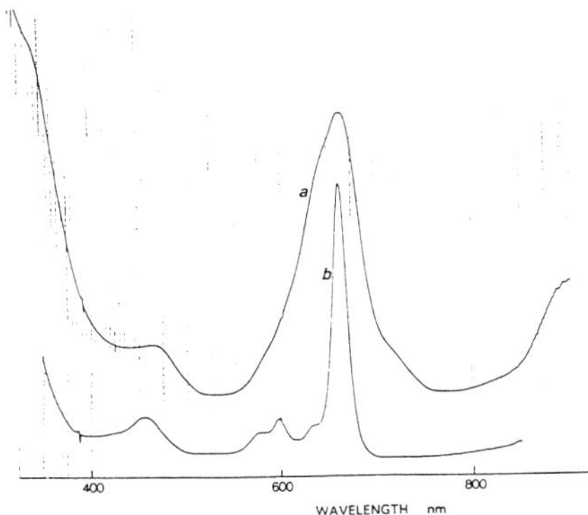

Figure 1.UV-Vis spectra for erbium diphthalocyanine in (a) DMF and (b) chloroform.

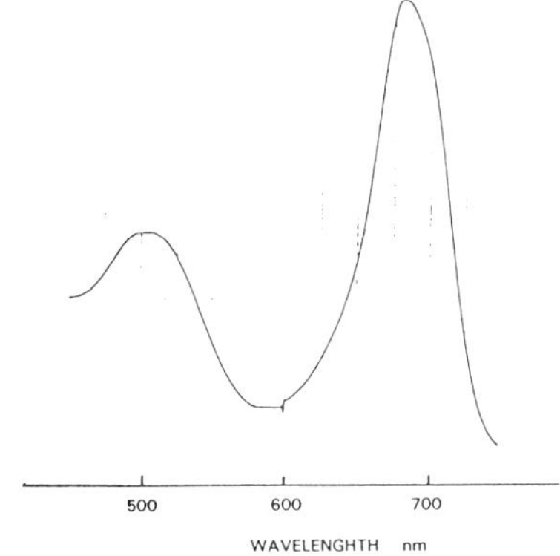

Figure 2. UV-Vis spectrum after exposure to chlorine.

Figure 2 shows a typical spectrum after exposure to chlorine (1000 ppm (vol) in N_2). It is clear that there is a significant change in the optical profile. Whilst this suggests that $ErPc_2$ might therefore be a candidate for an optical Cl_2 sensor the redox change over a period of time when exposed to oxidising or reducing agents, found in DMF, might also occur with prolonged exposure to air.

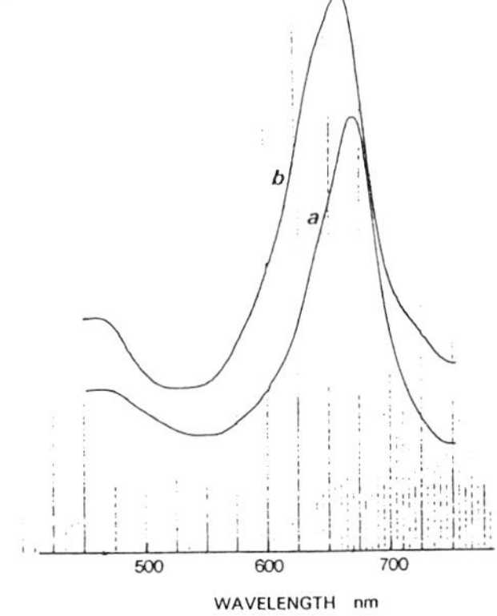

Figure 3. UV-Vis spectra after exposure to (a)NO_2 and (b)NH_3.

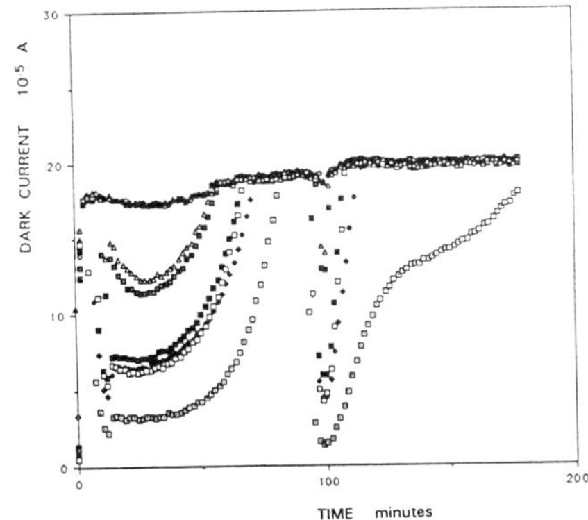

Figure 4. Effect of chlorine exposure on conductivity.

Figure 3 shows the effect of exposure to NO_2 and NH_3. There is very little change in the spectra in these cases. Table 1 shows the various changes observed for the different gases. Chlorine has the highest electronegativity and can thus oxidise the $ErPc_2$ to a greater extent in a given time than the other gases. Thus there are more molecules oxidised and a greater colour change in the case of chlorine. Despite this FTIR results showed no discernible spectrum changes after exposure to chlorine. Samples consisting of compressed KBr and $ErPc_2$ did not change colour on exposure to chlorine but some colour change did occur if the powders were exposed to chlorine before pressing. The FTIR results contrast with those of Schoch and Temofonte (1988) for $LnPc_2$. However, in the present work it would not be possible to observe changes in the vibrational frequency associated with the Cl - M bond as any form of such vibration would occur at frequencies less than our limit of 370 cm^{-1}. Moreover, Cl - N, H, C vibrations are difficult to observe.

	Cl_2	NO_2	NH_3
%change in optical transmittance.	- 42	- 12	- 5
change in λ_{peak} position nm.	+ 17	+ 6	0
%recovery in λ_{peak} position after 3 days in air.	83%	83%	No change.

Table 1. Effects of the different gases on optical properties.

4. ELECTRICAL RESULTS

$ErPc_2$ in common with the other di-Pcs exhibits a resistivity which is several orders below that of most of the mono-phthalocyanines. This is associated with the greater overlap for the π orbitals. Exposure to chlorine leads to a decrease in conductivity as shown in Figure 4 for a number of samples of different thicknesses. After removal of the chlorine ambient there is considerable recovery over a period of an hour, the dark current returning to within 20% of its pre-exposure value. This is without any form of heat treatment being applied to the samples. After re-exposure to Cl_2 the conductivity drops to the previous value but this time recovery is much reduced, the current reaching only about 50% of its initial value. These results are in contrast to other MPcs where conductivity is increased in the presence of oxidising gases. Figures 5 and 6 show for various sample thicknesses the effect of NO_2 and NH_3 exposure. NO_2 shows characteristic oxidising behaviour, creating $Er^{\oplus}Pc_2$ and thus increasing the carrier concentration. Removal of the gas leads to slow desorption of the chlorine and a gradual decrease in the current. Ammonia has no effect on the conductivity in agreement with the work of Jeffery (1992) on $DyPc_2$.

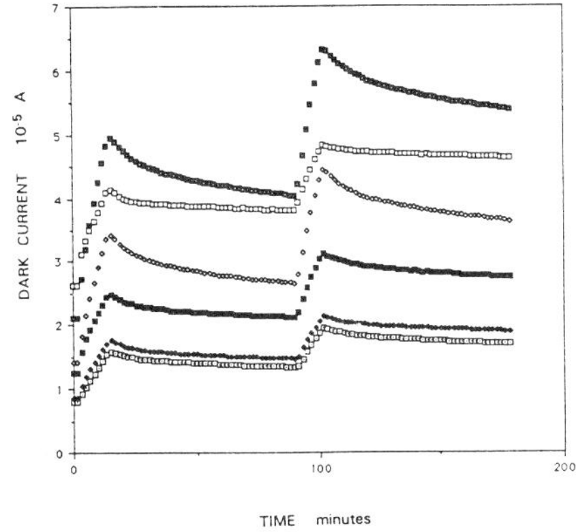

Figure 5. Effect of NO_2 on dark conductivity.

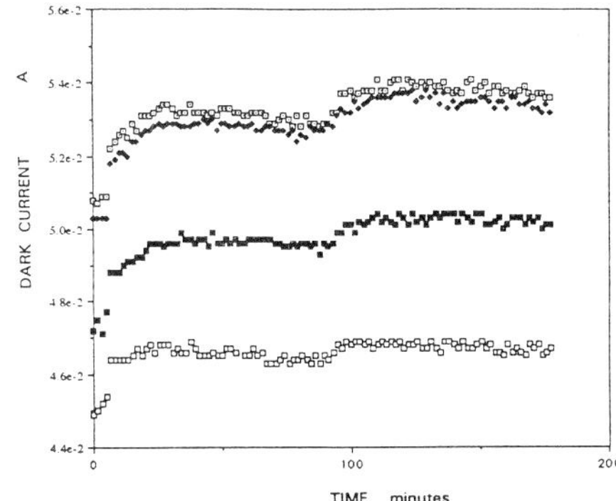

Figure 6. Effect of NH_3 on dark conductivity.

5. SUMMARY

Erbium diphthalocyanine has been shown to exhibit a measurable sensitivity to both Cl_2 and NO_2. Both optical and conductive properties are affected. However, considerable research is still required both to understand the processes occurring and their specific effects (e.g. current reduction in the presence of chlorine) and in the synthesis and characterisation of the material.

REFERENCES

Bott B. and Jones T.A. 1986 Sensors and Actuators 9, 19
Clarisse C. and Riou M.T. 1987 Inorg.Chim.Acta 130, 139
Even R. and Trometer S. 1992 Sensors and Actuators, B8, 129
Jeffery M.J. 1992 Ph.D. thesis, University of Lancaster
Jeffery M.J. and Collins R.A. to be published
Moskalev P.N. and Alimova N.I. 1975 Russian J.Inorg.Chem. 20, 1474
Moskalev P.N. and Kirin I.S. 1965 Russian J.Inorg.Chem. 10, 1065
Moskalev P.N. and Kirin I.S. 1970 Russian J.Inorg.Chem. 15, 7
Wright D.J. 1989 Prog. Surface Science 31, 1

Nanocrystalline WO_3-based H_2S sensor

Hong-Ming Lin, Huey-Yih Yang, Chi-Ming Hsu, Chih-Fu Yang and Pee-Yew Lee[+]
Department of Materials Engineering, Tatung Institute of Technology, Taipei, Taiwan, R.O.C.
[+]Marine Engineering Department, National Taiwan Ocean University, Keelung, Taiwan, R.O.C.

ABSTRACT : Nanocrystalline materials, exhibiting a large surface area, may be applied to gas sensors for which an excellent surface effect is required. In this study, tungsten oxide is synthesized by the gas evaporation method and the mean particle size about 21 nm is obtained. The step-heating sintering process is used to obtain the porous network-like structure of WO_3. The results indicate the nanocrystalline WO_3 excels thin-filmed WO_3 in sensing the reducing gases. The sensitivity of 9.9 can be achieved in 7.7 wt% Pt-doped WO_3 at 220 °C under 100 ppm H_2S/Air. The response time of 7.2wt% Pd-doped at 200 °C and 7.7wt% Pt-doped WO_3 at 220 °C is lower than 0.11 sec.

1. INTRODUCTION

In 1967, Shaver illustrated that the WO_3-based gas detector activated by platinum possesses the enhanced sensitivity to small amounts of airborne hydrogen and other hydrogen-containing gases like N_2H_4, NH_3, and H_2S. He got a tungsten metal film by vacuum evaporation and oxidized it by heating in air at 600-750 °C for several minutes. The oxide film was activated by sprinkling platinum black on the surface. It was observed that the resistance of the tungsten oxide in 0.1% H_2/Air at 350 °C changed abruptly in one second [1]. The further studies and developments of WO_3-based gas sensors had been accomplished by Barrett, Miura, and Akiyama [2-4]. Barrett[2] prepared the WO3-based gas sensors of the metal oxide semiconductor (MOS) type by the means of decomposing ammonium tungsten $(NH_4)_{10}W_{12}O_{41}\ 5H_2O$. It was found that the WO_3-based gas sensor was sensitive to H_2S even the concentration down to 50 ppm. The 90% response time approached 7-8 minutes when the concentration of H_2S was higher than 50ppm. The sensitivity of WO_3 to 200 ppm H_2S/Air got an ultimate value at 469 K. At the same time, Miura[3] developed a couple of Pt-loaded oxide electrodes as a proton-conductor gas sensor. It was selective and sensitive to CO even at room temperature. It was reported that the CO sensitivity of this type of WO_3-based sensor was 7 times higher than the H_2 sensitivity and the 90% response time was about 3 minutes in 1000 ppm of CO gas and 1 minute in 1% H_2. In 1991, Akiyama[4] found that the tungsten oxide-based semiconductor sensor was highly sensitive to NO and NO_2. The sensitivity, defined as the ratio of the resistivity in the detecting gases to that in air, was as high as 31 and 97 in 200 ppm NO and 80 ppm NO_2, respectively, at 300 °C.

The gas-evaporation method for fine oxide powders is essentially based on the evaporation and condensation of solid materials in an environment of oxygen gas or inert gases; in the later, powders must be oxidized by heat treatment after fabrication or the raw materials before evaporated is oxide powders [5-7]. It's a wide-used and popular technique because there is no contamination during the process and many kinds of metal oxide can be synthesized by this method. The particles are of the order of 1-100 nm in size and the size distribution is narrow. Nanocrystalline materials (NCM) with the diameter of the particle smaller than 100 nm exhibits many special characteristics which are not found in traditional materials[8-12]. In this study, the nanocrystalline (NC) of WO_3 particles has been synthesis to examine their sensing properties H_2S gas. Tungsten is used as the raw material and is synthesized into NC tungsten oxide by gas evaporation under oxygen atmosphere. Also the noble metal-doped tungsten oxides are fabricated to study the dopants effects on the sensitivity and the respond time in H_2S gas.

2. EXPERIMENTAL PROCEDURES

Al_2O_3 plate painted with electrodes is used as a substrate. It is at first cleaned by ultrasound in acetone and baked to get rid of humid gas. Then the substrate is attached to the collector in the gas evaporation equipment, which is shown in Figure 1. The pure tungsten sheet is hold by two electrodes and used as a raw material. The chamber is vacuumed by rotary and diffusion pump and always pumped down to 3 x 10^{-5} torr. Helium is purged into the chamber and pumped out in order to make sure that there is no more air or other gases in it. Pure oxygen is controlled at the pressure of 20 mbar and kept dynamic equilibrium. Tungsten is heated by an A.C. power supply and volatilities at about 1200 ^{o}C in the oxygen atmosphere.

To synthesize doped-tungsten oxide powders, the noble metals, Pt and Pd, have been made into nanocrystalline powders first to reduce their melting temperature so that they are allowed to evaporate at lower temperature. NC palladium and platinum, and gold are dispersed on the tungsten sheet and evaporated under 20 mbar of O_2 to form uniform distribution of dopants on the NC WO_3 substrate.

The principles of sintering in this study are to retain the porous structure of particles, to improve the adhesion between WO_3 particles and Al_2O_3 substrate and also to minimize the grain growth. To retain the porous structure of NC tungsten oxide, sintering by step-heating is used. The sample is first pre-heated at 50 ^{o}C for one hour to receive a strengthened structure, and then heated successively at higher temperature step by step.

The sensitivity is defined as the ratio of the resistance of WO_3 in dried air to that of WO_3 in detected gas which diluted by dried air. The measurement equipment is shown in Figure 2. The sample is settled a removable reactor which is put in a furnace. Thermocouple is attached to the sample holder for controlling the operating temperature. The detected gas, H_2S, flowing through the reactor is passed and dissolved in the water. The reactor is heated by the tubular furnace to achieve the operating temperature. LCR meter and thermometer are both linked to the computer to record the data directly. The measured frequency and voltage are 1 kHz and 1 V, respectively.

SEM is used to observe the morphology of the sample before and after sintering. The porosity and the thickness is also estimated by SEM. The dopant concentration is determined by EDS analysis. TEM is used for particle size estimation and crystal structure analysis.

3. RESULTS AND DISCUSSION

3.1 Analysis of NC WO_3 Powders

The TEM images and diffraction pattern of pure tungsten oxide powders is illustrated in Figure 3. The mean particle size is calculated from the image of TEM. The results indicate the average particle size of NC WO3 is about 21 nm under 20 mbar of oxygen pressure. The diffraction pattern of tungsten oxide is identified as WO3 with tetragonal structure. Several different shapes (rhombic, hexagonal or spherical) of particles have been observed in nanophase WO_3.

To maintain the same thickness of NC WO_3, every substrate collects the evaporated nanocrystalline particles in the equal time interval at constant evaporating rate. Several specimens have been examined by SEM and the thickness difference is within few μm range. Figure 4 shows the cross section of WO_3/Al_2O_3 specimen is approximately 40 μm.

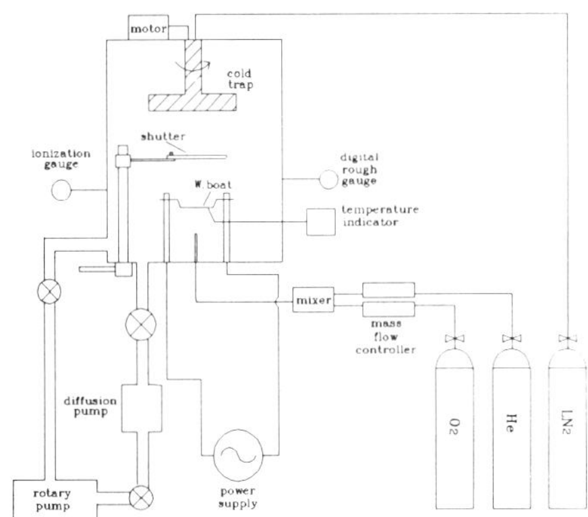

Figure 1 : Apparatus of gas-evaporator.

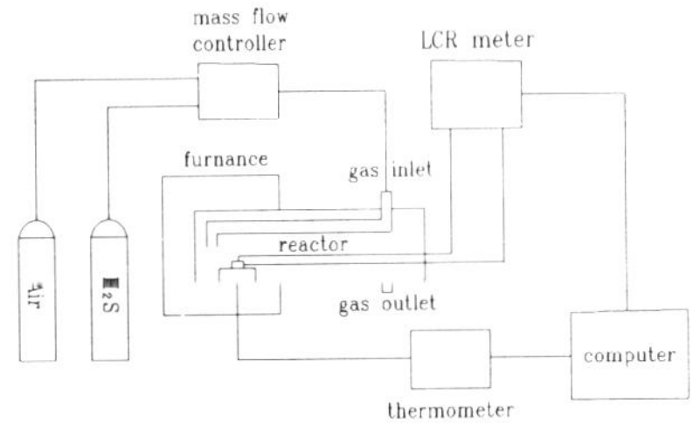

Figure 2 : Electronic measurement system.

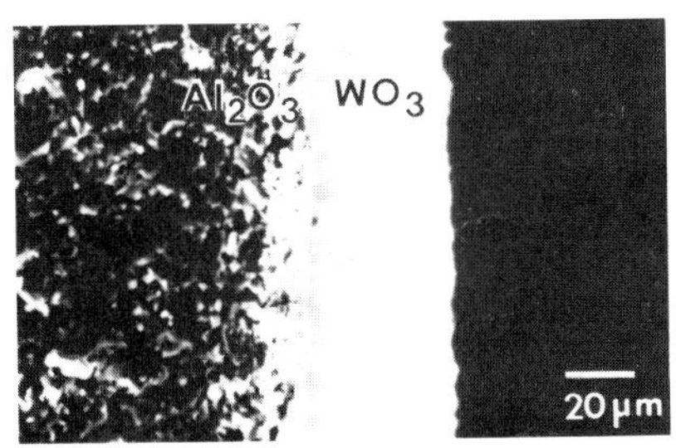

Figure 4 : The cross section image of pure WO_3 on Al_2O_3 substrate.

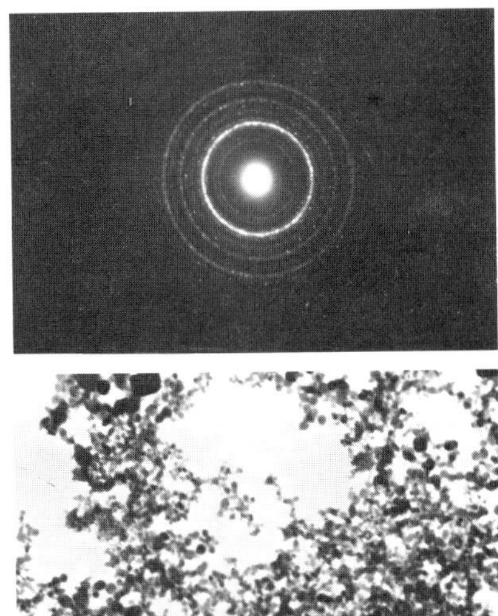

Figure 3 : TEM image and diffraction pattern of pure tungsten oxide prepare under 20 mbar of O_2 pressure.

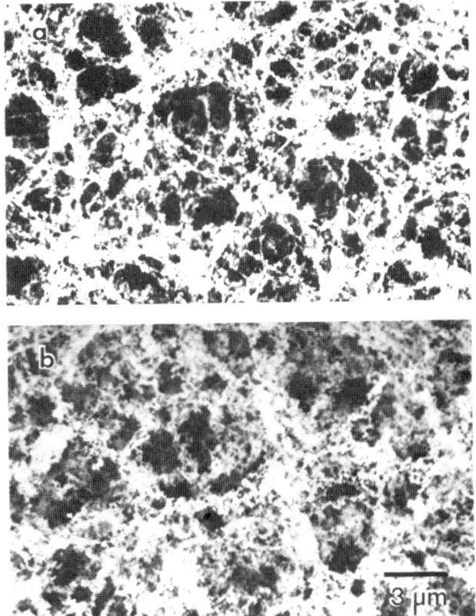

Figure 5 : The morphology change by step-heating (a) as received, (b) sintering by step-heating.

EDS Xray mapping indicates the distribution of the dopants, Pd, Pt and Au, in tungsten oxide is homogeneously distributed out through the specimen.
The morphology of the samples prepared by the gas evaporation is the porously network-like structure before sintering. If the specimens are directly heated at higher temperatures, the porous structure will be destroyed and turn into a thin film. Since the coherent force between the particles is weak, the structure becomes densified after sintering. To increase gas sensitivity, it is necessary to maintain the porous structure to increase the surface effect. A sintering process with step-heating is used to improve this phenomena. The sintering conditions with best porous structure is obtained by using step-heating Figure 5 is the SEM images of surface morphology in before and after step sintering. The results indicate the slow step-heating enhances porosity and strengthens of surface structure.

3.2 The Sensitivity of WO3-based Sensor

The dopant affects the sensitivity of WO_3 in 100 ppm H_2S/Air is shown in Figure 6. The weight percentage of dopant is determined by EDS analysis. It is clear that the doped WO3 reveals a superior sensitivity and decreases the temperature of the ultimate sensitive point. The highest sensitivity is observed in 7.7 wt% Pt-doped WO_3 and is close to 7.2 wt% Pd-doped WO_3. It reveals the "spillover" effects of Pt dopant is larger than Au dopant and is almost same as Pd dopant. The optimum operating temperature of pure WO_3, 10 wt% Au-doped, 7.7 wt% Pt-doped and 7.2 wt% Pd-doped WO_3 are 270 ^oC, 236 ^oC, 220 ^oC and 170 ^oC, respectively. According to the sensitivity and optimum operating temperature, Pd is a ideal dopant in WO_3 sensing system. The amount of the dopants also influences the sensitivity and the optimum operating temperature. Figure 7 indicates the increase weight percentage of Pd dopant will increase the sensitivity and lower the optimum operating temperature in 100 ppm H_2S/Air. This results indicate the effects of different dopants and doping concentration can be used to control the sensitivity and optimum operating temperature in WO_3 sensing system and will enhance the selectivity of detection in various reducing gases.
A step-sintering porous structure of 5.4 wt% Pd-doped WO_3 and a over sintering 5.4 wt% Pd-doped WO_3 thin-film specimen are prepared to study the structure effect on the sensitivity and optimum operating temperature. The distinction of these two kinds of structures is shown in Figure 8. The maximum sensitivity of the thin film is about 4.0, two third of the sensitivity of the porous structure. The optimum operating temperature of thin film structure is about 45 ^oC higher than that of porous structure.

3.3 The Response Time of WO3-based Sensors

The characteristics of sensors that must be studied well includes the sensitivity and the response time. The response time is affected by the temperature, the dopants, and the gas concentration. Figure 9 reveals the resistance change of 7.2 wt% Pd-doped WO_3 versus time in 100 ppm H_2S at 200 ^oC. The response is too fast to be measured by LCR meter. The instrument limit indicates the response time is lower than 0.11 second. The similar results is obtained in 10 wt% Pt-doped WO_3 in 100 ppm H_2S at 220 ^oC. For the pure WO3, the 90% response times as the function of H_2S concentration in different temperatures show in Figure 10. The response time is increased as the temperature decrease or H_2S concentration decrease. The comparison of the results in this study to others' is list in Table 1.

4. CONCLUSIONS

1. The nanocrystalline WO_3 sintering by step-heating can retain the porous structure and increase the gases sensitivity. The slower the heating rate is, the higher the porosity can be got.

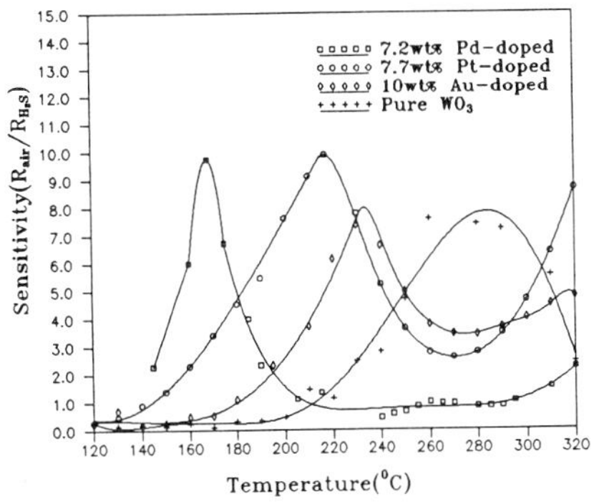

Figure 6 : The sensitivity of pure WO_3, 7.7 wt% Pt-doped, 7.2 wt% Pd-doped and 10 wt% Au-doped WO_3 in 100 ppm H_2S/Air.

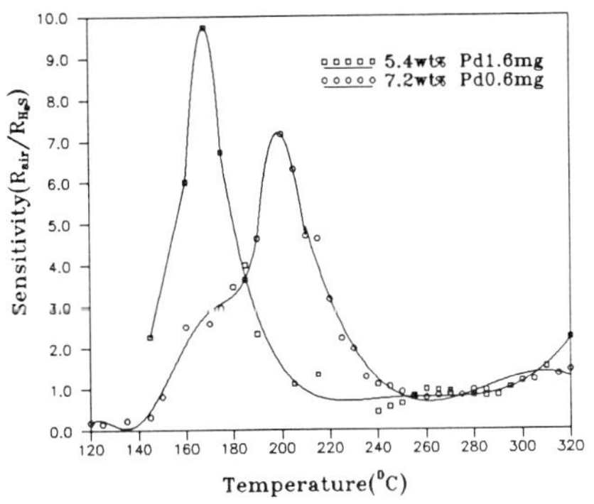

Figure 7 : The influence of different amounts of the Pd dopants on the sensitivity which are measured in 100 ppm H_2S/Air.

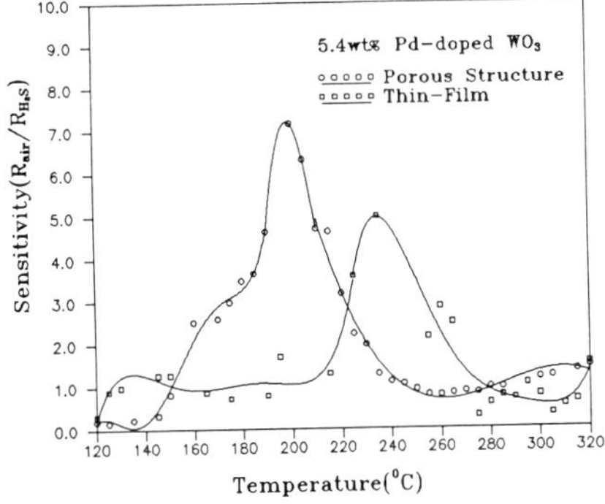

Figure 8 : The sensitivity of thin film and porous structure of 5.4 wt% Pd-doped WO_3 in 100 ppm H_2S/Air.

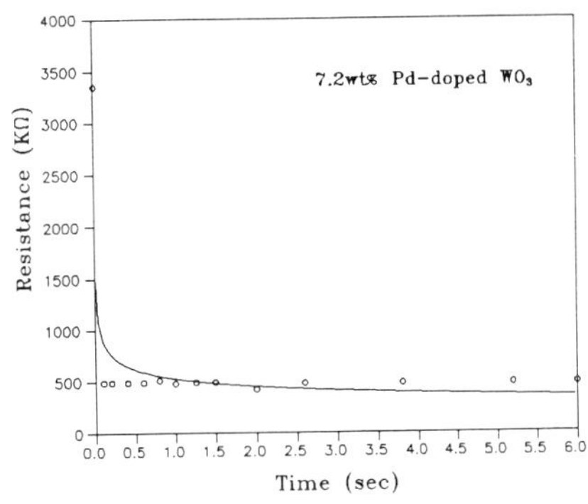

Figure 9 : The response time of 7.2 wt% Pd-doped WO_3 in 100 ppm H_2S/Air at 200 °C.

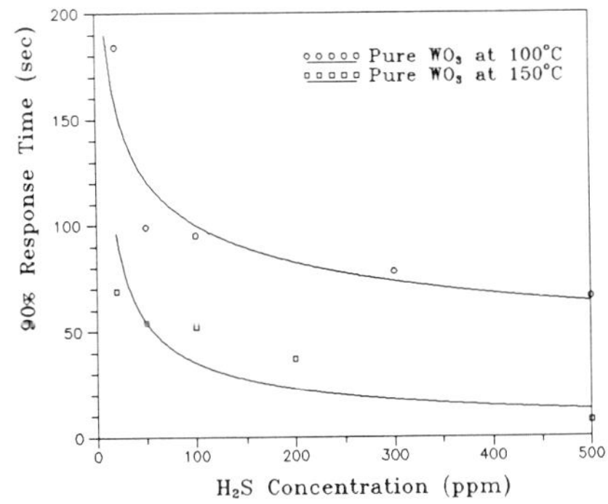

Figure 10 : The 90% response time of pure WO_3 in different concentrations of H_2S/Air at 100 °C and 150 °C.

2. The highest sensitivity is observed in 7.7 wt% Pt-doped WO_3 and is close to 7.2 wt% Pd-doped WO_3. It reveals the "spillover" effects of Pt dopant is larger than Au dopant and is almost same as Pd dopant.
3. The optimum operating temperature of pure WO_3, 10 wt% Au-doped, 7.7 wt% Pt-doped and 7.2 wt% Pd-doped WO_3 are 270 ^oC, 236 ^oC, 220 ^oC and 170 ^oC, respectively. According to the sensitivity and optimum operating temperature, Pd is a ideal dopant in WO_3 sensing system. The effects of different dopants and doping concentration can be used to control the sensitivity and optimum operating temperature in WO_3 sensing system and will increase the selectivity of detection in various reducing gases.
4. We can find that WO_3 doped with the noble metal yields a fast response time. The response time of 7.2 wt% Pd-doped and 7.7 wt% Pt-doped WO_3 is lower than 0.11 second at 200 ^oC and 220 ^oC, respectively. The response time is also influenced by the operating temperature and the concentration of the detected gas. The higher temperature and concentration both result in faster response time.

5. REFERENCES

1. P. J. Shaver, *Appl. Phys. Lett.*, **11**(8), (1967)255.
2. E. P. S. Barrett, G. C. Georgiades, and P. A. Sermon, *Sensors and Actuators*, **B1**, (1990)116.
3. N. Miura, K. Kanamaru, Y. Shimizu and N. Yamazoe, *Solid State Ionics*, **40/41**, (1990)452.
4. M. Akiyama, J. Tamaki, N. Miura, and N. Yamazoe, *Chemistry Lett.*, (1991)1611.
5. Chihiro Kaito, Kazuo Fujita, and Hatsujiro Hashimoto, *Japanese J. Appl. Phys.*, **12**(4), (1973)489.
6. Manabu Kato, *Japanese J. Appl. Phys.*, **15**(5), (1976)757.
7. R.W. Siegel, S. Ramasamy, H. Hahn, L. Zongquan, L. Ting, and R. Gronsky, *J. Mater. Res.* **3**(6), (1988)1367.
8. X. Zhu, R. Birringer, U. Herr, and H. Gleiter, *Phys. Rev. B*, **35**, (1987)9085.
9. J. Rupp and R. Birringer, *Phys. Rev. B*, **36**, (1987)7888.
10. R. Birringer, H. Hahn, H. Hofler, J. Karch, and H. Gleiter, "*Diffusion and Defect Data*", Trans Tech, Aedermannsdorf, (1988)17.
11. V. I. Torbov, V. N. Troitskii, and A. Z. Rakhmatullina, *Porshk. Metall.*, **12**, (1979)27.
12. R. Birringer, *Materials Science and Engineering*, **A117**, (1989)33.

Table 1 The Comparison of WO3-based Gas Sensors

	Shaver[1]	Barrett[2]	Miura[3]	Akiyama[4]	This Study
Type	Thin Film	Thin Film	Diode	Sintered-body	Sintered NCM
Dopant		Pt	-----	Pt	---- Pd, Pt
Detected Gas	H_2	H_2S	CO	NO, NO_2	H_2S
Gas Conc.	0.1 %	200 ppm	1000 ppm	200, 80 ppm	100 ppm
Operating Temp.	350 ^oC	196 ^oC	-----	300 ^oC	200, 220 ^oC
Sensitivity	-----	0.8	-----	31,97	9.7, 9.9
Response Time	<1 sec.	7-8 mins.	3 mins.	10 mins.	< 0.11 sec.

LED based fibre optic oxygen sensor using sol-gel entrapped dyes

G. O'Keeffe, B.D. MacCraith, C. McDonagh, School of Physical Sciences, Dublin City University, Glasnevin, Dublin 9, Ireland.

B. O'Kelly and J.F. McGilp, Department of Pure and Applied Physics, Trinity College, Dublin 2, Ireland.

Abstract: An LED-based oxygen fluorosensor is reported. Sensor operation is demonstrated using blue LED excitation and photodiode detection. The sensor employs a declad optical fibre coated with a thin microporous sol-gel-derived silica film containing an oxygen-sensitive luminescent ruthenium complexe. These easily manufacturable, low cost sensors are shown to have excellent oxygen sensitivity. Results establish a baseline for further developments leading to planar waveguide or distributed oxygen sensors.

1. Introduction

In this paper, recent results leading to the development of a portable low-cost optical oxygen-sensor are presented. The principal features of the work are operation under blue LED excitation with photodiode detection, excellent performance in an evanescent-wave configuration, and ease of fabrication of the sensor. The sensor fabrication technique developed by us is a generic one which involves coating an optical fibre or waveguide with a microporous glass film prepared by the sol-gel process. The porous coating contains an analyte-sensitive dye (in this case a ruthenium complex whose fluorescence is quenched by oxygen) which is trapped in the nanometre-scale cage-like structure. The combination of the simple, low-cost fabrication technique with a compact semiconductor source and detector offers the potential of a handheld inexpensive sensing device with the feasibility of disposable sensing elements. Furthermore, the results presented here establish a baseline for future work on planar waveguide devices and (quasi)-distributed fibre-optic sensors.

2. Optical Sensing of Oxygen

Optical sensors for oxygen are attractive in that they are fast, do not consume oxygen and are not easily poisoned. The most common method adopted is based on the quenching of fluorescence signal, I, as described by the Stern-Volmer equations

$$\frac{I_o}{I} = 1 + K_{sv}\, pO_2 \qquad \text{.................... (1)}$$

$$K_{sv} = k\tau_o \qquad \text{.................... (2)}$$

where I_o and τ_o are respectively the fluorescence signal and excited state lifetime in the absence of oxygen, K_{sv} is the Stern-Volmer quenching constant and k is the bimolecular quenching constant. The ruthenium complex, Ru(II) tris (4,7-diphenyl-1,10-phenanthroline)= $Ru(Ph_2\ phen)_3^{2+}$, was chosen for this work as it exhibits high oxygen sensitivity due to its long unquenched lifetime τ_o (approx. 5μs). Furthermore, in common with other ruthenium polypyridyl complexes, it absorbs strongly at blue-green wavelengths and has a relatively large Stokes shift.

3. Sensor preparation

Sol-gel-derived glass can be prepared from hydrolysis and condensation polymerisation of metal alkoxide solutions followed by a curing process at low temperatures [1]. The process parameters can be selected to produce a microporous glass which can act as a support matrix for analyte-sensitive dyes which are added to the precursor solution. Such material can then be used to coat waveguiding substrates. In this work a short length (~10 cm) of declad multimode PCS600 fibre was dipcoated with a thin (~300 nm) layer of sol-gel-derived silica containing the oxygen-sensitive $Ru(Ph_2 phen)_3^{2+}$ complex. A typical fibre configuration in a gas flow cell is shown in Fig. 1. This approach to sensor preparation is simple and avoids the complex chemistry often associated with other immobilisation methods. The coating is tough, inert and intrinsically bound to the silica fibre core. Although distal tip coating is feasible, the side-coated evanescent wave approach is more attractive because it offers much greater flexibility by enabling adjustment of sensitivity-determining parameters such as length, thickness, porosity and refractive index of the coating [2]. Furthermore, the problem of photobleaching is considerably reduced, especially when an LED source is used.

4. Sensor system and results

Fluorescence sensors based on sol-gel-derived coatings were reported previously by us [3,4]. Such results were obtained using an air-cooled argon-ion laser source emitting at 488 nm. With a view to producing a low-cost, hand-held device this source has been replaced by a blue LED (Ledtronics, USA) with a peak emission at approximately 470 nm. The low output of such LED's demanded considerable optimisation of the launch and collection optics in order to achieve adequate signal-to-noise ratios. The experimental system is shown in Fig. 2. The LED is pulsed for lock-in detection and its output is launched through a short-wave pass filter (cut off ≅ 500 nm) into the coated fibre which is mounted in a gas cell. The evanescent field of the guided light excites the entrapped ruthenium complex and a fraction of the resultant fluorescence is captured by the fibre. The forward-propagating portion of this fluorescence passes through a longwave pass filter (cut on ≅ 610nm) to the detector. The launch and detection filter combination minimises detection of the LED excitation signal without significant curtailment of the fluorescence signal. The output from the detector is passed to the lock-in-amplifier for synchronous detection with the LED pulsing signal. Measurements are made using either a photomultiplier or silicon photodiode as detector.

Precise gas mixtures of oxygen and nitrogen may be passed to the gas cell via mass-flow controllers. The response of the sensor to varying concentrations of oxygen is shown in Fig. 3. These data which were measured with a PMT detector, indicate the high sensitivity at low concentrations of oxygen with a limit of detection of <1% oxygen concentration. The sensor response time is approx 2s. Fig. 4 shows sensor response when measured with a standard silicon photodiode. This figure illustrates the sensor repeatability. These data were obtained with a 15 mcd source. The signal-to-noise ratio will be improved considerably with 150 mcd blue LED sources obtained at the time of writing. These preliminary results from an unoptimised sensor design indicate the potential for a LED-based sensor system. Furthermore, the long (5μs) unquenched excited-state lifetime of the $Ru(Ph_2phen)_3^{2+}$ complex in the sol-gel-glass environment facilitates sensor design based on the inherently referenced decay time measurement.

5. Conclusions

Although further characterisation and development work remains to be carried out this work has established the potential for low cost portable oxygen sensors. These early results from an unoptimised sensor indicate the viability of an LED-based system with photodiode detection. The continued improvement in the output power of blue LED's and the potential availability of blue laser diodes means that considerable improvement is achievable with the possibility of distributed sensing. The eventual implementation of this sensor is likely to be in the fluorescence-decay-time measurement mode because of the advantages of this approach over intensity-based sensing.

6. References

1. Brinker, C.J., and Scherer, G.W., *Sol-gel Science*, Academic Press, New York, 1990.
2. MacCraith, B.D., Sensors and Actuators B, 11, 1-3, 29-34, 1993.
3. MacCraith, B.D., Ruddy, V., Potter, C., O'Kelly, B., and McGilp, J.F., Electron Lett., 27, 1247-1248, 1991.
4. MacCraith, B.D., McDonagh, C.M., O'Keeffe, G., Keyes, E.T., Vos, J.G., O'Kelly, B., and McGilp, J.F., The Analyst, Vol. 118(4), 385-388, 1993.

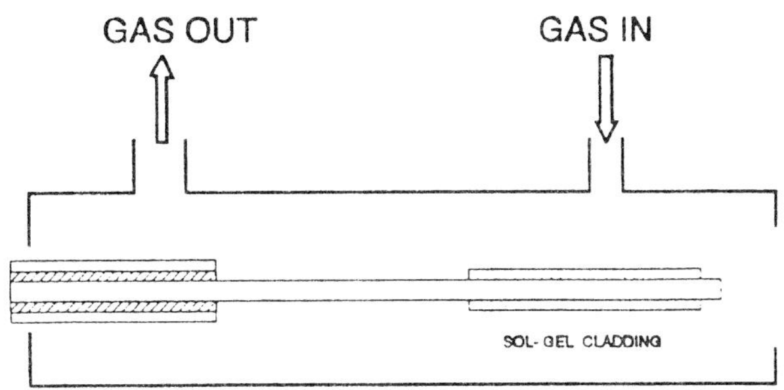

Figure 1 Coated fibre in gas cell

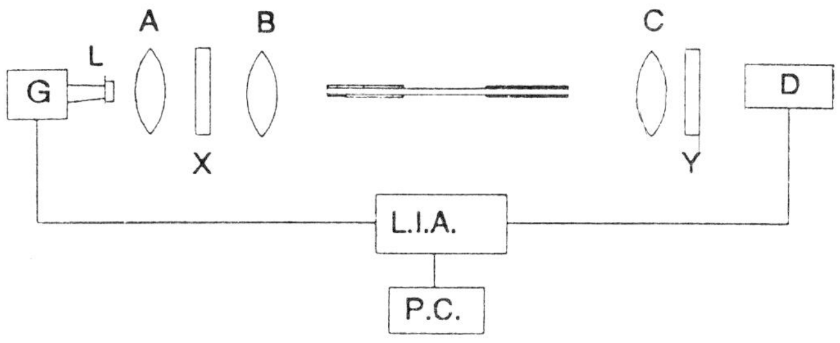

Figure 2 Experimental system

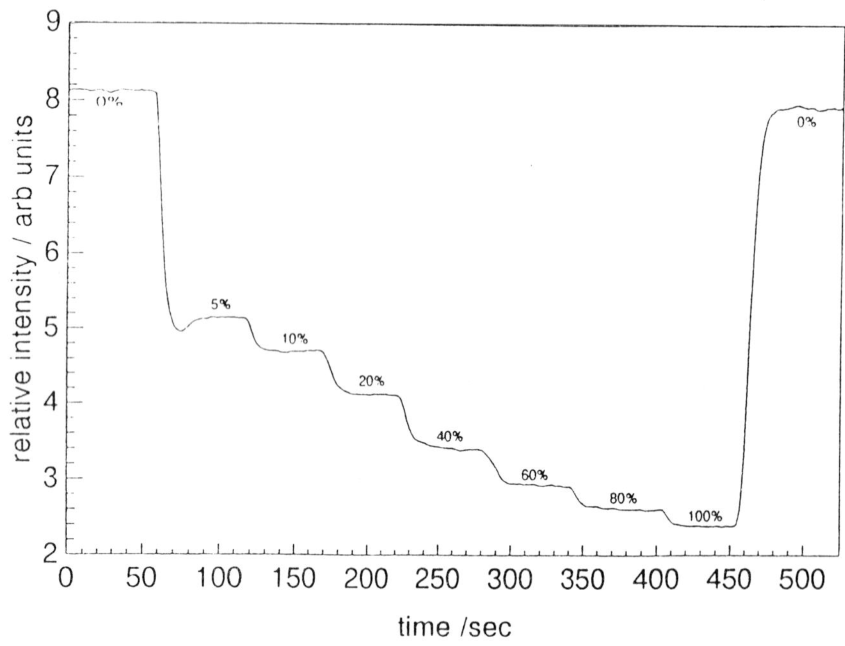

Figure 3 Sensor response with PMT detection

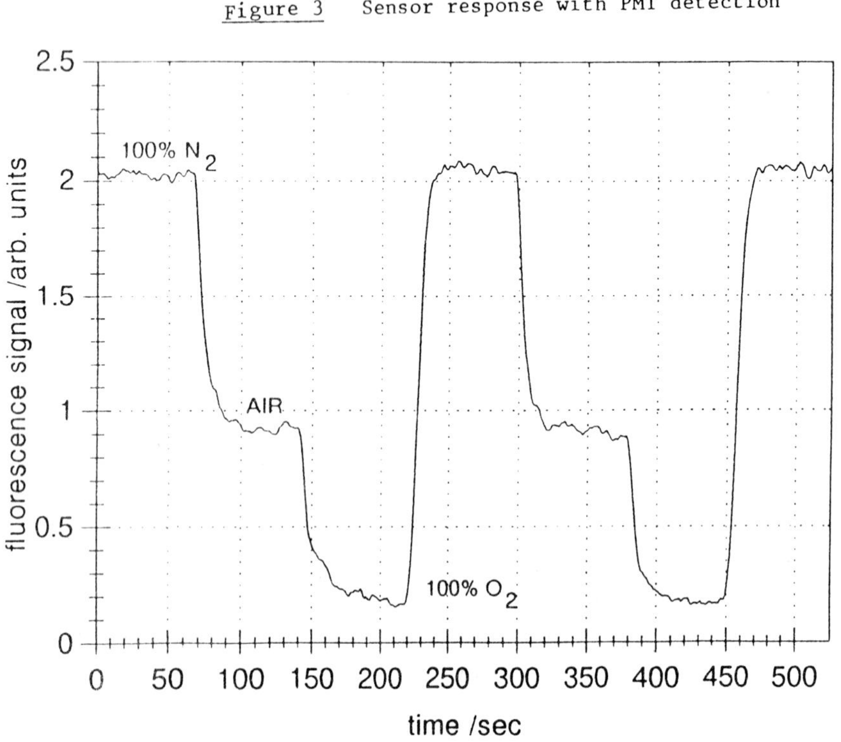

Figure 4 Sensor response with photodiode detection

An evaluation of thick and thin film tin dioxide based sensors for the selective detection of NO_2

Geraint Williams and Gary S.V. Coles

Department of Electrical and Electronic Engineering,
University of Wales,
Singleton Park,
Swansea, SA2 8PP, U.K.

ABSTRACT: Low resistance polycrystalline SnO_2 based thick film sensors exhibit a significant response to mixtures of nitrogen dioxide in air while remaining reasonably insensitive to reducing gases, especially at operating temperatures of <300°C. NO_2 sensitivity values are lower than those determined for thin film sensors produced by radio frequency sputtering.

1. INTRODUCTION

Most recent studies on SnO_2 based sensors for detecting the oxides of nitrogen have focused on the use of thin film devices produced via evaporation or sputtering techniques [1-3]. The sensing mechanism is believed to involve the extraction of electrons by the NO_x molecules from the semiconductor conduction band leading to an increase in film resistivity. This chemisorption process can only proceed providing that the electron affinity of the adsorbate is greater than the work function of the semiconductor. The high sensitivity of thin film SnO_2 devices arises because the thickness of the sensing layer (usually <100nm) is comparable to the width of the depletion layer caused by the decrease of electron concentration in the vicinity of the sensor surface. Thus, changes in surface conductivity are efficiently mirrored in the film bulk.

In several cases, researchers have modified the sensor composition with additives such as In or Cd in an attempt to improve NO_x sensitivity and also to eradicate cross sensitivity to reducing gases such as CO and CH_4 [4-7]. However, optimisation and fine tuning of sensor composition involving such systems appear to be laborious processes requiring the preparation of a range of Sn alloy targets or sequential multi-layer sputtering followed by thermal oxidation. In contrast, thick film sensors based on polycrystalline tin dioxide appear more versatile in this respect, though little work has appeared in the literature regarding their behaviour in atmospheres containing strongly oxidising gases such as NO_2. Leppavuori *et al* [8] observed that the presence of NO caused the carbon monoxide response of glass bonded SnO_2 thick films to be decreased considerably. Others have investigated the effect of NO on Al, Pt, Pd and Sb-doped devices and found that film conductance decreased by up to three orders of magnitude depending on temperature and oxygen content of the containing environment. In this paper we compare and contrast results obtained for both thick and thin film sensors produced at Swansea. The NO_2 response of each device was optimised with respect to operating temperature and the sensor cross sensitivity to various reducing gases such as CO, CH_4 and H_2 was established.

2. EXPERIMENTAL

The fabrication of thin film SnO_2 by means of a two step sputtering/oxidation process is described elsewhere [9]. Thick film sensors were prepared in the following manner. Tin dioxide (Keeling and Walker Superlite) was mixed along with the required additives in an organic medium consisting of ethyl cellulose stabilised α-terpineol. This paste was then applied over the Pt electrode array of a planar alumina substrate which was then heated to 100°C in order to slowly evaporate the organic vehicle. Finally the sensor was fired at 1000 °C for 2h employing a furnace heating rate of 8°C/min.

3.RESULTS AND DISCUSSION

3.1 Thick film sensors

Since, in general, the resistance of a tin dioxide sensor increases upon exposure to nitrogen dioxide then ideally the resistivity of a NO_2 sensing device should be low thus ensuring a high surface electron density available for reaction with the oxidising gas molecules. The majority of thick film sensors tested here possess low resistivity (resistance in clean dry air <1kΩ) and as a consequence appear relatively insensitive to the presence of reducing gases such as CO, CH_4 and H_2. It would seem likely that this lack of sensitivity is due to the high electron density of the SnO_2 sensing material which largely obscures any decrease in resistance caused by the presence of reducing gas. Figure 1 shows the results obtained for a thick film sensor prepared from a commercially available SnO_2 powder of low resistivity, presumably resulting from a high degree of non-stoichiometry.

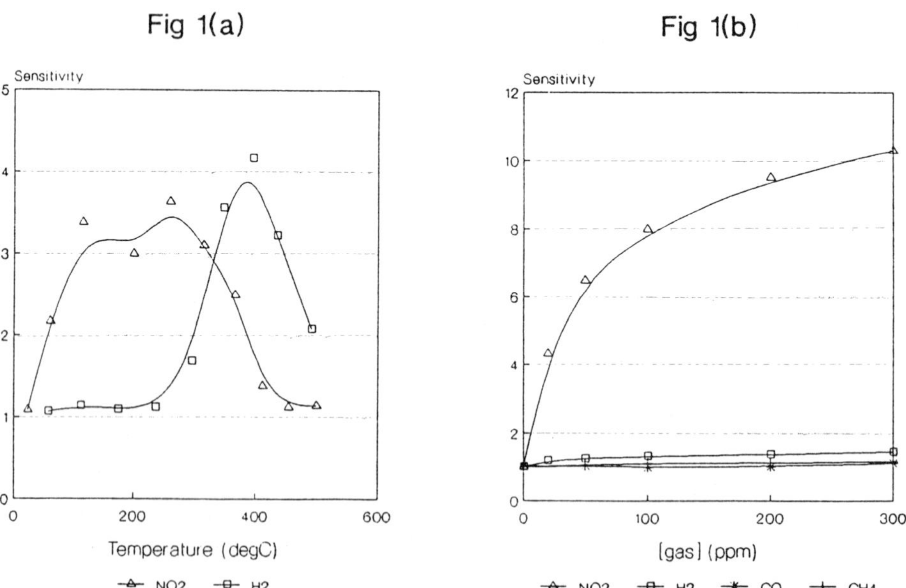

Figure 1 (a): Sensitivity versus operating temperature profiles obtained for a sensor prepared from a commercially available SnO_2 powder upon exposure to 100 ppm concentrations of NO_2 and H_2 in air.
(b): Sensitivity versus gas concentration for the same sensor operated at a temperature of 240°C.

Table 1: A summary of results obtained for a series of tin dioxide based gas sensors using a contaminant gas level of 100 ppm in dry air.

Sensor type	Optimum operating temperature (°C)	Base Resistance	R_{NO2}/R_{air}
1. Thin film SnO_2 (0.3μm)	200	31.5Ω	13.0
2. Thin film SnO_2 (1.0μm)	300	63.8kΩ	11.1
3. Polycrystalline SnO_2	230	0.92kΩ	8.0
4. Polycrystalline $SnO_2+Ta_2O_5$ (5% w/w)	250	0.31kΩ	5.96
5. Polycrystalline $SnO_2+Sb_2O_3$ (2% w/w)	330	9.6Ω	1.34
6. Polycrystalline $SnO_2+Cr_2O_3$ (1% w/w)	310	20.5MΩ	1.07
7. Polycrystalline $SnO_2+Cr_2O_3$(0.2 % w/w)+Sb_2O_3 (0.5% w/w)	220	1.0kΩ	2.25
8. Polycrystalline $SnO_2+In_2O_3$ (35% w/w)	210	0.55kΩ	4.99

Sensor type	Operating temperature (°C)	Hydrogen sensitivity	CO sensitivity	Methane sensitivity
1.	200	0.97	1.04	1.0
2.	300	1.42	0.94	0.98
3.	230	1.32	0.99	1.11
4.	250	1.44	1.13	1.03
5.	330	1.05	1.0	1.0
6.	310	0.90	1.0	1.0
7.	220	1.27	1.27	1.15
8.	210	1.43	1.24	1.11

N.B. For reducing gases such as CO, CH_4 and H_2, sensitivity is represented by R_{air}/R_{gas}, where R_{air} is the sensor resistance in dry air and R_{gas} is the resistance in a contaminant/air mixture. However, for an oxidising gas such as NO_2, which causes an increase in sensor resistance, sensitivity is given by the ratio R_{gas}/R_{air}.

A comparison of the plots of both H_2 and NO_2 sensitivity versus temperature (Figure 1(a)) reveals that hydrogen response is considerable at T>300°C. However, decreasing the temperature beyond this value leads to the disappearance of H_2 sensitivity rendering the sensor responsive to NO_2 only. Therefore, the judicious choice of operating temperature can lead to excellent selectivity as shown in Figure 1(b) where sensitivity is plotted as a function of contaminant concentration at a temperature of 240°C.

Table 1 shows the results obtained for a series of SnO_2 based thick film sensors upon exposure to NO_2, H_2, CO and CH_4 mixtures in air. Data acquired for two thin film sensors under the same conditions are also displayed for comparative purposes. The addition of antimony to a polycrystalline SnO_2 sensor reduces base resistance considerably but virtually eliminates any oxidising and reducing gas response while the inclusion of Ta(V) or In(III) appears to have little benefit over the undoped oxide both in terms of enhanced NO_2 sensitivity or selectivity. The effect of a Cr(III) additive was tested in the light of the findings of Solymosi and Kiss [10], who established that a SnO_2-Cr_2O_3 powder effectively catalysed to the reduction of NO. However, in agreement with the observations of others [7], the addition of 1% w/w chromium (III) oxide leads to an increase of 4 orders of magnitude in sensor resistance and inhibits any response to NO_2. As a consequence the Cr_2O_3 level was decreased to 0.2% w/w and antimony oxide was added in order to reduce resistivity. Disappointingly though, the NO_2 sensitivity of this device turned out to be much lower than that of the undoped SnO_2 sensor. Therefore, although the reducing gas response of the thick film gas sensors is low, the NO_2 sensitivity is significantly less than for the thin film sensors studied.

3.2 *Thin film sensors*

The gas sensing properties of SnO_2 films of different thicknesses (0.3 and 1.0μm) were investigated. Experiments showed that operating temperature has a large influence on NO_2 sensitivity and the specificity of the device. As in the case of thick film sensors, reducing gas detection is favoured by employing operating temperatures of 400°C or greater while maximum NO_2 response is attained below 300°C. The interfering effects caused by the presence of reducing gases in a flow of NO_2 (100 ppm) in dry air were studied. The dynamic response of a 1.0μm SnO_2 thin film sensor at two different working temperatures is shown in figure 2. At a temperature of 300°C the resistance change observed in NO_2 flow remains relatively unaffected by the presence

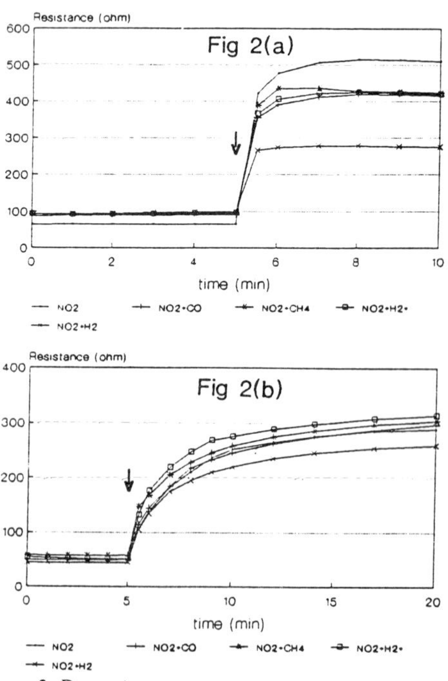

Figure 2: **Dynamic response of a thin film sensor (1.0μm) to a gas flow containing 100 ppm NO_2 along with 1000ppm levels of possible interferant reducing gases when operated at (a) 300°C and (b) 200°C. The gas flow marked * contains a 100 ppm H_2 concentration.**

of high levels of CO or CH_4 (1000ppm) and a low level of H_2 (100ppm). However, the inclusion of a 1000 ppm concentration of H_2 leads to a considerable reduction in the NO_2 signal. Decreasing the temperature to 200°C eliminates the interfering effects of the hydrogen but has the disadvantage of increasing the response time considerably.

4. CONCLUSION

Both thick and thin film SnO_2 based sensors display high sensitivity to NO_2 with little interference from reducing gases such as CO, CH_4 and H_2 when operated in the 200 - 300 °C range. The sensors produced from sputtered Sn [9] display the highest oxidising gas response, though their fabrication is far more time consuming than the production of thick film devices. Optimisation studies are now in progress to maximise NO_2 response by modifying fabrication parameters such as film thickness and oxidation conditions.

REFERENCES

1. S.C.Chang, Thin film semiconductor NO_x sensor, IEEE Trans., Electron. devices, *ED-26(12)*, (1979), 1875-1880.
2. K.D.Schierbaum, S.Vaihinger, W.Gopel, H.H.Van der Vlekkert, B.Kloek and N.F.De Rooij, Prototype structure for systematic investigations of thin film gas sensors, Sensors and Actuators B, *1*, (1990), 171-175.
3. W.Gopel, K.D.Schierbaum, D.Schmeisser and H.D.Wiemhofer, Prototype chemical sensors for the selective detection of O_2 and NO_2 in gases, Sensors and Actuators, *17*, (1989), 377-384.
4. G.Sberveglieri, P.Benussi and G.Coccoli, Radio frequency magnetron sputtering growth and characterisation of Indium-tin oxide (ITO) thin films for NO_2 gas sensors, Sensors and Actuators, *15*, (1988), 235-242.
5. G.Sberveglieri, P.Benussi, G.Coccoli, S. Groppelli and P. Nelli, Reactively sputtered indium tin oxide polycrystalline thin films as NO and NO_2 gas sensors, Thin solid films, *186*, (1990), 349-360.
6. G.Sberveglieri, S. Groppelli and P. Nelli, Highly sensitive and selective NO_x and NO_2 sensor based on Cd-doped SnO_2 thin films, Sensors and Actuators B, *4*, (1991), 457-461.
7. G.Sberveglieri, S. Groppelli, P. Nelli, V.Lantto, H.Torvela, P.Romppainen and S.Leppavuori, Response to nitric oxide of thin and thick SnO_2 films containing trivalent additives, Sensors and Actuators B, *1*, (1990), 79-82.
8. P.Romppainen, H.Torvela, J.Vaananen and S.Leppavuori, Effect of CH_4, SO_2 and NO on the CO response of an SnO_2-based thick film gas sensor in combustion gases, Sensors and Actuators, *8*, (1985), 271-279.
9. G.Williams and G.S.V.Coles, NO_x response of tin dioxide based gas sensors, accepted for publication in Sensors and Actuators.
10. F.Solymosi and J.Kiss, Adsorption and reduction of NO on tin (IV) oxide doped with chromium (III) oxide, J. Catal., *54*, (1978), 42-51.

Development of a thick film ammonia sensor combined with a ventilation rate sensor for livestock buildings

G. Huyberechts, M. Van Muylder, M. Honoré, J. Desmet, J. Roggen

Interuniversity Microelectronics Center (IMEC), Materials and Packaging Division, Micro-Systems Group, Kapeldreef 75, B-3001 Leuven

D. Berckmans, J.-Q. Ni

Katholieke Universiteit Leuven, Laboratory for Agricultural Building Research, Kardinaal Mercierlaan 92, B-3001 Leuven

ABSTRACT : A major problem is the environmental load as produced by agriculture. It has been shown that the main part of the ammonia emission into the atmosphere in Europe is coming from livestock wastes. To measure, and control, the total ammonia emission from individual livestock buildings, the combination of a thick film ammonia sensor with an accurate ventilation rate sensor is studied. An accurate ventilation rate sensor for livestock buildings (accuracy ± 60 m^3/h) and a prototype of a thick film ammonia sensor for a measuring range from 0 to 150 ppm are developed.

1. INTRODUCTION

Ammonia is the most abundant alkaline component in the atmosphere. It plays an important role in atmospheric chemistry and has given rise to serious environmental concern. A substantial part of the acid in the atmosphere generated by the oxidation of sulphur dioxide and nitrogen oxides is neutralised by ammonia. As a result ammonium is a major constituent of atmospheric aerosols [1]. On the other hand, ammonia promotes the oxidation rate of sulphur dioxide dissolved in hydrometeors. Airborne measurements in the United Kingdom indicate a 5 to 25 fold in oxidation rate of sulphur dioxide when sulphur dioxide laden air masses pass over major ammonia emission zones [2].

It is believed that the excess emissions of ammonia into the atmosphere are associated with environmental and ecological damage. It has been found that ammonia causes direct and indirect damage to the ecosystem, the indirect effect being more serious and widespread than the direct effect [3,4].

The influence of ammonia (ammonium) in contact with soil leads to acidification and imbalance of nutrificients [5] resulting in destruction or change in natural vegetation [6].

Another aspect from ammonia emission is odour hindrance. It has been shown that ammonia is not the main component responsible for odour hindrance, however it has been observed that relations between odour hindrance and ammonia emission exist. Data on ammonia emission factors for several animal categories have been compiled and reported earlier [7].

Obviously high ammonia concentrations in animal houses represent a potential hazard for the animals and the workers [8,9,10].

Given the environmental impact and the steady growth rate of anthropogenic ammonia emissions, together with more stringent future regulation on these emissions (e.g. the Dutch government insists on a 70 % reduction in 2000 with respect to the 1980 level [11]), an interest in the emission control of ammonia is justified.
The ammonia problem consists of two parts, on one side there is the in-building ammonia concentration, which has potential physiological effects on the animals and health effects on the workers. On the other side the emission of ammonia from the animal buildings into the atmosphere contributes much to the air pollution, both chemical and as offensive agricultural odour. In most of the animal raising systems, the air flow from the buildings to the atmosphere is forced by a mechanical ventilation system. The objective of the ventilation system is to control gas concentration, air humidity and inside temperature.
It is evident that the most effective way to measure total ammonia emission from livestock buildings is to combine an ammonia sensor with the ventilation system. In order to control both the in-building concentration and the total ammonia emission to the atmosphere, it is important to have quantitative information about ammonia drawn out by the ventilator. Usually agricultural ammonia emissions both from livestock buildings and from the field application of slurry are measured by wet chemical methods or gas chromatography. For atmospheric ammonia measurements other approaches have been reported. [12,13-16]. Unfortunately, most of these techniques are either expensive or inconvenient to manipulate. Therefore the aim is to develop low cost reliable sensors to eventually include measurement of ammonia concentrations at various locations in the stable as well as in the ventilation chimney in the control system. Such low cost systems are not available yet.

2. VENTILATION RATE SENSOR

A low cost ventilation rate sensor for application in animal housings has been developed at the Laboratory for Agricultural Building Research. Its accuracy is ± 60 m^3/h (versus about ± 600 m^3/h for previous sensors as used by other research teams measuring ammonia emission) [17], in a measurement range from 200-5000 m^3/h and for pressure differences from 0-120 Pa. In order to measure the global ventilation rate through the building, the air flow rate sensor is placed in the exhaust chimney with a diameter of about 0.5 m, which is standard in Belgium and the Netherlands. The ventilation rate sensor was carefully calibrated in a laboratory test rig and its installing position was studied [18]. Figure 1 gives a schematic representation of the principle of the ventilation rate sensor. Figure 2 gives the result of a linear regression analysis of the two-blade impeller prototype ventilation rate sensor.

3. THICK FILM AMMONIA SENSOR

At IMEC an ammonia sensor based on a doped semiconducting metal oxide is realised using conventional screen printing techniques. The resulting sensors consist of a patterned multi layer structure consisting of a 96% alumina substrate with a heater element, a dielectric layer, a contact layer and finally the gas sensitive semiconductive metal oxide layer. Each deposition step is followed by the appropriate heat treatment. All thick film inks, except for the metal oxide precursor ink, are commercially available. On completion of the production process, the individual sensors are isolated from the substrate by laser scribing and suspended in a modified TO-package by its contact wires. A metal cover with laser drilled holes to allow gas diffusion completes the sensor package.
The doped metal oxide as well as the precursor ink is home made from the appropriate starting materials. The optimum composition and thermal processing steps have been deduced by using experimental design techniques, as will be reported elsewhere.
The conductivity of semiconducting metal oxide films at a certain temperature is influenced by the presence of reducing gases in the surrounding atmosphere. In general the conductivity is described by :

Figure 1 - Schematic representation of the principle of the ventilation rate sensor.

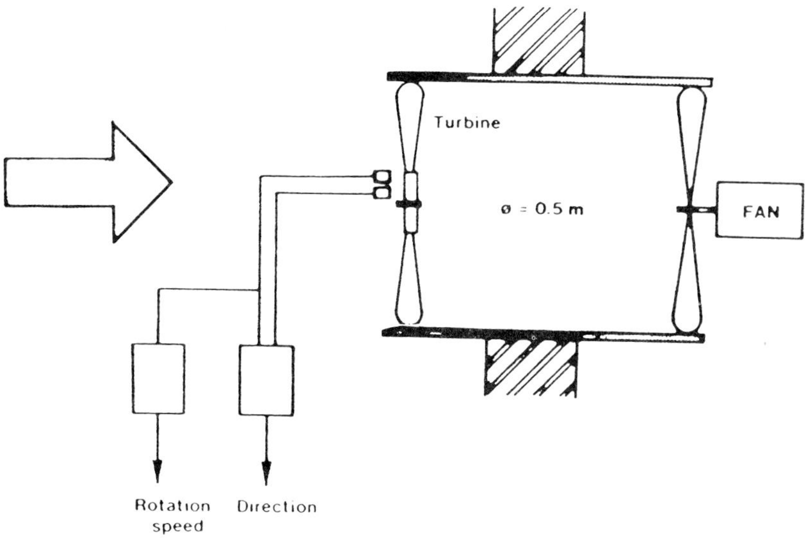

Figure 2 - Linear regression analysis of the two-blade impeller prototype ventilation rate sensor.

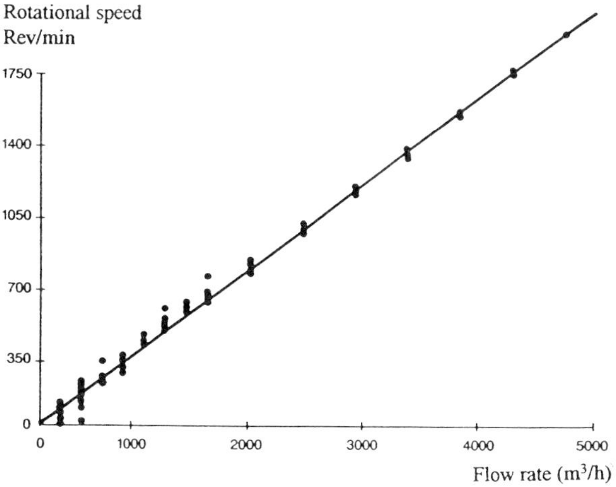

$$\sigma = \sigma_0 + a \cdot R^b$$

whereby σ and σ_0 are the conductivity of the sensor material at a given temperature in the gas mixture under investigation and in un-polluted air respectively, R is the pollutant concentration and a and b are constants depending on the composition and physico-chemical characteristics of the sensor material and the nature of the reducing gas (as well as eventually on the presence of other gases).

4. LABORATORY TESTS ON THE AMMONIA SENSOR

Since the sensor is designed to be usable in animal raising buildings, the selectivity towards ammonia has to be optimised with respect to other gases known to occur in these atmospheres. Laboratory tests have focussed on a decrease of the methane sensitivity. As shown in figure 3 a good response to ammonia in the required concentration range is observed in dry air at an operating temperature of 450 °C (obtained using the incorporated heating element). Under these conditions the discriminant value for ammonia versus methane can be defined as :

$$\frac{\sigma(150\ ppm\ NH_3) - \sigma_0}{\sigma(2.5\ \%\ CH_4) - \sigma_0}$$

and results in values up to five. It was also shown that the ammonia sensitivity is decreased in humid air, especially when changing from dry air to 50% relative humidity. Over the humidity range observed in livestock buildings (50 to 90 % RH) the sensitivity remains fairly constant, as can be seen in figure 4.

5. FIELD TESTS ON THE AMMONIA SENSOR

During field measurement the ammonia sensor was heated by application of a constant voltage over the incorporated heater resistor. At the first stage of this research the sensor temperature is not controlled, the main objective being the evaluation of the sensor before and after exposure to the aggressive environment in the pig stable. At this stage of the research the proper functioning of the ammonia sensor after prolonged operation (> 2 month) is confirmed, as well as the sensitivity towards ammonia. The sensor temperature has been shown to be air flow dependent, resulting in the need for appropriate control. Tests on cross sensitivity of other compounds present in the livestock building atmosphere are planned.

6. CONCLUSIONS

The continuous measurement of total ammonia emission from ventilated livestock buildings involves the measurement of two parameters, the air flow rate and the ammonia concentration. A good solution of air flow measurement has been obtained by using a sensitive ventilation rate sensor. A low cost tin dioxide based thick film sensor is developed for the measurement of ammonia concentrations. This sensor remains operative after more than two months of field testing.

The quantitative data of total ammonia emission from livestock buildings are significant both for the in-building climate control and for environmental protection. It is not only significant for research purpose but also for the practical use of a controlling and monitoring system in modern commercial livestock buildings.

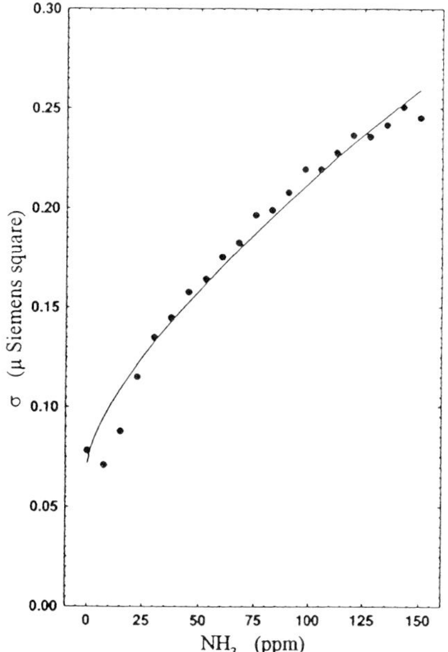

Figure 3 - Typical sensor response (PA07) in dry air at 450 °C

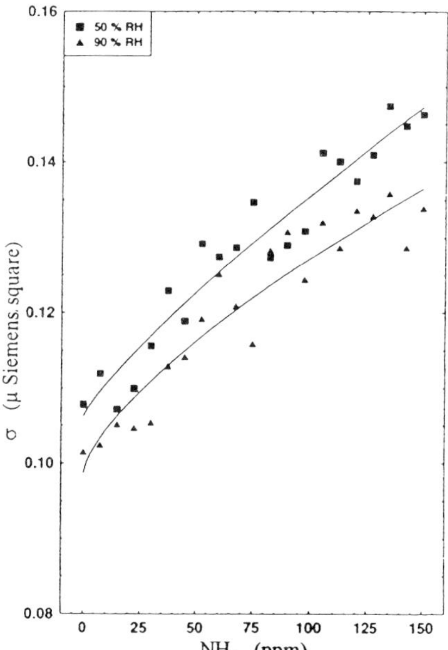

Figure 4 - Typical sensor response (PA07) in humid air at 450 °C

7. ACKNOWLEDGEMENTS

The Flemish Government is acknowledged for the financial support of part of the presented research under project VLIM/H/9032 of the "Vlaams Impulsprogramma Milieutechnologie".

8. REFERENCES

[1] W.A.H. Asman, J. Van Jaarsveld, Atmospheric Environment, 26A (3), 445-464 (1992)
[2] H.M. ApSimon, M. Kruse, J.N.B. Bell, Atmospheric Environment, 21 (9), 1939-1946 (1987)
[3] A.C. Stern, R.W. Boubel, C.B. Turner, D.L. Fox, "Fundamentals of air pollution", Academic Press, San Diego (1984)
[4] H.M. ApSimon, M. Kurse-Palss, in "Odor and ammonia emission from livestock farming" V.C Nielsen et Al (Eds) Elsevier, London, 17-22 (1991)
[5] E. Bijsman, J.-W. Erisman, Journal of Atmospheric Chemistry, 6, 265-280 (1988)
[6] J.G.M. Roelofs, A.L.M. Houdijk, ref. 4, pp 10-16
[7] D. Berckmans, J.-Q. Ni, J. Roggen, G. Huyberechts, International Winter Meeting ASAE, Paper No. 924552, Nashville (1992), and references therein.
[8] F.N. Reece, B.D. Lott, W. Deaton, Poultry Science, 59, 486-488 (1980)
[9] B. Crook, F. Robertson, S. Glass, E. Boothroyd, J. Lacey, M. Topping, Americam Industrial Hygiene Association Journal, 52 (7), 271-279 (1991)
[10] L. Carr, F. Wheaton, L. Douglass, Transactions of the ASAE, 33(4) 1337-1342 (1990)
[11] N. Verdoes, Research Institute for Pig Husbandry, the Netherlands, (1991)
[12] Z. Genfa, Environmental Science & Technology, 23, 1467-1474 (1989)
[13] K. De Praetere, W. Van Der Biest, Journal of Agricultural Engineering Research, 46, 31-44

(1990)
[14] R. Scholters, in "Ammoniak in der Umwelt : Kreisläufe, Wirkungen, Minderung", Braunschweig, Germany, Oct 1990,
[15] D.M. Pranitis, M.E. Meyerhoff, Analytical Chemistry 59, 2345-2350, (1987)
[16] ASHRAE Handbook 1977 Fundamentals, American Society of Heating, Refrigerating and Air-Conditioning Engineers, Inc., New York, (1980)
[17] D. Berckmans, P. Vandenbroeck, V. Goedseels, International Summer Meeting ASAE, Paper No. 914031, New Mexico (1991)
[18] D. Berckmans, P. Vandenbroeck, V. Goedseels, Indoor Air (3) 323-336 (1991)

Mercury vapour-sensitive MOSFET with gold gate membrane

A.Galdikas, V.Gečys, V.Jasutis, S.Mickevičius, H.Tvardauskas

Semiconductor Physics Institute, A.Goštauto 11, 2600 Vilnius, LITHUANIA

ABSTRACT: Mercury-sensitive silicon field effect transistors with gold gate membrane have been developed. MOSFETs with the fine-grained of Au-membrane show the considerable increase of the threshold voltage U_t due to mercury vapor in ambient air. Threshold voltage dependence on mercury concentration C trends to saturation at $C \approx 2.5 \times 10^{-3}$ g/m^3. The time constant of U_t change is about 10 minutes. XPS experiments show no-zero mercury concentration in the gold membrane as well at Au-SiO_2 interface.

1.INTRODUCTION

Since the 1975 when I.Lundstrom [1] introduced H_2-sensitive silicon field effect transistor (FET), a great interest has been shown in chemical sensors based on metal-oxide-semiconductor (MOS) structures. The MOS-capacitors and MOSFET's with catalytic metals (Pd,Pt,Ir...) or semiconductor (SnO_x) films as active membranes sensitive to various gases were developed [2].

On the other hand gold is known as a strong absorber of mercury. The parameters of the thin gold films (resistance, mass) change after its exposing to mercury vapor [3,4].Therefore Au-SiO_2-Si structure appears to be mercury-sensitive. The possibility of mercury detection with MOS-capacitors was proved by authors [5].

In this contribution the silicon field effect transistor with gold gate sensitive to mercury vapor is presented.

2.EXPERIMENTAL

The usual n-channel silicon field effect transistors (MOSFET) were used in our experiments. Transistors were fabricated on the surface of the p-type boron doped silicon wafers with (100) plane orientation and 12 ohm.cm resistivity. Thickness of the gate oxide SiO_2 was 50nm, the distance between drain and source regions-200nm. Hg-sensitive gold membrane (thickness 10nm) was deposited by a thermal evaporation.The lead wires and contact squares of the transistors were covered by epoxy resin in order to avoid an erosion due to Hg vapor. MOSFET's were placed in a can with silver inner walls at once after fabrication in order to avoid mercury adsorption on the gate surface.

The composition of the gold films before and after its exposition to a mercury vapor was investigated by X-ray photoemission spectroscopy (XPS).The spectra of the Au-SiO_2-Si samples with a square 5*5mm^2 were taken by a calibrated XSAM 800 electron spectrometer with a MgKα source at 15KV. The samples fabrication technology and the thickness of the SiO_2 and Au films were the same as of the MOSFET's gate.

The transistors and samples for XPS investigation were exposed to mercury vapor in a chamber at 20°C temperature. The certain concentration of mercury vapor in chamber were obtained by the diluting the saturated vapor with air.

The drain current of the MOSFET with gold gate slightly depends on air humidity (current in saturation region increases about 10% when humidity changes from 30% to 75%). Therefore relative humidity in the chamber was kept at a level 60%.

It should be mentioned that no influence of the mercury vapor on I-U characteristic of MOSFET with Al gate was observed.

3.RESULTS

Experiments have showed the decisive importance of the structure of Au membrane on MOSFET's response to mercury vapor in air. It seems likely the threshold voltage U_t of the transistors with the membrane consisting of the grains which size d=0.02-0.07μm (Fig.1) decrease due to Hg exposing. Mercury-induced U_t shift of these MOSFET's is very different: typical values of ΔU_t are scattered in the range 0-0.1 V.On the other hand, the behavior of MOSFET's with fine-grained gold membrane is well defined.

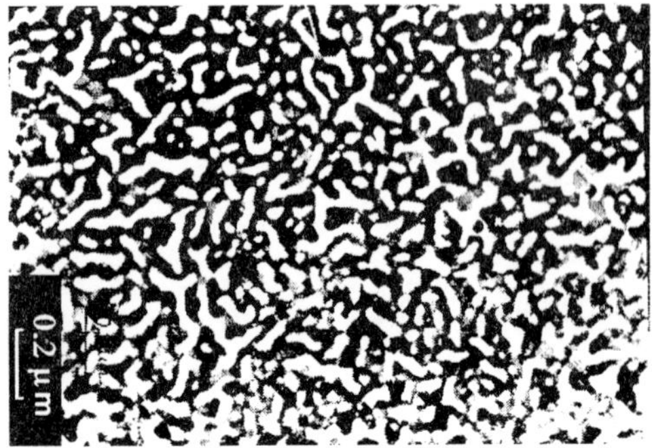

Figure 1. Transmission electron micrograph of the gold gate membrane.

Fig.2 shows the influence of mercury vapor on the MOSFET's with membrane grains size d<<0.01μm I-U characteristic, measured in diode regime (drain and gate are connected together). Mercury causes the parallel shift of I-U characteristic toward the higher values of drain voltage U, or, identically, the increasing of the threshold voltage U_t.The equivalent

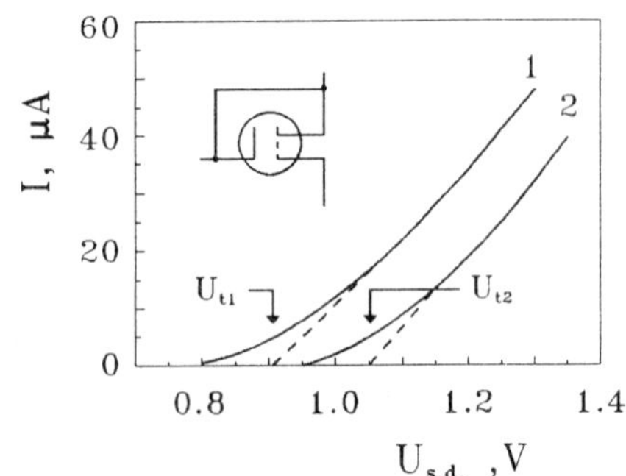

Figure 2. Diode characteristic of MOSFET with gold gate in air without (1) and with (2) mercury vapor (concentration 3.7x10^{-3}g/m^3).

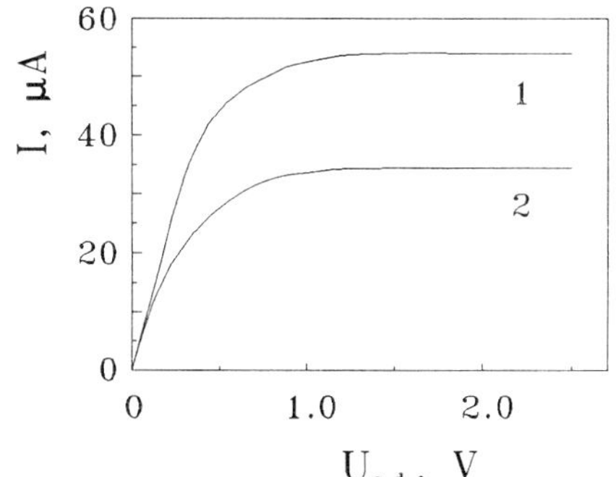

Figure 3. MOSFET with gold gate characteristic in air without (1) and with (2) mercury vapor (concentration $2x10^{-3}g/m^3$). U_{gate}=1.884V.

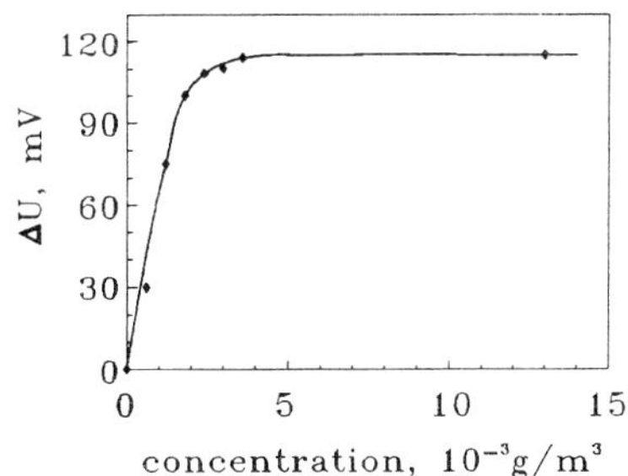

Figure 4. Dependence of the shift of threshold voltage on the mercury vapor concentration in air. The concentration $1.3x10^{-2}g/m^3$ corresponds to a saturated vapor at 20°C.

results were obtained when measurements were carried out in the case of constant gate voltage: mercury vapor leads to the decreasing of the drain current I (Fig.3). This effect trends to saturation at Hg vapor concentration $2.5*10^{-3}g/m^3$ (Fig.4).

The time constant of U_t or I change due to mercury vapor is about 10 minutes. The data shown in Fig.2, 3 and 4 correspond to the steady state. But when mercury vapor is removed from the chamber, the I-U characteristic remains the same as in the presence of mercury. The initial I-U dependence (as before exposing of MOSFET to Hg vapor) may be obtained, however, by annealing of the transistor in the air at the temperature 230-300°C for 10-30 min. The sensitivity of such annealed device to mercury vapor is almost the same as before annealing. But the parameters of MOSFET become significantly worse if these procedure is repeated five or more times.

The wide scan spectra of the no Hg-affected films shows only a peaks corresponding to $4f_{5/2}$ and $4f_{7/2}$ excitations of the gold atoms. The 12-hour exposition of the samples in the saturated mercury vapor at room temperature leads to the appearing of the mercury doublet $4f_{5/2}$ and $4f_{7/2}$ in the 100-102eV energy range.
In order to obtain the concentration of the Hg atoms in the Au-films and to determine Hg penetration depth, the samples were profiled by argon ion bombardment with energy of 2keV. The measured erosion rate was about $0.025nm*s^{-1}$.The profiling data (Fig.5) shows the ratio of Hg and Au concentration on the samples surface about 1:14. The greater part of Hg atoms is located in the near-surface layer of Au with thickness about 6nm. But the concentration of mercury at Au-SiO_2 interface is not zero: Hg:Au ratio near SiO_2 layer is about 1:100.

The presence of Hg at the Au-SiO_2 interface gives an opportunity to explain the sensitivity of MOSFET with gold gate by the influence of Hg on a work function of Au. The threshold voltage of MOSFET is proportional to the work function of the gate material [1]. The work functions of Au and Hg are 4.30 and 4.52eV, respectively. We assumed that the impurities of Hg cause the increasing of work function of Au, and therefore U_t rises (or I decreases) when MOSFET with gold gate is exposed to the mercury vapor. But the quantitative estimation shows smaller Hg influence to MOSFET parameters than observed experimentally. For an exact explanation of obtained effect it is necessary to take into account a possible change of Au layer's structure due to Hg, for example, formation of amalgam.

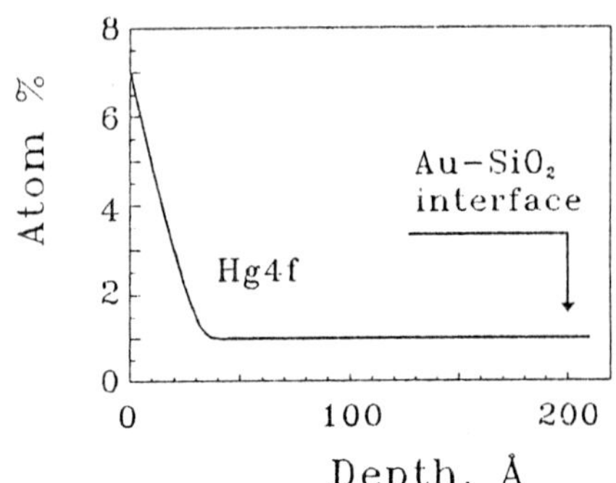

Figure 5. Profiling data of Au film exposed to saturated mercury vapors for 20 min.

In general, the presented results (MOSFET's sensitivity to mercury, time constants) are not in contradiction with those obtained for MOS capacitors [5]. Unfortunately, the sign of the capacitance change due to Hg was not indicated in [5].

REFERENCES

1. K.I.Lundstrom, M.S.Shivaraman, and C.M.Svenson. J.Appl.Phys. 1975. V46. No9. P.3876-3881.

2. W.Gopel, J.Hesse, J.N.Zemel. Sensors. A comprehensive survey. V.2 Chemical and biochemical sensors.1991.P.360-380.

3. A.N.Mogilevski, A.D.Mayorov, N.S.Stroganova, D.B.Stavrovski,I.P.Balkina, L.Spassov, D.Michailov, R.Zaharieva. Sensors and Actuators A.1991. V.28.P.35-39.

4. P.J.Murphy. Anal.Chem. 1979. 51.P.1599.

5. F.Winquist and I.Lundstroms. Sensor and Materials, 1991, V.2,N.4,P.229.

The effect on response due to the addition of glass frit to tin oxide to form materials suitable for the screen printing of thick-film gas sensor arrays

E.Sizeland and J.K.Atkinson

Faculty of Engineering and Applied Science
University of Southampton
Southampton SO9 5NH
United Kingdom

Summary

Addition of lead borosilicate glasses to tin oxide to form high temperature firing screen printable materials has been investigated. The mechanical strength of fritted and un-fritted materials has been compared and found to be similar. The effect on the conductivity response of these materials to methane however shows a reversal of the normal behaviour of the tin oxide above certain operating temperatures. Further studies of this behaviour pattern are expected to reveal a much better understanding of the mechanisms involved in the conductivity changes of semiconductors when exposed to gases which effect their charge carrier concentrations. The reversal of the response characteristic also offers increased scope for the attainment of orthogonal responses in individual members of a sensor array designed to be used with a pattern recognition approach to the detection of specific gases.

Introduction

Tin oxide has been researched extensively as a material well suited for use as the basis for a gas sensor[1]. With the addition of suitable dopants the sensitivity exhibited by tin oxide films in the form of measurable conductance changes in the presence of various reducing gases can be very large. A major drawback exists however in a pronounced lack of specificity which has chiefly been addressed through the addition of various promoters and catalysts in attempts at increasing the magnitude of response to specific gas species.

An alternative method for improving specificity lies in the use of arrays of tin oxide sensors variously doped with additives such that differing responses are obtained from the individual array elements. In this way the possibility of using signal processing, and in particular pattern recognition methods, can be exploited to achieve levels of specificity from the array not normally obtainable with individual sensors.

The work described here forms part of an on-going programme of research aimed at the development of tin oxide sensor arrays for the detection of flammable and toxic gases. As a necessary stage in this development a means of producing cheap and repeatable sensor arrays has been pursued through the use of thick-film screen printing. In order that the sensors themselves are sufficiently robust a means of producing the tin oxide as a high temperature fireable thick-film ink has been sought.

Several different approaches to the production of a fireable thick-film ink are possible including the addition of a glass frit which is capable of forming the bond to a supporting ceramic substrate such as alumina. Previous work concerning the use of thick-film arrays of gas sensors[2] resulted in the production of a supporting structure consisting of an alumina substrate carrying an array of buried platinum heating elements underneath and electrically isolated from gold electrodes onto which gas sensitive materials can be deposited. In this way the conductivity of films of gas sensitive materials, heated to controlled temperatures, can be conveniently and simultaneously measured.

Construction of the Gas Sensor Arrays

The use of the previously described supporting thick-film circuits involved some modification since the anticipated operating temperatures were considerably in excess of those used previously with organic materials for which the substrates were originally designed. The original sensor design incorporated Palladium Silver solderable terminations. This was necessary as a leadframe was attached to the sensor array which then would be dipped in a high temperature (250°C) solder. Any experimental use above 300°C resulted in the solder reflowing and the occurrence of silver migration across the solder dam into the remainder of the sensor, consequently a new packaging method had to be sought.

Experiments were undertaken using various mixtures of commercially available adhesives to attach the leadframe to the substrate and an extensive search into the possibility of using a high temperature (>350°C) conductive polymer epoxy was carried out. This was unsuccessful and the use of metal packages using wire bonds was seen as the only worthwhile alternative. The design of the sensor however was not substantially altered, although the use of gold terminating pads was used, as was gold 0.001 inch wire bonds and gold bond pads in the metal packages.

The use of the same metallurgy meant that temperatures up to 1000°C could easily be obtained without significant ageing of the wire bond. This point is further emphasised when observing the experimental usage of the sensor array. The sensors were suspended, inverted, and were purely held by the strength of the wire bonds. The bonds were stich bonded, employing a double bond on their tails thus creating a stronger bond. However, to bond adequately the substrate must initially be secured to the metal packaging. This was accomplished using a non-conductive heat activated adhesive which looses it's adhesive properties at 150°C with negligible outgassing. Thus the sensors once at elevated temperatures are suspended freely by the gold wire bonds only.

Another departure from the original sensor design comes in the form of the construction of the array. Each tin oxide sensor is individually "snapped out" from the substrate and hence there is no thermal conduction between adjacent sites through the alumina.

Preparation of the Active Material

The active components of the thick film inks were treated prior to being formed into a paste. The additives and Tin (IV) Oxide (SnO_2) were mixed together on a weight-to-weight basis and ground into a fine powder using a pestle and mortar. The mixed SnO_2 samples were constructed by sintering in a box furnace of 2730cm^3 capacity at 1000°C for two hours, using a temperature ramp of 12.5°C per minute, any colour changes observed were recorded. Following sintering the compounds were again ground in a pestle and mortar and incorporated into a thick film ink.

Commercially available organic solvents were used in the preparation of the pastes (*ESL 400 vehicle and ESL 401 thinner*). The easiest paste to make incorporated 100% vehicle as the only organic solvent, this however was thought to be too expensive. The cost of each ink was subsequently reduced by the addition of thinner on a 50% volume-to-volume basis. The other additions to the paste were the active components, SnO_2 and the binding agents, for example Lead Borosilicate glass.

The Tin Oxide and glass based pastes were prepared on a weight-to-weight (wt-wt) basis with the relative proportion of glass ranging from 0 to 10% wt-wt. The powdered constituents were weighed with 25 grams being placed into a jar. The vehicle/thinner mixture was then added slowly until a paste was formed that flowed from a spatula in an even, continuous stream. Once this had been completed the jar was sealed and labelled with the ink components and date of manufacture. The jar was then placed in a refrigerator prior to use.

After printing the tin oxide inks were fired in a single zone belt furnace at temperatures in the region of 850°C. The fired SnO_2 ink was found to be of a submicron particle size when viewed via a scanning electron microscope (SEM), as shown in figure 1. The resulting morphology emphasises the particular usefulness of SnO_2 for use as a gas sensor due to it producing a very high surface area, which helps improve the response time of the sensor.

Figure 1 : SEM of Tin Oxide thick film

A Lead Borosilicate (*ESL Glass 7*) having a peak firing temperature of 850°C was incorporated into the pastes to increase the mechanical strength of the fired film. Several pastes were fabricated using a percentage weight to weight (wt-wt) of the glass ranging from 0 to 10% wt-wt. An Energy Dispersive X-Ray Analysis System (EDAX's) was used to investigate the exact formation of the glass. This was necessary as the composition of the glass is proprietary information to the glass manufacturer. Boron however, one of the main constituents has a lower atomic weight than Carbon and cannot be quantitatively analyzed by this approach. The attempted use of several other techniques to analyze the Boron content including Fourier Transform Infrared (FTIR) also pointed out the difficulty in analyzing Boron. However conversations with the glass manufacturers suggested that the Boron content was between 1% and 10% wt-wt.

Gas Testing

The gas rig design employed in the experiments is shown in Figure 2. This set up facilitates the examination of multiple gases. The gas rig design has a complex switching arrangement which must be carried out in exactly the correct order to minimise cooling effects. Any cooling effects that may have been encountered were monitored via the platinum heaters by noting any resistance change when the switch from clean air to gas mixture is made and vice versa. The gas rig is semi-automated in that the data is collected automatically using an analog to digital converter (ADC), [Amplicon Liveline, PC30A].

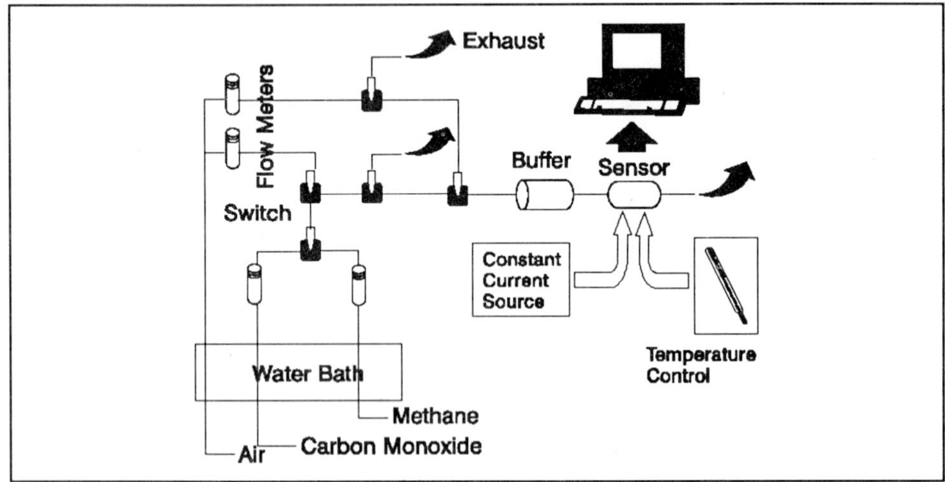

Figure 2 : Semi-automated gas rig (multiple gases)

Standard Gas Exposure Experiment

A standard gas exposure experiment was formulated to facilitate the testing of the sensor arrays. To allow the testing of the arrays using a range of gas concentrations over as short a period of time as possible a system had to be devised whereby the sensors could detect and recover from exposure to the analyte without the response being inhibited by the previous experiment. The formulation of the standard experiment, designed for twenty minutes, incorporated a three minute period of base line readings, seven minutes of exposure to analyte and ten minutes recovery. The final recovery period is designed to ensure that the active material recovers from the exposure with only the deep sites within the sensor still being bound by the analyte gas carriers.

An experiment to determine the flow rate necessary for the twenty minute time scale was undertaken using a range of operating temperatures from 240 to 530°Celsius. This experiment showed that using a flow rate of 1.0 litre per minute produced the greatest response (natural logarithm of conductance in gas divided by the conductance in air) to 50% of the lower explosive limit (LEL) of Methane. It was found that at flow rates above this the results become increasingly non-repeatable possibly due to turbulence. Below this rate a reduced response was observed possibly due to the there being an insufficient volume of gas passing over the sensor to maintain the diffusion profile.

Observed Gas Sensor Responses

Response versus Methane Concentration

This experiment was undertaken using the standard exposure time scaling. The operating temperatures of individual sensors were held constant and each sensor was repeatedly exposed to an increasing concentration of Methane. The results obtained showed that increasing the concentration of the analyte gas increased the response of the sensor as shown in figure 3.

Response versus operating temperature

The response of the Tin (IV) Oxide gas sensors at constant gas concentration showed a peak of response in the region between 450 and 530° Celsius. A plot of response versus operating temperature for 50% LEL is shown in figure 4.

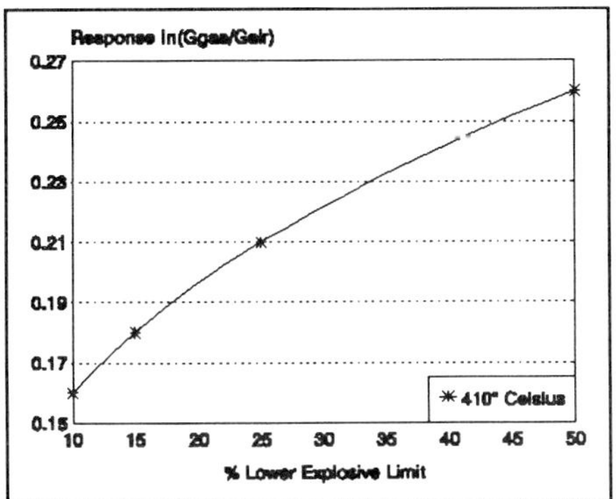

Figure 3 : Response versus % LEL Methane

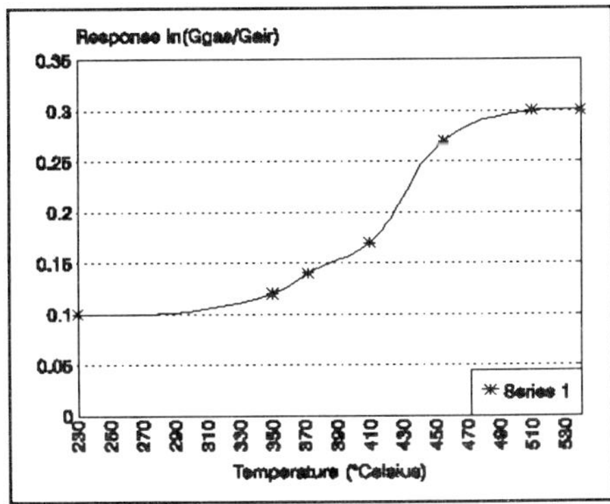

Figure 4 : Average response to 50% LEL Methane versus Operating Temperature

Effect of Added Glass Frit

As has previously been mentioned a lead borosilicate based glass *(ESL Glass 7)* was added to the pure tin oxide to increase the mechanical strength of the material. The proportions of added glass showed no increase in mechanical strength when added in quantities less than 5% weight-weight (wt-wt).

The addition of the glass dramatically increased the working resistance of the sensor material by a factor of approximately twenty in the case of 5% wt-wt glass (compared to that of pure tin oxide). As mechanical strength did not increase until 5% wt-wt all experiments incorporating glass were undertaken using the 5% and 10% wt-wt sample.

The mechanical strength of the inks was subsequently found to be sufficiently strong without glass and the inclusion of a glass frit was eventually deemed unnecessary. However the response of the glass based sensors provided some quite interesting results.

It was found that operating at temperatures below 300° Celsius Tin (IV) Oxide was the dominant active material and the classical n-type material response was observed. At temperatures above this the glass appeared to start to compete for dominance in the carrier response mechanism and above 500° Celsius become the dominant semiconductor mechanism. Typical responses to glass based sensors are shown in figures 6 and 8 with the corresponding plots for pure Tin (IV) Oxide films shown in figures 5 and 7.

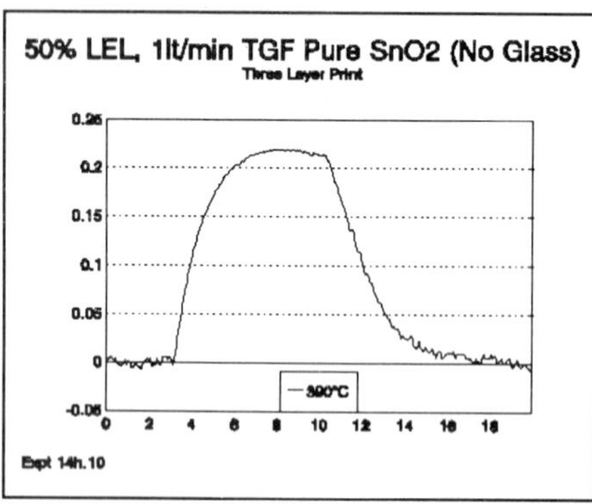

Figure 5 : Response to 50% LEL for Pure Tin (IV) Oxide @ 390°C

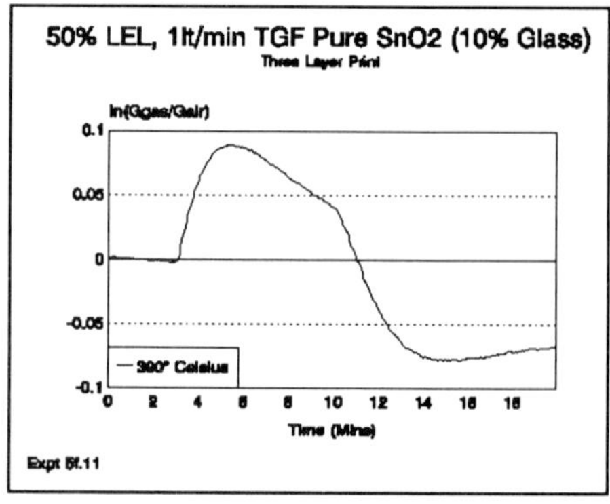

Figure 6 : Response to 50% LEL for 10% Glass @ 390°C

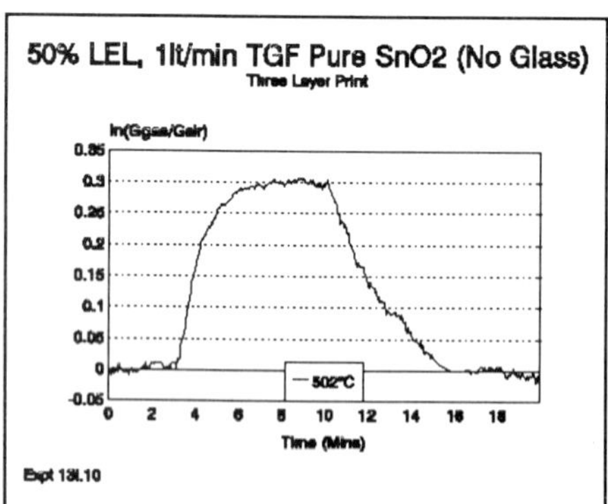

Figure 7 : Response to 50% LEL for Pure Tin (IV) Oxide @ 500°C

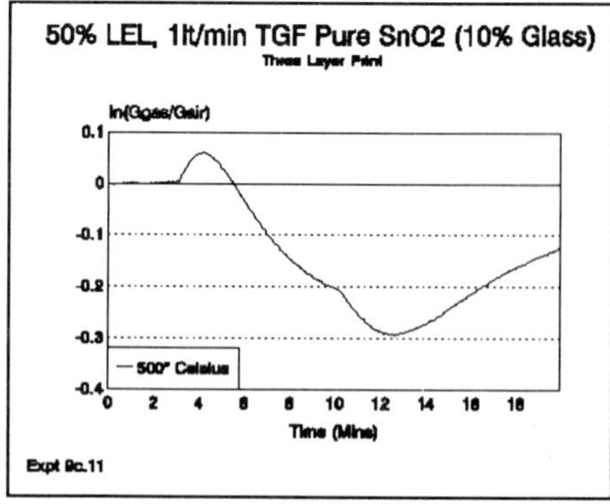

Figure 8 : Response to 50% LEL for 10% Glass @ 500°C

The fact that the presence of lead borosilicate glasses appears to induce p-type behaviour in the sensor material suggests that the boron content is producing a doping effect which only begins to operate at higher temperatures. This phenomenon is being studied in materials where controlled amounts of boron have been added. It is hoped that these experiments will lead to a clearer understanding of the exact mechanism by which the conductivity of the material is modulated by carrier exchange between the gas species and the semiconductor.

Discussion of Results

It was found that the addition of glass frit to tin oxide thick-film pastes was unnecessary for the purposes of enhancing the bonding of the material to alumina substrates. It was noted that at a suitable operating temperature the presence of a Lead Borosilicate additive produces a change in behaviour that would however appear to offer interesting possibilities in relation to improving specificity.

The effect of the glass additive is such that the response to the reducing gas (methane) is reversed whilst the response to other interferants (for example water vapour) is not. Hence it should prove possible to differentiate between these gases using an array containing sensors doped with promoters such as antimony and sensors doped with p-type additives such as boron.

References

[1] G.S.V. Coles, K.J. Gallagher and J. Watson, *Fabrication and preliminary tests on tin (IV) oxide-based gas sensor*, Sensors and Actuators, 7, 1985, pp 89-96.

[2] A.W.J. Cranny, *Ph.D. Thesis*, University of Southampton, 1992.

Highly selective low power Cu and Pt doped SnO_x film based CO gas sensor

V.Ambrazevičienė, A.Galdikas, A.Mironas, A.Šetkus

Semiconductor Physics Institute, A.Goštauto 11, 2600 Vilnius, LITHUANIA

ABSTRACT: The simultaneous influence of the Pt- and Cu-doping on the sensitivity and selectivity of the screen-printed SnO_x film has been investigated. The very remarkable dependence of the resistance response to exposure of H_2 and CO gases on the amount of Cu-additive was obtained at low temperatures ($\leq$150°C). The optimum value 0.16wt% of Cu-doping was established for highly sensitive and selective CO gas sensor. The optimum range of the low working temperature (70°C-100°C) and high stability in respect to variation of the temperature through the range have been gained also.

1. INTRODUCTION

Semiconductor gas sensors based on the tin oxide or other metal oxide offer the very attractive attributes such as high sensitivity, small size, simplicity and low cost. The lower working temperature (<300°C-500°C) and the rather high selectivity of those sensors are preferable for many cases of practical applications. The desirable characteristics could be gained sometimes by addition of foreign metals as has been reported by number of authors [1-6].

It is well known now that addition of small amount of Pd or Pt promotes gas sensitivity and decreases the working temperature [1-4]. The effect is related to the catalytic activities of noble metals and sufficient selectivity could not be obtained. On the other hand the number of authors found a variety of additive metals such as Al, In, Cu, Cd, etc. suitable to produce the selectivity in response to exposure of the certain gas (e.g. H_2, NO_x, H_2S) SnO_x-based elements. It must be noted however that the gases could be detected only at moderate temperatures (200°C-300°C). Though the nature of the effect of the "ordinary" additives is not clear yet, taking [2] into account the Fermi level and consequently the conduction electron concentration dependence on the doping level may be assumed as the main cause of the effect. More details are necessary before the model is complete. In order to distinguish the effect of the "ordinary" additives from other one the lower temperatures (50°C-100°C) seems preferable to a higher one. It is because the first donor level of the oxygen vacancy (0.03 eV [7]) should be fully ionized in commonly used SnO_x elements with electron concentration **n** of the order $10^{18}cm^{-3}$ at these temperatures and also the influence of the second donor level of the oxygen vacancy (0.15 eV [7]) on **n** could be neglected. Hence the Fermi level in the bulk of SnO_x should be affected by the foreign dopant only.

The simultaneous influence of the Pt- and Cu-additives on the sensitivity and selectivity of the SnO_x-based gas sensors has been investigated in this paper. The sufficient sensitivity of the elements due to catalytic activities of Pt at low temperatures is promoting the detection of the peculiar features of the foreign metal doping by a simple testing system. If consider that Cu acts as electron acceptor in SnO_x [2] the significant dependence of the Fermi level on small amount of Cu additive could be expected. On the other hand the quite different results on Cu doping in [2] and [3] were very suggestive because of different amount of the additive. In the present paper the quantity of Cu was varied through the range from 0.04wt% to 4wt%. The very first experimental results of the investigation are reported. The development of the model of the phenomenon is in progress now so it will be reported elsewhere.

2. EXPERIMENTAL

Gas sensitive SnO_x-based films were constructed on Si substrate with insulating SiO_2 layer by screen printing. The golden terminals were evaporated on the substrate and shaped before film deposition. Several coatings of the water solution of the $SnCl_2 \cdot 2H_2O$ and HCl based mixture were printed on the substrate. The dopants platinum acid $H_2PtCl_6 \cdot 6H_2O$ and copper chloride $CuCl_3$ were added to the solution. The temperature of the substrate was approximately 80°C. Sensor elements were annealed at 300°C in an ambient atmosphere for an 0.5 hour. In order to distinguish the Cu-doping effect from other one the experimental results were compared with those obtained on Pt-doped SnO_x films. These films were produced by the same method except for Cu-doping and are called the basic elements trough the rest of this report.

The amount of Cu-additive was varied trough the range from 0.04wt% to 4wt%. The amount of Pt was constant and equal approximately 1wt% for the all elements.

Electron microscopy and X-ray diffraction analysis showed the polycrystalline structure of a doped SnO_x films with the grain size of the order 0.1 μm. The thickness of the films depends on the solution amount and usually was about 1 μm.

All the tests were carried out in the test chamber with an ambient atmosphere containing the definite amount of a reducing gas (mainly CO or H_2). The sensor resistance was measured by a simple electric circuit. The main results are presented at working temperature 85°C though the experiments were carried out at different temperatures of the range 20°C-260°C. The temperature was measured by a common thermocouple being attached to the element.

3. RESULTS AND DISCUSSION

The dependence of the resistance change $\Delta R/R_0$ of the sensors due to CO and H_2 gases on amount of Cu additive are shown in Fig.1 at 85°C. The dependencies represent the average of the results obtained on the different SnO_x sensors with the different amount of Cu. One can see the exciting drop down of $\Delta R/R_0$ at 0.16-0.17 wt% of Cu in the presents of the H_2 gas. The opposite influence of the Cu on the resistance change due to exposure to CO gas was found. The same results were obtained at the gas concentrations up to approximately 10^4 ppm. We should point out that only certain amount of Cu of the 0.05wt%-0.5wt% range can be admitted in order the suppression of the gas sensitivity should be avoid. In accordance with this too

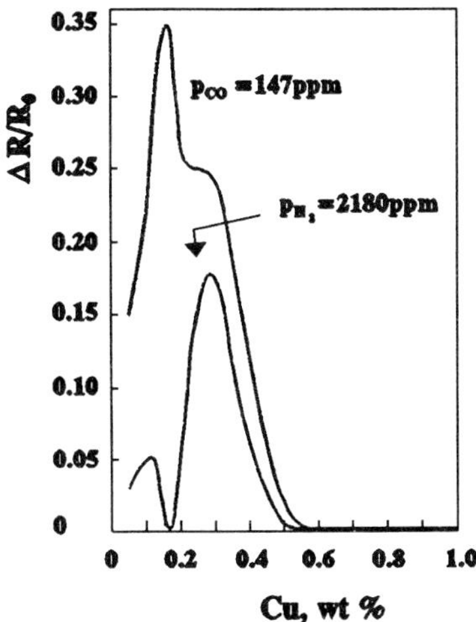

Figure 1. The dependencies of the sensor resistance change on amount of Cu-additive under exposure to H_2 and CO at 85° C.

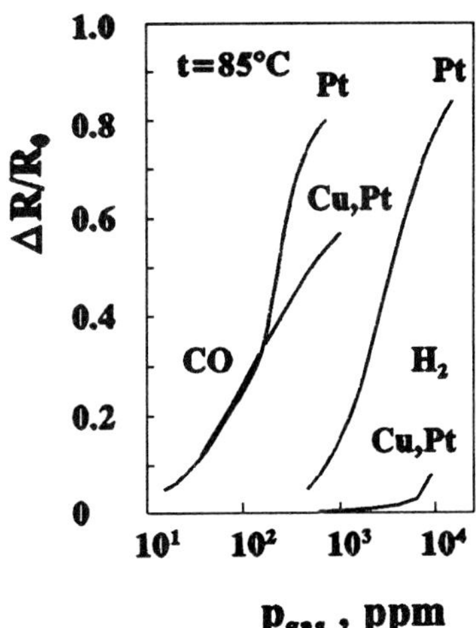

Figure 2. The dependencies of the sensor resistivity change on the partial pressure of CO or H_2 gases at 85° C. **Pt** marks the basic element and **Cu,Pt** - Cu-doped element.

large amount of Cu (by our estimation ~2.7wt%) should be the main cause of almost insensitive to all of the tested gases SnO_x-based element produced in [2].

The dependencies of the resistance change on CO and H_2 gases concentration in an ambient atmosphere are shown in Fig.2 for 0.16wt% Cu-doped element and for the basic element. The resistance change of Cu doped sensor due to the H_2 gas is less than 3% in the wide range of the H_2 gas concentration (up to $6 \cdot 10^3$ ppm) while that due to CO is significant and increases by power law $\Delta R/R_0 \sim p^{\alpha}_{CO}$ with increasing concentration. We found almost constant value $\alpha \approx 1.8 \cdot 10^{-3}$ at all the concentrations of the CO gas tested. The results on Fig.2 show almost the same resistance change due to exposure to H_2 and CO gases for the basic element. By comparing the dependencies of $\Delta R/R_0$ on gas concentration for Cu-doped and the basic elements shown in Fig.2 one can deduce the significant increase of the CO gas selectivity by Cu doping.

The resistance change of the Cu-doped element due to exposure to other gases tested was negligible (see Table 1).

Table 1.

Gas	CH_4	CO_2	SO_2	NO_x
Concentration,ppm	8700	6300	7000	8700
$\Delta R/R_0$	0.04	0.04	0.04	0

In addition we should point out the high stability of the Cu-doped elements in respect to variation of the working temperature. It could be illustrated by the comparison of the experimentally obtained resistivity dependance on temperature for the 0.16wt% Cu-doped and the basic elements plotted on Fig.3. The resistivity R_0(Cu,Pt) of Cu-doped element in ambient atmosphere is almost constant through the all range of the temperatures while the resistivity R_0(Pt) variation with temperature of the basic element is significant. The resistivity response of Cu-doped element to exposition of CO gas R_{CO}(Cu,Pt) ceases at temperature ~150°C almost abruptly. But the resistivity R_{CO}(Cu,Pt) variation with temperature is negligible at low temperatures (50°C-100°C). We have obtained the discrepancies of less than ±1.5% and ±3% between the exact values of the resistivity change $\Delta R/R_0$ and those in Fig.2 when the temperature was varied from the working temperature 85°C by the value through the range accordingly ±15°C and ±30°C. On the other hand the resistivity response of the basic element to Co gas R_{CO}(Pt) depends significantly on temperature through the same range (Fig.3). It must be pointed out also there was not found the noticeable influence of the temperature change on the selectivity of the Cu-doped elements.

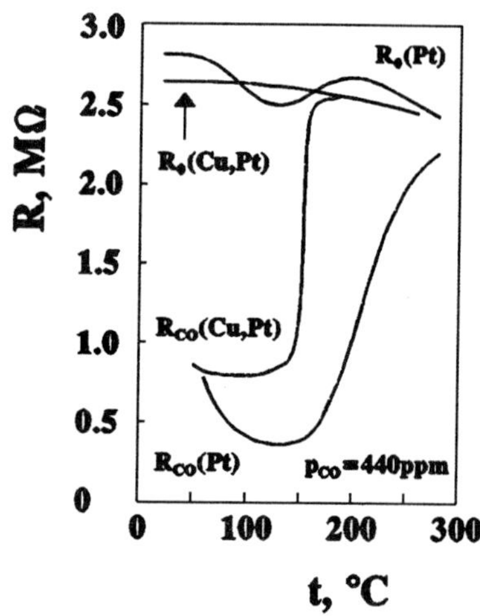

Figure 3. The dependencies of the sensor resistivity on temperature in ambient atmosphere (index **0**) and under exposure to **CO** gas (index **CO**). **Pt** marks the basic element and **Cu,Pt** - Cu-doped element.

REFERENCES

[1] D.Kohl, The role of noble metals in the chemistry of solid-state gas sensors,**Sensors and Actuators B, 1** (1990) 158-165.

[2] N.Yamazoe, New approaches for improving semiconductor gas sensors, **Sensors and Actuators B, 5** (1991) 7-19.

[3] N.Yamazoe, Y.Kurokawa and T.Seiyama, Effects of additives on semiconductor gas sensors, **Sensors and Actuators, 4** (1983) 283-288.

[4] Duk-Dong Lee, Byung-Ki Sohn, Dong-Sung Ma, Low power thick film CO gas sensors, **Sensors and Actuators, 12** (1987) 441-447.

[5] G.S.V.Coles, S.E.Bond and G.Williams, Selectivity studies and oxygen dependence of tin(IV) oxide-based gas sensors, **Sensors and Actuators B, 4** (1991) 485-491.

[6] G.Sberveglieri, S.Groppelli and P.Nelli, Highly sensitive and selective NO_x and NO_2 sensors based on Cd-doped SnO_2 thin films, **Sensors and Actuators B, 4** (1991) 457-461.

[7] S.Samson and C.G.Fonstand, Defect structure and electronic donor levels in stannic oxide crystals, **J.Appl.Phys., 44** (1973) 4618-4621.

Section B

SILICON SENSORS

Silicon micromachined sensors

Kurt Petersen
Lucas NovaSensor, 1055 Mission Court, Fremont, CA 94539

ABSTRACT: The use of Silicon Fusion Bonding (SFB) for the fabrication of silicon micromachined sensors is described. Basic device technology is reviewed and numerous examples are given to illustrate the power and versatility of wafer bonding when combined with other micromachining processes and etch-back techniques. Pressure sensors, acceleration sensors, microwave power detectors, resonant pressure sensors made with SFB surface micromachining, and thermally isolated structures all benefit from the availability of SFB as a core process technology applied to the design of silicon micromachined sensors, actuators, and other microstructures.

1. INTRODUCTION

Few silicon processing techniques, applied to solid-state sensors and actuators, have proven to be as versatile, powerful, and commercially successful as Silicon Fusion Bonding. SFB can be employed in both major microstructure fabrication methodologies, Bulk and Surface micromachining. Applied to "bulk" silicon micromachining principles and design methodologies, revolutionary new pressure sensors, acceleration sensors, and other devices have already been commercialized less than 5 years after the general introduction of the process. Applied to "surface" micromachining principles and design methodologies, microactuator and microresonator structures have been commercialized and the background has been established for more advanced and novel surface micromachined devices whose performance is not limited by the constraints and disadvantages of thin, deposited, polycrystalline films.

Described by Lasky in 1985 [1], Silicon Fusion Bonding (or Direct Wafer Bonding) is most applicable to sensors and actuators when combined with various controllable etch-back techniques [2,3]. Of particular interest here is the anisotropic electrochemical etching principles analyzed by Seidel [4]. All the devices described here employ these important processing methods.

2. SOI STRUCTURES

2.1 Ultra-Stable, High-Temperature Pressure Sensor

In addition to the SOI structures now under extensive development in the integrated circuit community, SFB offers special advantages for heavily-doped, dielectrically-isolated piezoresistors. Figure 1 shows a pressure sensor rated for very high temperature operation

(>250°C). The chip relies on dielectrically-isolated piezoresistors fabricated with silicon fusion bonding [5]. First, the sacrificial wafer is implanted with a patterned heavy boron dose to form a series of P+ resistors in the configuration of a wheatstone bridge. Next, the sacrificial wafer is bonded to an oxidized bottom substrate wafer. Now the top, sacrificial wafer can be removed by a combination of grinding and dopant-selective etching. The resulting wafer now consists of a series of implanted P+ piezoresistors bonded on top of an insulating oxide. These resistors can be thermally oxidized and interconnected with a high temperature metallization, followed by a conventional backside, anisotropic etch to form a pressure sensitive diaphragm.

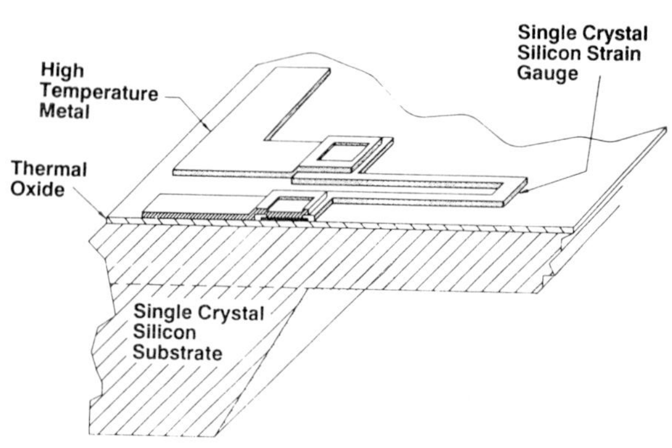

Figure 1

Cross section of a pressure sensor rated for high temperature operation (> 250C). The dielectrically isolated resistors are formed by silicon fusion bonding.

The dielectrically-isolated format provides three important performance advantages for pressure sensors. First, since the piezo-resistors are not junction-isolated, the sensor can operate at higher temperatures without the junction-leakage problems associated with conventional piezoresistive sensors. Second, since the isolated resistors have no associated space charge regions, they are very tolerant to ionic surface contamination, and the absolute values of these resistors are, therefore, extremely stable over long time periods. Third, since the resistors are heavily doped, the temperature coefficients are small and linear, allowing very accurate temperature compensation and calibration. These chips are currently in use on the US Space Shuttle fleet to monitor tire pressure.

3. PRESSURE SENSOR STRUCTURES

3.1 Absolute Pressure Sensor

Conventional micromachined pressure sensors have many mechanical disadvantages. First, because the diaphragm is defined by a cavity etched from the backside of the wafer, the alignment of the frontside piezoresistors to the diaphragm is inaccurate. Second, backside anisotropic etching causes the cavity sidewalls to slope "outward" from the diaphragm edge. Besides expanding the size of the chip, the resulting stress distribution causes inherent output non-linearities in the pressure response. Third, the anodically-bonded glass constraint" usually incorporated in conventional pressure sensors induces differential mechanical and thermal coefficients. This effect results in non-linear performance characteristics which are difficult to correct.

All these disadvantages can be avoided by employing silicon fusion bonding processes in the design of pressure sensors. See Figure 2. In a typical SFB process flow [6], a cavity is first etched in the bottom wafer. Of course, this cavity can be any shape or depth. Next, the top wafer is bonded over the etched pit, creating a vacuum cavity. After the top wafer is thinned back to the desired thickness (by a combination of grinding, polishing, and/or electrochemical etching), all the standard IC processing required to produce the familiar piezoresistive wheatstone bridge is completed, as shown in Figure 3. The final pressure sensor, together with its integral vacuum cavity, is now composed entirely of silicon, thereby eliminating differential thermal and mechanical coefficients. In addition, since the cavity as well as the piezo-resistors were both defined on the top surface of the wafer, the easy and accurate alignment results in improved and very reproducible sensor performance characteristics.

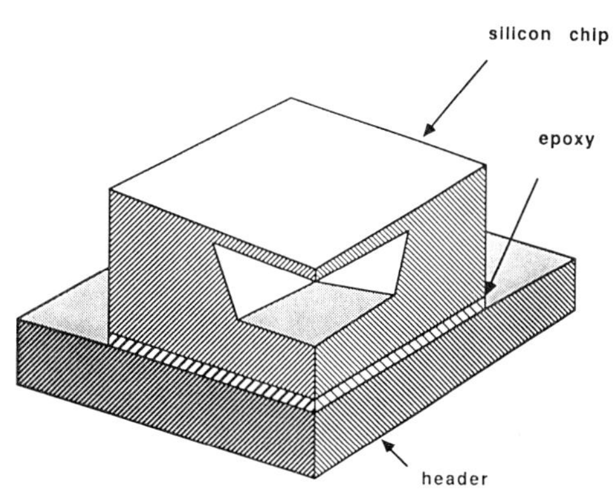

Figure 2

Cutaway view of an absolute pressure sensor formed by silicon fusion bonding. The diaphragm is formed by thinning of the top bonded wafer.

3.2 High Overrange Tolerance Pressure Sensor

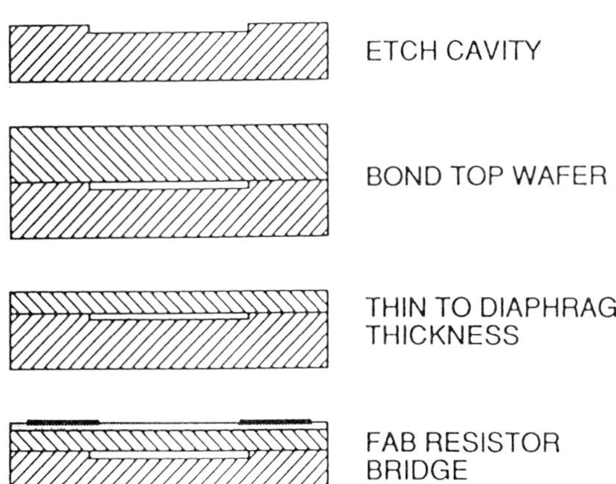

Figure 3

Process flow for an SFB Pressure Sensor. If the cavity is etched deeper, the device shown in Figure 2 is obtained.

Many unique and useful devices are made possible due to the fact that the vacuum cavity created during the standard SFB pressure sensor process can be any shape or depth. Figure 4 shows a pressure sensor with a vacuum cavity only a few microns deep [7]. The total full-scale deflection in the center of a typical pressure sensor diaphragm is only a few microns. For applied pressures greater than full-scale, the diaphragm will begin to contact the bottom surface of the etched cavity. The bottom surface therefore serves as an over-pressure stop. Conventional pressure sensors with full-scale ranges of only 15 psi typically burst at about 20x over-pressure. Similar devices built with this shallow over-range stop (made possible by SFB processing), not only survive over-pressures as high as 500x, but their performance, after repeated cycling to 500x, is degraded less than 0.2%. This dramatic improvement in over-pressure capability is very important for industrial process control applications.

Other types of pressure sensors have been realized using a similar process methodology. Shoji et al [8] demonstrated an absolute capacitive pressure sensor in which both plates were single crystal silicon. Huff et al [9] fabricated a structure in which the two plates were designed to touch as a specific pressure, thereby functioning as a pressure switch.

An important and far-reaching aspect of this SFB design concept is its similarity to polysilicon surface micromachining. In fact, the structure of the single crystal pressure sensor described here is virtually identical to the original polysilicon surface micro-machined pressure sensor [10]. As we will discuss below, instead of the sacrificial oxide underneath a polysilicon layer, SFB allows either a vacuum cavity or a sacrificial oxide underneath a single crystal silicon layer. The significance of this development is that many of the important surface micromachined structures and devices first demonstrated with polysilicon can now be made with single crystal silicon - exhibiting much improved performance. This will be discussed further below.

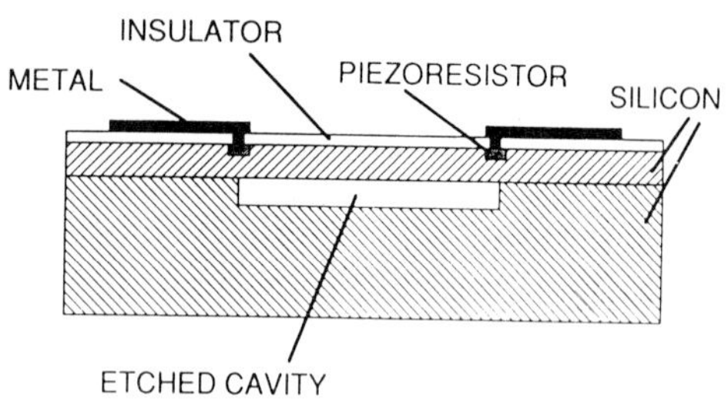

Figure 4

Cross-section of a pressure sensor with high over-range tolerance. This 15 psi sensor is capable of being cycled at 500 times over-pressure with less than 0.2% performance

4. HANDLE WAFER CONCEPT

An important concept in micromachining fabrication technology is the use of a "handle" wafer. Many of the types of devices built with micromachining techniques are required to be very thin and/or very small. Even today, such devices are sometimes fabricated by fully processing wafers as thin as 100 microns. In order to minimize breakage, 25 and 50 mm diameter wafers were used. The concept of the handle wafer allows most of the processing to proceed on a thin region bonded to a normal thickness substrate, which is ultimately removed at the end of the process. The bonded assembly must be strong, stress-free, and capable of withstanding high processing temperatures associated with such steps as diffusion and oxidation. Silicon Fusion Bonding is ideal for this application.

4.1 Catheter-Tip Pressure Sensor

The SFB process for pressure sensors makes it possible to create extremely small catheter-tip pressure sensors [6] for diagnosis of cardiac abnormalities. In the process described above, the vacuum cavity can be a 250 micron square anisotropically etched pit. Since the resulting vacuum cavity now has inward-sloping walls, the entire chip actually needs to be only slightly larger than the diaphragm itself, as shown in Figure 5. Also, since the final sensor chips must be very thin, the wafer is simply ground and polished back to its desired thickness (typically 150 microns) in the final process step. During this step, the anisotropically etched pit is opened to the atmosphere, creating the desired gage pressure sensor. In effect, the substrate serves largely as a "handle" wafer so that most of the processing can be completed on a normal thickness wafer.

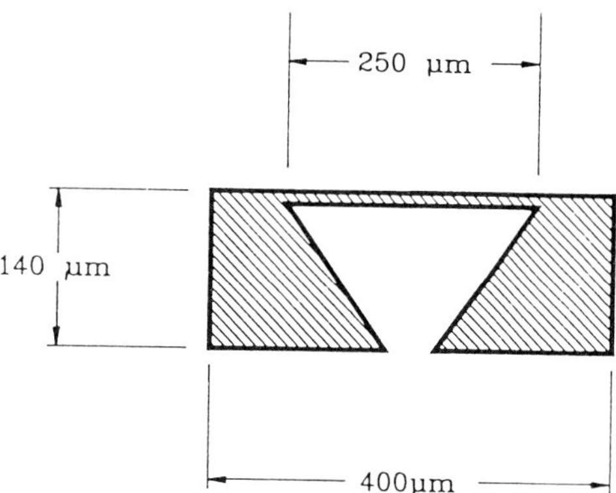

Figure 5

Cross-section of a catheter-tip pressure sensor. When the "handle" wafer is thinned sufficiently, the cavity is exposed to form a gage pressure sensor.

Such minute catheter-tip pressure sensors have now been in production for several years. They are 400 x 900 x 150 microns. Over 16,000 die are produced on a single wafer. The introduction of these extremely small pressure sensors has greatly increased the safety, the success rate, and the frequency of use for this diagnostic procedure.

4.2 Microwave Detector

Another example of the handle wafer concept is exemplified in the high bandwidth microwave power detector. Since these detectors incorporate minute, thermally isolated structures, the chips are usually very thin and small (1mm x 1mm x 0.1 mm). A typical microwave power detector consists of a thin diaphragm (5 microns), a surrounding frame (100 microns thick), and plated, gold beam-leads [11]. This part is manufactured by bonding 2 wafers together (the top wafer containing a patterned masking material at the bond interface), then grinding and polishing the top wafer to the required device thickness. At this point, the wafer assembly is robust enough for the growth of an epitaxial layer, oxidation, diffusions, and beam-lead metallization. After attaching the completed top surface to a soluble organic film, the bottom "handle" wafer is removed by etching, and the thin diaphragm is formed by electrochemical etching. Finally, the completed, individual chips are rinsed into a beaker of alcohol, cleaned, dried, and sorted.

4.3 Tri-Wafer Catheter-Tip Pressure Sensor

Assembly of a small catheter-tip pressure sensor into the catheter itself is a major challenge. Besides guiding 3-4 small diameter wires down a long, thin plastic tube and bonding them to the chip, the backside of the pressure sensing diaphragm must also be vented to the outside atmospheric pressure through the long, thin tube. The assembly procedure can be simplified if the vent to the back of the diaphragm is provided at the side of the chip, rather than the back.

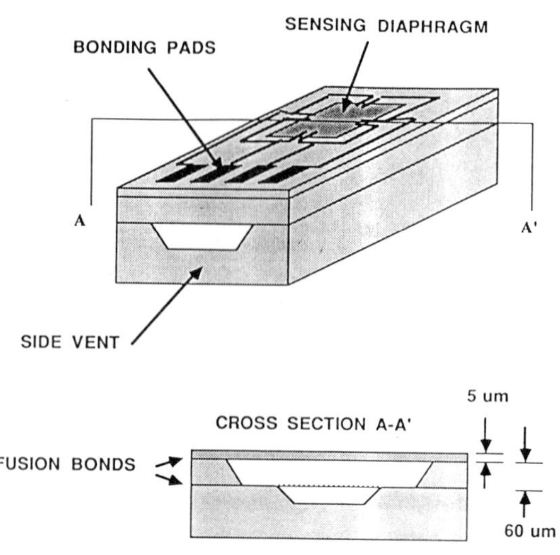

Figure 6

Tri-wafer bonding is employed to fabricate this catheter-tip pressure sensor. The tri-wafer concept makes it possible to create a built-in side vent.

This and other difficult three-dimensional structural problems can be solved by a three-wafer SFB process [12]. First a channel, nearly the length of the final chip, is anisotropically etched into the bottom wafer. Next, the middle wafer is bonded, then ground and polished back to about 60 microns thickness. After this point, the process sequence is identical to the sensor described above. A pit is etched into the middle wafer (aligned to penetrate the buried channel), the third wafer is bonded and electrochemically etched back to form the sensing diaphragm, piezoresistors are defined on the top surface, and the substrate (or "handle") wafer is polished back to its desired thickness. Finally, when the thinned, 3-wafer assembly is sawn, the end of the channel is exposed on the edge of the chip near the wire bond pads, as shown in Figure 6. This configuration greatly simplifies and reduces the assembly costs of catheter-tip pressure sensors.

5. ACCELERATION SENSORS

While the commercialization of high volume micromachined pressure sensors have dominated the micromachining industry for many years, the next high volume micromachined product will be acceleration sensors for the automotive suspension control and crash detection markets. SFB processing provides unique and valuable capabilities for designing high performance accelerometers. The device shown in Figure 7, for example, consists of a square, micromachined seismic mass, surrounded by the silicon frame [13]. The mass is attached to the frame by 2 cantilever beams located along one side of the mass. Piezoresistors on the 2 beams detect stresses induced as the seismic mass moves up and down in an acceleration field. In addition, over-range protection "tabs" are positioned around the mass to inhibit its motion under shock conditions.

Sense beams and over-range tabs are created by SFB processing. The beams and tabs are 10-15 microns thick, suspended over etched cavities about 8 microns deep. The process is similar to the high over-range pressure sensor described above except for 2 additional, final steps in which the seismic mass is formed by conventional anisotropic wet etching and the thin, bonded layer is etched through from the top surface to separate the mass from the frame and to define the beams and the tabs.

6. ACTUATORS

6.1 Self-Test Accelerometer

An important commercial actuator application is in self-test accelerometers. Crash detectors designed for automotive air-bag deployment systems must be assured of the highest standards of reliability. Developers of these systems demand that solid state acceleration sensors incorporate a full self-test capability. That is, when the system is tested each time the car is started, the electronics can fully simulate a crash on the acceleration sensor chip itself. This capability can be incorporated on the accelerometer described by the addition of a surface micromachined thermal actuator beam [14].

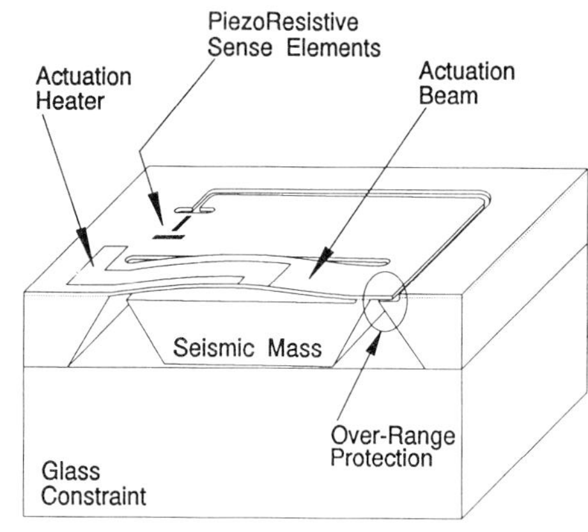

Figure 7

Cross-section of an acceleration sensor made with SFB. The surface micromachined sensing beams, actuator beam, and overrange tabs are formed from the top bonded wafer.

Indicated in Figure 7, the thermal actuator beam is fabricated at the same time and in the same process sequence as the sense beams and over-range tabs. It is attached to the surrounding silicon frame at one end, and to the far end of the silicon seismic mass at the other end. An implanted "heater" resistor, about 150 Ohms, is placed on the beam. When a current pulse is applied to the heater resistor, the actuator beam is locally heated. Since the rest of the chip remains at ambient temperature, the differential thermal expansion of the actuator beam creates a torsional force which pushes down on the end of the seismic mass. The deflection of the seismic mass is detected by the sensing resistors in exactly the same manner as a acceleration signal. In this way, a crash is simulated on the chip itself. Note that this thermal actuation mechanism is completely different from a bimetallic effect. Only one material is used - single crystal silicon. Indeed, a bi-metal would cause performance problems in the sensor because it would induce stresses even during ambient temperature changes.

This important new device concept, the first ever to incorporate a microsensor and a thermal microactuator on the same chip, will be installed on model year 1994 automobiles.

7. SURFACE MICROMACHINING USING SILICON FUSION BONDING

As we have seen, the powerful design capabilities of silicon fusion bonding are not limited to microsensors. In recent years, surface micromachining processing techniques have been used to demonstrate a variety of thin-film mechanisms and actuators, from laterally resonant devices, to microtweezers, to micromotors. While the traditional fabrication material for these devices has been polysilicon, the same principles can be exploited to fabricate these actuators with single crystal silicon [17]. In fact, the thermal actuator incorporated into the self-test acceleration sensor described a-bove is a surface micromachined actuator.

7.1 Resonant Beam Pressure Sensor

Two technical aspects of piezoresistive pressure sensors have required constant development and optimization over the past 20 years. First, piezoresistors exhibit large temperature coefficients which must be measured and cancelled in the interface circuitry. Second, resistor values can drift due to ionic or other contamination on the surface of the chip. Accurate temperature compensation techniques and new wafer processing methods to enhance resistor stability have improved piezoresistor long-term performance and overall accuracy to about 0.01% (100 ppm) per year in the best devices. A well recognized technology for obtaining even better pressure sensor performance, however, is the use of resonant devices [15].

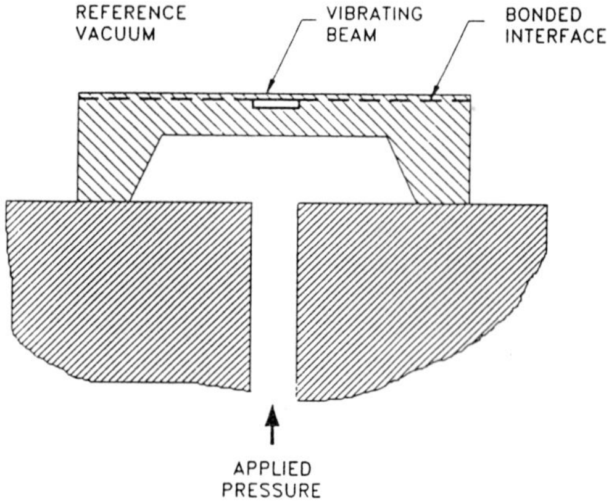

Figure 8

High-accuracy, resonant-beam pressure sensor fabricated with SFB surface micromachining. The resonant frequency of the beam increases with pressure applied to the backside.

Surface micromachined structures are being developed which consist of a small beam positioned on the surface of a flexible diaphragm [16,17,18], as illustrated schematically in Figure 8. The beam can be electrostatically excited into its fundamental resonant mode (typically 100kHz), and piezoresistors on the beam detect the resonant vibrations. When pressure is applied to the diaphragm, the beam is stretched and its resonant frequency increases. An external "maintainer" circuit tracks the resonant frequency, which is a direct measure of the applied pressure.

Since the resonant frequency depends only on the mechanical properties of silicon, and not its electronic properties, extremely stable, accurate, and reproducible measurements have been realized. Overall accuracy of such devices is now approaching 0.001% (10 ppm) per year [18]. Although similar resonant beams have been fabricated from polysilicon [17], the resonant beam in the pressure sensor structure described here is fabricated by SFB processing in a sequence identical to the accelerometer described above. Most of the sacrificially etched structures demonstrated during the late 80's could also fabricated with the same SFB processing sequence.

8. CONCLUSION

Few new processing technologies for silicon sensors have been successfully commercialized as quickly as silicon fusion bonding. The reasons are that the process is easy and reliable, the material (single crystal silicon) is very well understood, and the technique is so versatile. An additional advantage is the attention SFB has received from the IC industry. In contrast to other micromachining technologies, silicon fusion bonding (or direct wafer bonding) has been enthusiastically embraced by the IC industry for applications in dielectrically isolated, radiation tolerant circuits. The SFB technology, equipment, expertise, and bonding methodologies developed by the IC industry in the coming years (for dielectrically isolated circuits) will directly impact the microsensor and microactuator industry. Exploitation of SFB technology for sensors and actuators has barely begun.

9. ACKNOWLEDGEMENTS

The authors would like to acknowledge the technical assistance and influence of Farzad Pourahmadi, Joe Brown, Janusz Bryzek, Joseph Mallon, Youssof Fathi, Rose Scimeca, Aniela Bryzek, Ted Vermeulen, Jay Jani, Philip Parsons, Andrew Glendinning, Mike Skinner, John Tudor, Huntley Millar, and Luke Rose. Without these co-workers, the work described here would not have been possible.

10. REFERENCES

[1] J.B. Lasky, Applied Physics. Letters, Vol. 48, pg 78, (1986).

[2] C. Harendt et al, Journal of Electronic Materials, Vol. 20 (3), pg 267 (1991).

[3] C. Harendt, H. Graf, E. Penteker, and B. Hofflinger, in the Proceedings of the 5th International Conference on Solid-State Sensors and Actuators, pg 927, June (1989).

[4] H. Seidel, "The Mechanism of Anisotropic, Electrochemical Silicon Etching in Alkaline Solutions", in the Technical Digest of the IEEE Solid-State Sensor and Actuator Workshop, pg 86, Hilton Head Island North Carolina, June 1990.

[5] Kurt Petersen, Joe Brown, Ted Vermeulen, Phillip Barth, Joseph Mallon, and Janusz Bryzek, "Ultra-stable, High-temperature Pressure Sensors Using Silicon Fusion Bonding", Sensors and Actuators, A21-A23, pg 96 (1990).

[6] Kurt Petersen, Phillip Barth, John Poydock, Joseph Mallon, Janusz Bryzek, "Silicon Fusion Bonding for Pressure Sensors", in the Technical Digest of the IEEE Solid-State Sensor and Actuator Workshop, pg 144, Hilton Head Island North Carolina, June 1988.

[7] Shuichi Shoji, Takemi Nisase, Masayoshi Esashi, and Tadayuki Matsuo, "Fabrication of an Implantable Capacitive Type Pressure Sensor", in the Proceedings of the 4th International Conference on Solid-State Sensors and Actuators, Tokyo, pg 305, June 1987.

[8] Lee Christel, Kurt Petersen, Phillip Barth, Farzad Pourahmadi, Joseph Mallon, and Janusz Bryzek, "Single-Crystal Silicon Pressure Sensors with 500X Overpressure Protection", Sensors and Actuators, A21-A23, pg 84 (1990).

[9] H. Guckel and D.W. Burns, "A Technology for Integrated Transducers", in the

Proceedings of the International Conference on Solid-State Sensors and Actuators, Philadelphia, pg 90, June 1985.

[10] Lee Christel and Kurt Petersen, "A Miniature Microwave Detector Using Advanced Micromachining", in the Technical Digest of the IEEE Solid-State Sensor and Actuator Workshop, pg 144, Hilton Head Island North Carolina, June 1992.

[11] Lee Christel and Kurt Petersen, "Tri-Wafer Catheter Pressure Sensor Fabricated with Silicon Fusion Bonding", to be published in the Proceedings of the International Conference on Solid-State Sensors and Acutators, Yokohama, June 1993.

[12] Phillip Barth, Farzad Pourahmadi, Robert Mayer, John Poydock, Kurt Petersen, "A Monolithic Silicon Accelerometer with Integral Air-Damping and Overrange Protection", in the Technical Digest of the IEEE Solid-State Sensor and Actuator Workshop, pg 35, Hilton Head Island North Carolina, June 1988.

[13] Farzad Pourahmadi, Lee Christel, Kurt Petersen, "Silicon Accelerometer with New Thermal Self-Test Mechanism", in the Technical Digest of the IEEE Solid-State Sensor and Actuator Workshop, pg 122, Hilton Head North Carolina, June 1992.

[14] Kurt Petersen, Dale Gee, Farzad Pourahmadi, Russell Craddock, Joe Brown, Lee Christel, "Surface Micromachined Structures Fabricated with Silicon Fusion Bonding", in the Proceedings of the International Conference on Solid-State Sensors and Acutators, pg 397, San Francisco, June 1991.

[15] R.T. Howe, "Resonant Microsensors", in the Proceedings of the 4th International Conference on Solid-State Sensors and Actuators, Tokyo, pg 843, June 1987.

[16] K. Ikeda, H. Kuwayama, T. Kobayashi, T. Watanabe, T. Nishikawa, and T. Yoshida, "Silicon Pressure Sensor with Resonant Strain Gauges Built into Diaphragm", in the Proceedings of the 7th Sensor Symposium, Japan, pg 55 (1988).

[17] H. Guckel, J. Sniegowski, T.R. Christenson, F. Raissi, "The Application of Fine-Grained, Tensile Polysilicon to Mechanically Resonant Transducers", Sensors and Actuators, A21-A23, pg 346 (1990).

[18] Kurt Petersen, Farzad Pourahmadi, Joe Brown, Philip Parsons, Mike Skinner, and John Tudor, "Resonant Beam Pressure Sensor Fabricated with Silicon Fusion Bonding", in the Proceedings of the International Conference on Solid-State Sensors and Acutators, pg 664, San Francisco, June 1991.

Silicon sensors in Western Europe

R W Bogue
Robert Bogue & Partners, Kingston House, Bere Alston, Devon PL20 7HB

ABSTRACT: As a consequence of American research into the physical properties of silicon, undertaken in the 1950s and early 1960s, a multi-million pound market has emerged for silicon sensors. This paper is based on a study that analysed and quantified this market in Western Europe. Following a discussion of the market, the issue of silicon sensor research is addressed and the following question posed: In view of the extensive commercialisation of silicon sensors that has occurred and the strong technological expertise resident within the major supply companies, why is the academic research community still so deeply involved in this field?

1. INTRODUCTION

American research into the physical properties of silicon during the late 1950s was stimulated by the emerging electronics industry and resulted in a family of sensors that today, constitute a significantly sized market.
The commercialisation of silicon sensors started modestly in the mid-1960s but accelerated in the following two decades as the technology improved and applications emerged which required small and rugged sensors that offered high performance but which were inexpensive by prevailing price standards.
This is widely known but despite various estimates, the size and value of the European silicon sensor market has never been accurately defined. This presentation is based on a study, undertaken by the author on behalf of Frost & Sullivan, which aimed to rectify this deficiency.
Quantifying and analysing this market allows us to judge the degree of success achieved by silicon sensors and to place them in a true commercial perspective, but more importantly to the research community, it raises a fundamental issue: the fact that despite almost three decades' commercialisation, silicon sensor research by the academic community is probably stronger today than ever. At first sight, this might be justified on the grounds of the significant forecasted growth in the market but it can be argued that much of this research is of uncertain relevance; a view that emerges when one examines the structure of the market in detail. Following such an examination, I will attempt to justify this view.

2. SCOPE

The following classes of silicon sensor are considered:

* Low cost pressure sensors
* Gas sensors
* Flow sensors
* Temperature sensors
* Radiation sensors
* Pressure transducers and transmitters
* Hall effect sensors and magnetoresistors
* Accelerometers and vibration sensors
* Chemical sensors
* Biosensors

Markets for the above are quantified for 1992 and 1997, in Western Europe; a region defined as the twelve EC nations, together with Scandinavia, Austria and Switzerland.

3. THE MARKETS

The markets for each of the sensor types listed above are considered in the following subsections.

3.1 Pressure sensors

Silicon technology has been outstandingly successful in the field of pressure sensing, both for precision measurements (transducers and transmitters) and in the low cost, lower accuracy sector. Many older electromechanical technologies have been displaced from the market and new applications have been created, for instance, manifold absolute pressure (MAP) sensing in car engine management systems and disposable medical blood pressure sensors.
Several different technologies and sensing mechanisms are adopted in silicon pressure sensors and each are at differing stages of commercialisation:

* Bonded piezoresistive (PR) silicon strain gauge (Rapidly declining)
* Planar PR silicon (Declining)
* Micromachined PR silicon (Steady)
* Deposited thin-film PR silicon (Growing)
* Micromachined capacitive silicon (Growing)
* Micromachined silicon resonator (Growing slowly)

In view of the greatly differing prices and applications served by these sensors, it is useful to split the market into low cost sensors (prices of $1-7 each in high volumes), and transducers and transmitters ($400-1500). The markets, categorised in this manner, are as follows:

	1992	1997
Low cost sensors	14.5 million	35.0 million
Transducers and transmitters	420,000	650,000
TOTAL	14.9 million	35.7 million

The dramatic forecasted growth in the low cost sector largely reflects an expansion of use in the automotive sector, stimulated mostly by environmental emission legislation. The transducer and transmitter market will expand as an ever-growing number of manufacturers adopt silicon sensing technology in their products.

3.2 Accelerometers and vibration sensors

In contrast to the above, the dominance of piezoelectric sensors is such that silicon technology has made few inroads into this market. Small numbers of precision silicon accelerometers are used for DC/steady-state measurements, based on bonded PR strain gauges and the more sophisticated micromachined PR and variable capacitance technologies. As with pressure sensing, the two latter technologies are rapidly displacing the former from the market.

However, emerging automotive uses, including crash sensing/air bag inflation and the control of adaptive suspension systems, will create numerically massive markets for low cost silicon devices. The markets for these sensors and precision silicon accelerometers are summarised below.

	1992	1997
Precision accelerometers	3,500	5,000
Low cost automotive sensors	10,000	10.5 million
TOTAL	13,500	c. 10.5 million

3.3 Flow sensors

Again, silicon technology has exterted little impact on the large flow sensing market and indeed, silicon flow sensors have only recently become available (from Honeywell). These are micromachined devices that exploit a heat transfer principle and presently find uses in HVAC systems and hospital equipment. As with low cost silicon accelerometers, these sensors are poised to satisfy emerging mass markets in the automotive sector as mass air flow measurement starts to replace MAP sensing in the engine management systems of larger cars.

Present-day and forecasted markets are summarised below.

	1992	1997
Industrial/medical	53,000	100,000
Automotive	-	500,000
TOTAL	53,000	600,000

3.4 Hall effect sensors and magnetoresistors

Silicon sensors that exploit the Hall effect have been exceptionally successful in penetrating the industrial position, proximity, current and speed sensing markets and in addition, they are used to measure magnetic field strength. Applications are particulary numerous in the automotive industry, where these sensors are used in engine management systems (crank-shaft speed, throttle position and cam-shaft speed sensing), in ABS systems, in electronic speedometers and to control the position of electric seats, sun roofs and windows etc. Around 45% of the Hall effect sensor market is presently taken by the automotive sector and this proportion is expected to rise to 55% by 1997.

Silicon magnetoresistors only constitute a small proportion of the overall magnetoresistor market and comprise thin films of permalloy deposited on silicon substrates. Major present-day uses are similar to those for Hall sensors (with which they compete) and include industrial and automotive position sensing.

The markets for Hall effect sensors and magnetoresistors are summarised below.

	1992	1997
Hall effect sensors	27.5 million	50.0 million
Magnetoresistors	4.0 million	9.5 million
TOTAL	31.5 million	59.5 million

3.5 Temperature sensors

Silicon temperature sensors satisfy only a small part of the overall temperature measurement market and have not penetrated the precision sector which remains dominated by thermocouples and PRTs. Major uses are in white goods, cars and HVAC systems, where low unit costs are vital. These applications jointly constituted 80% of the market in 1992.
Despite limited overall market penetration, these sensors nevertheless sell in very high volumes, as illustrated below.

1992 market: 18.7 million sensors
1997 market: 31.0 million sensors

3.6 Radiation sensors

As with the above, silicon technology has succeeded in penetrating only a small part of the total radiation detection market and devices such as scintillation counters, Geiger tubes, film badges and thermoluminescent detectors (TLDs) dominate this business. Most of the silicon radiation sensing market comprises high cost ion-implanted detectors but smaller numbers of RADFETs are also sold. The former are used widely in medicine, environmental monitoring and nuclear physics research and the latter, mostly in the defence and space sectors. The markets are illustrated below.

	1992	1997
Ion-implanted detectors	3,000	4,500
RADFETs	500	1,000
TOTAL	3,500	5,500

3.7 Gas sensors

An extensive research literature exists on the subject of silicon gas sensors and most work has concentrated on the FET structures, dubbed GASFETs. Despite pioneering efforts by Thorn EMI in the UK and the international research community, significant markets have failed to materialise for GASFETs; a fact that reflects the wide range of other, well proven gas sensors that are now available. It is therefore ironic that the class of silicon gas sensors now starting to exert a significant market impact is based on one of the oldest and often overlooked gas detection techniques, that of differential thermal conductivity. Sensors that exploit this effect but which benefit from state-of-the-art silicon micromachining and fabrication technologies are now finding uses in health and safety, environmental monitoring and as replacements for conventional sensors in industrial binary gas analysers. Known European GASFET applications are almost wholly limited to detecting leaks in telecommunications cables (hydrogen detection).
The markets for silicon gas sensors are summarised below.

	1992	1997
Micromachined thermal cond.	2,800	30,000
GASFETs	200	300
TOTAL	3,000	30,300

3.8 Chemical sensors

As with the above, extensive research efforts have been directed towards the development of silicon chemical sensors based on FET technology (ISFETs). ISFETs that respond to many ions and cations have been demonstrated in the laboratory and during the last few years, pH-responsive types have entered the European market.
In contrast to their gas-responsive counterparts, these sensors are expected to achieve widespread market acceptance by virtue of their low unit cost and rugged, glass-free construction. Present-day applications are mostly in portable pH meters which are used in the water, environmental monitoring, food and drink and related process industries. As ISFETs appear on the market that respond to further chemical quantities (eg. sodium, potassium, magnesium and calcium), markets are expected to expand dramatically, as illustrated below.

1992 market: 25-30,000 sensors (pH only)
1997 market: c.100,000 sensors (pH and other species)

ISFETs that respond to the many chemical species of interest to the water industries (chlorine, ammonia, heavy metals, iron, aluminium and dissolved oxygen etc.) are expected to enjoy widespread use in the longer term in water quality monitoring instruments.

3.9 Biosensors

As yet, no silicon biosensors are available commercially in Western Europe but it is likely that some will be launched during the next five years, probably by Japanese and US companies. The overall biosensor market is poised for strong growth but silicon must compete with the many other sensing technologies under investigation and is not expected to emerge as a dominant force within this business. Devices at the research stage include various planar biosensors, where silicon is simply a passive substrate, and FET-based types, frequently referred to as BIOFETs, ENFETs and IMMUNOFETs. Markets are most likely to comprise the former type, certainly in the short term, and may possibly reach tens-of-thousands of units/annum by 1997.

3.10 Market summary

The total Western European silicon sensor market for the years 1992 and 1997 is summarised in the table overleaf, expressed as numbers of units. This illustrates that the market is poised for very considerable growth and will rise from 65.2 million sensors in 1992 (valued at around $320 million) to reach around 137.4 million sensors by 1997; an increase of 110.7% in the annual market. From this table and the preceeding sections we can readily identify the major, high growth areas, as follows:

- Low cost (mostly automotive) pressure sensors
- Low cost automotive accelerometers
- Automotive air mass flow sensors
- Hall effect sensors
- Magnetoresistors
- Temperature sensors
- Chemical sensors (ISFETs)
- Gas sensors

	1992	1997
Hall effect sensors	27.5 million	50.0 million
Temperature sensors	18.7 million	31.0 million
Low cost pressure sensors	14.5 million	35.0 million
Magnetoresistors	4.0 million	9.5 million
Pressure transducers & transmitters	420,000	650,000
Flow sensors	53,000	600,000
Chemical sensors	30,000	100,000
Low cost (automotive) accelerometers	10,000	10.5 million
Precision accelerometers	3,500	5,000
Radiation sensors	3,500	5,500
Gas sensors	3,000	30,300
Biosensors	-	10s-of-'000s?
TOTALS	65,218,000	137,415,000

4. THE SUPPLY COMPANIES

The companies supplying silicon sensors to the high growth areas listed above, together with their market shares, turnovers and R&D expenditures are as follows.

Sensor type	Company	Market share	Turnover	R&D spend
Hall effect	Allegro	18%	$105m	$18m
	Honeywell	74%	$6.2bn	$300m
	Siemens	8%	$48.6bn	$5.3bn
Magnetoresistors	Philips	98%	$33.8bn	$2.3bn
Automotive pressure	Delco	)	$3.9bn	$635m*
	Motorola	) c.80%	$11.3bn	$1.1bn
	Honeywell	)	$6.2bn	$300m
Auto. accelerometers (companies with products under development)	Analog Devices	-	$538m	$89m
	Delco	-	$3.9bn	$635m*
	Motorola	-	$11.3bn	$1.1bn
	Lucas	-	$1.4bn	$170m
Automotive flow	Honeywell	?	$6.2bn	$300m
Temperature	Philips	65%	$33.8bn	$2.3bn
	Siemens	25%	$48.6bn	$5.3bn
	Texas Insts.	6%	-	-
Chemical sensors (ISFETs)	Several small European university spin-offs Leeds and Northrup			
Gas sensors	Hartmann & Braun (T/O = c.$800m) Small European university spin-offs			

(* R&D spend of GM Hughes, the holding company)

5. THE END-USER INDUSTRIES

Whilst silicon sensors are used by numerous industries, the 1992 market was dominated by the process, aerospace and automotive sectors and in terms of value consumption, these industries jointly took around 56% of this market. By 1997, this figure will have risen to approximately 64%. However, in numerical terms the automotive sector dominates this business and consumed around 35-40% of the 1992 market and is forecast to take over 60% by 1997. The key point to arise from this is that the bulk of the market will comprise sensors that must meet stringent operating requirements, be exceedingly reliable yet cost very little. Typical automotive prices for a complete tested, calibrated and packaged sensor range from less than £1 to maybe £10 in high volumes. Few companies are able to achieve this and compete in this business.

Indeed, the price issue is such that to meet these restraints, several companies operate complex product development, manufacturing and sales strategies. One of the world's leading suppliers of automotive silicon pressure sensors adopts the following:

Research and development	- US (various locations)
Silicon processing and micromachining	- US (another location)
Packaging and testing	- Korea
Centralised European sales and marketing	- France
National sales and distribution	- Various countries and regions

6. SILICON SENSOR RESEARCH

Those familiar with the recent literature will be aware that Western Europe has a strong community of academic and government groups involved with silicon sensor research. Whilst it is beyond the present scope to review this in any detail, some topics that have received particular attention in recent years include:

- Micromachined capacitive, resonator and PR pressure sensors
- Micromachined capacitive and other accelerometers
- Poly-silicon sensors
- Magnetotransistors, magnetoresistors and magnetodiodes
- GASFETs and GASFET arrays
- ISFETs
- Silicon biosensors (FETs and planar types)
- Micromachined silicon microphones and hearing aids
- Silicon resonators (for various mechanical quantities)
- Silicon radiation sensors
- Advanced micromachining technologies
- Packaging and encapsulation techniques
- Electronically integrated silicon sensors

A government-funded study by the present author identified forty university or government silicon sensor research groups in the UK, France, Germany, Italy, Sweden, the Netherlands and Switzerland alone, and the total number in Western Europe exceeds fifty. This study revealed the relative weakness of the UK compared with certain other European countries and this was felt by the DTI to be a significant national deficiency. This assumes that because markets are developing rapidly a strong research effort is important. At first sight, this view, the research topics listed above and the number of active groups appear justified on the grounds of forecasted market growth but it can be argued that this view is not altogether valid and that much academic research is of uncertain value, for the following reasons:

1. In many areas silicon sensor technology is now developed to a point where it is capable of yielding devices whose sophistication exceeds known market requirements (particularly micromachined sensors);

2. Most high volume markets are or will be served by small numbers of large companies with extensive in-house R&D facilities and state-of-the-art manufacturing technologies;

3. Much volume growth is in the automotive market, one which is again served by small numbers of large companies who can manufacture, test and package vast numbers of sensors with exceptionally low unit-costs.

The first point requires little expansion except to refer the more sceptical to the sensor research literature and the IOP's own publication, the "Journal of Micromechanics and Microengineering".
The arguments for research on the grounds of market growth weaken considerably when points two and three are taken into consideration, as it is clear that the size of the companies dominating the major rapidly growing markets (and their technical resources) are such that future product generations will emerge from corporate rather than academic research laboratories. Furthermore, the market dominance by large companies allows little scope for new market entrants or small companies exploiting university research. Markets for sensors aimed at the automotive sector, in particular, will remain in the hands of the large companies geared to meet the price restraiints imposed by this industry.
However, some markets are at an earlier state of development and are not yet dominated by large companies with extensive in-house resources, and other fields still require some basic science. It is in these areas that academic research should be directed and I therefore tentatively propose that silicon sensor research should be conducted as follows.

Fields for academic research	Fields to leave to the manufacturers
- ISFETs	- Pressure sensors
- Silicon biosensors	- Accelerometers/vibration sensors
- Silicon gas sensors (but not FETs)	- Hall effect sensors
- Silicon radiation sensors	- Magnetoresistors, magnetodiodes and magnetotransistors
- Silicon microphones	- Micromachined sensors for most physical quantities
- Integrated silicon optical sensing	- Flow sensors
- Silicon analytical instruments	- Electronically integrated sensors
- Silicon microactuators	- Temperature sensors
	- Fabrication and packaging technologies

7. ACKNOWLEDGEMENTS

I hereby express my thanks to Frost and Sullivan Market Intelligence for allowing me to reproduce the market data included in this presentation.
I should also like to thank the numerous individuals from the silicon sensor business who freely shared their knowledge of this business with me during the initial research.

8. REFERENCES

The European Market for Silicon Sensors, Report E.1760, Frost & Sullivan Market Intelligence, 1993.

A Comparative Review of Advanced Sensor Research in the UK and Mainland Europe, Bogue R W, 1991, LGC/DTI.

Low-stress polysilicon process compatible with standard device processing

P.J. French, B.P. van Drieënhuizen, D. Poenar, J.F.L. Goosen, P.M. Sarro+ and R.F. Wolffenbuttel

Laboratory for Electronic Instrumentation, Department of Electrical Engineering,
Technical University of Delft, Mekelweg 4, 2628 CD, Delft, The Netherlands.
+DIMES, Technical University of Delft, PO Box 5053, 2600 GB Delft, The Netherlands.

ABSTRACT: If surface micromachined devices are to be combined with on-chip circuitry, any high- temperature processing must be avoided in order to minimise the effect on active device characteristics. High-temperature annealing, often used to reduce intrinsic stress, cannot therefore be applied to these structures. It is furthermore advantagous to be able to use film thicknesses commonly used in standard processing. In this work a low stress polysilicon process without resorting to high temperature annealing has been developed. After polysilicon deposition, an 850°C for 30 minutes is used to activate the dopant. No further annealing was used to reduce stress. By careful consideration of the processing a low- temperature polysilicon processing, which can be used to fabricate thin micromachined structures, has been developed.

1. INTRODUCTION

In order to fabricate free-standing surface-micromachined structures, it is usually necessary to construct the structures in materials which have low tensile intrinsic stress. Films in compressive stress tend to buckle causing severe limitations on the dimensions of free-standing structures. High tensile stress is also unsuitable as it may result in fracturing of the structures.

Processing parameters have a considerable effect on the electrical and mechanical properties of polysilicon [1]. Intrinsic compressive stress can be released by high temperature annealing (>1000°C) [2], but this is not compatible with standard silicon processing. It is therefore necessary to develop a process where low tensile stress can be achieved without any high-temperature steps.

The studies presented here have included the effect of deposition temperature, *in-situ* [3] and post implant annealing. The studies have also included the effect of changes in the processing for the underlying, sacrificical layer. From these studies, processes yielding the desired strain and free standing structures for thin polysilicon films (<0.5μm), with a small air gap (~0.5μm) have been developed. Both the annealing temperatures and the film thicknesses are compatible with standard bipolar processing.

2. PROCESSING

Sacrificial layers, 4000Å - 6000Å, were formed either by wet oxidation of polysilicon or by phosphor-silicate glass (PSG) deposition. Polysilicon films of 4000Å were deposited, to form the mechanical structures, by LPCVD with a reactor pressure of 150 mtorr and the silane flow of 43 sccm, followed by, in some tests, an *in-situ* anneal in nitrogen. After deposition, wafers were doped using phosphorus implantation at 80keV with doses of $1x10^{15}cm^{-2}$ or $5x10^{15}cm^{-2}$. Doped samples were annealed in N_2, at a range of temperatures (850°C - 1100°C), for 30 minutes. The polysilicon was patterned by plasma etching and the sacrificial oxide removed using 20% HF. The under-etch rate of the sacrificial oxide was 0.2μm for oxidised polysilicon and 5.5μm for the PSG.

In order to reduce the problems of sticking, wafers were given a final rinse in IPA and dried on a hot plate. Care was taken not to allow the samples to dry until the final bake. This is of particular importance when such thin sacrificial oxides are used.

3. STRESS MEASUREMENT TECHNIQUES

Several techniques have been developed and tested to measure the stress in micromachined structures and these have been described elsewhere [4]. For this work a pointer structure was used where the expansion of two beams is converted into the rotation of a third beam. This has the advantages over buckling beams [5] that a single structure can be used to measure both compressive and tensile strain. The structure of this device is shown in figure 1.

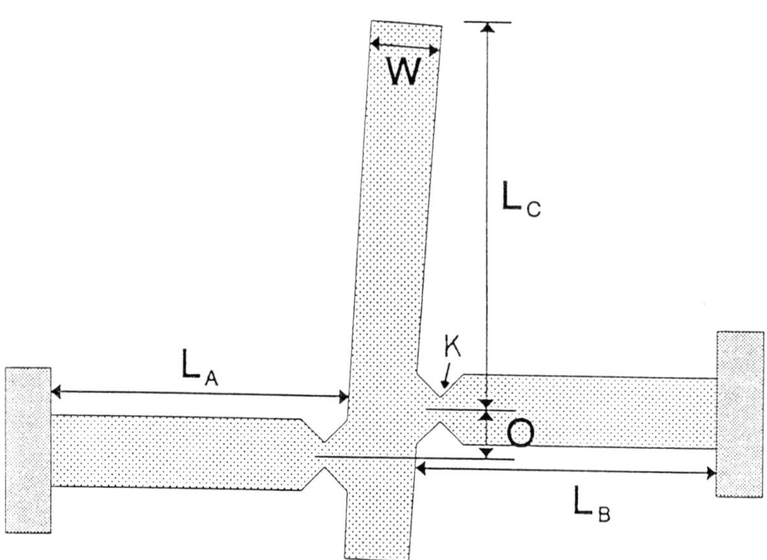

Fig. 1. Structure used for measuring both tensile and compressive stress.

Computer simulations have been used to calculate the effect of the turning point thickness, K, on the measured value of rotation. Through these simulations a correction factor has been calculated to compensate for this error.

4. MEASUREMENT RESULTS

Each stage of the processing was examined in order to achieve the desired strain levels. Suitable stress characteristics were found for a deposition temperature just below the transition between amorphous and polysilicon deposition. The effect of deposition temperature on intrinsic strain is shown in figure 2.

By depositing as amorphous ensured that the thin grain boundary was formed by annealing and not during deposition, which reduced the stress contribution from the grain boundary. Furthermore, the resulting small grains lead to improved stress distribution, greater repeatability and smoother surface (the last factor is of particular importance where moving parts are required).

Films deposited below the transition and subsequently doped by implantation and annealed at 850°C were found to be in tensile strain, whereas those deposited above this temperature were in compressive stress.

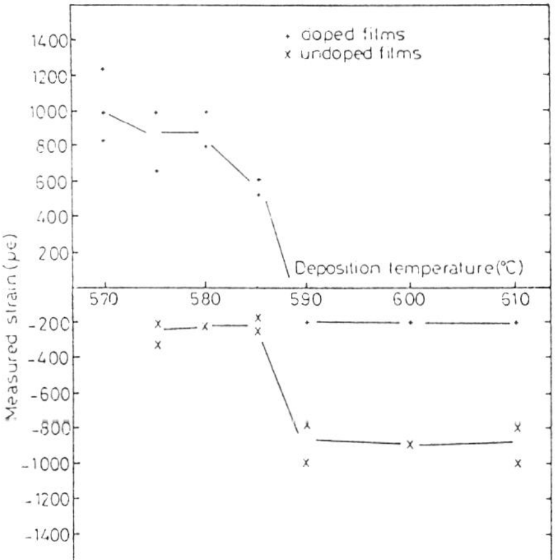

Fig. 2. The effect of deposition temperature on the stress levels in both doped and undoped films deposited on oxidised polysilicon.

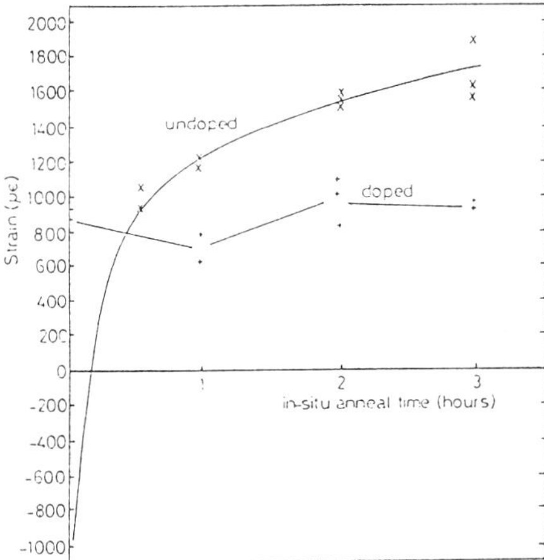

Figure 3. The effect of *in-situ* anneal on film stress levels, deposited on oxidised silicon.

Even films deposited near the transition temperature, where the grain size was small, were found to have compressive stress, indicating that grain size alone does not determine the stress levels. The sharp change in strain coinciding with the transition between polysilicon and amorphous deposition, but not changing greatly either side of this transition, was a further indication of the importance of the grain boundary formation and crystal orientation. Undoped and unannealed films were in compressive strain for all deposition temperatures. A low temperature, 600°C, *in-situ* anneal as short as 30 minutes was sufficient to have a significant effect on the stress levels in the films resulting in a change from a compressive strain of $\sim$-1000$\mu\epsilon$ to a tensile strain of $\sim +1000\mu\epsilon$ for undoped films. As expected doped films were less influenced by the use of the *in-situ* anneal. The measured stress for these films is shown in figure 3. No significant differences between stress levels for the two doping concentrations used could be found.

A range of post implantation annealing temperatures (850°C-1100°C) were investigated. These showed a further reduction in strain levels with increasing anneal temperature. However, for the structures envisaged for this work the stress levels for the 850°C anneal were satisfactory and the most compatible with other processing. For depositions above the transition temperature, the films were in compressive stress for all anneal temperatures, although the stress approached zero as the anneal temperature was increased. Figure 4 shows the measured strains for films deposited at 570°C (below the transition) and 610°C (above the transition) as a function of anneal temperature after doping by phosphorus implantation at a dose of $1x10^{16}cm^{-2}$. This figure shows that if the deposition parameters are not optimised, high-temperature annealing as reported in the literature [2,6] is indeed required to remove the compressive stress.

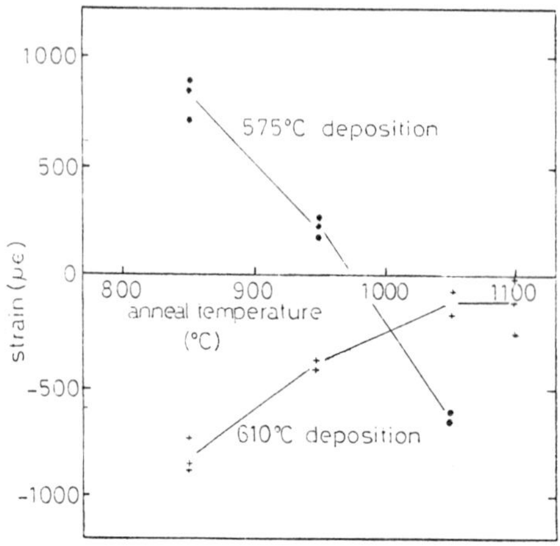

Fig. 4. Effect of anneal temperature on stress in films deposited above and below the transition temperature.

In this work both PSG and oxidised polysilicon have been used as a sacrificial layer. In most cases the PSG samples yielded slightly lower stress levels. However, both layers were found to be suitable sacrificial layer materials.

5. CONCLUSIONS

All aspects of processing have been studied and further materials examinations are underway to establish a more detailed relationship between the film structure and the in-built stress. Initial test indicated that the formation of the grain boundary and the crystal orientation are major factors in determining the stress. Therefore, by careful consideration of the processing, satisfactory levels of strain have been achieved with anneal temperatures and film thicknesses compatible with on-chip circuitry processing. Further investigations are presently underway to verify the origins of stress in the polysilicon in order that improvements can be incorporated into the process.

6. REFERENCES

1) P.J. French and A.G.R. Evans, "Piezoresistance in polysilicon and its applications to strain gauges", Solid State Electronics, 32, (1989), pp 1-10.

2) K.L. Yang, D. Wilcoxen and G. Gimpelson, "The effects of post processing techniques and

sarcificial layer materials on the formation of free standing polysilicon microstructures", Proc. IEEE Micro-mechanical systems, Utah, USA, 20-22 February, 1989, pp 66-70.

3) H. Guckel, J.J. Sniegowski, T.R. Christenson, S. Mohney and T.F. Kelly, "Fabrication of micromechanical devices from polysilicon films with smooth surfaces", Sensors and Actuators, 20, (1990), pp 117-122.

4) B.P. van Drieënhuizen, J.F.L. Goosen, P.J. French and R.F. Wolffenbuttel, "Comparison of techniques for measuring both compressive and tensile stress in thin films", To be published in Sensors and Actuators.

5) H. Guckel, T. Randazzo and D.W. Burns, "A simple technique for the determination of mechanical strain in thin films with applications to polysilicon", J. Appl. Phys., 57, (1985), pp 1671-1675.

6) R.S. Hajab and R.S. Muller, "Residual strain effects on large aspect ratio micro-diaphragms", Proc. IEEE Micro Electro mechanical systems conference (MEMS), Salt Lake CIty, Utah, USA, 20-22 February, 1989, pp 133-138.

ACKNOWLEDGEMENTS: The authors wish to thank the members of DIMES, in particular R. Mallée, for their help in fabrication of the devices. This project is supported in part by FOM/STW under the project numbers DEL 99.2136 and DEL 11.2524.

An inductively coupled high temperature silicon pressure sensor

Robert S. Okojie and William. N. Carr
Microelectronics Research Center, New Jersey Institute of Technology
Newark, New Jersey 07102 USA

ABSTRACT: We report results of a silicon-based pressure sensor suitable for operation in a high temperature environment of 650°C. The device utilizes magnetic induction as the basis for pressure sensing. The operating temperature range is extended above that obtainable with piezoresistive silicon. Problems of leakage current and redistributed impurity concentration profile are virtually eliminated. The problem of TCE mismatch is greatly reduced by symmetrical layering, resulting in a Temperature Coefficient of Offset (TCO) of 200ppm/°C. The proposed device operated with a 10mA and 2MHz primary current, provides an output voltage of 167mV at full-scale(FS) pressure of 120kPa. Under the same conditions the output voltage increases to 190mV at 650°C. A unique feature of this device is an increased sensitivity at low pressures. This unique non-linearity, often considered a disadvantage, does extend the dynamic range of the sensor considerably at low pressures.

1. INTRODUCTION

The quest for energy conservation, waste management, and process optimization, has created a growing need for more accurate and precise measurement of energy consumption rates inside combustion chambers and reactors. To achieve these goals, it is imperative for sensing devices to interact directly with the sensed environment. Secondly, data obtained should be transmitted with minimum distortion through the interface circuitry. The major difficulty encountered in attempting to make these sensors interact directly is the fact that typical temperature of the chamber is usually above 125°C [1]. Beyond this temperature, device reliability and survivability becomes questionable for most conventional piezoresistive sensors. We are proposing a new generation of devices that could survive temperature environments as high as 700°C. Several papers[2,3,4] have reported pressure sensors suitable for temperatures of around 350°C. Because most of these reported devices still retain the high temperature sensitive parameters, they are essentially an upgrade of existing devices such as piezoresistive and capacitive pressure sensors. In these devices, the problems of redistributed impurity concentration profile, leakage current, and strain-inducing TCE mismatch have at best been reduced but not eliminated. We have attempted to solve these problems by applying different technology that does not involve resistors or capacitors in diffused membranes. Sensitivities due to TCE mismatch is reduced by symmetrical layering for similar materials (silicon). We have selected a design using single silicon wafer with $TaSi_2$ inductors.

2. PRINCIPLE OF OPERATION

The microsensor described in this paper is a Linear Voltage Differential Transformer. It operates by the principle of magnetic induction. Fig. 1 shows two square planar conducting coils 1 and 2 defined as primary and secondary coil, respectively. The two coils are separated by a gap g (not seen). An alternating current passes through the primary coil, generating a magnetic field that creates a flux in the secondary coil. By integrating the Biot-Savart relation [5],

$$dB(r,t) = \frac{\left[\mu J(r,t)dV \times \hat{r}_{pp'}\right]}{4\pi\left|r - r'\right|^2}, \tag{1}$$

we are able to calculate the magnetic flux density at points on the plane of the secondary coil contributed by each primary coil turn. We then evaluate the magnetic flux density in the plane of the secondary coil. The flux yield over the planar surface of the secondary coil of *N*-turns is expressed as

$$\varphi(t)_N = \oint_S B \bullet ds = \sum_i^M B_{ij} \sum_j^N b_j^2 \tag{2}$$

where M is the number of primary coil turns. By taking the derivative of (2) with respect to time, the induced voltage in the secondary coil is obtained as

$$V_{out} = \frac{4\omega\mu I\cos(\omega t)}{N} \sum_{i=1}^{M} \sum_{j=1}^{N} U, \text{ where}$$

$$U = \frac{b_i^2}{\left(b_i^2 + g^2\right)\left[2b_i^2 + g^2\right]^{\frac{1}{2}}} b_j^2 \tag{3}$$

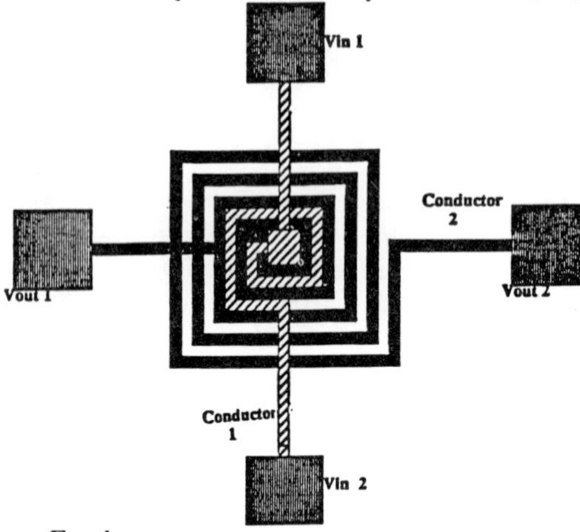

Fig. 1.
Exagerated top view showing coil windings. V_{in} contacts top coil. V_{out} contacts lower coil.

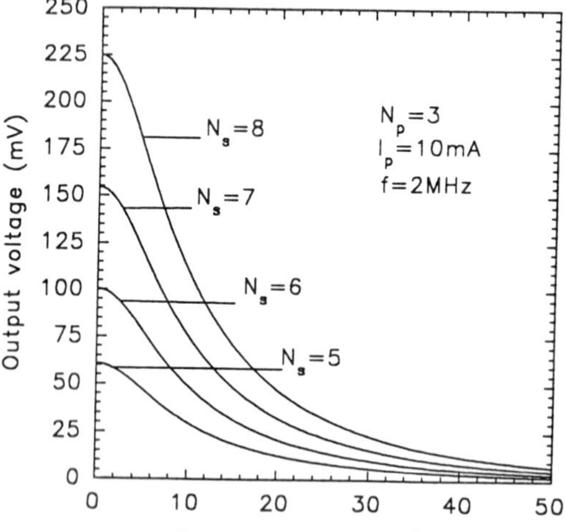

Fig. 2.
Output voltage versus coil gap for various secondary coil turns.

Note in the above that the output voltage V_{out} varies as g^{-3}. Hence, at fixed current *I* and frequency *w* the output voltage of the secondary coil V_{out} is a function of the coil gap *g*. To achieve high output voltage V_{out}, more secondary turns *N* are used. Fig. 2. shows the output voltage response to coil gap for various secondary/primary turns ratios *N/M*. We also note that applications-viable output voltages of hundreds of millivolts are obtained within a coil gap range of 0 to 10μm for the present device. Sensitivity, reliability, and physical considerations make it necessary to reduce the number of primary turns *M* without degradation of performance. The sensitivity of the output voltage V_{out} varies with the gap as follows:

$$\frac{dV}{dg} = \frac{4\omega\mu I\cos(\omega t)}{N}\sum_{i=1}^{M}\sum_{j=1}^{N} b_i^2 b_j^2 QZ \tag{4a}$$

$$\text{where } Q = \frac{2}{b_i^2 + g^2} + \frac{1}{\left(2b_i^2 + g^2\right)^3} \quad \text{and } Z = \frac{-g}{\left(b_i^2 + g^2\right)\left(2b_i^2 + g^2\right)^{\frac{1}{2}}} \tag{4b, c}$$

From the plot, it turns out that the coil gap g for minimum linearity error is between 0 and 8 μm.

3. MICROELECTROMECHANICAL DESIGN

The deflection w(w>0.5a) at the center of the circular diaphragm as a function of pressure, P from [7,8] is

$$w = 0.662a\sqrt[3]{\frac{Pa}{Eh}} \tag{5}$$

where

w, a, P, E, and h are deflection of membrane (μm), radius of membrane (μm), applied pressure (Pascal), Young's modulus of membrane (Pascal), and membrane thickness(μm), respectively.

The maximum stress occurs at the center of the membrane. For a 3 μm center deflection with a polysilicon diaphragm of diameter 200 μm and thickness of 1 μm, the maximum stress is 110MPa. This stress, corresponding to 120kPa of pressure differential on the diaphragm is less than 2% of the yield limit stress as we desire.

The sensor output voltage V_{out} is

$$V_{out} = \frac{4\mu\omega I\cos(\omega t)}{N}\sum_{i=1}^{M}\sum_{j=1}^{N} b_i^2 b_j^2 QZ \tag{6}$$

where QZ now becomes a function of the gap g and the gap expressed as $g = g_o - w$, g_o being the initial gap.

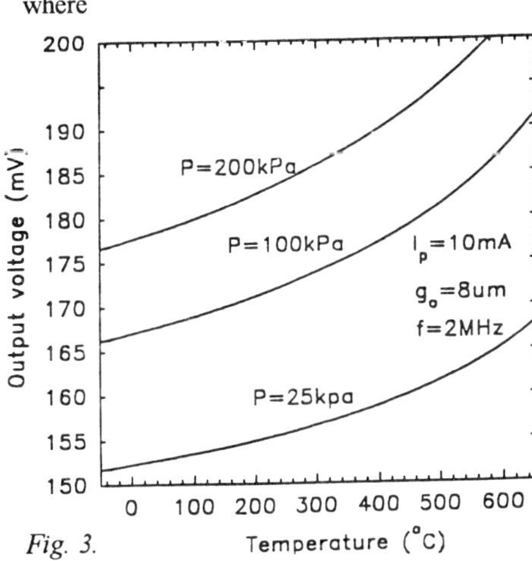

Fig. 3. Output voltage versus temperature at low, medium, and high pressures

TEMPERATURE EFFECTS

The output voltage V_{out} is a function of the TCE of the device materials. A plot of output voltage V_{out} versus temperature T for low, medium, and high pressures is shown in fig.3., for a coil gap of 8 μm taking into account the TCE of materials within the device sandwich. From this we are able to evaluate the Temperature Coefficient of Offset as

$$TCO = \frac{1}{V_o}\frac{V_f - V_o}{T_f - T_o} \tag{7}$$

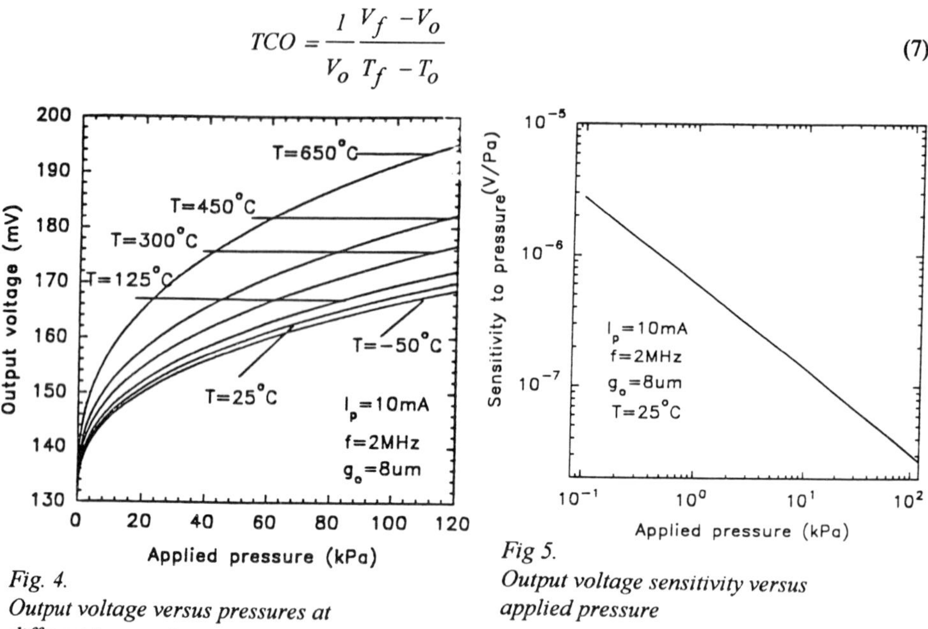

Fig. 4.
Output voltage versus pressures at different temperatures

Fig 5.
Output voltage sensitivity versus applied pressure

The output voltage as function of pressure P at various temperatures is evaluated and plotted in fig. 4. From fig's 3 and 4, we deduce the output voltage sensitivity to pressure at different temperatures and also sensitivity to temperature at fixed pressures. The Best-Fit-Straight-Line[11] technique is used to calculate voltage output linearity error. Output voltage V_{out} sensitivity to pressure at 25°C is shown in fig. 5. We note with interest the increased sensitivity at extremely low pressures, thereby extending the device dynamic range.

5 DESIGN CONSIDERATIONS

The design of the device follows the design rule of the Microelectronics Research Center of the New Jersey Institute of Technology. The design decision took into consideration problems of high temperature operation. Typical of such problems are degradation of piezoresistors at elevated temperatures and also TCE mismatch. To resolve the problems, the device uses tantalum silicide ($TaSi_2$) as a coil conductor for its excellent properties. It has a high melting point that makes it stable to 650°C. The low resistivity (50μΩ-cm) makes it viable for low power dissipation devices. The starting material is a <100> silicon wafer (n or p-type). An 8μm cavity is etched anisotropically into the wafer. A 200nm silicon dioxide is grown on the wafer. This is followed by sputtering 250nm $TaSi_2$ over the oxide. It is defined and dry-etched to form the secondary coil link. The next step is to deposit 100nm thick Si_3N_4 by PECVD over the conductor. The nitride essentially isolates the wires. A via is defined and etched through the nitride to the conductor, followed by another $TaSi_2$ sputtering. It is defined and dry-etched to pattern the secondary coil at the base of the cavity. The next step involves another nitride depositon similar to above. This nitride layer, together with the previous one, sandwiches the secondary coil. At elevated temperatures $TaSi_2$(TCE=9ppm/^oC) expands faster than Si_3N_4(TCE=0.8ppm/^oC). The two nitride layers absorb the $TaSi_2$ stress and therefore provide an improved TCE-match to the polysilicon.

The cavity is then filled with sacrificial PSG and planarized at the rim. This is followed by 100nm thick PECVD nitride. $TaSi_2$ is sputtered, defined, and patterned to form the primary coil. A 100nm PECVD. nitride deposition follows, sandwiching the primary coil . An array of 2μm diameter holes are etched through the nitride layers (coil avoided) to the PSG. The glass is subsequently selectively etched to empty the cavity. The vias are sealed with 100nm PECVD[6]. A via is created through this nitride to expose the center of the primary coil. This is followed by a $TaSi_2$ sputter. The silicide is defined and dry-etched to pattern the primary coil link. The link is then covered with the final 50nm PECVD nitride. A 1μm polysilicon is then deposited to form the diaphragm. Four holes are etched to expose the contact pads. Fig. 6 shows the cross section of the final device.

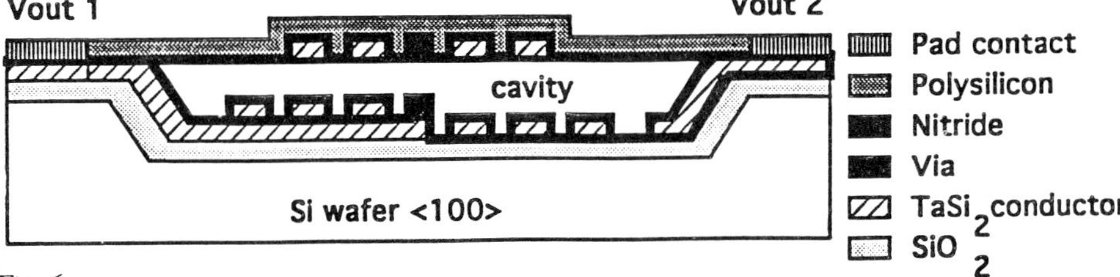

Fig. 6.
Cross section of inductively-coupled high-temperature pressure sensor (not to scale; primary coil terminals cannot be seen from this perspective. Refer to fig. 1)

6 CONCLUSION

We have described a high temperature pressure sensor design that uses inductive coupling. This design reduces temperature sensitivity problems encountered by most existing pressure sensors at high temperature. The problem of degradation in PZ resistivity is solved by avoiding the diffusion process. The problem of TCE mismatch is solved by symmetrical layering. Table 1 summarizes the results.

Table 1: Simulation Summary

Parameters	25° C	650° C
Full Scale(FS) Pressure sensitivity(V/Pa)	.125x10^{-6}	0.1x10^{-6}
Temperature Sensitivity(V/°C) (FS)	.06x10^{-3}	.09x10^{-3}
Temperature Coeff. of Offset (ppm/°C)	-	200
Output voltage range (mV)	40	62
Output voltage Offset(mV)	130	130

References

1. Solouf,Jnr., R E; Silicon Sensors for Automotive Applications; Transducers '91 Intl. Conf. on Solid State and Actuators Digest of Technical Papers; p170
2. Chung, G. S, Kawahito, S, Ishida, M, and Nakamura, T.; Temperature-independent Pressure Sensors Using Epitaxially Stacked Si/Al_2O_3/Si Structures; Sensors and Actuators A, 29 (1991) 107-115
3. Kurt Petersen, Joe Brown, Philip W. Barth, Joseph Mallon and Janusz Bryzek; Ultra-stable High Temperature Pressures Using Silicon Fusion Bonding; Transducers '91, page 90.
4. French, P.J., Muro, H.,Shinohara, T., Nojiri; SOI Pressure sensor; Transducers '91, Intl. Conf. on Solid State and Actuators Digest of Technical Papers; p181
5. Seshadri, S. R; Fundamentals Of Transmission Lines and Electromagnetic Fields; Addison Wesley, 1971 pp264
6. Sugiyama, S., Shimaoka, K. and Tabata, O; Surface Micromachined Micro-diaphragm Pressure Sensors; Transducers '91, Intl. Conf. on Solid State and Actuators Digest of Technical Papers; p188
7 Timeshenko, S and Young, D.H; Elements of Strength of Materials, Fifth Edition, 1968
8 Timeshenko, S and Woinowsky-Krieger, S; Theory of Plates and Shells, Second Edition, 1959
9 Patel, J.R & Chaudhuri,A.R; Macroscopic properties of Dislocation-free Ge; J. App. Phys. Vol.34 No9, September 1963, p2788.
10 Huff, M.A, Nikolich, D. and Schmidt, M. A; A Threshold Pressure Switch Utilizing Plastic Deformation of Silicon; Transducers '91, Intl. Conf. on Solid State and Actuators Digest of Technical Papers; p177.
11 Technical Apps Note; Pressure Sensors; Motorola, Inc., 1992

Micromachined silicon resonant sensors with robust optical interrogation

R. A. Pinnock

Sensors and Actuators Department,
Lucas Advanced Engineering Centre,
Dog Kennel Lane, Shirley, Solihull, West Midlands, B90 4JJ (U.K.)

ABSTRACT: In recent years, the development of micro-fabrication techniques in silicon has led to the introduction of a new generation of miniature resonant sensor elements. In addition to their small size and frequency-based output, the sensor-heads may be made electrically passive by interrogating the resonant elements optically. This paper describes some of the practical issues associated with the development of optically-interrogated micromachined silicon pressure sensors, for use in hostile environments and temperatures up to 300°C.

1. INTRODUCTION

The micromachining of silicon is an extremely precise mechanical fabrication technique, which is becoming widely exploited in the transducer industry for the fabrication of a new generation of miniature resonant sensor elements. Such elements can be fabricated from single silicon crystals, resulting in high accuracy, good long-term stability and low hysteresis. Resonant structures fabricated using these techniques can therefore offer significant performance benefits over conventional sensing methods, and recently reported devices of this type are certainly achieving excellent levels of performance.

In general, these devices can be driven into resonant oscillation via electrostatic means, and maintained at resonance through the use of suitable detection electrodes monitoring the resonator motion, and providing feedback to the drive signal. These electrical drive and read techniques currently place an upper limit of about 150°C on the temperature at which the sensors can be operated. Many applications, particularly in the aerospace field, but also in some cases in the industrial field, require sensors which can operate at substantially higher temperatures than this, whilst maintaining their excellent performance characteristics. For example, pressure and temperature sensors are required for control purposes on gas turbine engines in the aerospace field, capable of operating with high accuracy at temperatures up to, and in some cases beyond, 300°C. One method by which this can potentially be achieved is through optical, rather than electrical, interrogation of the microresonator.

During the last decade, several groups have reported sensors based on the optical driving and reading of resonant mechanical structures. In general, these devices have been fabricated in silicon, using photolithographic and anisotropic etching techniques to produce tiny, but in some cases relatively complex, sensor structures. Similar structures based on other materials (e.g. quartz) have also been reported. The use of optical fibres to transmit optical signals to and from the microresonator enables totally electrically-passive sensor-heads to be produced. This, combined with their small size and frequency-based outputs, makes them particularly suitable for a number of specialist measurement requirements in thermally, electromagnetically and chemically hostile environments.

The work described in this paper is aimed at developing optically-interrogated silicon microresonator pressure sensors, capable of operating with high accuracy at temperatures up to 300°C. It is hoped to be able to demonstrate the operation of the devices on a gas-turbine engine demonstration rig in the near future. The main emphasis of this paper will be on the practical issues associated with fabricating and packaging the devices, since this an area which will be vitally important in their commercial application.

2. PRINCIPLE OF OPERATION

Figure 1 shows the basic principle of operation of an optically-interrogated silicon microresonator:

The microresonator is driven at its resonant frequency (approximately 100kHz to 200kHz) by light transmitted from a pulsed source by an optical fibre. The oscillations induced in the resonator by the pulsed optical drive signal are detected through the use of a second (c.w.) optical source (at a different optical wavelength) and associated detector. The device is maintained at resonance by locking the detected output frequency to the drive frequency of the pulsed source. For a pressure sensor, the microresonator is attached to a diaphragm to which the pressure is applied. This deforms the diaphragm and strains the microresonator, thereby changing its resonant frequency. Thus, a frequency-based output is obtained from an electrically-passive sensor head, with only a single optical fibre addressing the microresonator.

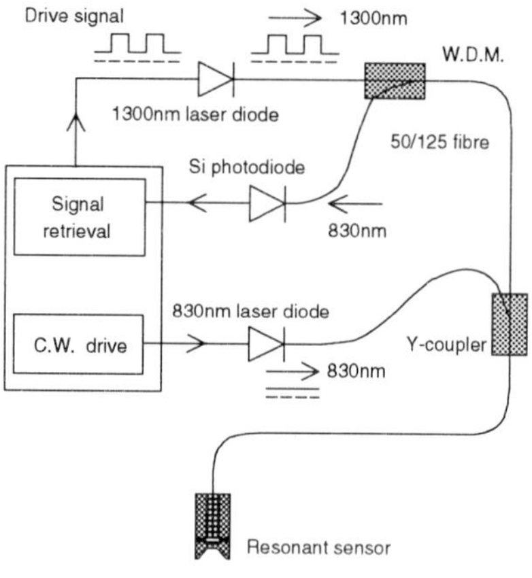

Figure 1 : Sensor System Configuration

The interrogation arrangement described above is one of a number of possible configurations which have been proposed. Some of these enable "self-oscillation" of the microresonator from a single c.w. source to be achieved. For example, Langdon and Dowe [1] have demonstrated self-oscillation effects in optically-interrogated cantilevers made from thin polyester film coated with aluminium. Ultimately, such techniques may prove to be simpler to implement, but further work remains to be done in order to integrate them into viable sensing systems.

The performance of the system shown above will be dependent on several factors. These include:

- the design of the silicon micromachined resonators and associated structures;
- the mechanical design of the optical fibre positioning / alignment system;
- the choice of technology for and design of the optical detection system;
- the optical systems design (choice of source, detector, fibre, source and detection electronics, etc.).

These issues will be considered in the following section.

3. SYSTEM DESIGN

3.1 Sensor Element

The basic sensor element consists of a "bridge" of silicon, supported on pillars at each end, fabricated on the underside of a silicon diaphragm. This structure can be manufactured as a whole in single-crystal silicon, using anisotropic etching and other micro-fabrication techniques. The bridges themselves are typically 800μm long, 60μm wide and 2μm thick: the size and thickness of the diaphragms will depend upon the required pressure range, but typically, they are approximately 1300μm square and 25μm thick.

Various designs for the bridges are possible; some simple ones are shown in Figure 2. The first is a simple rectangular bridge supported at both ends; The second design is a dual resonator arrangement, as proposed by Willson [2], where the second (cantilever) resonator may be used for temperature compensation purposes in the pressure sensor; the third has the supporting pillars positioned at the natural nodal points of the oscillatory motion, in order to minimize the energy lost to the supporting structure. Other, more complex arrangements are of course possible.

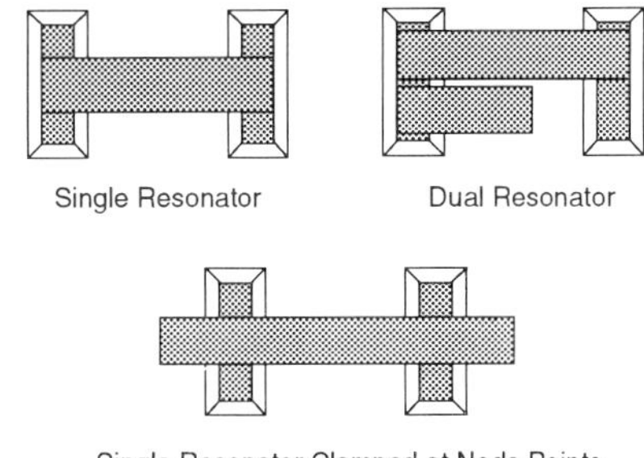

Figure 2 : Simple Resonator Bridge Designs

In this work, the bridges are initially defined in the silicon by boron-doping. The introduction of the boron prevents these regions of the silicon from etching during the anisotropic etching process. However, it has also been shown that non-uniform stresses can be set up within the resonator bridge, due to the difference in size of the boron and silicon atoms, and the non-uniform distribution of boron in the doped region. These stresses can severely affect the performance of the resonator, particularly in terms of its temperature coefficient. In order to help to overcome this problem, the bridges are annealed after fabrication, thereby making the stress distribution uniform throughout the bridge.

3.2 Interfacing of Optical Fibres

The accurate alignment and positioning of the interrogating optical fibre with respect to the microresonator is crucial to the successful operation of these devices. This aspect assumes even greater importance when considered in terms of the eventual manufacturability of such sensors. The physical dimensions of the resonator elements are of the same order of magnitude as the addressing fibres, and precise alignment, particularly with respect to the width of the resonators, is obviously critical in order to maximize the amount of light impinging on the bridges. Equally important, however, is the positioning of the fibre in terms of the distance of its end from the surface of the resonator. As will be discussed below, it is desirable to use a LED source (rather than a laser diode) to both drive and read the sensor. The oscillatory movement of the bridge, however, will only be of the order of tens of nanometres, therefore demanding the use of an interferometric detection technique. The white-light interferometry scheme outlined below can meet this requirement, but again requires precise positioning of the fibre relative to the bridge.

Two basic constructions are being investigated in the current work programme. In the first, shown in Figure 3, the fibre support structure is a concentric assembly, positioned above the resonator and diaphragm. The fibre is held in a glass or ceramic ferrule, which is in turn fixed into a silicon "collar". This assembly is attached to a glass (or silicon) spacer, whose dimension in the vertical direction defines the fibre-to-resonator separation. The devices are in fact fabricated at the wafer level, these being subsequently sawn up into individual sensors.

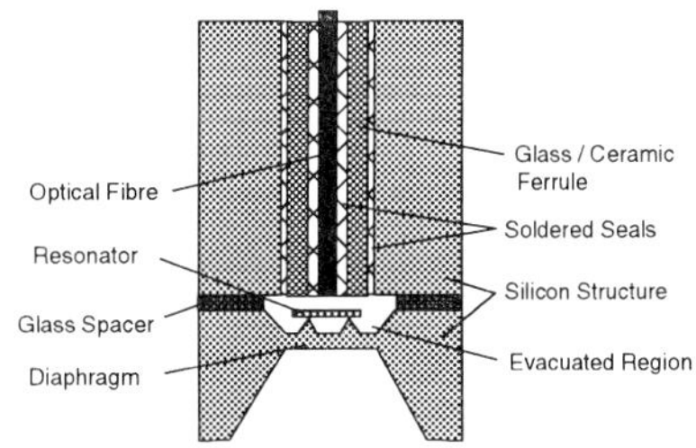

Figure 3 : Microresonator with Concentric Interrogation Fibre

The second construction is shown in Figure 4. Here, the fibre is fixed into a v-groove, defined by anisotropic etching in a silicon base. Also defined is a 45° face, against which the end of the fibre butts. Light emerging from the fibre is reflected upwards by the angled facet, to impinge on the resonator which is positioned directly above as shown. Thereafter, the light follows the return path back into the fibre. Once again, these devices are fabricated at the wafer level.

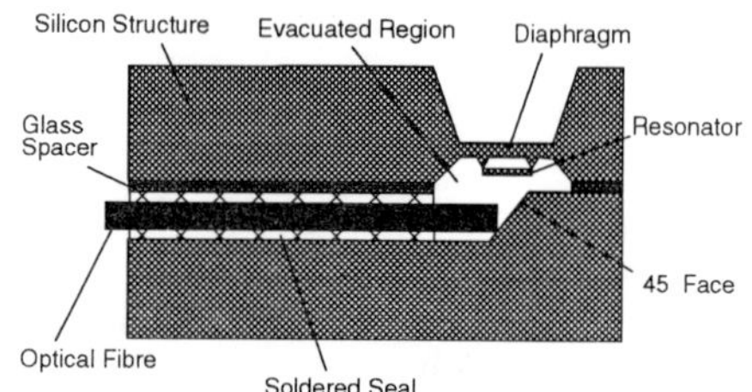

Figure 4 : Microresonator with Side-interfacing Fibre

This arrangement has been developed at Lucas Advanced Engineering Centre in order to greatly simplify the problem of accurately positioning and aligning the optical fibre with respect to the resonator. The position of the fibre is accurately defined by the v-groove and 45° facet, and the distance of its end from the microresonator can be set by careful control of the etching process. The full procedure by which these devices are manufactured has been described by Jacobs-Cook and Bowen [3]. An example of such a device (without a diaphragm, in order to show the microresonator) is shown in Figure 5. In this case, the fibre is 50/125, and the bridge has approximately the dimensions given earlier.

Figure 5 : Fibre Alignment with Angled Interface

In order to achieve the large Q-values required for high-accuracy measurement, the region around the resonator must be evacuated. This in turn means that for both the constructions shown above, the interrogating fibres must be hermetically sealed into the surrounding silicon structures. In order to satisfy the requirement for reliable operation at high temperatures, solders are being investigated as a means for fixing and sealing the optical fibres (see Figures 3 and 4). These offer the best compatibility with both the manufacturing processes and the desired performance specification of the devices (particularly in terms of the high-temperature requirement). This again is a critical aspect of the manufacturing of these sensors.

3.3 Optical Detection Scheme

As indicated above, the movement of the resonator, induced by the pulsed optical drive, will be of the order of tens of nanometers, and this therefore dictates the use of an interferometer to detect the oscillations. In the present case, a "white-light" interferometry arrangement is being used, since this allows the read-source to be a LED rather than a laser. The sensor-head effectively operates as a two-beam interferometer, one beam being derived from the light signal reflected from the end of the optical fibre, and the other reflected from the resonator itself. Assuming that the distance between the end of the fibre and the resonator is greater than the coherence length of the read-source, the two beams will propagate back along the fibre, without interference, to a second (detection) interferometer. In the usual manner of white-light interferometry, this has an optical path difference which matches the "built-in" path difference of the sensor, and allows the coherent recombination of the two optical beams. The output of this interferometer is a visibility function with varies with the oscillation of the microresonator.

A significant problem which needs to be overcome is the fact that the microresonator is fixed to a diaphragm which will move by several microns under the applied pressure measurand. This movement is equivalent to several fringes of the interferometer transfer function, and is many times larger than the oscillatory movement of the bridge. The operating point of the sensor interferometer will therefore be continually changing, and may easily be altered to a point where the sensitivity is much reduced, or almost zero (i.e. at one of the maxima or minima of the sinusoidal interferometer transfer function). This is the well-known signal fading problem, which in the present case could prevent reliable phase-locking of the resonance. A number of techniques have been proposed which can overcome this problem (e.g. [4], [5]), and these can be made compatible with the white-light detection system outlined above.

3.4 Systems Issues

There are many inter-related systems issues which need to be considered in the design of these sensors. Such matters as the choice of optical fibre (multimode or single mode), the choice of the optical sources and detector, the optical drive (and read) power, the characteristics of the loop-closing electronics, and so on, all depend on the desired sensor performance, the availability of components, and the eventual manufacturability of the devices.

In the present work, 50/125 fibre is being used to interface to the microresonators. This is because the core size closely matches the physical dimensions of the resonator, and also because the use of multimode fibre allows LEDs to be used as the optical sources. In turn, the use of LEDs is advantageous primarily because they are simpler to implement than laser-diodes.

In addition, the relatively long coherence length of laser diodes can be problematic when detecting the oscillations of the microresonator. This seems to be because the effective cavity length between the end of the interrogating optical fibre and the resonator can sometimes be matched to the wavelength of the laser in such a way that "self-oscillation" effects can be induced by the read-source. This is a highly unstable effect, and leads to considerable instability in the measured resonance. Under soft vacuum (0.07mbar), stable resonances with very high Q-values can be obtained. In the example shown in Figure 6, the resonant frequency is approximately 178.6kHz, and the Q-value is more than 230,000. This was measured with a device similar to that shown in Figure 5.

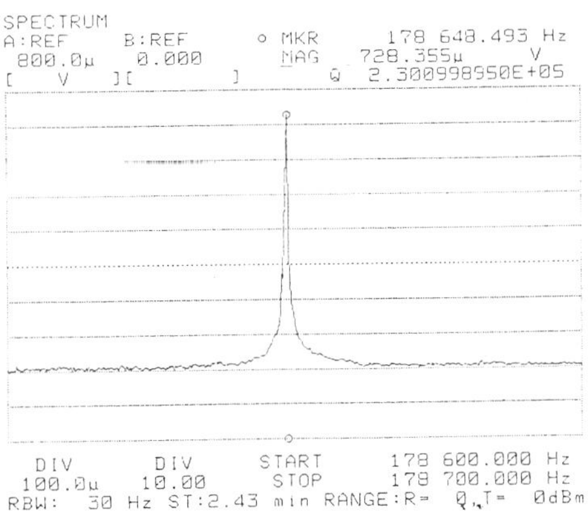

Figure 6 : Resonance of Optically-interrogated Resonator

A further potential problem with the use of optical interrogation techniques to drive and read the resonators is the fact that the resonant frequency is dependent upon the optical power delivered to the resonator by the drive and read sources. The resonant frequency of the bridge is dependent on its temperature, and this will in turn be affected by the average optical power delivered to (and subsequently absorbed by) the resonator. Unfortunately, the result of this is that the frequency-output of the device effectively becomes sensitive to simple intensity changes in the light delivered by the optical fibre. This effect can be seen for both the drive-source and the read-source in Figures 7 and 8. In these cases, the change in delivered optical power is relatively large, and in reality, better control of the delivered power should be possible. However, for highly accurate sensors (i.e. accuracy better than 0.1% f.s.), the effect may still be problematic.

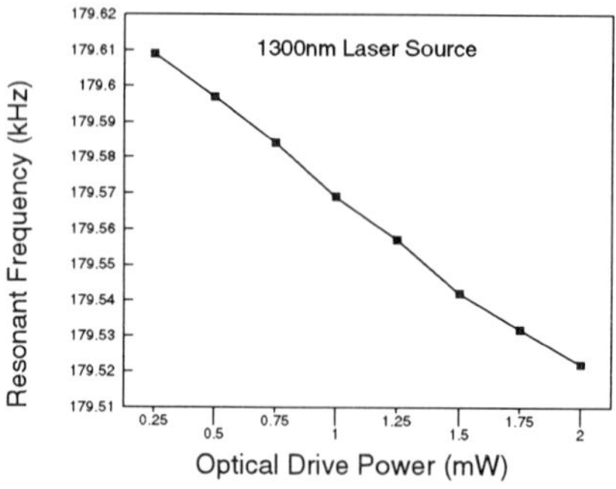

Figure 7 : Resonant Frequency Variation with Drive Power

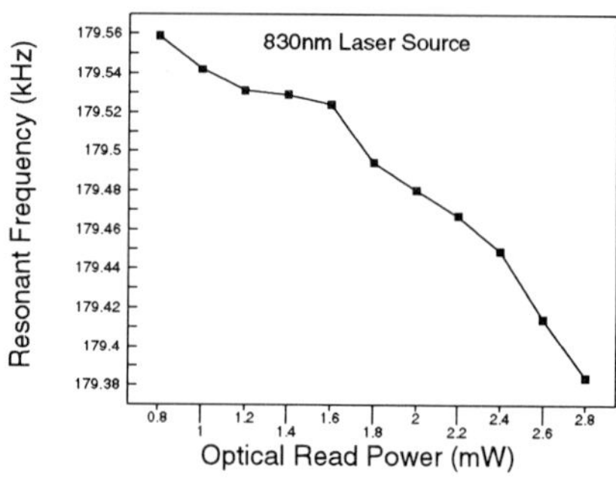

Figure 8 : Resonant Frequency Variation with Read Power

One way to help overcome this problem is to keep the delivered optical powers as low as possible. In the present work, successful operation has been demonstrated with drive powers, for example, as low as 25μW. This is again compatible with the use of LEDs as the optical sources. Keeping the drive power low will also help to overcome spurious hysteresis effects associated with over-driving the resonator [6].

4. CONCLUSIONS

This paper has described some of the work currently being undertaken at the Lucas Advanced Engineering Centre, aimed at addressing important practical issues associated with the design and manufacture of optically-interrogated silicon microresonator pressure sensors. Many of these issues are not fully solved as yet, but solutions will need to be found if these devices are going to fulfil their potential as accurate, frequency-based sensors.

5. ACKNOWLEDGEMENTS

The author wishes to thank the many colleagues who have contributed to the work described in this paper, particularly Messrs A. J. Jacobs-Cook, M. E. C. Bowen, N. Hart, R. J. Hazelden and R. E. Jones; also, Dr. S. J. Prosser and Dr. P. Extance for their support in publishing this work.

6. REFERENCES

1 R. M. Langdon and D. L. Dowe : "Photoacoustic Oscillator Sensors"; SPIE Vol. 798 (Fibre Optic Sensors II) (1987), p. 86 - 93

2 J. P. Willson : "Sensors"; U.K. Patent GB 2,223,311 B

3 A. J. Jacobs-Cook and M. E. C. Bowen : "Novel Optical Fibre / Microresonator Interfacing Technology"; Eurosensors VI, October 1992

4 K. Weir, W. J. O. Boyle, A. W. Palmer, K. T. V. Grattan and B. T. Meggitt : "The Measurement of Vibration Using a Fibre Probe and a Michelson Interferometer"; "Sensors: Technology, Systems and Applications", Ed. K. T. V. Grattan, The Adam Hilger Series on Sensors, 1991, p. 269 - 274

5 R. Jones, M. S. Hazell and R. J. Welford : "A Generic Sensing System Based on Short Coherence Interferometry"; CCL White-light Interferometry Seminar, March 1991

6 M. V. Andres, K. W. H. Foulds and M. J. Tudor : "Nonlinear Vibrations and Hysteresis of Micromachined Silicon Resonators Designed as Frequency-out Sensors"; Electronics Letters, Vol. 23, No. 18, 1987, p. 952 - 954

Planar fibre-optic interface for silicon microresonators

A. J. Jacobs-Cook & M. E. C. Bowen

Lucas Advanced Engineering Centre,
Dog Kennel Lane, Shirley, Solihull, West Midlands B90 4JJ (U.K.)

ABSTRACT

Novel fibre-optic interfaces have been developed using anisotropic etching of <110> silicon. An interface comprises an alignment v-groove and, at one end, a 45° angled mirror face to redirect light from a fibre in the groove. Both an interface structure and a silicon microresonator may be integrated as a single component. One such device has been tested under conditions of soft vacuum using optical drive and detection. At a pressure of 7×10^{-2} mbar the device was observed to resonate at just under 180kHz with a Q factor of over 32 000.

1 INTRODUCTION

Optically addressed microresonant structures have attracted much interest in recent years and many papers have been published on the subject e.g. [1,2,3,4]. Most of this effort has however concentrated on either the resonator design or the particular means by which the optical drive and interrogation may be employed.

One area of great importance is the interfacing of optical fibres to such silicon structures in order to produce viable sensors. Optically-addressed silicon diaphragm and microresonator devices have been proposed as potential solutions for a number of sensor applications (e.g. pressure, temperature, etc.), and the alignment and fixing of optical fibres on the silicon structures is a crucial aspect of the manufacture of these devices.

At present, it is a difficult and time-consuming process to achieve the required alignment tolerance. Also, the fibres generally have to approach the silicon structure from above or below (i.e. out of the plane of the silicon), which may be inconvenient in some cases, and can make the structures difficult to seal. Fig. 1 illustrates a conventional approach to interfacing an optical fibre to a microresonator.

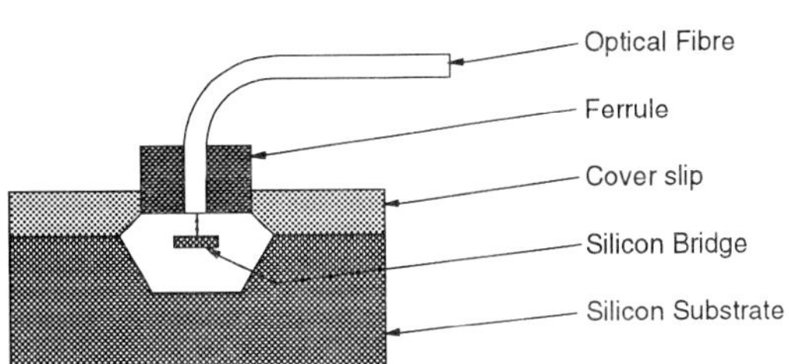

Fig. 1: Conventional method of interfacing an optical fibre to a resonant sensor

Limited work in this area of technology has been reported, although similar ideas to those in this paper have been proposed [5]. Also, the use of integrated optic waveguides could provide a possible solution to the problems of packaging fibre optic sensors [6].

The main aim of the work described in this paper was to investigate potential processes for producing silicon structures which would enable "in-plane" addressing fibres to be interfaced to the structures. In particular, the aim was to develop devices with suitably angled integral facets, which could be used both to position optical fibres with respect to the structures, and to direct the emergent light beams onto silicon based optical sensors. It was decided to make the interfaces with integral resonant bridges, in order to assess the potential for using the structures as sensors, with the interrogating light signals directed on to the resonators via the angled facets. It was considered that such devices would be of use in the further development of temperature and pressure sensors based on silicon microresonators.

2 INITIAL DESIGN

One of the prime requirements of the design is a v-groove, to allow accurate 'in-plane' positioning of an optical fibre. A reflective facet, angled at 45° to the v-groove axis, is therefore required to redirect the light from the fibre resting in the groove. In addition, it is advantageous to be able to integrate silicon beams (as resonant sensors) with the interface structure. Fig. 2 gives an outline sketch of the envisaged interface structure in cross-section.

An initial study of the properties of silicon and the process of anisotropic etching indicated that the required angled facets could be achieved using suitably orientated <110> silicon. A 45° face could be fabricated in order to redirect light from an optical fibre (e.g. on to a resonator) and (if necessary) back into the fibre again.

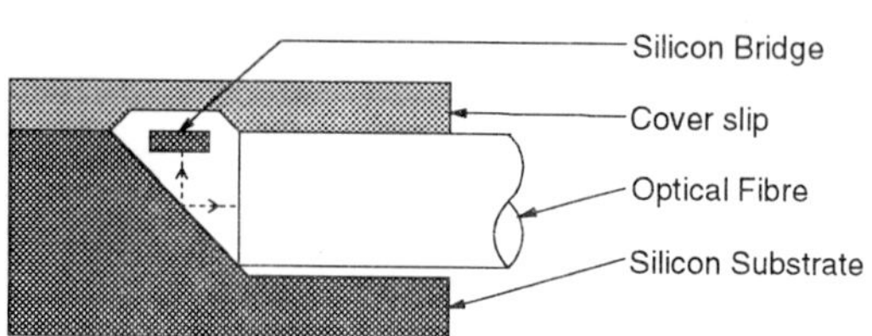

Fig. 2 : Envisaged structure for interface

Simultaneously, a pair of angled facets at 35.7° to the horizontal could be produced in order to form a v-groove in which to position the fibre. Since the positions and form of these facets could be accurately controlled by the mask design and etching process, alignment of the fibre within the structure should be considerably eased.

3 DEVICE FABRICATION

In addition to the fabrication of the essential components of the device (i.e. the resonator with optical interface), several other process steps were necessary to produce a robust package. The various fabrication stages are detailed as follows.

3.1 Fabrication of interfaces/resonators

A two layer mask set was required for the fabrication of the silicon devices. Fig. 3 shows the combined areas of detail from both masks for a single device having an optical interface combined with a microresonator.

The silicon must be <110> type in order that the structures form correctly in the final step of anisotropic etching. Having oxidised the surface of the silicon, mask layer 1 is applied to photolithographically define windows in the oxide (as shown in Fig. 3). A high concentration boron doping step follows and the areas of unprotected silicon are doped to a sufficient etch stop concentration for KOH/H_2O solution (~ $1x10^{20}$ atoms cm^{-3}). Later, during etching, these areas will be under-etched to form bridge structures. Alternatively these areas may be left undefined by the mask so that bridges are not included.

Fig. 3 : Mask layout to define bridge structures and interfaces

The next step is to define the areas of silicon which are to be etched in the final stage. Again, referring to Fig. 3, the shaded areas of surface oxide are removed after being defined by mask layer 2. All other areas are left protected by the oxide and will therefore not be etched.

The silicon is now immersed in a solution of 30 wt.% KOH/H_2O @ 70°C and left until the structures completely form. During this time the v-groove, lying perpendicular to the bridge, is formed by {111} planes lying at 35.7° to the surface plane. The area between the bridge and the v-groove etches and in doing so a {100} plane, angled at 45° to the surface plane, is revealed which progresses under the bridge as it is etched away. Thus the bridge is under-etched and therefore freed by the progressing {100} plane. Etching is complete when the top edge of the progressing {100} plane just breaks into the corners of the small diamond-shaped 'timing marks'.

Although successful fabrication relies on a timed-etch process, the etch rate is fairly reproducible. Also, the position of the 45° angled face with respect to the bridge need only be accurate to ± 5μm, in order that most of the light from a fibre illuminates the bridge. Thus both these factors mean that the etch time is not critical and must only be controlled to within ± 5 min.

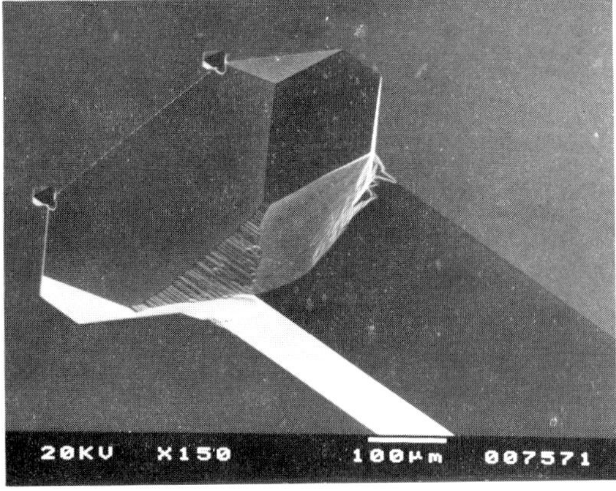

Fig. 4 : Basic interface structure (without bridge)

Figs. 4 and 5 show a basic interface structure and an interface with integral bridge resonator respectively. The v-groove and 45° angled mirror face can be clearly seen.

3.2 Preparation of glass seals

From Fig. 2, it can be seen that, in order protect the device with a glass cover, a shallow pit must be etched in the surface of the glass. This ensures freedom of movement of the resonating bridge and also prevents the bridge from bonding to the glass during anodic bonding.

In order to do this the surface of the glass is appropriately defined, by photolithography, so that a pattern of square openings exists in the photoresist.

Fig. 5 : Basic interface structure (with bridge)

The glass is then immersed in a solution of concentrated HF acid (48 wt.%) for approximately 15 mins after which time square pits (~50μm deep) will have etched into the surface. It may be necessary to coat the surface of the glass to be etched with a thin layer of nichrome. This improves adhesion of the photoresist and prevents uneven etching.

The glass used for the protective cover was Pyrex 7740 since various samples were available and its thermal expansion coefficient is well matched to that of silicon.

3.3 Preparation of optical fibres

Multimode optical fibre (50/125 graded index) is suitable as a means of directing light onto the resonators since the core diameter is of the same order as the width of the resonators. Thus if an optical interface is incorporated in the sensor, one will find that there is a gap of 125μm between the end of the fibre and the resonator bridge structure. This is due to the 45° face introducing the gap which is determined by where the point on the fibre circumference touches the angled face (see Fig. 2).

In order to maximise the coupling efficiency of light between fibre and resonator a lens may be formed on the end of the fibre. More light will not only be coupled to the resonator but also back into the fibre. The latter is important for interferometric detection of the resonator motion.

Lensing of the fibre end may be achieved simply as follows. After stripping the buffer coating back, the fibre end is cleaved and then wiped clean using isopropyl alcohol (IPA). A BHF solution is made up, comprising ten parts ammonium fluoride 40 wt.% to one part hydrofluoric acid 48 wt.%. The end of the fibre is left in the solution for approximately 60 mins after which time a convex lens is formed in the core. The convex etching profile arises due to the dopant variation from the centre of the core to its outer extreme.

Fig. 6 shows an SEM of a 50/125 fibre which has been cleaved but not etched and, for comparison, Fig. 7 shows the same type of fibre having been etched in BHF for approximately 60 mins. The lens can be clearly observed. The etch time is again not critical in this application and approximately 5 mins more or less would be allowed.

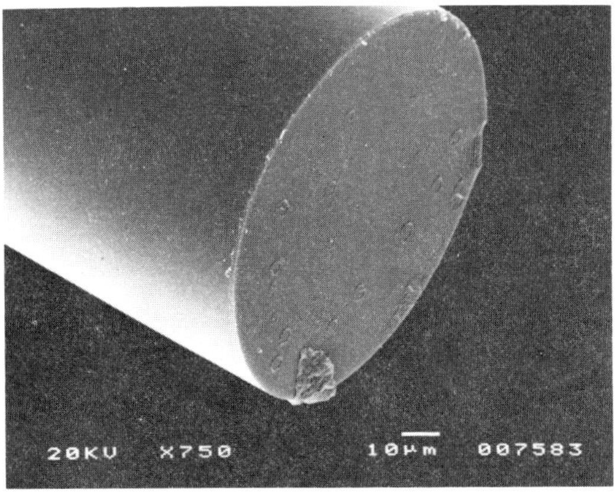

Fig. 6 : Unlensed optical fibre

3.4 Final assembly

The first stage of assembly requires the silicon interface/resonator structure to be anodically bonded to the glass cover. After cleaning the silicon and glass samples, they are inserted into an anodic bonder and aligned one over the other. When the alignment is satisfactory, the chamber is evacuated and the samples are heated to 400°C. At this point a voltage of around 500V is applied across them which is sufficient to initiate a permanent bond.

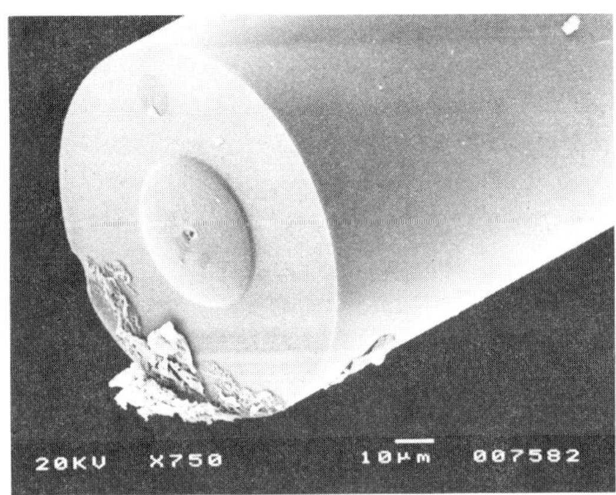

Fig. 7 : Lensed optical fibre

The second stage of assembly requires the optical fibre to be inserted along the v-groove until it abuts with the 45° angled face. When this occurs the fibre is bonded in the groove using adhesive. In this case UV-curing glue was used.

An optical fibre can be seen, in Fig. 8, bonded into the v-groove of an interface with integral resonator.

The glass cover was not in place when the SEM was taken since glass tends to become charged by the electrons and so the image would be distorted.

4 RESULTS

4.1 Analysis of unlensed and lensed optical fibre

Fig. 9 shows a photomicrograph (100x mag.) of an optical fibre having been inserted along the v-groove of an optical interface structure and bonded in position. Note the spot of light illuminating the bridge.

Two photomicrographs were taken: one with an unlensed fibre, the other with a lensed fibre. However, after photographic reduction, the difference in spot size would be less obvious and so only one is shown as an example. The measured spot sizes give a much better indication of the improvement a lensed fibre provides. The spot size at the bridge from the unlensed fibre was measured as 75μm. The spot size at the bridge from the lensed fibre was measured as 62μm. The coupling efficiency (for light illuminating the bridge) is inversely proportional to the cross-sectional area of the optical spot. Therefore, taking the ratio of the respective spot areas, to compare the coupling efficiency, shows an improvement of over 46% for the lensed fibre over the unlensed one. If one considers that reflected light must be collected again by the optical fibre, the improvement should be better still, since the light from the unlensed fibre will continue to diverge comparatively more than that from the lensed fibre.

Fig. 8 : Completed interface structure

4.2 Evaluation of resonators

In order to test the interfaces with integral resonator bridges, a number were fabricated as described above.

Fig. 9 : Incident light spot from an optical fibre

Initially, a straight-cleaved 50/125 fibre was inserted into one of the interface devices, its position being defined by the v-groove in the silicon substrate and the base of the 45° angled face, as described previously (see also Fig. 2). At this stage, the fibre was held in position using UV-curing glue. Using a microscope, it was possible to see the light from the end of the fibre reflected onto the resonator via the angled face. On inspection of the first completed device, the resonator was only partially illuminated by the spot of light. A closer inspection of the device showed that this was due to a slight over-etching having caused the 45° angled face to be displaced longitudinally by ~10μm from its desired position. Approximately 60% of the light spot was actually hitting the resonator in this case.

The arrangement used to test the operation of the devices is shown in Fig. 10. Pulsed light from a 1300nm laser diode is used to drive the resonator, whilst c.w. light from a 830nm laser diode is used to detect the oscillations. A wavelength multiplexer couples the light signals into the single fibre interrogating the interface device.

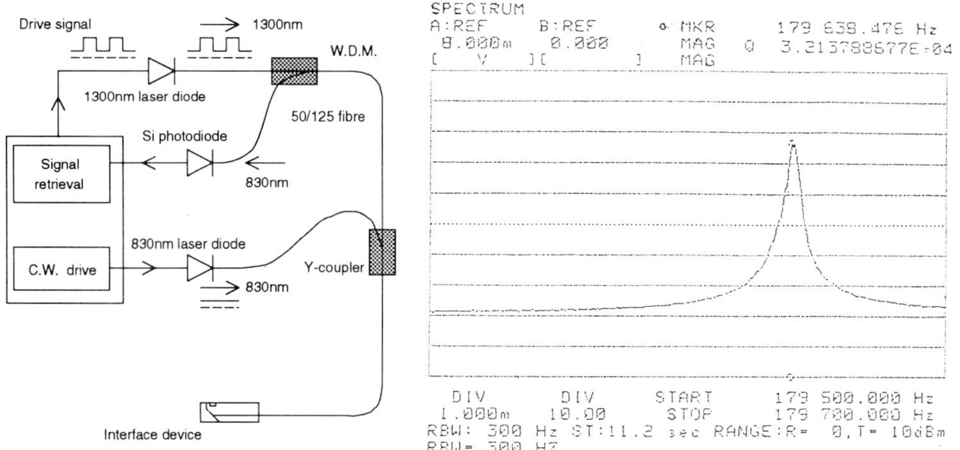

Fig. 10 : Arrangement for testing interface operation

Fig. 11 : Initial test results

When an attempt to operate this device at atmospheric pressure was made, no resonance could be detected. Silicon resonators (without optically absorptive coatings) have inherently low absorption and air damping significantly reduces the Q factor. Optical excitation is therefore not very efficient under these conditions. Metallic coatings improve the optical absorption of resonators, at the expense of reducing the Q factor, but this does make operation at atmospheric pressure possible. However no metallic coatings were used in this case.

Since the uncoated device would only be able to operate under vacuum, it was inserted into a small vacuum chamber attached to a pump, with the interrogating fibres emerging via a seal from the chamber. The chamber was then evacuated to 7×10^{-2} mbar, and a further attempt made to drive the resonator. Sample results are shown in Fig. 11. The device is seen to have a resonance at just under 180 kHz, with a Q of more than 32 000. Dimensions of the resonator are 800μm x 56μm x 1.4μm (l x w x t).

5 CONCLUSIONS

This programme of work has resulted in the successful production of novel interface structures, combining silicon microresonator sensors with a package which allows the simple alignment of the interrogating optical fibre. The devices have been shown operating, using wavelength-multiplexed optical driving and reading signals transmitted down a single optical fibre and, under conditions of vacuum, high-Q resonance has been achieved.

At present, the devices are configured as temperature sensors (although they have not been tested for temperature response as yet). They may equally be configured for the measurement of other parameters (e.g. pressure), and do not necessarily need to have an integral microresonator sensor element. In the latter case, the devices may be used as interfaces to other structures and devices (e.g. silicon diaphragms, electro-optic components, etc.).

The main issue, which must be addressed, is the sealing of the fibre into the interface, so as to allow a permanently evacuated region to be created around the microresonator bridge. This would then improve the Q factor and thus the drive efficiency. Possible ways of achieving this will be investigated in future work.

6 ACKNOWLEDGEMENTS

The authors wish to thank Mr N. Hart for his many technical contributions and advice on fabrication of the silicon devices. Acknowledgements must also go to Messrs R. A. Pinnock and J. Logan for useful technical discussions and information provided. Finally the authors must thank Drs S. J. Prosser and P. Extance for their support in publishing this work.

7 REFERENCES

1 K. E. B. Thornton, D. Uttamchandani and B. Culshaw, Novel optically excited resonant pressure sensor, *Electronics Letters, Vol. 24, No. 10 (1988) 573-574.*

2 N. A. D. Stokes, R. M. A. Fatah and S. Venkatesh, Self-excited vibrations of optical microresonators, *Electronics Letters, Vol. 24, No. 13 (1988) 777-778.*

3 R. E. Jones, J. M. Naden and R. C. Neat, Optical fibre sensors using micromachined silicon resonant elements, *IEE Proceedings, Vol. 135, Pt. D, No. 5 (1988) 353-358.*

4 J. Suski, D. Largeau, A. Steyer, F. C. M. van de Pol and F. R. Blom, Optically activated ZnO/SiO_2/Si cantilever beams, *Sensors and Actuators, A, 24 (1990) 221-225.*

5 R. A. Buser, N. F. de Rooij, Silicon Microetching and FOS: properties, compatibility and synthesis of transducers, *Proceedings EFOC/LAN 88, Amsterdam, Netherlands, 29th June - 1st July 1988, pp. 223-228*

6 H. Bezzaoui and E. Voges, Integrated optics combined with micromechanics on silicon, *Sensors and Actuators A, 29 (1991) 219-223.*

Low-cost fabrication technology for IC-compatible backside contacted ISFETs

A. Merlos, J. Esteve, M.C. Acero, C. Cané and J. Bausells.

Centre Nacional de Microelectrònica - CSIC.
Universitat Autònoma de Barcelona, E-08193 Bellaterra, Barcelona, Spain.

ABSTRACT: A simple and fully IC-compatible microelectronic technology has been developed for the fabrication of back side contacted ISFET's. Problems related to photolithographic steps on microstructured wafer surfaces have been avoided and only standard processes have been used. In order to avoid IC compatibility problems, the anisotropic etching of silicon has been carried out in TMAH-based solutions. This technology allows the fabrication of low cost and high yield BSC-ISFET. Therefore this kind of devices are very suitable for industrial and environmental applications.

1. INTRODUCTION

ISFET based chemical sensors have a very large number of potential applications. In the field of environmental control, for example, this kind of devices are very suitable to monitor the quality of drinking water, waste water and rivers. Microelectronic fabrication technologies lead to devices with a very robust structure (all solid state) with low maintenance requirements and allows a great reduction in the cost of unitary sensors due to their inherent high volume production. Additionally, by means of the deposition of the adequate ion selective membrane over the ISFET gate area, it is possible to change their natural pH sensitivity into other chemical sensitivities to different ions: Na^+, K^+, Ca^{2+}, NH_4^+, ... [1,2,3].

The most important problem that affects ISFETs functioning is their high drift rate. This drawback, however, can be easily circumvented by using them in flow injection analysis (FIA) systems [4]. Another important drawback is packaging. Then, conception, design and fabrication of ISFET based chemical sensors are strongly related to the packaging system [5]. The lifetime of ISFET's is mainly dependent on the integrity of the encapsulation of electrical connections, and packaging automatization is difficult. In order to solve the packaging problem, back side contacted (BSC) ISFET's seem to be more appropriate. Moreover, BSC-ISFET's are also very promising for the automatization of the deposition of ion sensitive membranes to obtain different types of ChemFET's.

In order to make the encapsulation of this kind of devices easier and more reliable, different technologies for the fabrication of back side contacted ISFET's have been proposed [6-8]. However, there are still some problems not fully solved in these technologies which make difficult their compatibility with standard CMOS technologies.

This work focuses the attention on the development of a technology, simple and fully IC-compatible for a high yield, volume production and low cost fabrication of back side contacted ISFET's. The design has been conceived in order to make use of only standard processes (photolithography, metallization, ...). Except for the wet anisotropic etching of silicon, all the fabrication processes are taken from the standard CMOS technology at the CNM clean room facilities. The excellent electrical characteristics of the BSC-ISFET obtained as well as the high reliability and yield achieved for the whole BSC-ISFET fabrication process, makes a future inclusion of some electronics into a "smart sensor" very promising.

2. ANISOTROPIC ETCHING.

In the fabrication of BSC-ISFETs, a silicon anisotropic wet etching process is commonly used in order to define a silicon membrane thinner than 25-30 μm, where the electrical connection between the top and bottom drain and source ISFET diffusions can be achieved. The most commonly used wet anisotropic etchants are KOH-based aqueous solutions, but they have important IC-compatibility problems due to possible mobile ion (K^+) contamination.

In recent years a special effort has been made to find a new anisotropic silicon etchant that fulfils CMOS compatibility requirements and easy handling. NH_4OH-based solutions have been proposed [9] as ion-free IC-compatible anisotropic etchants, but it is difficult to obtain high quality and hillock free surfaces. The most promising results have been reported by using TMAH-based (tetramethyl ammonium hydroxide) solutions [10-13]. TMAH is fully IC-compatible, nontoxic and has very good anisotropic etching characteristics.

The silicon anisotropic etching in TMAH-based solutions has been investigated in a previous work [12] in order to select the best experimental conditions: TMAH wt.% and IPA vol.% solution concentrations and etching temperature. (100)-silicon etch rates were found to range between 10-45 μm/h depending on the exact experimental conditions. The etch rates of the most commonly used masking layers in IC technologies have also been determined: 10-100 Å/h for thermally grown SiO_2; less than 20 Å/h for Si_3N_4 and up to 350 Å/h for LTO (low temperature deposited silicon oxide). The quality of etched surfaces is mainly dependent on TMAH wt.% concentration. The addition of IPA (isopropyl alcohol or 2-propanol) to TMAH solutions reduces the undercutting ratio by a factor of more than 2 and leads to smoother surfaces of the etched sidewalls. The etching selectivity with respect to high boron doped silicon (p^{++} etch stop characteristics) is improved by the addition of IPA.

In order to corroborate the CMOS compatibility of the commercially available TMAH aqueous solutions that we were using, a series of MOS capacitors were fabricated onto silicon wafers previously etched in TMAH-based solutions. Electrical results have confirmed that the flat band voltage related to these MOS structures was not affected by the previous silicon etch and therefore, the claimed ion-free etching characteristics of TMAH solutions were verified.

3. BSC-ISFET FABRICATION

One of the most important technological problems related to the fabrication of BSC-ISFET devices is the step coverage with resist when photolithographic processes have to be done onto deeply microstructured wafer surfaces, since the depth of anisotropic etched holes is greater than 200 μm. Step coverage problems can affect drastically the yield of the fabrication process.

The conception and design of the BSC-ISFETs allows us to avoid the step coverage problems by placing the ISFET gate area at same wafer side where the silicon anisotropic etched holes are made (see figs. 1a-1e). The channel dimensions and the geometry of the drain and source ISFET areas can be fixed before the anisotropic etch step by using a double dielectric layer, consisting of a silicon nitride film over a silicon oxide film. Both materials, SiO_2 and Si_3N_4, can be used as masking layers when etching silicon in TMAH solutions [12]. After the anisotropic etching all the remaining photolithographic steps are made on the wafer side without holes allowing the use of standard processes.

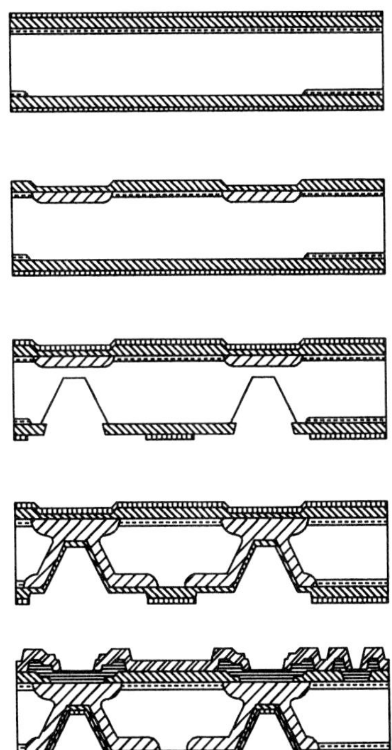

Fig. 1 Fabrication process of back-side contacts ISFETs.

The starting material for the fabrication of the back side contacts ISFETs were double-side polished, 300 μm thick, <100>-oriented, boron doped (p-type) silicon wafers with nominal resistivity of 12-17 Ω·cm (which corresponds to a doping level of $9x10^{14}$ cm^{-3}). This low doping level of the silicon substrate was selected in order to obtain depletion transistors. The front to back wafer alignment steps were carried out with a K&S infrared aligner. This technique allows a precision in the double side alignment better than 10 μm. The wafer is thick enough to assure that the substrate rigidity is not drastically affected.

The fabrication process sequence is schematically shown in figures 1a-1e. The process starts with a short silicon dry etch used to define the marks (2000 Å deep) needed in all the subsequent double side wafer alignment processes. Then, a stop channel boron implant is made to avoid the activation of parasitic transistors. After that, a layer of 9000 Å thick field silicon oxide layer is thermally grown and a 1000 Å layer of LPCVD Si_3N_4 is deposited on both sides of the wafer (fig. 1a).

The next steps are the opening and the doping of the drain and source ISFET areas which are located at the top wafer surface. The n-type (phosphorus) doping step is carried out at 1050 °C in a $POCl_3$ ambient (fig. 1b). The doping process has been optimized in order to achieve a final diffusion junction depth greater than 15 μm from both sides, top and bottom, of silicon wafers. After that, another 1000 Å LPCVD Si_3N_4 layer is deposited and the openings for the silicon anisotropic etch are defined at the bottom wafer surface. The etching time is carefully controlled to obtain a membrane thickness of 20-25 μm (fig. 1c). The optimized etching parameters in TMAH solutions allow to assure reproducibility and a very good uniformity in the hole wafer.

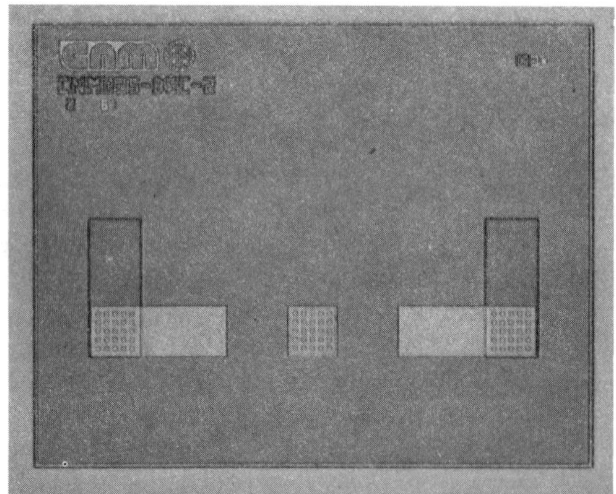

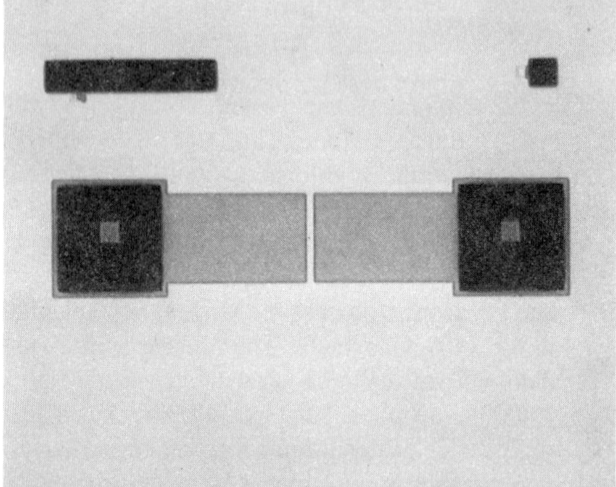

Fig. 2 Photographs of the top and bottom surfaces of the BSC-ISFET devices.

In figure 3 an optical photograph of a vertical cross-section of an anisotropically etched hole for the electrical contacts of drain and source is shown. The n-type (phosphorus) diffusions have been selectively etched to show clearly the junction depths.

After the silicon anisotropic etch, the bottom drain and source phosphorus doping is performed and followed by a long thermal drive-in process to achieve the final diffusion junction depths (fig. 1d). Then after the removal of the silicon oxide and silicon nitride bottom layers previously used as masking films during the anisotropic etch process, the ISFET gate structure is formed. It consists of a 1800 Å thick LPCVD Si_3N_4 layer deposited over a 780 Å thick thermally grown SiO_2 layer.

The BSC-ISFET fabrication process concludes with the contact window opening, the aluminum metallization and patterning, the deposition of a silicon oxide passivation layer and the opening for the external electrical contacts (fig. 1e). All these photolithographical steps can be made without any patterning problem because of the planarity of the top wafer surface. Optical photographs of the BSC-ISFET top and bottom surfaces can be observed in figure 2.

4. DEVICE CHARACTERIZATION

The electrical characteristics of the BSC-ISFET are very good and comparable to those obtained for our standard front-side ISFET devices [4]. A typical drain current versus drain voltage, Id-Vd, characteristics for the BSC-ISFETs are shown in figure 4. They have been

Figure 3 Optical photograph of a vertical cross-section of an anisotropically etched hole for diffusions electrical contacts.

measured at T=25°C for a solution of pH=7. The typical Id-Vg characteristics for different drain voltages and temperatures in the same solution are shown in figure 5. It can be seen that the threshold voltage is negative corresponding to depletion devices. We can also observe, that in the case where these sensors have to work in environments that are not thermally controlled, the thermal sensitivity can be reduced by a proper selection of the ISFET working point: the athermic drain current Id_{at} depends on the drain voltage, Vd. The chemical sensitivity of these devices to the pH value of a solution is determined by the Si_3N_4-gate membrane and a value of 55 mV/pH was measured, as for the standard front side ISFET devices [4].

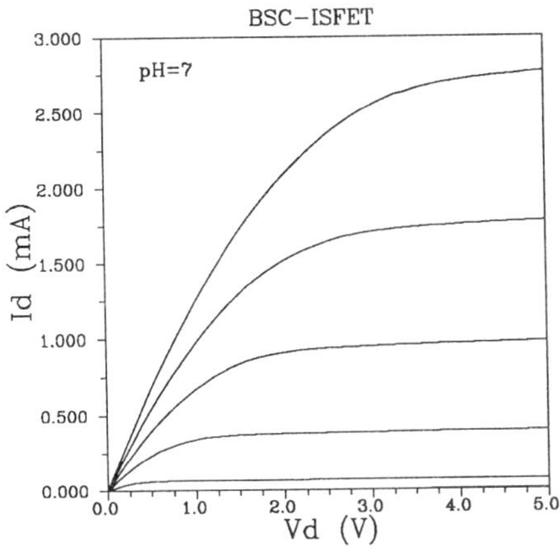

Fig. 4 Typical Id-Vd characteristics of BSC-ISFETs in pH=7, for different gate voltages: Vg=-2, -1, 0, 1, 2, 3 V.

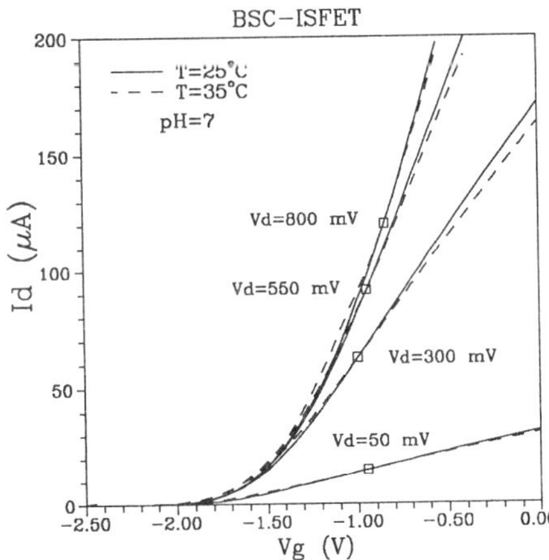

Fig. 5 Typical Id-Vg characteristics of BSC-ISFETs in pH=7, for T=25,30°C and different drain voltages: Vd= 50, 300, 550, 800 mV.

5. CONCLUSIONS

A fully IC-compatible technology suitable for low cost and high yield production, has been developed for the fabrication of back side contacted ISFETs. The most important problems usually found in the fabrication of this kind of devices has been circumvented. IC-compatible

TMAH solutions have been extensively characterized for their application in the silicon anisotropic etching processes. The presented design based on back side contacts avoids the yield problems related to the non-standard photolithographic processes on deeply microstructured wafer surfaces. The fabrication technology and the electrical and chemical characteristics of these devices make them very suitable for industrial purposes.

6. ACKNOWLEDGMENTS

The authors would like to acknowledge Dr. Francesca Campabadal for the characterization and analysis of MOS capacitors and for various interesting discussions on TMAH compatibility. Authors would also like to acknowledge MBT Tecnología Ambiental for the financial support of this project.

7. REFERENCES

[1] A.Sibbald, P.D.Whalley and A.K.Covington, A Miniature Flow-Through Cell with a Four-Function CHEMFET Integrated Circuit for Simultaneous Measurements of Potassium, Hydrogen, Calcium and Sodium Ions, *Anal. Chim. Acta*, **159** (1984) 47-62.

[2] U.Oesch, S.Caras and J.Janata, Field Effect Transistor Sensitive to Sodium and Ammonium Ions, *Anal. Chem.*, **53** (1981) 1983-1986.

[3] S.Alegret, J.Alonso, J.Bartrolí, J.Domènech, N.Jaffrezic-Renault and Y.Duvault-Herrera, Flow-Through pH-ISFET Detector for Flow-Injection Analysis, *Anal. Chim. Acta*, **222** (1989) 373-377.

[4] A.Merlos, I.Gràcia, C.Cané, J.Esteve, J.Bartrolí and C.Jiménez-Jorquera, CMOS flow-through pH-ISFET, *5th Conference on Sensors and Their Applications, Edinburgh, UK*, September 1991.

[5] A.Grisel, C.Francis, E.Verney and G.Mondin, Packaging Technologies for Integrated Electrochemical Sensors, *Sensors and Actuators*, **17** (1989) 285-295.

[6] H.H.van den Vlekkert, B.Kloeck, D.Prongué, J.Berthoud, B.Hu and N.F.de Rooij, *Sensors and Actuators*, **14** (1988) 165-176.

[7] D.Ewald, A.van den Berg and A.Grisel, Technology for Backside Contacted pH-sensitive ISFETs Embedded in a p-Well Structure, *Sensors and Actuators*, **B1** (1990) 335-340.

[8] T.Sakai, I.Amemiya, S.Uno and M.Katsura, A Backside Contact ISFET with a Silicon-Insulator-Silicon Structure, *Sensors and Actuators*, **B1** (1990) 341-344.

[9] U. Schnakenberg, W. Benecke and B. Löchel, NH4OH-Based Etchants for Silicon Micromaching, *Sensors and Actuators*, **A21-A23** (1990) 1031-1035.

[10] O.Tabata, R.Asahi, H.Funabashi and S.Sugiyama, Anisotropic Etching of Silicon in $(CH_3)_4OH$ Solutions, *Transducers'91, San Francisco, CA, USA, June, 1991*, pp. 811-814.

[11] U. Schnakenberg, W. Benecke and P. Lange, TMAHW Etchants for Silicon Micromachining, *Transducers'91, San Francisco, CA, USA, June, 1991*, pp. 815-818.

[12] A.Merlos, M.C.Acero, M.H.Bao, J.Bausells and J.Esteve, A Study of the Undercutting Characteristics in the TMAH/IPA system, *J. Micromech. Microeng.*, **2** (1992) 181-183.

[13] A.Merlos, M.C.Acero, M.H.Bao, J.Bausells and J.Esteve, TMAH/IPA Anisotropic Etching Characteristics, *EUROSENSORS VI, San Sebastian, Spain*, October 1992. (to be published in *Sensors and Actuators*).

A multi-function sensor interface circuit (MFSIC) for use in chemical or biochemical sensors

G. Williams* and C. D'Silva

University of Sunderland, School of Health Sciences, Centre for Studies in Chemistry, Galen Building, Sunderland, Tyne & Wear SR1 3SD.

*University of Wales, Bangor, Institute of Molecular & Biomolecular Electronics, Dean Street, Bangor, Gwynedd LL57 1UT.

ABSTRACT: The commercial development of portable chemical or biochemical sensors have been hampered by the lack of availability of a suitable low power sensor interface integrated circuit. We have recently investigated this problem with the view to identifying the minimum circuit requirements needed for the development of such a device and report here our unique solution. The integrated circuit developed has been designed to be modular in character, so that the final device can be configured for a wide variety of applications, in chemical or biochemical sensing.

1. INTRODUCTION

A significant amount of research effort over the past decade has concentrated on developing sensors for environmental monitoring applications, health care, veterinary medicine, food, pharmaceutical and the agrochemical industries. These devices usually utilise a wide range of sensing principles but can share common instrumentation features which if identified, could form the basis for the development of a Multi Function Sensor Interface Circuit (MFSIC). The need for such an integrated circuit (IC) becomes apparent when we explore the advantages afforded by such a device of low power consumption, reduced manufacturing costs and minimal circuit size. Low power consumption and minimal size are an important consideration when developing portable instrumentation for on site testing, especially in the fields of environmental monitoring or health care. The major requirement of portable sensors is that they be lightweight, user friendly, require minimum personnel training and be reliable enough to perform analysis under a wide variety of hostile environmental conditions. Although there has been a tendency to develop instrumentation around microprocessor based systems this approach can be expensive requiring large amounts of software development and may not have the flexibility of application demanded of a wide range of sensing principles or provide the user friendly characteristics which could be designed into an application specific integrated circuit (ASIC). The design of an IC requires the circuit designer to first identify the specifications necessary for the device, then develop and test the necessary instrumentation circuitry using discrete components before laying out the IC, testing of the design and manufacture at a silicon foundry. There are many classes of sensors which could form the basis for the development of an IC chip but of these different types, electrochemical sensors have proved successful in a wide variety of applications including health care, environmental monitoring and the food industry. Therefore it was decided that the development of a MFSIC should be targeted initially to satisfy the needs of sensors in this area.

2. INSTRUMENTATION

Electrochemical sensor instrumentation in principle requires five main functions, an input voltage source, potentiosat, signal conditioning, storage and readout circuitry. An example of the type of instrumentation that has been used in a portable electrochemical environmental monitoring sensor (1), designed to detect and display the concentrations of four heavy metal ions, is shown in Fig. 1. The major differences in circuitry between this device and that required for an electrochemical enzyme sensor is the replacement of the waveform generator and timer by an adjustable voltage source and the addition of an integrator and timer between the I/V converter and the auto ranging logic. Figure 2 shows the schematic of such a sensor configured for the measurement of four enzyme electrodes. Additional circuit changes required for such a sensor would be the addition of a calibration block to correct for the non-linearity of enzyme characteristics which generally obey a hyperbolic relationship which in its

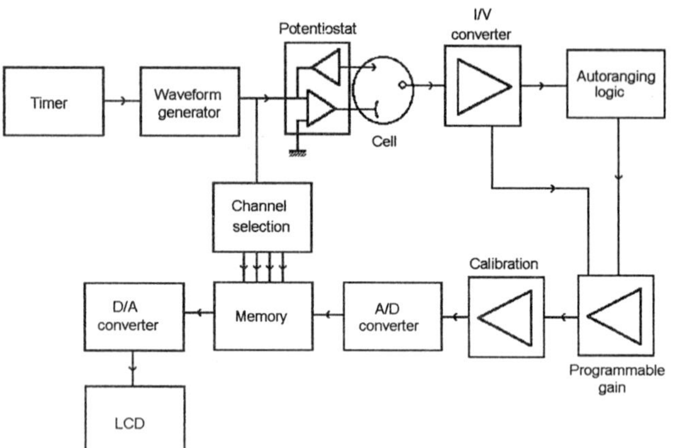

Fig. 1. Schematic diagram of electroanalytical instrumentation developed for environmental monitoring.

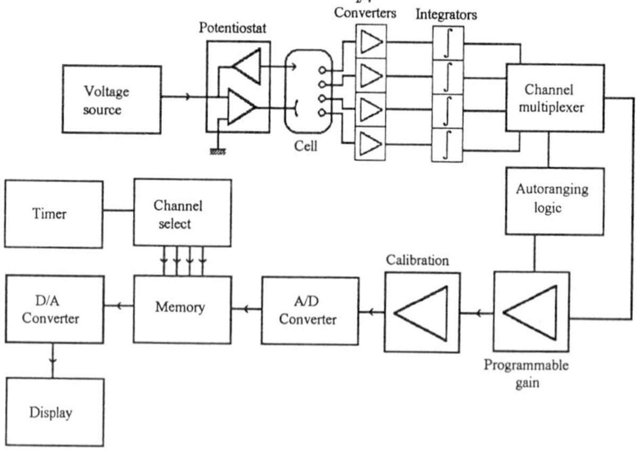

Fig. 2. Schematic diagram of instrumentation required for a four function enzyme sensor.

simplest form can be described by the Michaelis-Menton equation [1] where v is the rate of product formation, [S] the substrate concentration, V_{max} the maximum rate of product formation dependant on enzyme concentration and K_m the Michaelis constant.

$$v = \frac{V_{max}[S]}{K_m + [S]} \qquad [1]$$

For environmental monitoring applications compensation circuitry would also have to be added to the calibration block to correct for the marked effect of temperature on the performance and accuracy of the enzyme sensor.

3. INTEGRATED CIRCUIT DESIGN

Analogue-digital circuit design is traditionally a combination of standard IC components connected on a printed circuit board. However this route is wasteful in space and power consumption due to the large number of IC's required to produce complex functions. Integrated ASIC development affords an efficient alternative, making possible the use of high level analog building blocks (such as op-amps and related circuitry), digital gates (and more complex digital functional operations) and A/D and D/A converters all on the same custom chip. The ASIC design route is inherently a process of integrating several systems onto a chip. Both types of electrochemical sensors described share several common instrumentation features, namely a requirement for a potentiostat, memory module, channels select and auto ranging logic. The potentiostat was not included onto the ASIC because the quality of commercially available low power quad operational amplifiers were far superior to those on the ASIC, which were slightly better in performance than a 741 device. The final instrumentation modules chosen to be incorporated onto the ASIC are shown in Fig.3. Each module was fabricated to work independently of other modules or connected as a MFSIC. In this way it was reasoned the performance characteristics of one failed module would not jeopardise the performance of the remaining modules. A total of five main modules were finally

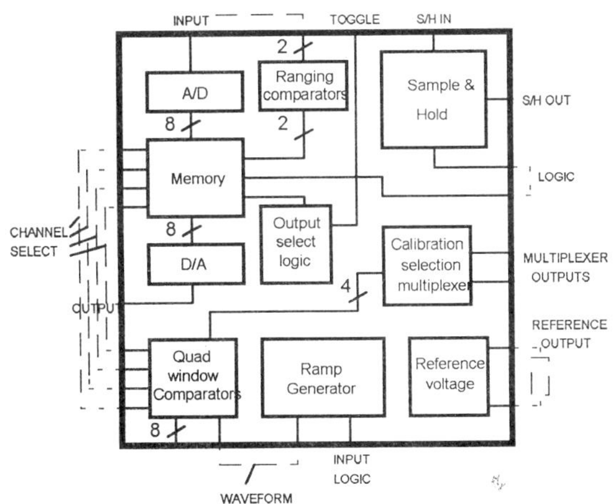

Fig 3. Schematic diagram of instrumentation modules incorporated onto the integrated circuit.

identified and incorporated onto the device, these were a data acquisition module, quad window comparator, ramp generator, scalable voltage reference and sample and hold.

3.1 Data acquisition module

Data acquisition (DA) in a non-microprocessor based systems usually requires several discrete IC's. Available DA IC's cater exclusively for microprocessor applications, where several analogue inputs are multiplexed and converted to a digital representation for storage on a chip and subsequent presentation to a data bus. There are no IC's on the market capable of storing several analogue values in digital memory and then converting the digital data back to analogue. The MFSIC DA module is unique, in that it implements 8-bit analogue to digital conversion, storage of four data values and subsequent digital to analogue conversion (see Fig. 4). This is achieved by close interaction between analogue and digital functions through the exchange of control and data signals. Matched A/D and D/A converters are used to ensure accuracy. Other useful control and information functions have also been included within the DA module. A simple state machine enables sequential output of the four memory location values by clocking one input pin. Two additional memory bits, connected to two comparators, are used to record ranging information, for use in auto-ranging systems.

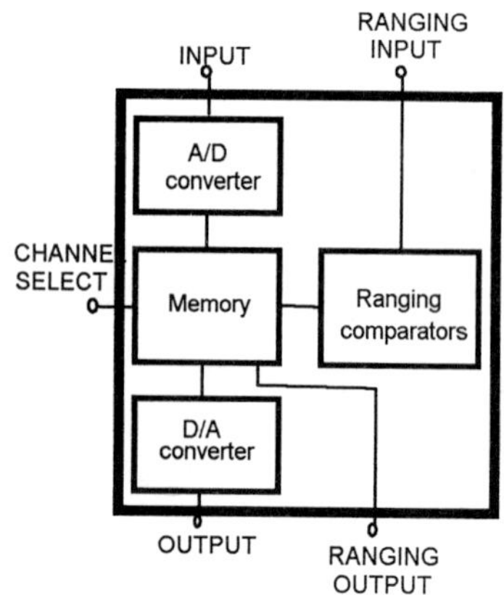

Fig. 4. Data acquisition module

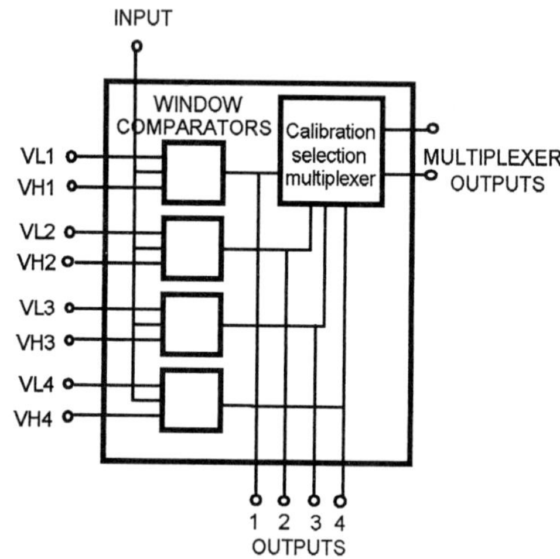

Fig. 5. Window comparators

3.2 Window comparators

Four window comparators have been implemented with a common input pin, and provide output logic high when the input is within the comparator range (see Fig. 5). The main purpose of this module is to provide channel select inputs for the DA module as well as driving multiplexer outputs for calibration curve selection. The two multiplexer outputs provided are for use with the LM604CN four channel op-amp to provide selectable calibration curves.

3.3 Ramp generator

A ramp generator is of general use in electrochemical instrumentation. It is also of use in circuit timing since a time dependant voltage is produced which can be combined with the quad window comparator and DA modules for channel selection. The ramp generator module has adjustable sweep rate and time delay before ramp commencement, set by clock speed and monostable delay respectively (see Fig. 6).

3.4 Other functions

Sample and hold is provided for use with the DA module for rapidly changing input values. The DA module provides a logic output, synchronised with the acquisition cycle of the analogue to digital converter. A scalable reference voltage, stable to within 0.5mV, is provided for use with the DA module to select the resolution of values held in memory.

4. ASIC EVALUATION

During development all MFSIC modules were thoroughly simulated using PSpice (analogue) and BX Software (digital), software, but it was still thought prudent to implement each module totally separately to allow for any unforseen errors. The final MFSIC design produced was on a 3 micron ISO-CMOS mixed analogue/digital gate array and packaged for use in a 68

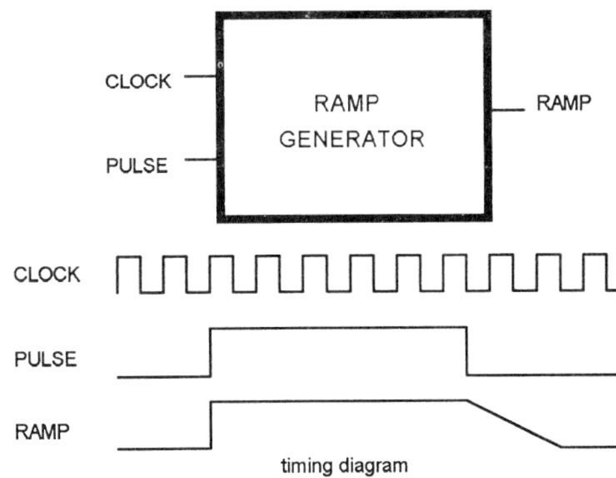

Fig 6. Ramp generator

Fig. 7. Photograph of a manufactured device.

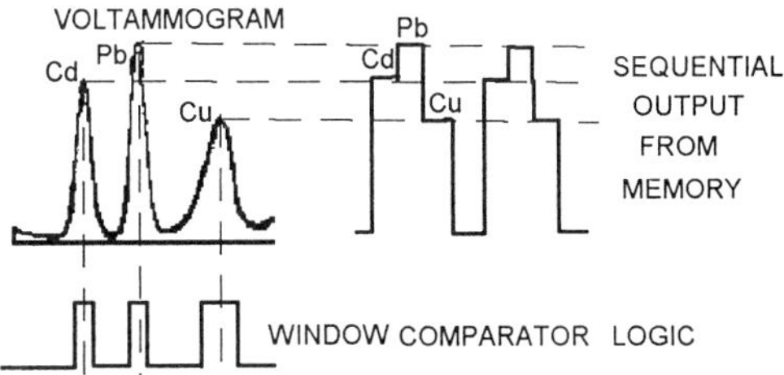

Fig. 8. Oscilloscope trace of voltammogram obtained with 100ppb Cd^{2+}, 100ppb Pb^{2+} and 200ppb Cu^{2+}

pin cerquad-J package as shown in Fig. 7. The manufactured device had a mean supply current of 10 mA and required a differential power supply of 5V. Testing of the MFSIC was effected by its application in the circuitry shown in Fig. 1. for the detection of heavy metals by stripping voltammetry (1). The oscilloscope trace of a voltammogram (I-V curve) obtained with 100ppb Cd^{2+}, 100ppb Pb^{2+} and 200ppb Cu^{2+} is shown in Fig. 8. The window comparators on this occasion were configured to allow for the detection of peaks within set limits and used to latch the maximum values into memory. Sequential outputs are shown on the right.

5. CONCLUSIONS

The MFSIC is a self contained data acquisition system which has all the logic necessary for inputting and sequentially displaying acquired values. The device is useful for one shot data logging applications and the use of a 68 pin cerquad-J package and 3 micron ISO-CMOS mixed analogue/digital gate array allows for portable applications where power consumption and size are at a premium. The device has been successfully tested by its application in circuitry developed for use in an environmental monitoring unit (EMU) and further work will aim to combine this circuitry with an integrated disposable thin film multi-electrode sensing element (2) designed for chemical or biochemical sensing applications to produce a completely portable device for field use. It is foreseen that the MFSIC developed here will be used soon in a multi-function amperometric enzyme sensor to be developed for applications in the food, health and sport fields. Other uses being considered for the device include its applications to ISFET's, conductimetric and piezo-electric sensors (3) for environmental, industrial or medical uses.

The successful development of this first MFSIC has validated the circuitry and strategy used in the development of this device. Further device work will aim to reduce pin count, by inter-connecting modules internally, so increasing functionality, since a lack of pins limited the overall number of functions that could be implemented, even within a 68-pin package.

6. ACKNOWLEDGEMENTS

The authors thanks the SERC for support under Grant GR/F 16424 & GR/H 26130. Gwyn Williams acknowledges the SERC for the award of a studentship and CD'S thanks GEC / Fellowship of Engineering for the award of a Senior Fellowship during part of this work.

7. REFERENCES

1. G. Williams and C. D'Silva. (1993)., Analyst. (submitted).

2. G. Williams and C. D'Silva., Fourth European Conference on Organised Thin Films, Bangor (UK). 1992, Sept 10th-12th. OC4.

3. C. Barnes., C. D'Silva., J. P. Jones., T. J. Lewis., (1991) ., Sensors and Actuators B, 3, 295-304.

Processing of silicon ICs using alpha irradiation—meeting a challenge of 'smart' sensing?

Sergey P Kostenko

Albedo 0.39, Research-and-Production Centre, 12 10th April Street, Apt 8
270020 Odessa, Ukraine

ABSTRACT The report presents a general approach to a sensor chip treatment which is based on the routine production process of the silicon IC. As a finishing treatment, alpha-irradiation from plutonium sources is introduced.

1. INTRODUCTION

The lag of smart sensor technology behind the technology of automated systems and actuators [1] results from the conflicting requirements of the multisensor chip. Pressure measurements, for instance, need a thermostabilized crystal or, at least onc showing minimum temperature fluctuations. However, to measure flow rates in the same spatial position, the temperature of the chip must exceed that of the medium to be measured.

Determination of gas component concentrations in the same environment necessitates the direct contact of the medium to be measured with the planar surface of a semiconductor IC. However, reliability and time stability standards specify the isolation of a "smart" sensor chip surface from the medium to be controlled. There is a problem with the use of such technology in sensing elements, yet retaining the above stability factors. The question remains how such a sensor chip can be used in what may be harsh conditions, without violating the other requirements for its production.

2. CHIP DESIGN

A general approach to a multisensor chip has been developed in reference [2], which involves the use of microelectronic IC technology. However, the reverse side of a silicon substrate undergoes anisotropic etching to produce a system of hollows, shaped like reversed polyhedra, producing pyramids, prisms, etc.

Fig. 1 represents one such system. The IC planar surface is protected by a hermetically sealed body with openings in the bottom.

The pyramid top is separated from the IC active zones by a distance of 4-6 μm. There may be several such hollows in a chip, whether discrete or adjoining, forming a complex profile. Such chips can be used to measure simultaneously such parameters as pressure, flow rates in two coordinates, temperature, and magnetic and optical components. The latter is possible when the hollow tips approach the chip surface. A layer preventing contact is formed by silicon nitride.

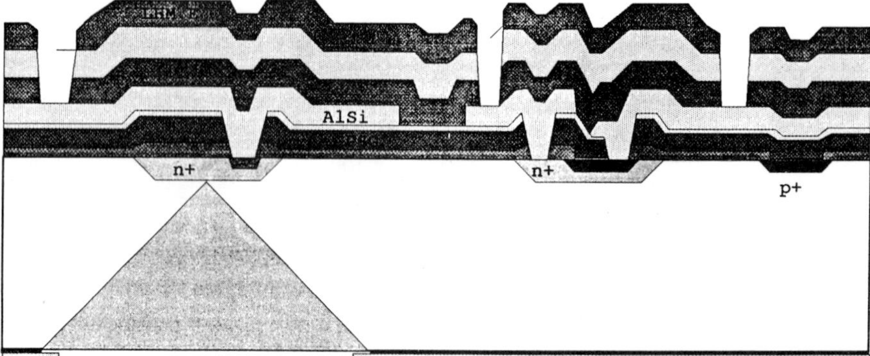

Fig. 1. Cross-section of one of the structures

The question remains how these chips can incorporate the potential for chemical measurements. The clue is to synthesize surfaces with a pre-determined ad-behaviour in the tops of the prisms and pyramids formed on the chips. Within the routine silicon IC technology, a finishing treatment by alpha-irradiation from ^{238} Pu sources is introduced to enable the production of a uni-chip "smart" sensor.

3. EQUIPMENT

Fig. 2 presents a fragment of the alpha-irradiation unit placed inside a bell jar. Four sources provide an exposure power of 8.10^{10} A/kg and 12 sources 4.10^{10} A/kg. Activity of the nuclides in the radioactive source is 15.2 and 7.44 Gbq, respectively. The sources are designed so as to exclude fission product emission, except for alpha-particles.

Fig. 2. The irradiation arrangement. When not in use, Pu sources are screened by shutters

This process was followed by 5keV electron-irradiation of the samples with *in situ* repeated Auger-analysis throughout the electron irradiation.

3. DIRECTED SYNTHESIS IN LOCAL VOLUMES OF THE SEMICONDUCTING CRYSTALS

The equilibrium of both mono- and multicomponent systems is governed by interactions between atoms, or atoms and defects (like vacancies, or interstitial atoms). The interaction modes of these interaction types may be different, - deformational electron-electron, dipole-dipole, spin-spin, or magnetic. It is only natural that under certain thermodynamic conditions one of the modes may appear dominant, for example the deformational one - due to the varying sizes of the atoms which constitute the system. Further interaction results in overlapping of electronic shells and turns out to be secondary electron-electronic in nature. Other variants may represent other interaction modes.

Thus, choosing which interaction mode is dominant, we obtain the major criterion for the specifying of synthesis conditions in the localized volumes of semiconducting crystals.

The method introduced here relates to the dominant direct electron-electron interaction mode. It is achieved by implanting a charge density into the surface layers which is achieved by means of alpha-particle irradiation from ^{238}Pu sources. The results exhibit such effects as the development or disappearance of surface clusters with preferential absorption, or modification of the surface layer composition, thus displaying the required properties [3], such as changes in crystallography of the surface layer [4], etc.

4. EXPERIMENTAL

Previous results on synthesis of surface structures with an enhanced response have been published elsewhere [2,5-7]. In ref. [6] we have presented our findings on chlorine molecule-specified absorption on a $CdGa_2Se_4$ single-crystal surface. It represented an attempt to deposit this material on silicon by cathode sputtering. The use of alpha-irradiation was shown to generate a prototype-like structure with preferential ad-behaviour.

4.1 SiC surface: the creation of C_x and S_i clusters

In this paper,the concept that small aggregates (carbon (C_x) or silicon (Si_y clusters, in particular) can provide elementary components with preferential sorption, for smart sensor electronics, is discussed. Under certain conditions (alpha-irradiation from ^{238}Pu sources) the SiC surface can produce these microclusters with a specified level of sorption. Micrographs (Figs 3-6) illustrate the effect of the processes.

The above property is related to the specific electronic structure of clusters, resulting both from their small sizes and geometry, and the influence of the underlying matrix atoms. Moreover, the fact that the 2s and 2p orbitals of carbon have lower energies than the 3s and 3p orbitals of the silicon [8] supports the view that some charge transfer is occurring from the carbon orbitals, with the resulting redistribution of electronic density and a further polarization of the material microfragments. This allows atomic forces to "feed" the process

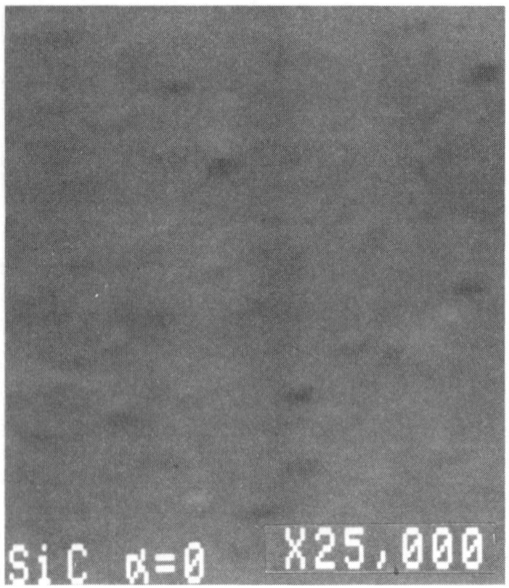

Fig. 3. The initial sample of SiC

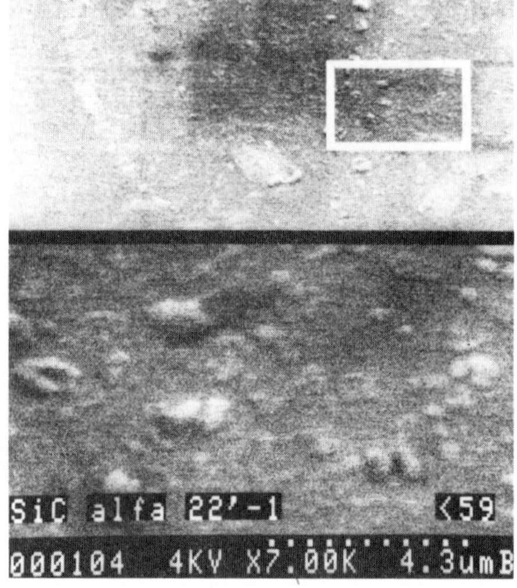

Fig. 4. The SiC sample after 22 min alpha-irradiation

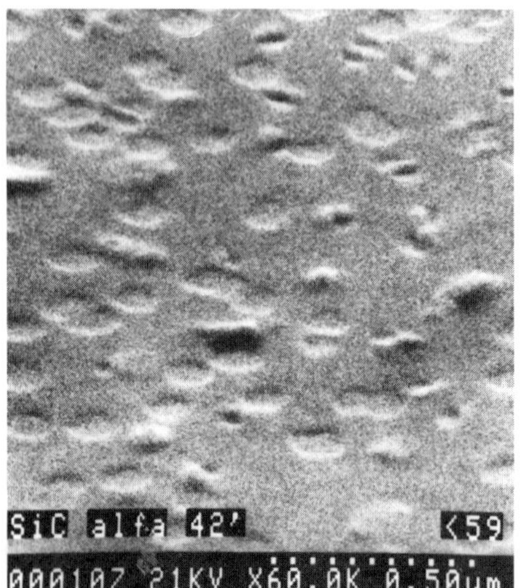

Fig. 5. The SiC sample after 42 min α - irradiation

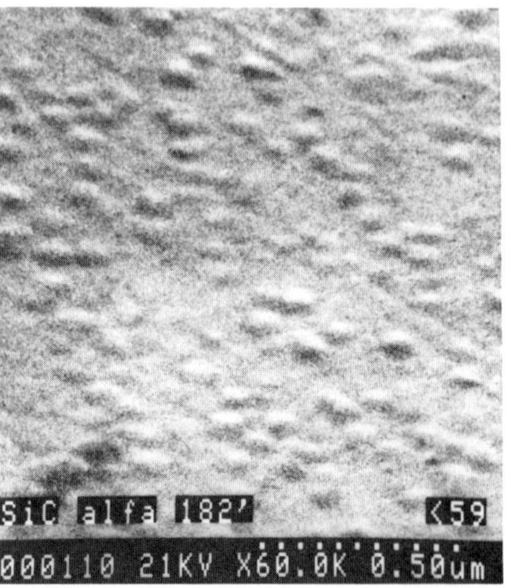

Fig. 6. The SiC sample after 182 min α - irradiation

of the system self-formation in the desired direction. Some aspects of such self-organization in semi-conductors are discussed in the work by Scholl [9] and in ref [10], for example.

In parallel with the above, migration processes of these clusters are observed to bring about restructuring of the affected surface layer of the matrix structure. The silicon atom quantity increases and that makes the probability of interaction between silicon atoms and alpha-particles higher. These phenomena were of a cyclical nature: under certain doses of alpha-irradiation Si-clusters appeared, an increase of the dose eliminated them. At the same time SiC surface sublayers give rise to C-clusters. By terminating the alpha-irradiation, a fixed surface pattern is obtained. Structures, dimensions and, consequently, the resulting properties of the clusters are related directly to alpha-irradiation doses.

5. CONCLUSIONS

The report presents a general approach to the production of a sensor chip which is based on the routine production process of the silicon IC. As a finishing treatment, alpha-irradiation from ^{238}Pu sources is introduced.

REFERENCES

1. J H Huijsing. Sensors and Actuators A, 30 (1992) 167-174
2. S P Kostenko et al. Sensors and Actuators B, 8 (1992) 13-19
3. D I Bidnyk and S P Kostenko. J Mater Chem, 1992 2(1) 19-21
4. S P Kostenko et al, -in the Book of Programme and Abstracts of 11th General Conf of the Condenced Matter Division, 8-11 April 1991, Eds by Profs K Bethge, G Thomas, J L Beeby, Exeter, England, 276
5. S P Kostenko and G D Tischchenko, -in the Book of Abstracts of 6th Intern Conf iib92, Thessaloniki, Greece, June 21-26, 1992, Eds A Rocher and Ph Komnimou, 1992, 154
6. S P Kostenko et al, in the Book: Sensors. Technology. Systems and Applications, Ed: K T V Grattan, Adam Hilger, Bristol, Philadelphia and New York, 1991, 57-62
7. S P Kostenko et al, in the Book: Sensors. Technology, Systems and Applications, Ed: K T V Grattan, Adam Hilger, Bristol, Philadelphia and New York, 1991, 37-41
8. M A Van Hove et al, in the Book: The Structure of Surface III, Proc 3rd Intern Conf ICSOS III, Milwaukee, Wisconsin, July 9-12, 1990, Usa, Eds S Y Tong, M A Van Hove, X D Xie, K Takaynagi, Springer-Verlag, Berlin, Heidelberg, New York, 1991, 550-554
9. E Schnoll. Nonequilibrum Phase Transformations in Semi-conductors. Self-Organization Induced by Generation and Recombination Processes, Springer-Verlag, Berlin, Heidelberg, New York, 1987, 459
10. S Janusonis and V Janusoniene. Self-Formation in Semiconductor Technology, Mokslas, Vilnius, 1993, 192 (in Russian)

Section C

ACOUSTIC AND ULTRASONIC SENSORS

A new approach to ultrasonic imaging using a novel multi-element integrated transmit/receive transducer

J V Hatfield, P A Payne[†], Q X Chen[†], A D Armitage, P J Hicks and N R Scales
Departments of Electrical Engineering & Electronics, and [†]Instrumentation & Analytical Science, UMIST, PO Box 88, Manchester M60 1QD

ABSTRACT: Much progress has been made towards integrating the transmit and receive circuitry associated with either linear or phased ultrasonic array scanning into the hand-held case of the transducer. The number of wires required to connect the transducer back to the display system is reduced dramatically and the path length between transducer elements and the pulser and receiver circuits may be kept to a few millimetres. The problems to be overcome in achieving this new structure are described under the following headings: transmit system design; receive system design; physical packaging and layout of the transducer.

1. INTRODUCTION

High resolution focused beams of ultrasonic energy are used for imaging purposes for example by the medical profession and in non-destructive testing. High resolution beam focusing is achieved by pulsing the elements of an ultrasonic array transducer in some predetermined manner. This may take the form of a phased array, which allows for an electronically steered beam (sector scanning), or a linear array which typically would allow groups of elements to be successively pulsed along the array, or even a combination of the two. Conventional multi-element ultrasound transducer arrays are connected by long multi-wire cables to the system's electronics and display unit. This can cause numerous problems including interference and reflections along the cables. The net effect of these is to make the system signal-to-noise ratio poorer. These effects are magnified if the system is designed to operate at higher than conventional ultrasound frequencies (eg, 20MHz) [1].

2. TRANSMIT SYSTEM DESIGN

Currently an Application Specific Integrated Circuit (ASIC), capable of providing sixteen programmable channels, each of 1nS time resolution and with a dynamic range of 2^{19} (easily extendable), has been developed in 1.5µm CMOS technology under the Eurochip initiative. The rôle of this chip is to provide delays of programmable length which determine the firing order of the high voltage pulse circuitry of the array elements. It is envisaged that this ASIC will form part of a generic chip set for operating high frequency arrays in any mode. The functionality of the chip could conceivably be achieved using programmable counters clocked at 1GHz until a terminal count is reached. This would require high speed, high power dissipation, technology, when the objective is to produce a low cost, compact, hand-held device and thus power consumption is of prime concern.

2.1 Circuit architecture

Instead of using a single high speed clock to meet the resolution requirements a novel adaptation of the time/memory cell approach of Arai *et al* [2] has been utilised. In the present scheme sixteen clocks, derived from a single source, but separated by one sixteenth of the clock period from each other are used. To achieve a 1nS time resolution each of the clocks has a period of 16nS (62.5 MHz) but are separated by 1nS from each other. In this way high resolution is obtained using standard CMOS technology. In order to set a pulse delay by this method the most significant bits of a 19-bit delay value are loaded into a 15-bit counter of which there is one per pulse channel. The appropriate counter clock is selected by the four least significant bits, by means of a 16-to-1 clock multiplexer, to give the fine resolution.

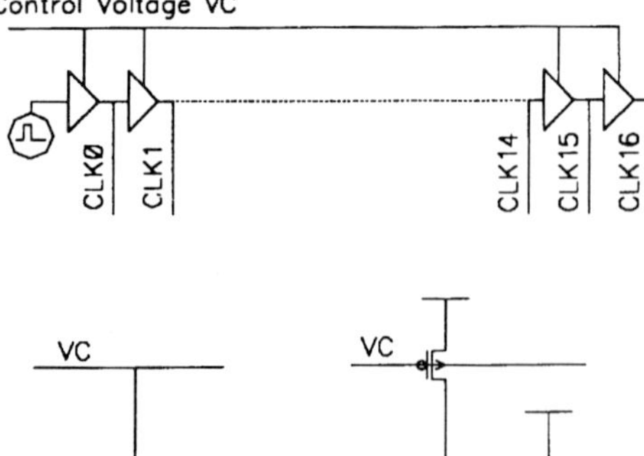

Fig. 1. Delay line schematic diagram and individual delay element

A string of delay elements is used to produce the 16 phased clocks (Fig. 1). The delay element is simply two inverters in series. The speed of the zero-to-one transition of the first inverter is determined by the extra pMOS pull-up transistor controlled by V_C. Thus the propagation delay of a falling edge from IN to OUT can be set by the voltage on V_C. An external reference clock supplies pulses to the delay line. As these progress down the delay line they are stretched due to the difference in propagation times for negative and positive edges. A form of Phase Locked Loop (PLL) circuitry has been developed to control the CMOS delay elements so that the negative edge of the zeroth clock lines up with the negative edge of the sixteenth clock. If these are the same then the delay for 16 elements is exactly the clock period and the delay for each element is exactly one sixteenth of the clock period [3].

2.2 Test methodology and results

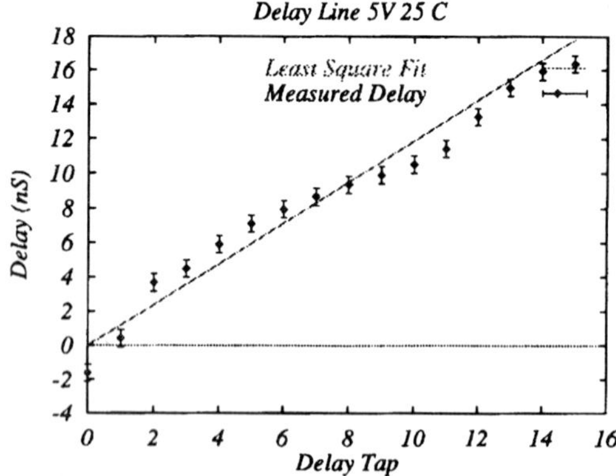

Fig. 2. Delay along the delay line as a function of delay element. Any one of the 16 "delay taps" can be selected as clock inputs to programmable down counters.

Delay line test results are shown in Figure 2. They were obtained using a Hewlett Packard 8180 data generator, Hewlett Packard 8182 data analyzer and a Stamford Research DG535, 5pS resolution pulse generator. This equipment was controlled, via an IEEE 488 bus, by computer programs written in the 'C' programming language. Briefly the output pulse from the DG135 was connected to the 'D' input of an on chip flip-flop. The DG135 was triggered by the delay line clock. The delay between the trigger and the input to the flip-flop could thus be varied until a change of flip-flop output was observed. This was done with each of the delay line taps connected to the flip-flop clock input by means of an on-chip multiplexer. The phase locking mechanism should, of course, make the IC self-compensating with respect to temperature and power supply variations, providing that the delay-line input clock remains stable. All tests indicate that over all conceivable ranges of interest this is indeed the case.

2.3 Miniaturisation of pulse generating circuitry

Work has also been successful in miniaturising the high voltage pulse generator components. ZETEX FMMT 415 surface mount avalanche transistors provide very fast high voltage pulses whilst occupying < 10% of the circuit area of conventional MOSFET pulsers. An array of these surface mount components is currently being mounted on a p.c.b measuring 12cm X 8cm which will also accommodate one of the 16-channel pulser ASICs. One p.c.b is mounted on either side of the 32-channel transducer of Figure 4.

3. RECEIVE SYSTEM DESIGN

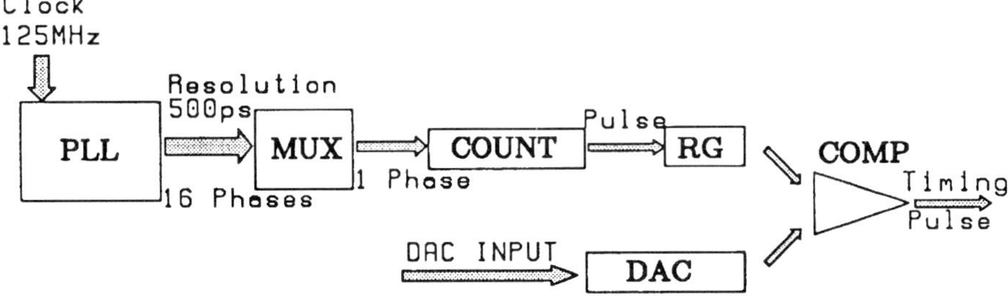

Fig. 3. Phased lock loop DAC and comparator system

The same array that transmits the ultrasonic energy will also act as a receiver of reflected ultrasound. To reconstruct the image analogue signals from the array elements are digitised using "flash" A to D converters and the temporal information recovered; again using programmable digital delay lines.

For imaging reconstruction purposes a step time as low as a few hundred picoseconds is desirable, for clear, high resolution, imaging. This makes a CMOS approach, with the technology currently available, more difficult for the receiver ASIC. For the receiver bipolar technology offers a better resolution potential. Using the same technique outlined above, however, the available bipolar technology only results in a 500ps resolution. In an attempt to increase the resolution still further fine resolution circuitry has been designed to be triggered by the 500pS pulses. This circuit consists of a Digital-to-Analogue converter (DAC), a ramp generator and a high speed comparator, (Figure 3). The ramp generator, when initiated by a 500pS pulse, produces a slope output which is one input to the comparator. The second comparator input is connected to the DAC. The comparator output toggles as the ramp output crosses a predetermined voltage set by the DAC. This has the effect of the delay circuitry providing the coarse time resolution and the DAC/ramp generator providing the fine time resolution within a single 500ps segment. Resolutions down to 30pS are possible using this technique. The main limiting factors are the stability of the PLL phase control and the DAC resolution. The DAC can also be used to calibrate the device and compensate for any constant timing errors.

4. PHYSICAL PACKAGING AND LAYOUT OF THE TRANSDUCER

It is extremely difficult and expensive to construct high frequency, high resolution ultrasonic arrays using conventional ceramic piezoelectric materials, and virtually impossible when the transducers are to operate above 10MHz centre frequency. Most of the problems encountered can be solved by employing piezoelectric polymer materials. This is due to the unique characteristics of these polymers, including: ready availability in thin film form, ranging in thickness from under 10 μm to several hundred micrometres; flexibility, mechanical robustness, and chemical stability; very high internal acoustic and dielectric losses, making them intrinsically wide band with very low inter-element interference or cross-talk, both acoustically and electrically.

4.1 Array construction methods

Two methods have been developed: the Direct Bonding method and the Rear Filling method, both being suitable for the arrays under study. The Direct Bonding method requires the electrode pattern to be applied to the surface of the backing stub. A non-metallised polymer film is then bonded to the stub and a metal layer is coated on top of the polymer film as the ground electrode. Relatively high temperatures can be used when preparing the electrode pattern since the piezoelectric polymer film is not involved at this stage. This can expedite the electrode pattern making process.

For the Rear Filling method the electrode pattern is applied to the upper surface of the piezoelectric polymer film with the ground electrode at the rear. Both methods enable backing materials having various acoustic impedances to be employed at the rear of the film to control the acoustic performance of the transducer. Fabrication of focused arrays is also possible with this second construction method since electrode patterns can be formed on a curved surface using the etch-back technique. The Rear Filling method has proved better for making high frequency transducers because there is no bonding layer between the polymer film and the backing and it gives better uniformity in performance between elements.

The electrode patterns are created using photolithographic techniques and were designed using a CAD package, AutoCAD, which produces files that can be used as input to commercial photolithographic equipment.

4.2 The integrated piezoelectric transducer array

The main concern is to interface the array with the on-board ASIC chips in a reliable and simple way. The present design is based on the Rear Filling method, and alumina substrates with screen printed gold patterns have been used to achieve the required interfacing with on board electronics. The main part of the array transducer consists of the array stub and the bonding pad extension, as shown in Figure 4. The array stub comprises an alumina substrate, the piezoelectric polymer film and the backing. For an initial feasibility study, a 32-element array was designed and fabricated. Each array element is 0.2mm wide and 7mm long with inter-element gap of 0.05mm. The alumina substrate has 32 bonding pads on two of the four sides, and two bonding pad extensions were used for easy connection to the electronics. The piezoelectric polymer film is glued to the top surface of the substrate and electrode patterns are produced using standard photolithographic techniques.

One of the critical areas of construction is connecting electrode patterns on the film to the electrode pads on the alumina substrate. This was done by using silver-loaded conductive paint. The ground electrode was formed by vacuum coating the other side of the piezoelectric film through the central opening of the alumina substrate, (Figure 4). Conductive epoxy or paint was used to attach an electrical lead to this ground electrode. A square tube was glued to the lower side of the alumina substrate and epoxy based backing materials having acoustic impedances close to the polymer were filled into the hollow thus formed.

Azimuthal focusing effects can be achieved by pressing the front face of the stub structure against a suitable cylindrical surface during the curing of the epoxy based backing materials, and different focal length can be obtained by changing the cylinder diameter. The two bonding pad extension

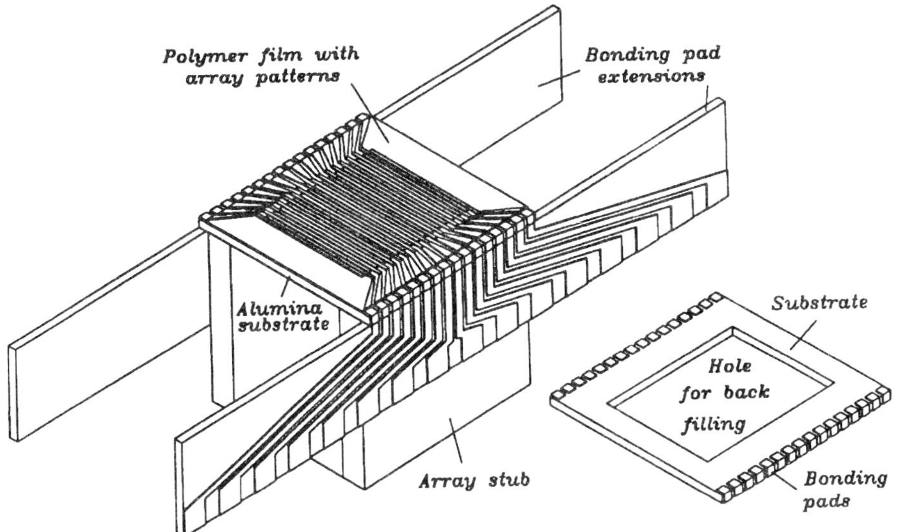

Fig. 4. Packaging and interface structure showing assembly prior to connecting the ASIC chip set and substrate with bonding pads prior to fixing polymer film

boards extend the bonding pads on the alumina substrate from 0.5 x 0.5mm to 1.5 x 1.5mm, making direct soldering to the pads possible. They also make it possible to evaluate the array element using an independent pulser and receiver, without relying upon the on board electronic circuitry. Since the main part of this 32-element array assembly is only about 12 x 12 x 10mm, there is plenty of room for the on board electronic circuitry and the whole IC array can be encapsulated in a hand held case.

5. PERFORMANCE MEASUREMENTS

Uniformity: The uniformity test was carried out in a water tank. A reflected echo from a plane steel target is obtained from each element, using the same pulser and receiver. Because of the high precision of the electrode pattern made with photolithographic techniques, the uniformity between the array elements is excellent. The responses from different elements are virtually identical.

Cross-talk: There are two forms of interference between elements in an array transducer: the electrical and the acoustic cross-talk. Electrical cross-talk arises from the dielectric coupling between elements and the acoustic cross-talk is caused by the propagation of acoustic waves along the surface of the array. For transducer arrays made from ceramics, cross-talk between elements is a major problem due to the high dielectric constant and low internal loss of these materials. For transducer arrays made from piezoelectric polymer materials, these problems are largely non-existent. The dielectric constant of PVDF is only about 1% of that of lead zirconate titanate (PZT) and the internal mechanical loss is about 25 times larger. The low dielectric permittivity reduces the electrical cross-talk and the high mechanical loss reduces the acoustic coupling. As a result, the cross-talk between elements for a polymer array is very small. This greatly simplifies the construction of polymer transducer arrays and there is virtually no design constraint on the dimensions of the array elements from the mechanical point of view.

Experiments have been carried out to measure the cross-talk between adjacent array elements.

A 15MHz tone burst with 10 V peak-to-peak value was used to drive one element of the array transducer and the output from six adjacent elements was monitored by a high input impedance oscilloscope. Typical cross talk is observed to be in the 100mV range. As there is no time delay between individual signals we can conclude that mechanical cross-talk is too small to be measured.

Pulse echo measurements: These were carried out using the pulser and receiver unit in the DIAS ultrasonic group's laboratory. The time domain waveforms were digitised by a Tektronix digital oscilloscope (model 7854), and then sent to a computer for frequency spectrum analysis. Figure 5 shows a typical example of the measurement results. It can be clearly seen that the transducer gives a very short pulse (80nS) and wide bandwidth.

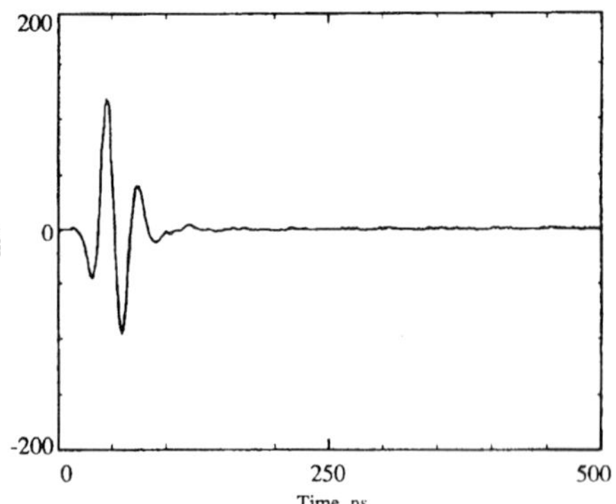

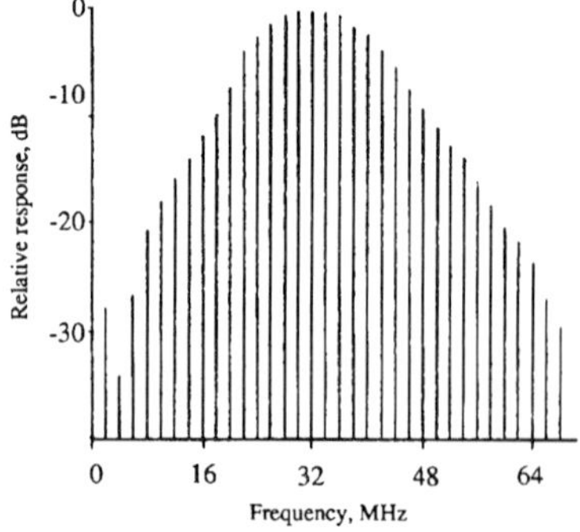

Fig. 5. Pulse echo response from a single array element, together with a frequency spectrum showing a centre frequency of 32MHz, bandwidth (3dB) 20MHz, pulse duration <80nS

6. CONCLUSIONS

New polymer transducer array designs have been produced and an interconnect technology allowing ASICs to be connected to the polymer transducers has been devised. However, further work on long-term reliability of this approach is required. Work is well advanced towards this objective. Individual components have been fabricated and tested and come up to specification in every respect. Most importantly a generic technique has been developed, and successfully demonstrated, which will enable both transmit and receive multi-channel ASICs to be realised.

REFERENCES

[1] P A Payne, 'Medical and industrial applications of high resolution ultrasound', *J.Phys,E: Sci Instrum.*, 1985, 18, pp.465-473

[2] Y Arai and T Oshugi, 'TMC - A CMOS time to digital converter VLSI', *IEEE Trans. Nuclear Science*, 1989, **36** No.1, pp.528-531

[3] J V Hatfield, P J Hicks, P A Payne and N R Scales, 'High-resolution multichannel pulse generator CMOS chip to drive ultrasonic array transducers', *Electronics Letters* 1992, **28** No. 23, pp.2113-2114.

ACKNOWLEDGEMENT

We acknowledge with gratitude the Science and Engineering Research Council of the UK for providing funding for this project.

Detection of photoacoustic waves in liquids by fibre optic interferometry

D.P.Hand, P.Hodgson, T.A.Carolan, K.M.Quan, H.A.Mackenzie, J.D.C.Jones
Department of Physics, Heriot-Watt University, Edinburgh EH14 4AS

ABSTRACT: A Michelson-based fibre interferometer was used to make absolute measurements of the sub-nanometre liquid surface displacements created by the photoacoustic process. This measurement system is both broad-band and non-contacting. Photoacoustic waves, generated using pulses from a Nd:YAG laser operating at 1.06μm, were measured in methanol, water, and various concentration mixtures. The wave amplitude demonstrated a quadratic relationship with concentration.

1. INTRODUCTION

Photoacoustic spectroscopy is an important analytical technique, particularly for liquids with low optical absorptions, where a much greater sensitivity can be obtained than with standard absorption spectroscopy. In addition, the problem of scattering is avoided. The photoacoustic process is a phenomenon where acoustic waves are generated as a result of optical absorption of laser pulses. The absorbed optical energy causes rapid localised heating and thermal expansion, resulting in an acoustic pressure wave, which may be detected with a suitable sensor. One application is the sensing of hydrocarbon pollutants in water [1]. This and other applications rely on accurate measurement of the photoacoustic pressure waves, which remains a significant problem in liquids. Contacting transducers, both piezo-electric ceramic [1] and Polyvinylidene difluoride (PVDF) film [2], have been employed, but each suffers from the problem of an acoustic impedance mismatching at the boundary with the liquid, which distorts the detected signal, and makes absolute calibration difficult. In addition, the bandwidth of piezo-electric ceramic transducers is limited by their resonant nature, whilst PVDF film, although non-resonant, has poor sensitivity. An alternative approach is to use optical interferometry which can, by contrast, measure the absolute displacement at the surface of the liquid resulting from the acoustic wave. This measurement is both non-contacting and intrinsically broadband. Measurement of acoustic waves on the surface of solid materials has been achieved with bulk optic interferometers [3], but to our knowledge this has not been reported with liquids. Although such interferometers are highly sensitive to small surface movements, they do suffer from stability and alignment problems.

Fibre optic interferometry offers a much more versatile and rugged solution, allowing remote detection in difficult environments. Indeed, such a Michelson-based fibre optic interferometer has recently been developed by us for detection of acoustic waves produced by a markedly different process: metal machining [4,5]. Successful operation has been achieved in the hostile machine tool environment. Although designed for a specific application, this interferometer is a versatile instrument capable of broad-band detection of small amplitude, high frequency movements in the presence of substantial lower frequency ambient perturbations, and so lends itself equally well to detection of photoacoustic pulses in liquids. In this paper we report measurements, made with this interferometer, of photoacoustic pulses generated in both water and methanol, together with various concentration mixtures. Comparative measurements are made between two techniques: optically probing a 25μm thick membrane cell wall containing the liquid, and probing the surface of the liquid itself.

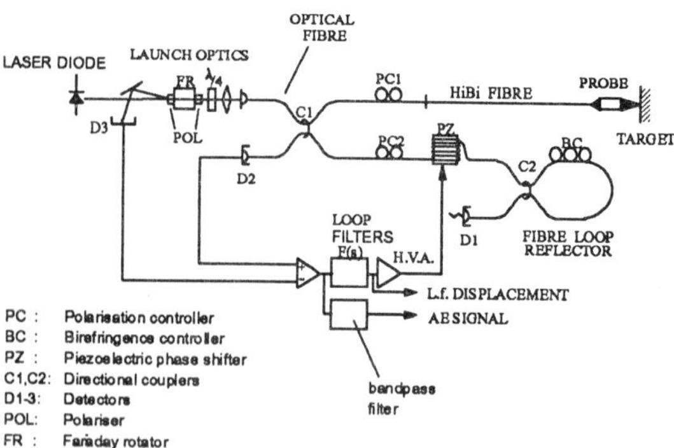

Figure 1: Schematic of optical fibre interferometer

2. DESCRIPTION OF FIBRE INTERFEROMETER

A schematic diagram of the non-contact fibre interferometer is shown in figure 1. It is based on a Michelson interferometer, but incorporates important modifications. These give anti-phase outputs and allow successful operation in the presence of the large low frequency vibrations (up to 10kHz) typically encountered when probing the metal machining processes for which the system was originally developed. The environment in which photoacoustic wave detection is required is generally much more benign than that associated with metal machining, but low frequency environmental vibrations and drifts are still considerably larger than the desired high frequency (~MHz) signal. An additional feature of the interferometer is the long fibre downlead (10m) which connects to a remote, ruggedised probe, allowing surfaces to be interrogated at a considerable distance from the main part of the interferometer. The probe itself is mounted inside a small rugged kinematic mount to provide beam alignment, and has a 25mm working distance. It is heavily shielded for protection from a hostile environment and to ease handling. The addressing fibre downlead is also armoured with protective tubing (Seicor Fanout Tubing), consisting of a low friction PTFE sleeve surrounded by a Kevlar strength member, and a hard plastic outer tube.

Light from the laser diode source (780nm) is launched into the single mode fibre, and is split equally between the reference and signal arms at the directional coupler. The signal arm consists of a length of single mode fibre spliced onto the downlead, which is polarisation-preserving fibre, making it insensitive to bending. The fibre polarisation controller before the splice ensures excitation of a single polarisation eigenstate. At the probe, a graded index (GRIN) lens, index matched to the cleaved fibre end, focuses the emergent light onto the test surface, where it is partially backscattered and re-collected by the fibre. Light in the reference arm, meanwhile, is returned by fibre loop reflector; the magnitude and polarisation state of this returned signal may be independently selected to match that reflected from the probe, simply by adjusting the birefringence controller contained within the loop [6]. Interference of the signal and reference occurs on recombination at the coupler, with guidance to detectors D2 and D3. The combination of the Faraday rotator, polarisers, and quarter-wave plate in the launch optics create an output at D3, with no insertion loss [5]. The signals detected at D2 and D3 are therefore sinusoidally dependent (in anti-phase) on the optical path difference between the signal and reference arms. This generation of antiphase outputs has two benefits. It provides compensation to first order of both the laser source intensity noise, and return intensity variations from the target.

To cope with the problem of low frequency movements caused by vibrations and thermal drifts, an active homodyne phase control system is employed, in which a piezo electric cylinder fibre stretcher is used as a phase modulator element in the reference arm, as part of a servo loop to maintain the interferometer at phase quadrature. This is driven by an error signal derived from the low frequency component of the anti-phase outputs. With the servo loop closed, the high frequency component of the target motion is directly proportional to the difference between the outputs D2 and D3, with a bandwidth set only by that of the detectors. To minimise the effect of laser frequency noise the fibre arm lengths are balanced to within 1mm, easily achieved using a frequency modulation technique.

3. THE PHOTOACOUSTIC EFFECT

The photoacoustic effect arises from a complex interaction that takes place when a modulated light source is incident upon an absorbing medium. The radiation causes heating of the sample, due to relaxations from optically induced excited states via non-radiative transitions. This heating of a limited volume causes pressure changes in the body, due to thermal expansion. An acoustic wave is generated which propagates away from the illuminated region, and which may be detected by a suitable transducer. An photoacoustic spectrum of the specimen is obtained by measuring the photoacoustic response as a function of the wavelength of the incident light. In addition, time and frequency domain analysis of the acoustic waveform provides information about the sample constituents, and the photoacoustic process itself.

Theories to predict the photoacoustic waveform generated in liquids have been developed, in particular by Lai and Young [7], and by Patel and Tam [8]. The photoacoustic signal is dependent upon the optical conditions causing its generation. For the case of a weakly absorbing sample, and an optical pulse duration τ which is much greater than the time for the acoustic wave to travel across the illuminated region, the peak of the generated pressure wave, P, is given by

$$P = \frac{k\alpha\beta\upsilon}{C_p}.E \tag{1}$$

where υ = acoustic velocity; α = optical absorption coefficient; β = thermal expansion coefficient; C_p = specific heat capacity; E = optical pulse energy; k = system constant

The peak pressure at position r can be approximated to the peak acoustic displacement s(r) by

$$P(r) = \frac{\upsilon\rho s(r)}{\tau} \tag{2}$$

where ρ = density [9]. More strictly, the pressure is related to the displacement velocity in the medium. Nevertheless, equation (2) allows reasonable estimates of the peak pressure to be calculated, and an absolute measurement of s(r) will lead to a calibrated estimate of P(r).

4. EXPERIMENTAL

A schematic of the optical arrangement is shown in fig. 2. Nd:YAG laser pulses at a wavelength of 1064nm and a repetition rate of 200Hz were focused into a quartz cuvette containing the sample. The pulse duration was 180ns (FWHM), and the pulse energy was 1.6mJ. The optical beam at the cuvette was approximately 25μm in diameter, and was a good approximation to a cylindrical source for photoacoustic generation, given the low optical absorptions of water and methanol (~0.1cm^{-1}).

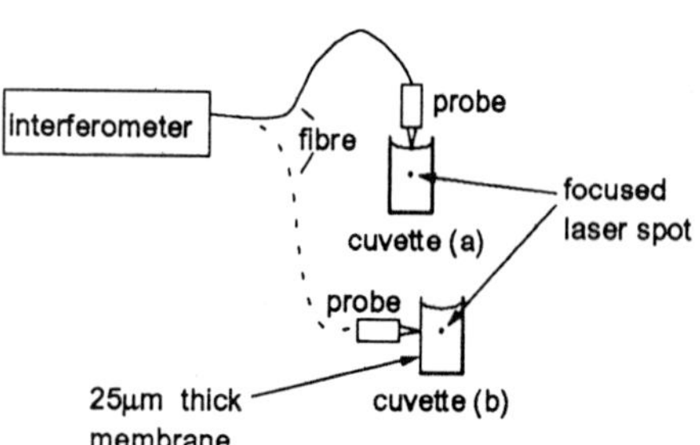

Figure 2 : Experimental arrangement

The acoustic displacements generated by the photoacoustic process were measured by probing the sample with the fibre interferometer. Two arrangements were used as shown. With cuvette (a) (5mm pathlength), the top surface of the liquid sample was directly probed with the interferometer, and the photoacoustic pulses detected as surface movements. Cuvette (b) (2mm pathlength), meanwhile, had one side wall replaced with a 25μm thick membrane of high optical reflectivity (>80%). In this case, the membrane was monitored by the interferometer, whilst acoustic pulses in the liquid sample caused the membrane to flex. The interferometer signal was averaged over 256 acquisitions on a digital oscilloscope, and transferred to a PC for further analysis.

5. RESULTS

The interferometer probe was aligned to obtain a reflection from the surface of the liquid placed in cuvette (a). Both water and methanol have refractive indices of ~1.33, giving a reflectivity of ~2% at the air interface. This resulted in a 1% return coupled back into the fibre. Despite this low optical power, the system operated satisfactorily, but with a higher noise floor than previously observed with a high reflectivity target [5]. This primarily resulted from additional small reflections within the probe, which are relatively insignificant with a higher reflectivity test surface. Signal averaging was employed to minimise this problem, and with 256 averages, a noise floor of 0.1nm was achieved over a 1MHz bandwidth, corresponding to 0.1pm/√Hz. Absolute calibration was difficult, however, as this relies on measuring the overall fringe depth of the interferometer, which has extra components resulting from the additional interference with the intrinsic probe reflections, giving an estimated 25% error in this case. Measurements of pulse shape are valid, however, and photoacoustic pulses produced in both 100% methanol and 100% water are shown in figures 3(a) and 4(a) respectively. In addition to the initial photoacoustic wave, reflections from the cuvette walls may be seen.

The probe was subsequently aligned with the high reflectivity membrane of cuvette (b), and results for both methanol and water are plotted in figures 3(b) and 4(b) respectively for comparison with those obtained by directly probing the liquid surface in cuvette (a). The membrane response can be clearly observed superimposed on the signal. In this case the noise

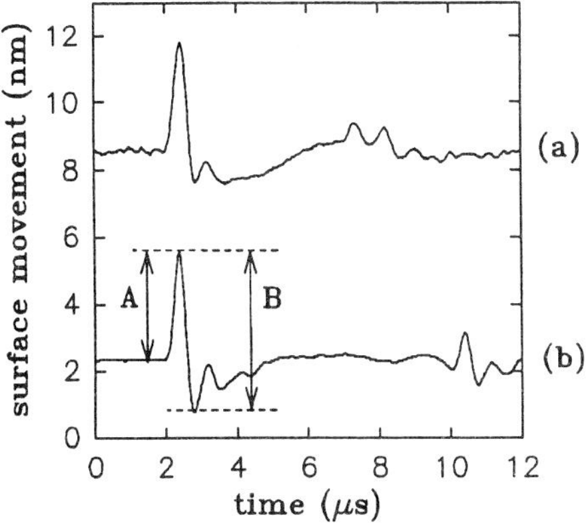

Figure 3 : Photoacoustic pulse produced in 100% methanol, measured by (a) probing liquid surface (b) probing reflective membrane

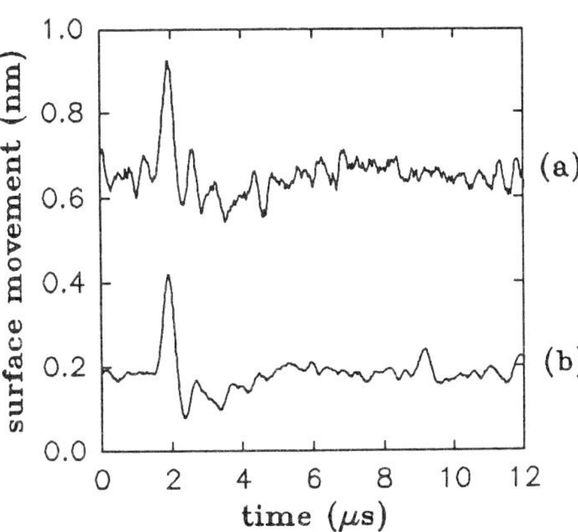

Figure 4 : Photoacoustic pulse produced in 100% water, measured by (a) probing liquid surface (b) probing reflective membrane

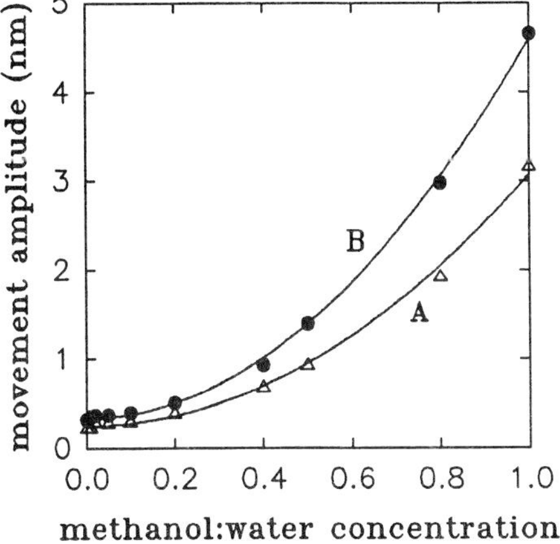

Figure 5 : Surface displacement amplitude as a function of methanol:water concentration

floor was reduced, and after 256 averages was measured to be 0.01nm, corresponding to 0.01pm/√Hz. Comparison between the two figures illustrates that the peak displacement generated in methanol is approximately 10 times that in water. From these data, and using equation (2), the peak pressure in methanol was estimated to be 1.6×10^4 Nm^{-2}, and in water 1.9×10^3 Nm^{-2}.

To further explore the sensitivity of the technique, a range of concentration mixtures of water and methanol were tested. Cuvette (b) was used to ensure independent calibration of each measurement. The results obtained are shown in figure 5, with both the amplitudes A and B, as defined in figure 3, plotted against concentration. A close fit to a quadratic curve is obtained in each case, as shown.

6. DISCUSSION

The measurements obtained with the fibre interferometer are of the displacement caused by a photoacoustic wave, in contrast to the more common detection of pressure, usually achieved via a piezoelectric transducer. The peak values of the measured displacement and calculated pressure generated in methanol are approximately an order of magnitude greater than those in distilled water. This may be attributed to the physical parameters of these samples, in particular β and C_p [10], and the role of these terms in photoacoustic generation, as summarised by (1). The dependence of the photoacoustic signal with methanol concentration in water agrees well with a quadratic curve, as shown in figure 5, fitted under the assumption that α,β,υ, and C_p vary linearly with concentration.

Although piezo-electric devices are the most sensitive available, piezo-electric ceramics generally suffer from being highly resonant detectors, and have a poor acoustic match to liquids. The electronic signals obtained from such detectors are strongly influenced by the properties of the detector itself. The fibre interferometer offers a rugged, broadband, non-contact detection scheme which allows a calibrated signal representative of the actual acoustic waveform to be measured.

Applications requiring a broadband calibrated measurement of vibrations or displacements in liquids are envisaged as being appropriate for this technique, especially in environments where a non-contact measurement is important. Spatial resolution of the displacement field can also be achieved, by scanning the probe across the reflecting surface. The calibration problem encountered when directly probing low reflectivity liquid surfaces could be easily solved by redesigning the probe optics to suppress intrinsic reflections. A further development would be a probe capable of measuring displacements in the liquid bulk, rather than at the surface. This would provide a measurement free from the effects of surface tension.

7. CONCLUSIONS

A ruggedised optical fibre interferometer has been used to make an absolute measurement of the acoustic displacements generated by the photoacoustic effect in liquids. Using this technique, sub-nanometre resolution of the displacement field is possible. These data can be utilised in studies of the fundamental photoacoustic generation process, and other hydrodynamic phenomena.

REFERENCES

[1] P.Hodgson, H.A.Mackenzie, G.B.Christison, K.M.Quan: in Near Infra-red Spectroscopy: Bridging the Gap between Data Analysis and NIR Applications, publ. Ellis Harwood (1982) 407.
[2] S.J.Komorowski, E.M Eyring: J. Appl. Phys. **62** (1987) 3066.
[3] P.Cielo, F.Nadeau, M.Lamontagne: Ultrasonics, **March**, (1985) 55.
[4] R.McBride, T.A.Carolan, J.S.Barton, W.K.D.Borthwick, J.D.C.Jones: Meas. Sci. Tech., in press (1993).
[5] D.P.Hand, T.A.Carolan, J.S.Barton, J.D.C.Jones: Opt. Comm., in press (1993).
[6] R.McBride, J.D.C.Jones: J. Mod. Opt., **39** (1992) 130.
[7] H.M.Lai, K.Young: J. Acoust. Soc. Am., **72** (1982) 2000.
[8] C.K.N.Patel, A.C.Tam: Rev. Mod. Phys., **53**(3) (1981) 517.
[9] A.C.Tam: Rev. Mod. Phys., **58**(2) (1986) 381.
[10] G.W.C.Kaye, T.H.Laby: Tables of physical and chemical constants and some mathematical functions, publ. Longman (1986) 15th ed.

Ultrasonic NDT with non-uniform transducers

Richard Stacey & John P Weight
School of Engineering, City University, Northampton Sq, London EC1V 0HB

Abstract: Conventional ultrasonic transducers generate both plane and edge waves in solids, producing signals which are complex and frequently difficult to interpret. Recently constructed Plane Wave Only (PWO) and Edge Wave Only (EWO) devices have simpler outputs giving simpler signals. The PWO transducers produce echo signals that are easily interpreted in terms of the defects. The EWO transducers are valuable for imaging applications. Our conclusions are backed by experimental results and calculations made using finite difference methods.

1. INTRODUCTION

Ultrasonic transducers are currently used in a range of non-destructive testing applications, for instance medical ultrasound, or the search for defects in nuclear reactor pipes. The transducer generates the ultrasonic waves, these are scattered by the anomaly or defect, and the resulting signal is then measured and interpreted. In pulse-echo scattering the transducer that generates the pulse also measures the echo.

A problem for all these applications is that the correct interpretation of the signal can be far from straightforward, particularly when the target lies in the near field of the transducer. The main reason for this problem is the complexity of the ultrasonic waves produced in a solid by a conventional transducer, which generates a uniform input over the whole transducer/solid interface. Figure 1(a) illustrates this complexity, with a cross-section through the wave field produced by such a transducer in aluminium, taken 2μs after the initial short input pulse. The transducer diameter is 19mm. The material surface is denoted by the line along the bottom and the transducer is positioned with its edges at the centres of the outgoing edge-wave semicircles. The data used were generated using finite difference methods [1].

Several wave-types are evident in Figure 1(a). As well as the expected longitudinal plane wave, propagating in front of the geometrical region of the transducer, there are both longitudinal and transverse edge waves radiating from the transducer edge, together with surface and "head" waves. The edge waves are produced by diffraction, and are significant well away from the solid surface. Now it is clear that the scattering of these plane and edge waves will give rise to multiple signals (for instance multi-pulse echo responses) which will vary greatly with both target position and size. The interpretation of the signals is therefore extremely difficult, and conclusions based on simple plane wave analyses (for example) can lead to inaccurate conclusions. The work reported here is part of a long-term project at the City University aimed at extracting more and better quality data from ultrasonic measurements, by taking into account these real-world effects.

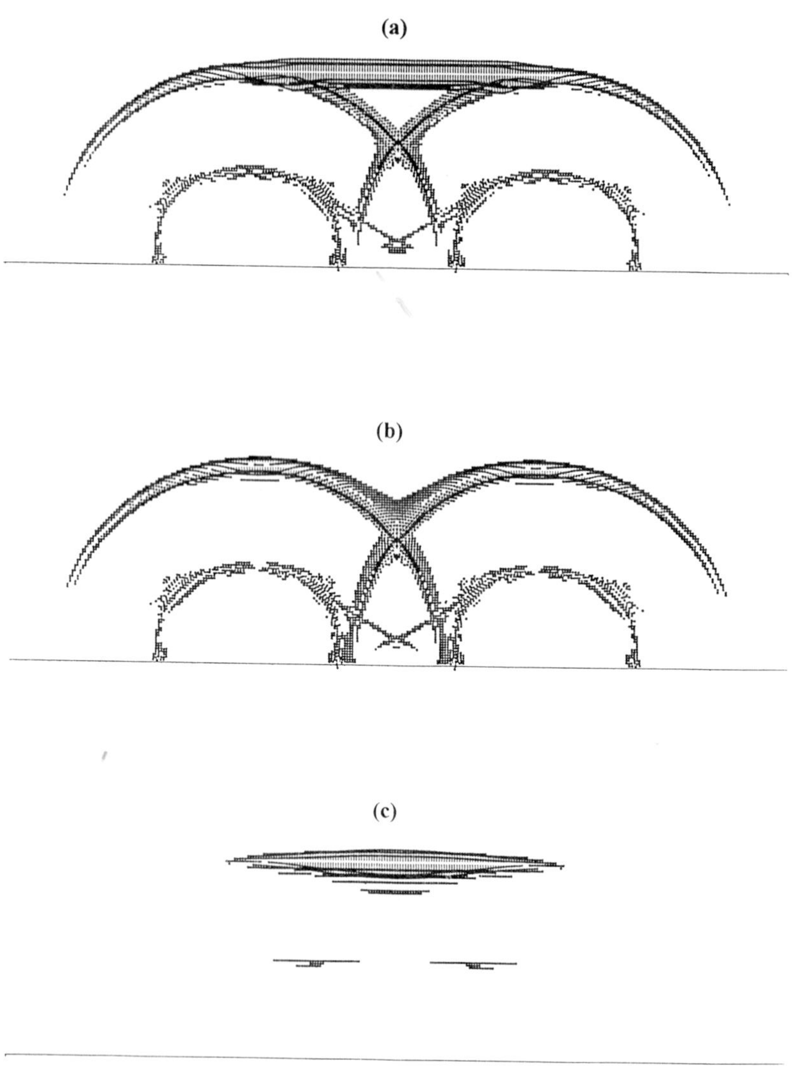

Figure 1: Outgoing wave profiles for (a) uniform (b) EWO and (c) PWO transducers

2. NON-UNIFORM TRANSDUCERS

It is possible to simplify the waves produced by a transducer by making its output non-uniform over the transducer radius, while preserving the angular symmetry [2]. The non-uniform transducers considered here were fabricated at City using non-uniformly polarised ceramic elements [3]. A concentric pattern of electrode rings was laid down on one face and used to apply a radially varying but axisymmetric polarising field. The polarised elements were backed by a tungsten-loaded epoxy composite in order to achieve a wide-band response. Two particular types of non-uniform transducer were constructed:

(i) Plane Wave Only (PWO) devices, where the field is strongest in the transducer centre and falls away smoothly to zero at the edge, and

(ii) Edge Wave Only (EWO) devices, where the field is strongest at the transducer edges and negligible in the centre.

Figures 1(b) and 1(c) show the wave profiles produced by such transducers after 2μs in aluminum. It is clear that the transducers' names are justified. The EWO device (Figure 1(b)) produces no plane wave, while the PWO device (Figure 1(c)) produces no edge waves and only a very small planar transverse wave component.

3. PULSE-ECHO SCATTERING WITH PWO TRANSDUCERS

The output from the PWO transducer is particularly simple, and as a result the echo signals it generates from defects in the solids are much simpler than those generated by a uniform transducer. Figure 2 shows the signals from an on-axis point target at depth 15mm in aluminium that are generated by (a) a uniform transducer and (b) a PWO transducer. The signals were again calculated using finite difference methods, with the transducers the same as those in Figure 1. In the PWO case there is essentially a single signal, while for the uniform transducer there is a whole complex of signals. Results for other defects, both calculated and experimental, show that this is a general pattern.

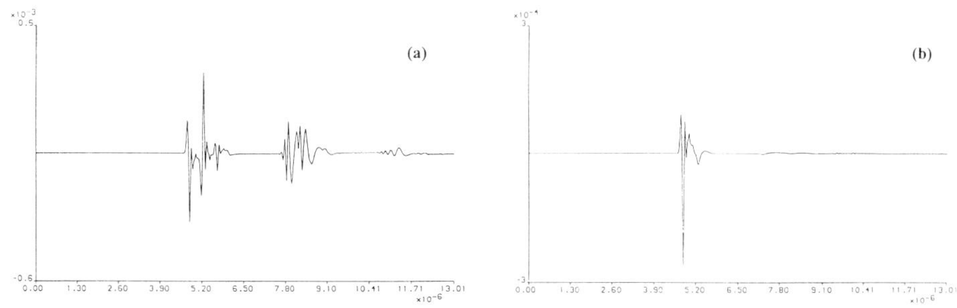

Figure 2: Calculated echo signals from a point defect on-axis at depth 15mm using (a) a uniform transducer and (b) a PWO transducer

The PWO device produces a single signal from each defect, up to multiple reflections, and the shape of the signal does not vary greatly with the position and size of the defect. Only the amplitude varies.

Figure 3 shows the experimental signal from a 1mm radius on-axis target at depth 15mm in a 35mm deep block of mild steel. Three signals can be seen: (i) the leading signal from the defect after about 5.2μs, (ii) the small double-reflection signal after about 10.4 μs, and (iii) the back-wall echo after about 12.4μs. It is notable that there are no other signals (as there would be with a uniform transducer) and that the back-wall echo has much the same structure as the defect echo, in line with our assertion about signal shape.

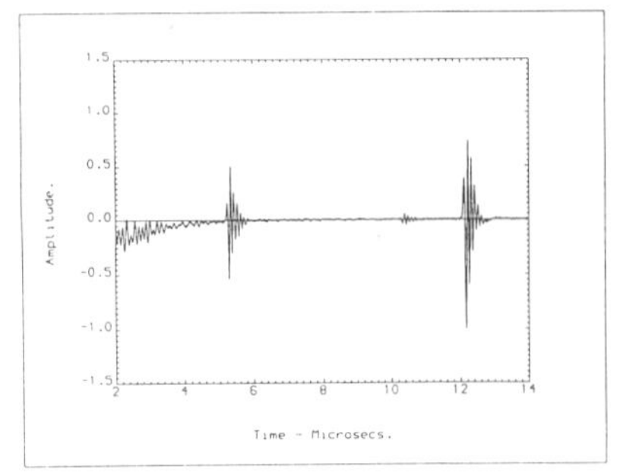

Figure 3: Experimental signal from a 1mm target at depth 15mm in steel using a PWO transducer

The PWO transducer can also produce valuable results when it is water-coupled. Figure 4 shows experimental measurements obtained using (a) conventional and (b) PWO devices, where the target was a flat-ended 2mm-diameter cylindrical brass rod with a 1mm diameter 1mm deep flat-bottomed hole drilled in it. The target was at a depth of 25mm in water, and both the transducers were 19mm diameter 5MHz devices. For the PWO case the double echo from the front of the rod and the bottom of the hole gives reliable information on the physical situation, even to the extent of the two signals' relative amplitudes reflecting the 3:1 area ratio of the two targets. For the conventional transducer the signal from the bottom of the hole is completely masked by the multiple pulse structure of the overall echo response, and much less good quality information can be extracted.

(a)

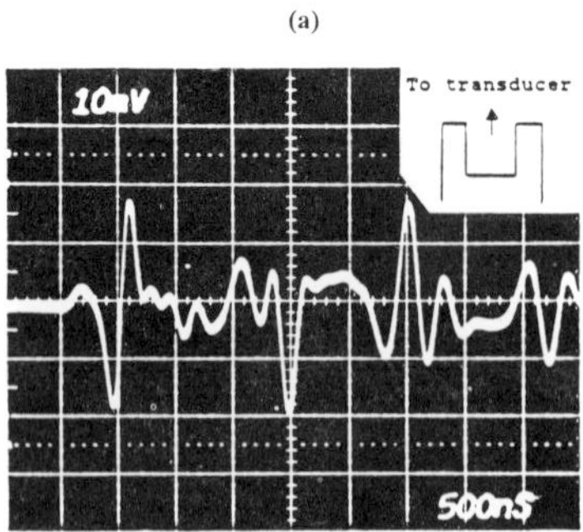

(b)

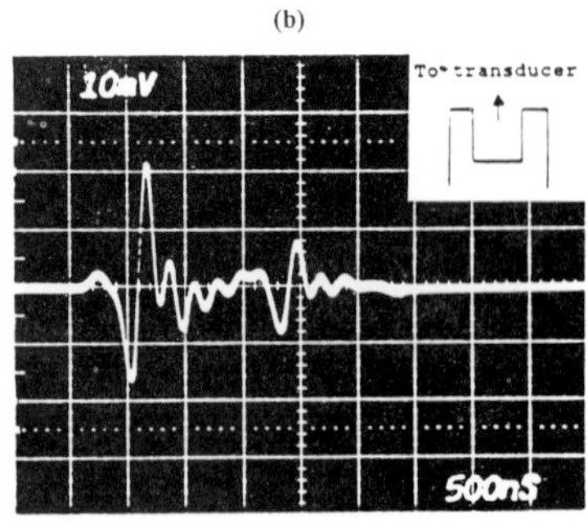

Figure 4: Echo signals from a water-coupled brass target (insert) using (a) a conventional transducer and (b) a PWO transducer

4. EWO TRANSDUCERS

The echo signals generated by an EWO transducer in a solid are more complex than those from a PWO transducer, because of the presence of two types of non-planar wave (longitudinal edge and transverse edge). However

(i) The echo signal shapes are generally simple, because they do not suffer from the interference between the effects of the longitudinal plane and longitudinal edge waves that complicate the signals from a conventional transducer.

(ii) The EWO transducers can also provide very precise positional information. Consider a small target on-axis. The edge waves from all parts of the transducer edge reach the target together and combine to produce the signal. If the target is moved very slightly off-axis, however, then the edge waves from different parts of the edge will reach the target at different times. Their effects will cancel or smear out, and the signal will reduce markedly.

(iii) This sharp axial focussing means that the echo signal amplitudes will not vary greatly with target size.

(iv) The precise angular alignment of the transducer is also less significant.

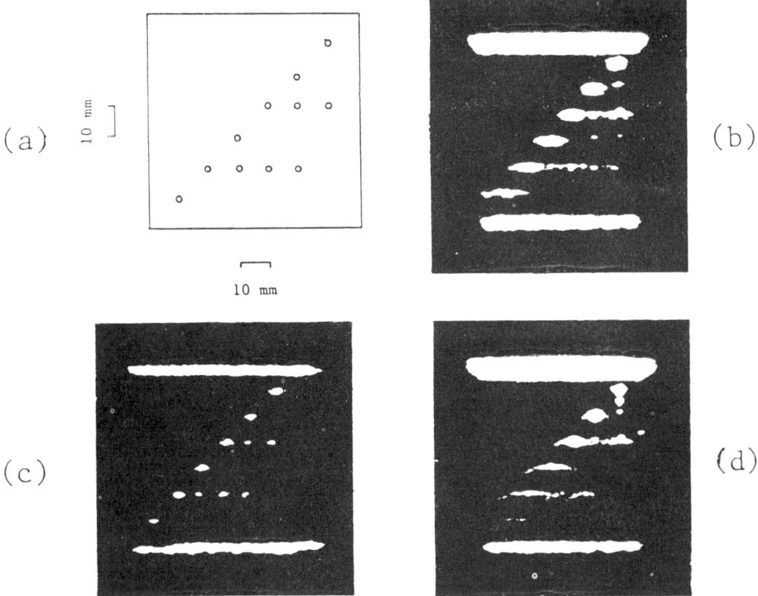

Figure 4: B-scan images of an aluminium test piece containing 11 side-drilled holes (shown schematically in (a)). The images were obtained with (b) a conventional transducer, (c) an EWO transducer, (d) a sharply focussed conventional transducer

All these properties are extremely valuable when we want to image targets using standard "B-scan" imaging techniques. This is particularly the case when the test-piece is water-coupled, which eliminates the effects of the transverse edge wave. We have produced images of metal blocks containing a range of defects, using conventional and EWO transducers. An example is given in Figure 5. Measurements were made using the test piece shown schematically in Figure 5(a), with 19mm diameter transducers and a water coupling path of 30mm. Results are shown for (b) a conventional transducer (Panametrics V3829), (c) an EWO transducer and (d) a sharply focussed conventional transducer. The imaging obtained with our prototype EWO transducer is clearly much the best.

5. FINITE DIFFERENCE MODELLING

Figures 1 and 2 were both produced using the finite difference methods developed in reference [1]. These have been shown to give extremely accurate reproductions of experimental results in pulse-echo scattering. The reason for this accuracy is that the full equations of motion and boundary conditions are approximated, without any extra simplifying assumptions. The finite difference modelling therefore unlocks the true content of the fundamental equations, in a way that is generally impossible analytically, and thereby provides a source of accurate "pseudo-data", on a scale that is impossible to match experimentally. We note that new extensions of the finite difference modelling of [1] allow us to separately calculate the various longitudinal and transverse wave contributions to all the scattering processes and echo signals [4].

6. CONCLUSIONS

The non-uniform ultrasonic transducers considered here (PWO and EWO) have been shown to be extremely useful for practical NDT problems, particularly for investigating defects in the near field. The PWO transducer produces single echo signals from single defects, and variations in defect position and size are reflected only in amplitude variations in the signal. This contrasts with the complexity of the signals produced using a conventional transducer. The EWO transducer has been shown to produce much better B-scan images of solid targets than a conventional transducer.

REFERENCES

1. Stacey, R. & Weight, J.P., "Ultrasonic echo responses from targets in solid media using finite difference methods", IEE Proceedings A: Science, Measurement & Technology, (in press) (1993).
2. Stacey, R. & Weight, J.P., "Pulse-echo scattering from solids with non-uniform transducers" (1993).
3. Brittain, R. & Weight, J.P., "Fabrication of high-resolution ultasonic transducers", Ultrasonics **25**, 101-6 (1987).
4. Stacey, R., "Separation of longitudinal and transverse waves in elastic scattering using finite differences" (1993).

Machine diagnosis by acoustic signal processing

R. Ohba*, Y. Tamanoi**, T. Ohtsuka**, M. Komatsu*** and T. Maeda***

*Department of Applied Physics, Faculty of Engineering,
Hokkaido University, N-13 W-8, Sapporo 060, Japan

** Division of Equipment Maintenance, Department of Engineering,
Marifu Oil Refinery, Koa Oil Co. Ltd., Iwakuni 740, Japan

*** Technical Centre, Ono Sokki Co. Ltd., Hakusan 531-1,
Midori-ku, Yokohama 226, Japan

ABSTRACT: An experiment on machinery diagnosis is performed on a model plant of oil refinery equipment. A hand held data processor and a specially designed sound collecting microphone used in the experiment are described as well as the principle of the basic experiment and the present diagnosis method.

1. INTRODUCTION

In Japan in recent years, it is a most remarkable feature that both the decrease in the population of young people and the tendency to recruit them to soft and service industries has brought about a reduction in the number of technicians or engineers in production industries. The numbers who want to be maintenance engineers are also decreasing rapidly today, so that those responsible for maintenance in industries are keen to develop automated maintenance/diagnosis methods for machines. We have been working on developing acoustic diagnosis methods to detect possible defects of ball bearings in practical machines in oil refining plants[1-3]. In the present paper, an experiment on machine diagnosis performed on a pilot model of oil refinery equipment is described. The data processing and sensors used in the experiment are also described, together with the idea and principle of the present diagnosis method.

2. PRINCIPLE OF THE METHOD

2..1 Background to the Problem

Machines in operation are always emitting some acoustic or vibration signals. A vibration signal is restricted in its source to those elements that are connected/contacted mechanically to each other. It is easy to separate vibration sources and to identify the specific source for a particular vibration signal. It might be the only practical means for machine diagnosis, although some sensors may be pre-fitted in order to automate the monitoring of the machine.

On the other hand, using an acoustic signal does not require any specific sensor to be pre-fitted and, it is easy to prepare a sensor for its detection. Acoustic means may be preferable to vibration to automate machine diagnosis. It has, however, one serious defect due to the necessity to separate/identify the specific source among several interfering sources, because acoustic signals are apt to interfere with one another. Human beings have the ability, known as the "cocktail party effect", to pick-up specific sounds one wants to hear among interfering plural sources. It would be possible to automate the maintenance of the system if it were possible to simulate this effect by some artificial means. The present method may be a step in such an attempt.

2.2 Autoregressive Model

Any time sequential signal can be regarded as the output from an appropriate linear system when its input is a white random signal. The linear prediction method[4] is well known to determine the corresponding system of any given time sequential signal and there exist established methods to solve the problem. An autoregressive model (AR model) is one of those which may be obtained from the methods. This is given as follows for a time sequence X(n) of n-th sampling time using M preceding values of sampled data, X(n-i), i= 1, 2, ... M.

$$X(n) = -\sum_{k=1}^{M} A_k X(n-k) + e(n) \tag{1}$$

where e(n) is white noise and is a virtual input to the linear system. An autoregressive model for the system is obtained by determining the set of coefficients $\{A_k\}$ using given sampled data.

2.3 Signal "Whitening" by Inverse Filtering

Define Y(n) as follows using the set of $\{A_k\}$ obtained, corresponding to the time sequence $\{X(n)\}$. Y(n) is called the linear prediction of X(n).

$$Y(n) = -\sum_{k=1}^{M} A_k X(n-k) \tag{2}$$

Now a residual, the difference between Y(n) and X(n), can be calculated as (3) using (1) and (2)

$$e(n) = X(n) - Y(n) \tag{3}$$

That is, the residual is white noise. If the linear prediction Y(n) based on preceding M sampled data of a time sequence is substituted from the n-th sampled data of the sequence X(n), the virtual input white noise of the system is obtained. The operation defined by (3) is called inverse filtering.

A white noise signal can be obtained by inverse filtering for any given time sequential signal only if it is represented by the same autoregressive model that is determined by the data used to design the inverse filter. In other words, any time sequence from a linear system can be whitened by the inverse filtering if, and only if, the sequence has the same statistical characteristics as those of the filter design data. In this case, any time sequence is whitened even if it is not the same data as that are used to design the inverse filter, if only its autoregressive model is the same that of the design data. This is true for any linear system so long as its dynamic characteristic remains unchanged.

A time sequence, however, cannot be whitened if it is an output of the system whose dynamic characteristics are changed or are different from those of the design data and no white noise may be obtained by the inverse filtering process.

Thus, it can be possible to detect any abnormal operation of machines by monitoring their output operational sound or vibration, based on an appropriate inverse filter that has been designed using those signals under normal operation. It may be possible even to diagnose the abnormal operation if we analyze more precisely, in detail, the residual signal. This is the basic idea of the present method.

3. MONITORING OF BALL BEARING DEFECTS

3.1 The Basic Experiment

Basic experiments are carried out in an ordinary laboratory room. Figure 1 shows a schematic diagram of the basic experiment. A flywheel directly connected to an electric motor is used for the experiment. The flywheel is supported by a pair of ball bearings, one of which is always normal and the other, is used for an experiment having 60 specimen bearings either being normal or with defects. A ball bearing with defects has artificial scratches either on 1) the outer race, on 2) the inner race or on 3) the ball. An acoustic signal is picked up by a precision condenser microphone with a parabolic sound collector at a point 600 mm distant from the bearing under test. The signal is input to a computer after pre-amplification, power amplification with an anti-aliasing filter and analog to digital conversion. The input signal is processed by the computer and the results are shown on the computer display.

At first 12 bit 1,024 data points are sampled with a 20 μs sampling period from the operational sound of a normal ball bearing and an inverse filter is designed, based on its autocorrelation, by calculating a set of linear prediction coefficients $\{A_k\}$ applying Levinson's algorithm[4]. Then by changing the specimen bearings, acoustic data are sampled with the same conditions that are used to design the inverse filter. The sampled data are processed by the inverse filter and we compensate spurious noise components not needed for the diagnosis, including the operational sound of the other bearing in a normal state. Abnormal bearings are detected by monitoring a moving average of the inverse filtered signal either if it exceeds a certain predetermined threshold value or if its peaks repeatedly lie above the threshold value. Resulting data processing and diagnosis are given on the computer display.

Figures 2 and 3 show some typical examples of those displays. Fig. 2(a) is sampled acoustic data of a normal ball bearing that is of the same type as one used to design an inverse filter. Fig. 2(b) is the residual signal of the data in Fig. 2(a) after the inverse filtering. Fig. 2(c) shows a moving average of the residual signal corresponding to the signal in Fig. 2(b). The moving average is calculated at every 50 th point from the M+lst sampling time (total 18 points) using Hanning windowed 128 points of the inverse filtered data. The 20 dB line in the figure shows a threshold level and which is +20 dB relative to peak power of the moving average calculated by the data

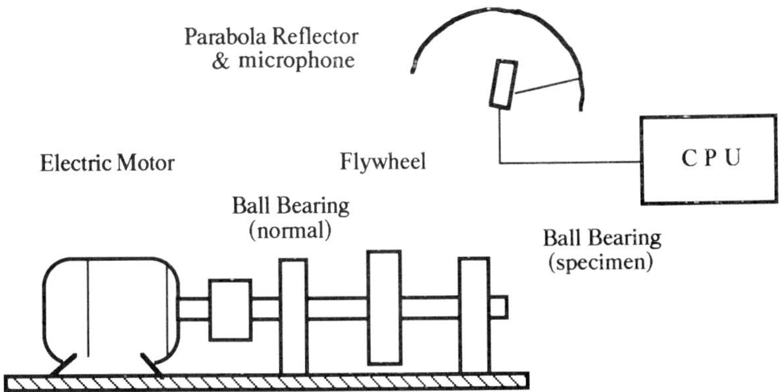

Fig. 1 Schematic diagram of the basic experiment.

used to design the inverse filter. Fig. 3(a) is sampled acoustic data of an abnormal ball bearing of the same type but it has artificial scratches on the outer race. Fig. 3(b) is the residual signal and Fig. 3(c) is its moving average. It can be seen that average power of Fig. 3(c) is more than the 20 dB threshold line and large peaks, nearly 10 dB over the threshold line, also repeat periodically. It is easy to diagnose an abnormal ball bearing, say, by finding those peaks or by monitoring the average power level of the moving average.

3.2 Experiments on a Model Plant

Experiments on a model plant were performed by using the newly developed hand held data processor (Ono Sokki: CF-1200). Figure 4 shows an overview of the model plant used in the experiment. The plant is located in the engineering yard near real oil refining plants and no special compensation is adopted for the ambient noise, so that the sound pressure level (SPL) of the stray noise at the site is about +60 dB. The SPL of the target operational acoustic signal is about 75 dB at the measuring point. Some additional noises, up to 88 dB, which is the maximum possible SPL of the total noise in the practical oil refinery, are used in the experiments by replaying recorded operational sounds of real oil refining plants. A special sound collecting microphone that is composed of a half-inch precision condenser microphone and a parabolic reflector made of glass fibre strengthened epoxy, whose focal length and length from the bottom to mouth are 7.5 mm and 200 mm, respectively, is used.

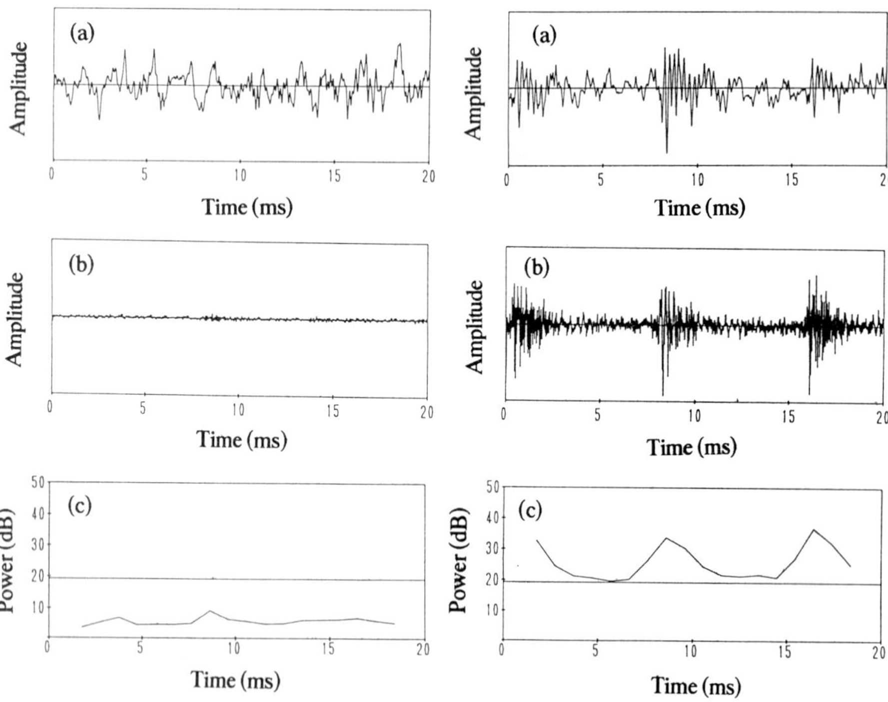

Fig. 2 Example of system operation. (a) operating sound of a normal bearing, (b) inverse filtered signal of (a) and (c) a moving averaged power of (b).

Fig. 3 Example of system operation. (a) operating sound of an abnormal bearing, (b) inverse filtered signal of (a) and (c) a moving averaged power of (b).

Fig. 4 Over view of the model plant.

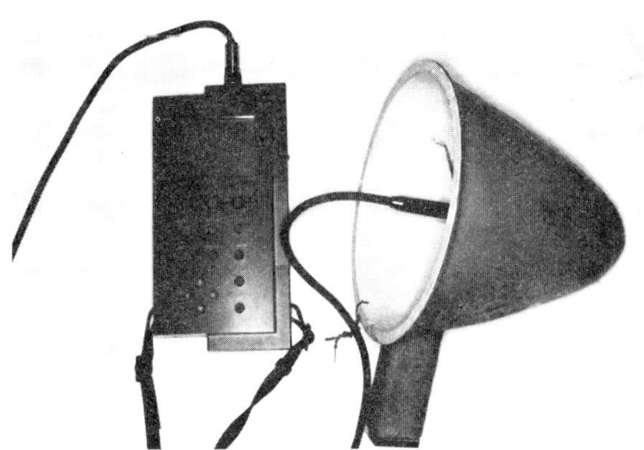

Fig. 5 CF-1200 hand held data processor and sound collecting microphone.

Figure 5 shows the hand-held data processor CF-1200 with the parabola reflector used in the experiments. Figure 6 illustrates the flow chart of the experiments. Operational sounds from a normal ball bearing are detected just above the bearing housing by setting the special sound collector at its focal point, 700 mm distant from the top of the housing. Four sets of design data are used for inverse filters: one is without additive noise and the other three are with additive noise of 80, 85 and 88 dB, respectively. Each set of data is sampled and A/D converted (12 bit at 1,024 points with 20 μs sampling period) by the hand-held processor and transferred to an off-line process computer. Inverse filters or the linear prediction coefficients $\{A_k\}$ are designed on the basis of the operational sounds of the normal bearing for each additive noise condition by the process computer. Calculated values of $\{A_k\}$s are again transferred back to the hand-held processor. The processor then performs the inverse filtering operation in real time. Abnormal ball bearings are successfully detected based on the sound signal detected, with the same conditions as the normal bearing by monitoring a moving average of the inverse filtered signal: in this case the threshold is the +3 dB level relative to the peak power of the moving average of the data used to design the inverse filter.

4. CONCLUSIONS

A method to diagnose an abnormal operating of a machine is proposed by applying an inverse filtering technique. A pilot model of the diagnosis hand-held type data processor is developed and it is successfully applied to detect abnormal ball bearings with three kinds of defects on either the outer race, inner race or the ball. A specially designed sound collecting microphone shows excellent performance to separate the target acoustic signal from ambient noise. The pilot model can detect those abnormal ball bearing specimens under 80 to 88 dB (SPL) additive noise conditions. Those results suggest a high possibility of the applicability of the present method in real plants. It is left for the future studies to confirm this, and we are continuing to confirm the performance of the pilot model in a real oil refinery.

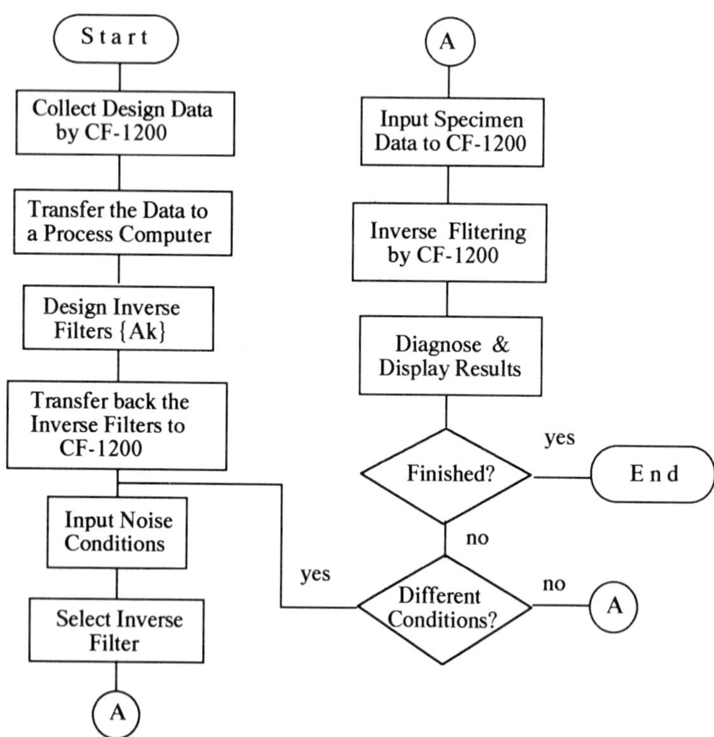

Fig. 6 Flowchart of the experiment.

REFERENCES

1) Ohba, R. and Tamanoi, Y.: Method and apparatus for machine diagnosis, Jpn. Pat. File No. H4-138681 (1992)

2) Tanaka, S. et al: Bearing Monitoring by Acoustical Signal Processing, Proc. of SICE '92, pp. 185-186(1992)

3) Ohba, R.: Acoustic Methods for Machinery Diagnosis, Proc. of the 27th SICE Hokkaido Branch Meeting, pp. 1-4 (1993)

4) Oppenhime, A. V. and Schafer, R. W.: Digital Signal Processing, Prentice-Hall Inc., Englewood Cliffs, N. J., 585pp.(1975)

Torque measurement by means of a SAW resonator

D.A. Hall, H.W. Wang, and A. Lonsdale*

Materials Science Centre, University of Manchester, Manchester, UK
*Sensor Technology Ltd., Banbury, UK

Abstract

The paper presents a new concept in torque measurement based on the use of a SAW resonator as a strain sensor. It is shown that the observed torque sensitivity is dependent on the adhesive used for bonding the SAW element to the test shaft and on the geometry of the element itself. The latter observation is explained qualitatively in terms of the strain distribution within the element. Proposals are made for a temperature compensated SAW-based torque transducer system based on the use of 2 SAW elements in a 'half bridge' configuration.

Introduction

The measurement of torque in rotating machinery is one of the most difficult problems facing mechanical development engineers. Torque is also one of the most useful parameters, since a knowledge of torque and rotational speed can be used in order to assess the power and, therefore, efficiency of a gearbox, transmission, or system as a whole. Within the context of the current drive towards more efficient, actively controlled systems it is widely recognised that a reliable, low-cost torque transducer is needed in a large number of diverse applications(1).

Conventional resistance strain gauge systems require excitation signals of a few volts, with the resulting output in the millivolt range. Careful signal processing is necessary in order to avoid ground loops and to ensure that the amplifier used is linear and has low drift. Furthermore, a bandwidth of 50 kHz or greater may be required for time-varying signals. In rotating mechanisms, an electrical connection to the gauge or gauges presents difficulties which may be insurmountable. For example, the use of slip rings is not feasible for large diameter shafts rotating at high speed, low-frequency AC carrier techniques have low bandwidth and high inertia, while add-on FM telemetry techniques can suffer from battery and balance problems. Other methods have been investigated in order to overcome the limitations of resistance strain gauges (e.g. optical or magnetic techniques), but most of these suffer from high cost, difficulties in attachment to existing assemblies, and calibration problems. The present work was carried out in order to evaluate the potential of a radical approach to torque measurement based on the use of a SAW (Surface Acoustic Wave) resonator as a strain sensor.

SAW devices have their origins in the mathematical description put forward by Lord Rayleigh in 1885(2), which showed that waves could propagate along the surface of an isotropic, elastic medium. However, it was not until 1965 that Voltmer and White demonstrated the excitation and detection of surface acoustic waves in a piezoelectric quartz substrate through metallic interdigital electrodes or transducers (IDTs)(3). The IDTs provide excitation and detection of the surface acoustic wave, while arrays of shorted metal strips or machined grooves serve as reflectors, allowing the production of SAW delay lines, filters, convolvers, and resonators(4). In RF signal processing, SAW devices allow a substantial miniaturisation relative to equivalent electromagnetic wave devices as a result of the relatively low velocity of the surface acoustic wave (~3 x 10^3 ms^{-1} compared with ~3 x 10^8 ms^{-1}).

It was recognised over 20 years ago that the disturbing influence of the environment on the characteristics of a SAW delay line or resonator could serve as the basis for a sensor, and subsequently SAW-based sensors have been proposed for monitoring stress/strain, acceleration, vibration, electrical and magnetic fields, displacement, gases/vapours, and temperature(5,6). For SAW strain sensors, a combination of variation in the periodicity of the electrodes (i.e. the wavelength) and the surface wave velocity (via variation in the density of the media) gives rise to a variation in operating frequency. The production of a sensor response in the form of a frequency shift has great advantages over conventional analogue sensors in terms of greater sensitivity and accuracy, high signal to noise ratio, potentially low cost, and is well suited to interfacing with digital systems.

Besides the improvements offered in existing applications, many novel sensing applications are made possible as a result of the high-frequency (typically 100-500 MHz) operation. For example, non-contact or non-invasive sensing may be achieved using simple capacitive or inductive links which require little modification to existing electronic systems and will find use in novel applications where some motion is required between the sensing element and detection system. One possible drawback is the temperature-dependence of the operating frequency, which may mean that long-term changes in strain are masked by temperature variations. A recent report of the use of SAW devices as torque sensors concluded that the variation of the resonant frequency with temperature was such that this type of transducer was felt to be unsuitable for industrial application, unless the temperature of the shaft could be measured accurately(7).

The present work reports results obtained as part of a study aimed at developing a temperature compensated torque transducer system based on the use of two SAW devices in a 'half bridge' configuration, as illustrated in Fig. 1. With such a configuration, it is possible in principle to use the difference frequency for measurement of strain and the sum for temperature(8).

Experimental Methods

The SAW devices used in the present work were commercial one-port SAW resonators, based on quartz, operating at frequencies of approximately 200 and 400 MHz. In practice, it is intended that the frequency of the 200 MHz element would be doubled by the supporting RF electronic package in order to achieve the desired temperature compensation. It is important to note that the elements should exhibit equal relative strain sensitivity (i.e. $\Delta f / f_0$) if a reliable measurement is to be achieved. The dimensions of the resonators were 3.50x1.20x0.61 mm (200 MHz element) and 2.10x1.00x0.61 mm (400 MHz element).

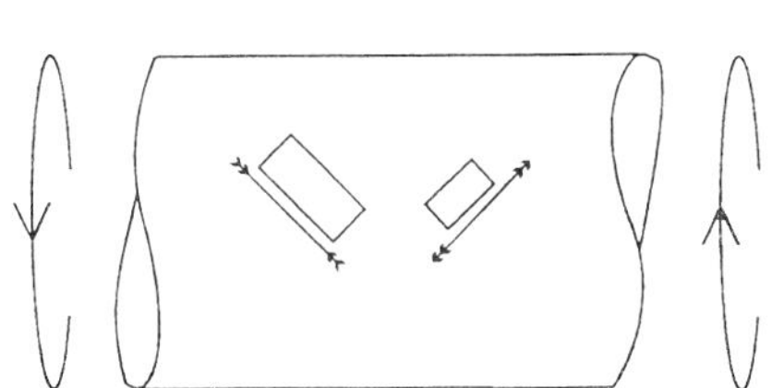

Fig. 1. Illustration of proposed technique for temperature compensation using two SAW resonators.

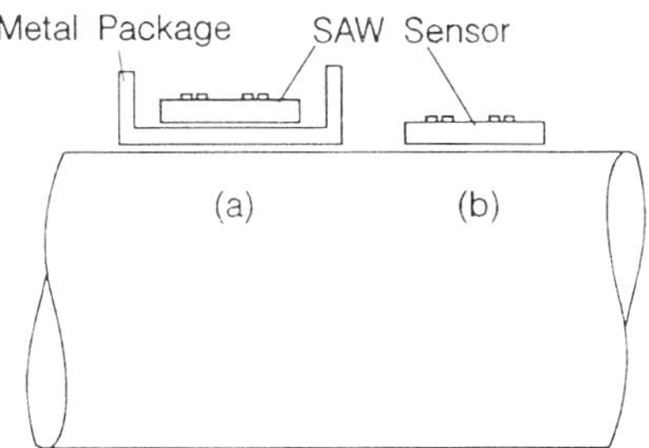

Fig. 2. Element configurations.
(a) packaged device
(b) directly bonded device.

Preliminary work involved investigating the torque sensitivity of packaged elements (using a TO220 package) bonded to a 10mm square flat on 0.5" diameter FV 520B stainless steel shafts, as shown in Fig. 2(a). However, as a result of poor linearity and hysteresis, subsequent work focussed on directly-bonded elements (Fig. 2(b)). Electrical connections were established by fine gold wires microbonded to the elements and fixed to adjacent solder pads on the shaft by conducting silver paint (Acheson Electrodag 915). The resonator characteristics were measured using a Hewlett Packard HP 8751A Vector Network Analyzer.

It was recognised in the early stages of the work that a relatively thick (~0.6 mm) quartz element would be much less compliant than a conventional epoxy-backed strain gauge and that this would impose particular demands on the adhesive used for bonding an element to a test shaft. As a result, three different adhesive types were investigated (epoxy, cyanoacrylate and methacrylate based) in order to evaluate their influence on torque sensitivity, creep under load, and hysteresis. Analysis of the strain distributions within bonded SAW elements was carried out using the LUSAS finite element package (FEA Ltd., Kingston upon Thames) in order to assess the effects of element geometry on the resulting strain sensitivity.

Results and Discussion

The packaged SAW elements were found to exhibit poor linearity and sensitivity, as shown by the results presented in Fig. 3. It is evident that the relative sensitivity of the low frequency element was greater than that of the high frequency element (1.7 ppm/N.m compared with 0.49 ppm./N.m). Directly-bonded elements gave improved linearity and sensitivity, as shown in Fig. 4. The unloading curves were obtained by carrying out measurements under a reducing load immediately after loading up to 22 N.m. It is clear that the observed hysteresis is introduced by time-dependent creep of the adhesive bonding layer, since both quartz and FV520B stainless steel are expected to behave in a perfectly elastic fashion at these relatively low levels of strain (up to ~300 μstrain). Again it is evident that the low frequency element exhibited a higher relative strain sensitivity than the high frequency element (3.6 ppm/N.m compared with 0.80 ppm/N.m).

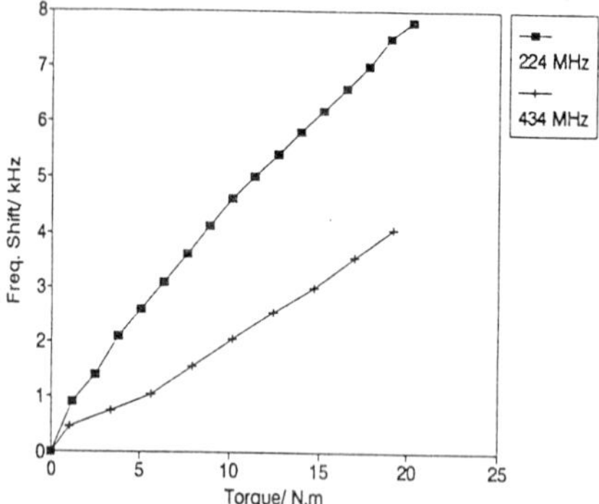

Fig.3. Torque sensitivity of packaged SAW resonators (bonded using methacrylate adhesive).

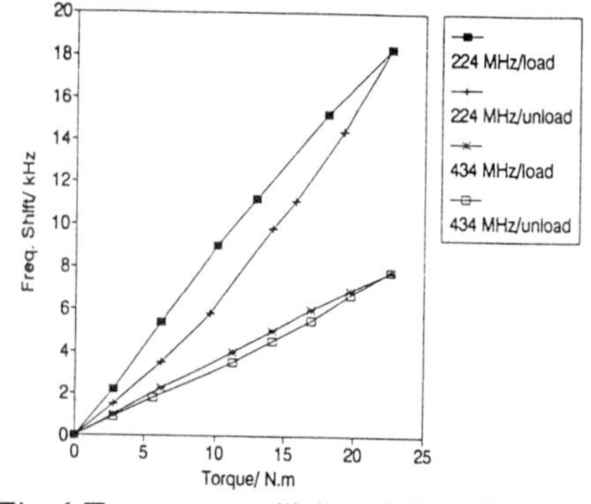

Fig.4. Torque sensitivity of directly bonded SAW resonators (methacrylate adhesive).

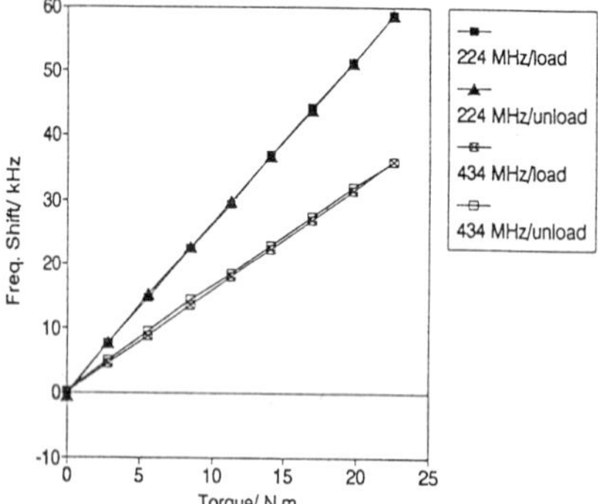

Fig.5. Torque sensitivity of directly bonded SAW resonators (cyanoacrylate adhesive).

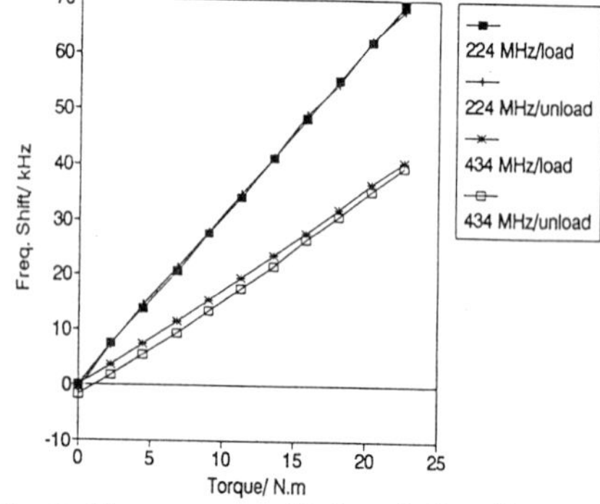

Fig.6. Torque sensitivity of directly bonded SAW resonators (epoxy adhesive).

Further studies on directly-bonded elements revealed that the observed torque sensitivity was dependent on the adhesive type, as shown in Figs. 4-6. Reasonable sensitivities and linearities were obtained for the epoxy and cyanoacrylate adhesives, while the results obtained using the methacrylate-based adhesive were poor. This was further underlined by tests of the longer-term stability under load, where creep in the adhesives led to gradual changes in the observed frequency shift, as shown in Fig. 7. The lower torque sensitivity of the high frequency element is evident in all these results. On the basis of these results, it appears that the cyanoacrylate adhesive should be most suitable for use in bonding SAW strain sensors, although it should be noted that some doubts remain regarding the long-term performance of such materials, particularly at elevated temperatures.

Finite element modelling of the stress distributions within quartz SAW resonators was undertaken in order to identify the cause of the low strain sensitivity of the high frequency element. A simple tensile model was chosen as being the simplest to implement, since it was found experimentally that elements bonded to 2mm thick steel plates, tested in tension, exhibited similar differences in sensitivity. For the purposes of the present model, it was assumed that the adhesive layer had a thickness of 15μm, this being an average of the values determined experimentally. The adhesive material was assumed to behave in a manner typical of a polymer below its glass transition temperature (E~1GPa, ν~0.35), since the detailed characteristics of the cured adhesives used in the study were not available. Conditions of plane strain were assumed in order to reduce the complexity of the model, although it is recognised that the real strain in the element would be greater if it were allowed to contract in the lateral dimension.

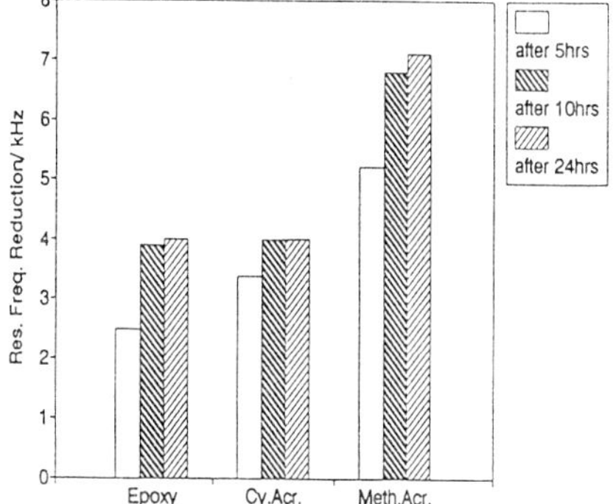

Fig. 7. Observed shift in resonant frequency under constant load.

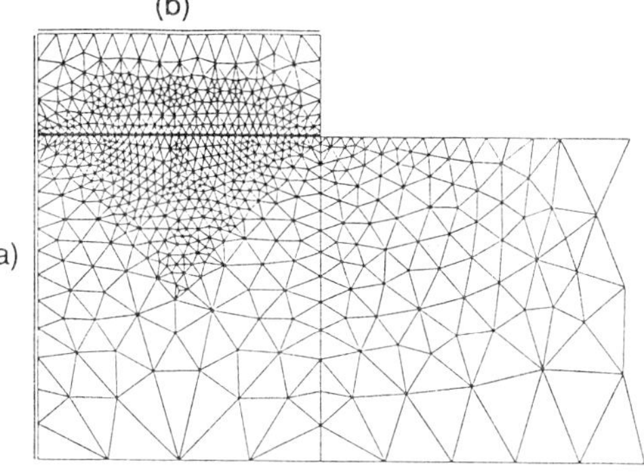

Fig. 8. Section of finite element model for low frequency SAW element.

A section of the model used in the analysis of the low frequency element is illustrated in Fig. 8. Analyses were carried out in order to determine the strain distributions resulting from an applied (horizontal) tensile stress of 1 MPa. The results obtained for the low frequency element are presented in Fig. 9 in terms of the distribution of strain along line section (a) (vertically, through the centre of the element). It was clear that substantial reductions in strain would occur both through the thickness of the element and along the surface, and therefore that the 'measured' strain, at the surface of the element, would always be substantially reduced relative to the strain in the underlying material.

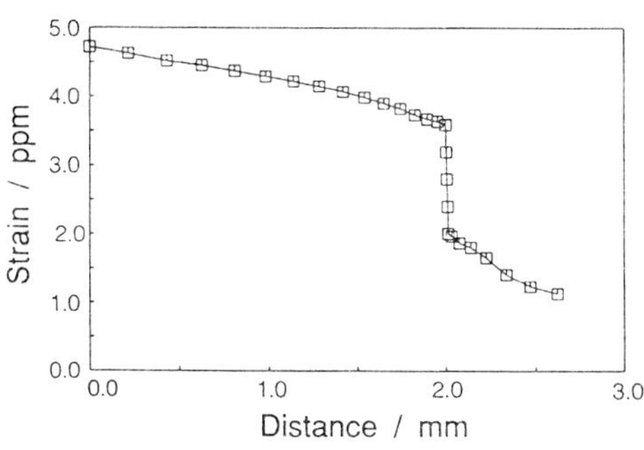

Fig.9. Results obtained for strain analysis of low frequency element along line section (a).

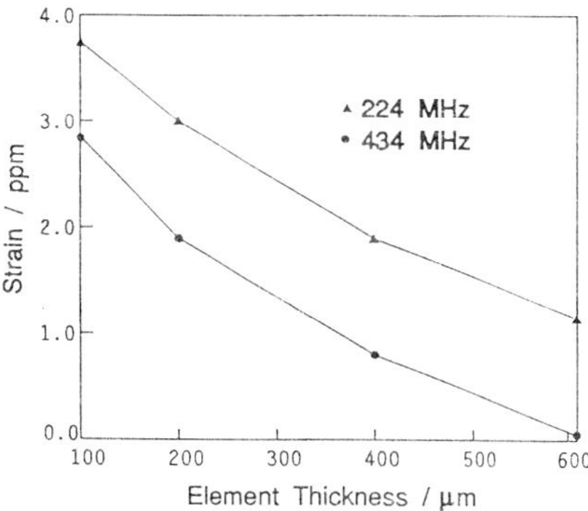

Fig.10. Summary of results obtained for induced strain at the surfaces of elements having different geometries.

By comparing the results obtained for different element geometries, it became clear that this 'dilution' of strain would become more severe if the aspect ratio (length/thickness) of the element were to decrease and that the strain sensitivity should be expected to improve if the element thickness were reduced, as shown in Fig. 10. This would also have the effect of reducing the shear stress and therefore reducing hysteresis caused by creep within the adhesive layer. The effect of the compliant adhesive layer can be seen as a discontinuity in the curve in Fig. 9. Therefore, the relatively low torque sensitivity exhibited by the high frequency element can be explained in terms of an inefficient transfer of strain from the test specimen to the sensing region at the surface of the element.

Conclusions

The present work has demonstrated that a reasonable torque sensitivity ($\Delta f/f_0 \sim$ 14ppm / N.m using epoxy adhesive) can be obtained using a passive quartz SAW resonator directly bonded to a test shaft. The torque sensitivity was found to be dependent on the adhesive type employed and on the geometry of the SAW element itself. Improved sensor characteristics may be expected if the aspect ratio of the element were increased. Further work remains to be done in order to produce elements showing equivalent relative strain sensitivities and operating at different frequencies, and hence to demonstrate the feasibility of producing a temperature-compensated SAW based torque transducer system.

It is clear that the potential of SAW strain sensors is immense. A substantial effort will be required in order to allow the production of devices which can be handled and employed in a manner similar to conventional resistance strain gauges, but this should be well rewarded. Particular areas requiring attention include the sensor itself, where thin film piezoelectrics (e.g. ZnO, AlN, $PbTiO_3$) hold great promise, bonding techniques, packaging, coupling methods, and the signal processing system.

Acknowledgements

The work reported here was carried out as part of a collaborative SERC/DTI Link project, "High Speed Torque Transducer", co-ordinated by Pera International. The authors would like to thank the SERC (D.A.H, H.W.W) and the DTI (A.L) for financial assistance, and all of the collaborating partners for practical help and support.

References

1. P. Kleinschmidt and F. Schmidt, Sensors and Actuators A 31, 35-45 (1992).
2. Lord Rayleigh, Proc. London Math. Soc., 7, 4-11 (1885).
3. R.M. White and F.W. Voltmer, Appl. Phys. Lett. 17, 314-316 (1965).
4. C.K. Campbell, Proceedings of the IEEE 77, 1453-1484 (1989).
5. R.M. White, 1985 IEEE Ultrasonic Symposium Proceedings, 490-494 (1985).
6. A. D'Amico and E. Verona, Sensors and Actuators 17, 55-66 (1989).
7. W. Baldauf, Technisches Messen 58, 329-334 (1991).
8. A. Lonsdale and B. Lonsdale, UK Patent Application 9 004822.4, 3/3/90.

Section D

OPTICAL SENSORS

Optical fibre marine fluorosensor

D. McStay and R. Milne

School of Applied Sciences , The Robert Gordon University, St Andrews St, Aberdeen, AB1 1HG

Abstract

A white light optical fibre fluorosensor for the measurement of marine phytoplankton population is reported. The excitation light is amplitude modulated and a phase sensitive detection system is employed. The spectral properties of two typical samples of phytoplankton obtained off the coast of Scotland have been characterised and used in laboratory tests of the fluorosensor. Initial tests show that the system is capable of resolving fluorescein concentrations in water of 10^{-8}mol/l and phytoplankton populations of 0.1μg/l

Introduction

The worlds oceans account for 99% of the space which is regularly inhabitated by living organisms on earth and is surprisingly enough the least known about. In order to obtain a better understanding of the oceans effect on global processes such as climate, the effects of local pollution and the contribution of environmental factors to the productivity of marine organisms, there is a need for new and improved techniques and instrumentation to process and analyse the extremely complex Biology, Chemistry and Physics of the marine environment e.g. a system to determine the abundance of phytoplankton and zooplankton, the first links in the food chain on which fish depend, is a typical requirement. Phytoplankton (algae) consist of small plants, mostly microscopic in size and unicellular. Two orders of algae commonly predominant in the phytoplankton are Diatoms (Bacillariophyceae) and Dinoflagellates (Dinophyceae). The presence of fluorescent pigments, e.g. Chlorophyll-a in the algae make it possible to investigate the algae properties and populations using optical techniques. This information is of vital importance to marine scientists in the future prediction of fish stocks and the determination of how pollution effects the food chain.

Optical techniques such as fluorescence, attenuation and light scattering measurements are routinely used in the determination of a range of marine parameters, e.g. phytoplankton, dissolved organic material concentration and particle size distribution, and are increasingly used in pollution monitoring e.g oil pollution [1,2,3,5]. The instrumentation used to perform these in-situ measurements are usually purpose built for the marine environment, but are essentially adaptations of traditional spectroscopic instruments, i.e. they contain a source, wavelength selective element, detector and sample chamber through which the water is passed [4]. These components along with the electronics package and the modifications (water tight housings etc) required for operation in a marine environment constitute the main bulk and expense of the device. The simplest optical device for in-situ fluorescence measurements uses a standard fluorometer equipped with a flow cell and a tube and pump system. This technique is, however, not strictly in-situ and suffers from problems associated with the fluid velocity in the sampling hose, mechanical agitation caused by the pump and a change in the light level which may change the fluorescent yield when examining phytoplankton. In the second type of in-situ system the whole instrument is submerged. The water tight housings, power and data cables required for such a device results in a large and expensive system, which render them ineffective when trying to resolve features such as phytoplankton patchiness [7] which occurs on a sub meter scale. Another drawback is that costly damage can occur to such devices, particularly when performing measurements near the sea bottom.

An alternative technique is to employ an optical fibre or fibre bundle to perform the subsea fluorescence measurements while having the source and detector system safely installed on a ship [7]. The optical fibres used in such a system are low cost and therefore, in contrast to conventional systems, even complete loss of the optical fibre probe results in minimal expense. Additionally there is no need for the water tight housings etc required in conventional in-situ devices. Finally the small size of the sensor offers the potential of resolving population variations occurring in less than a meter.

Marine Phtyoplankton

Phytoplankton samples where provided by the Scottish Office Department of Agriculture Food and Fisheries, Marine Laboratory, Aberdeen. Two types of phytoplankton were used, the first sample was a Microflagellate called Tetraselmis Chui (Butcher), class Prasinophyceae strain CCAP 8/6, grown in an f/2 medium (Guillard and Ryther), obtained from a culture collection of algae and protozoa off the west coast of Scotland near Oban in March 1993. These are minute organisms ranging in size from 1-20μm, as shown in figure 1(a). The second sample, was cultured at the Marine laboratories and consisted of mixed Dinoflagellate phytoplankton from Stonehaven Bay, near Aberdeen (North sea). Dinoflagellates are unicellular, biflagellate organisms which range in size from 15-400μm, a typical example is shown schematically in figure 1(b).

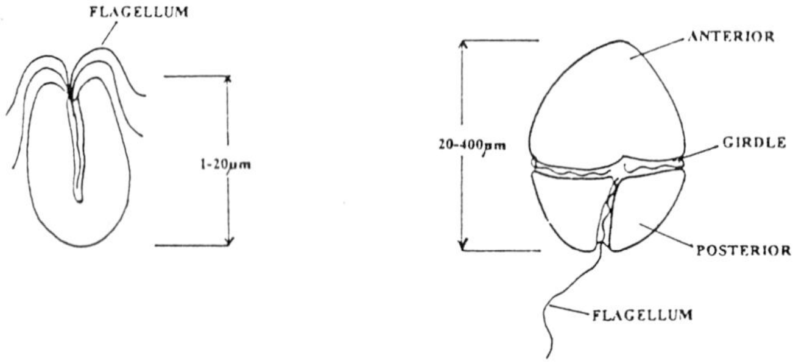

Figure 1: Diagram of phytoplankton species investigated in this work (a) Microflagellate (b) Dinoflagellate

The optical properties of both samples, under typical North Sea environmental conditions, were characterised using standard spectrophotometers. The measured emission characteristics for varying excitation wavelengths for each sample is shown in figure 2. Seawater was used at all times during the fluorescence tests to make up the different phytoplankton concentrations from the original sample stock. From the figures it is apparent that both samples possess a similar excitation and fluorescence spectrum both exhibit a broadband fluorescence extending from around 490nm to 600nm, with that of sample (a) centred around 510-517nm while that of sample (b) is centred around 515-520nm. Sample (a) also exhibits a small secondary peak around 540nm. These results are consistent with previous studies of marine microbiology which have shown that for excitation in the 400nm region there are three main bands associated with marine micro-organisms [6,12]:

1) a broad band emission centred around 470-490nm and extending to around 600nm, due to an unknown pigment [6]. Heterotrophic dinoflagellates, which have been observed to account for 40% of the dinoflagellate population have, however, been shown to exhibit such blue-green fluorescence [12].

2) an emission band centred around 570-585nm due to phycoerythrin [6].

3) the 685nm emission of chlorophyll-*a*

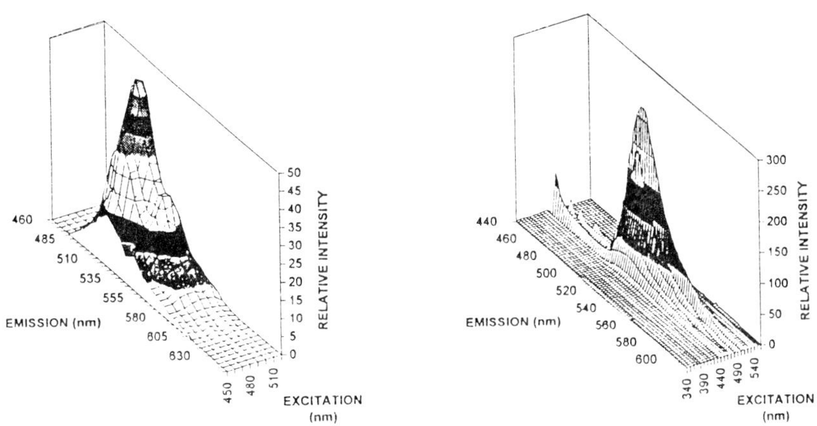

Figure 2: The emission characteristics for varying excitation wavelengths for: (a) Microflagellate (Tetraselmis Chui (Butcher)), class Prasinophyceae strain CCAP 8/6, obtained from a culture collection of algae and protozoa off the west coast of Scotland near Oban in March 1993. (b) mixed North Sea Dinoflagellate obtained from Stonehaven Bay (North Sea).

The actual concentration of the phytoplankton in the samples used in this work was determined by extracting the chlorophylls and carotenoids from replica samples in organic solvents using standard techniques [11]. Chlorophyll-*a* is a pigment universally found in plants and consequently is commonly used as a standard for measuring plant populations. The fluorescence spectrum of the Chlorophyll-*a* for the mixed dinoflagellate sample is shown in figure 3. The phytoplankton populations found, using the above technique, for the original stocks of samples 1 and 2 were 470.μ6g/l and 1664.5μg/l respectively. Separate dilutions of these samples were made and used in tests of the optical fibre marine fluorosensor.

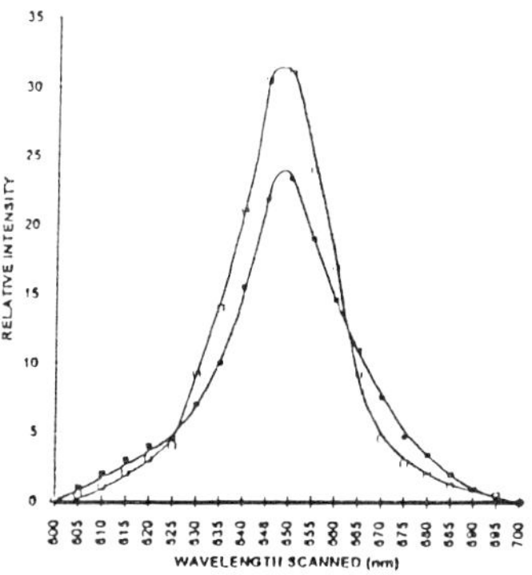

Figure 2: The fluorescence spectrum of the chlorophyll-*a* obtained for the mixed dinoflagellate sample used in the determination of the plant population

Optical Fibre Marine Fluorosensor

The marine fluorescence sensing system we report here utilizes a lightweight, low cost optical fibre probe. The sensor consists of two optical fibres. The first conveys the excitation light which emerges from the far end of the fibre in a cone determined by the numerical aperture of the fibre and the refractive index of the water. The second (detection) fibre is adjacent to the excitation fibre at an angle of 20° to collect as much of the fluorescence generated in this cone as possible, as shown schematically in figure 4. The use of two fibres helps to minimise the potential problem of fluorescence from the fibre core degrading the signal [7]. The wide fluorescence spectrum of the phytoplankton under investigation allow the use of a simple optical filter arrangement as the wavelength selection system for both the excitation and detection, resulting in a low cost spectroscopic system with both a high rejection and throughput. The excitation light is passed through a short-pass filter (cut-off 500nm) and an optical chopper. The amplitude modulated light (0.9kHz) is then launched into the excitation fibre. Repeatable launching of light from the source is noncritical due to the large beam diameter and core size of the optical fibres employed (600-1000μm). The fibres were 20m in length and held in a v-shaped perspex mould. Spectral filtering of the detected light emerging from the near end of the detection optical fibre was achieved by having a band-pass filter arrangement (520-590nm) at the near end of the detection fibre, in front of the photodiode (R.S. Components BPW 21). The output from the photodiode was connected to a lock-in amplifier (EG & G Brookdeal 9503-SC) which took a reference from the chopper. The use of a phase sensitive detection system coupled with the acceptance angle of the optical fibre help minimise the problem of light contamination e.g. that due to the down-welling of surface illumination.

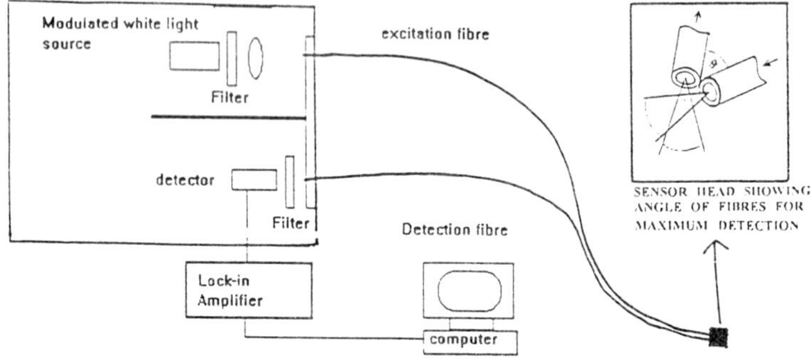

Figure 4: Schematic diagram of the optical fibre marine fluorosensor

Results

The optical fibre marine fluorosensor was initially tested using varying concentrations of the dye fluorescein in water. As can be seen from figure 5 (a), the system was able to resolve concentrations down to 10^{-8}mol/l. The sensitivity of the present system is limited by light contamination from the source caused by the limited rejection of the optical filter system employed. The fibre fluorosensor was then used to examine varying concentrations of phytoplankton in typical North Sea coastal water. In order to prevent the phytoplankton settling at the bottom of the sample container the solution was continually stirred during the measurement using a magnetic stirrer. Figure 5 (b) shows the typical variation of the detector output with concentration of the Dinoflagellate sample. From this figure it is clear that the system can readily resolve phytoplankton populations of the order of 0.1ug/l. This sensitivity is appropriate for measurements of phytoplankton populations in the open ocean where concentrations of 0.1 - 1ug/l are typical [6,12]. The performance of marine optical sensors can be seriously degraded by light contamination, particularly that due to the downwelling of surface illumination. Laboratory tests on the present system have shown that such background light has little effect on the signal. A series of field trials is, however, planned to determine the effect of such downwelling on the sensor performance.

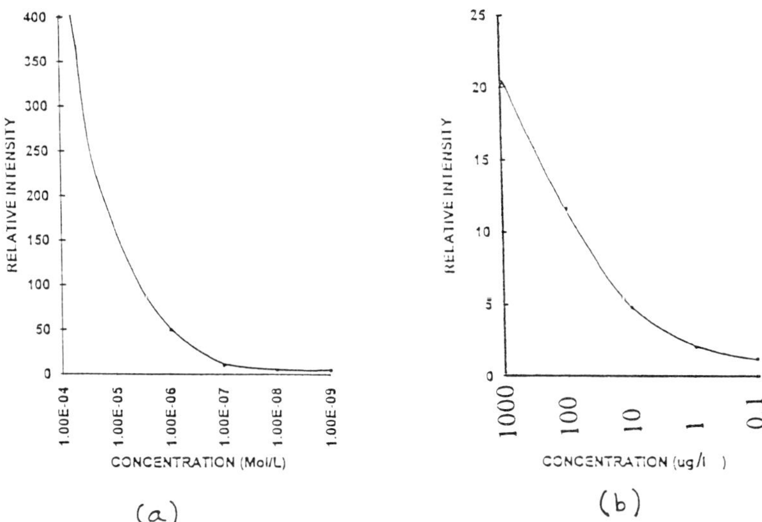

Figure 5 : (a)The variation in the fibre fluorosensor detector output with concentration of (a) the dye fluorescein in water (b)North Sea Dinoflagellate phytoplankton

Conclusions

A simple optical fibre fluorosensor for marine applications has been demonstrated. The system is capable of resolving fluorescein concentrations in water of 10^{-8}mol/l and phytoplankton populations of 0.1μg/l The system has the advantage of being simple, mechanically rugged and has a small sensor head, which allows variations in concentration occurring in less than a meter to be investigated. As well as allowing the use of a single optoelectronics package (containing the source and detection system), which does not have to be submerged, optical fibres have the advantage of being inexpensive, lightweight and chemically inert (and are thus immune to problems associated with corrosion).

Acknowledgements

This work was facilitated by a collaborative research agreement with the Scottish Office, Department of Agriculture, Food and Fisheries Marine Laboratory, Aberdeen as part of the Aberdeen Research Consortium. The authors are grateful to J. Dunn, E. MacDonald and M. Heath of the Marine Laboratory for providing the samples of phytoplankton and for their generous help.

References

1) Blizard M.A. (1988) SPIE Vol.925, Ocean Optics IX
2) Smith R.C. et al (1991) J. Geophys Res, 96, 8665-8686
3) Briggs R. and Grattan K.T.V. (1990), Trans.Inst. MC,12(2), 65-84
4) de Sa E.S. and Desa B.A. (1991), Opt. Eng., 30(10), 1576-1582
5) Smith T.E. (1992) Sea Tech., February, 10-13.
6) Yentsch C.S. and Phinney D.A. (1990) SPIE Vol. 1302 pp328-333
7) Cowles T.J. et al (1989) Applied Optics 28 (3), 595-599
8) Mignani A.G. and Brenci M. (1991) SPIE Vol. 1524, Bioptics: Optics in Biomedicine an Environmental Sciences, 272-289
9) Willams R. and Aiken J. (1990) SPIE Vol. 1269, Environment and Pollution Measurement Sensors and Systems, 186-194
10) Exton R.J. et al (1983) Applied Optics, 22 (1),54-64
11) Boney A.D.(1983) Studies in Biology No. 52 Phytoplankton (Edward Arnold), pp90-92
12) Mazel C. (1990) SPIE Vol 1302, pp320-327.

Optically powered hybrid current measurement system

N.A. Pilling*, R. Holmes+ and G.R. Jones+.

*The National Grid Company plc, Guildford, GU2 5BH.
+Dept. of Elec. Eng. & Elect., University of Liverpool, Liverpool L69 3BX.

ABSTRACT: An optically coupled and optically powered hybrid current measurement system for use in high voltage power systems is described. The transmitter, which uses pulse frequency modulation, is designed for minimum power operation of 2.5 mW. The signal to noise ratio is 54 dB corresponding to a resolution of 1 A. System accuracy is determined by the temperature sensistivity of the transmitter which is approximately 400 ppm°C^{-1}.

1. INTRODUCTION

A high performance optically coupled current measurement system for use on high voltage systems has been the objective of many investigations. Basically two approaches have been followed in this research. The first utilises the Faraday effect which describes the rotation of the plane of polarisation of light travelling through a medium subjected to a magnetic field, Sawa et al (1990).

The second approach is to use a transmitter incorporating active electronic components sited on the current carrying conductor, Berkebile et al (1981). This hybrid transmitter generates an optical signal encoded with information describing the measured current. Hybrid transmitters used for current measurement usually employ a current transformer (CT) to develop a secondary current proportional to the line current. The main problem with hybrid transmitters is that of powering the active components within the transmitter.

One method of powering the transmitter is from the line itself via an auxiliary CT and although it is possible to operate at line currents of less than 1 A, Pilling et al (1992), the system is not fail safe.

An alternative method of powering the transmitter is to use an optical fibre power transmission system. This technique has not been successful previously due to the large power requirement of hybrid transmitters.

It is apparent that the principal objectives in designing a hybrid transmitter are high performance and low power consumption. The transmitter described herein, achieves new standards in this respect, which has allowed optical energisation to be implemented successfully.

2. PRINCIPLE OF OPERATION

A block diagram of the hybrid current measurement system is shown in figure 1.

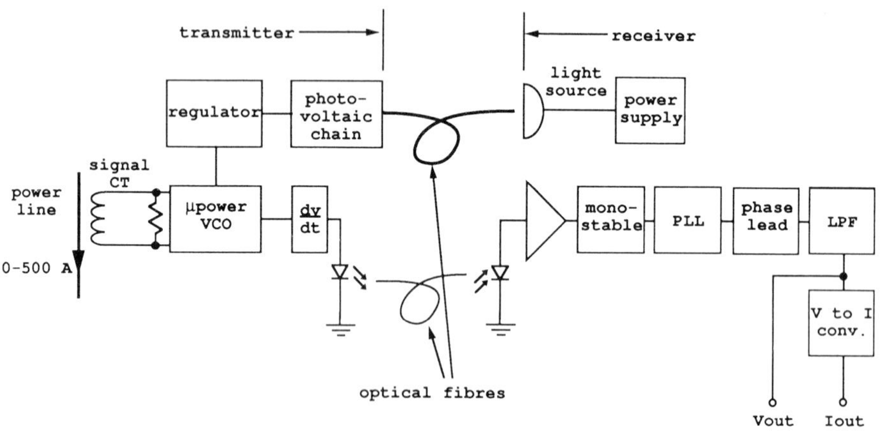

Figure 1 Hybrid system schematic diagram

Approximately 80 mW of optical power is launched into the optical fibre bundle remote from the power line. At the transmitter photovoltaics are used to convert this light into electrical power which is voltage regulated by micro-power circuitry. The supply provides power to a voltage controlled oscillator (VCO) which has been designed to operate with a minimal power consumption of 2.5 mW. The VCO input is driven from a semi-conventional bar primary CT and burden. The high input impedance of the VCO allows this CT to be made much smaller than a conventional CT of the same accuracy. A frequency modulated square wave is generated by the VCO in which the frequency deviation corresponds to the amplitude of the line current. Following differentiation and rectification of the square wave a pulsed frequency modulated waveform is produced to activate the LED. The PFM optical signal is conveyed to a remote point from the power line by multimode optical fibre. The receiver uses a photodiode to drive an amplifier to reconstruct the narrow pulses. A monostable circuit widens these pulses to regenerate a nominal square wave. This has the effect of redistributing the power in the PFM higher harmonics towards the fundamental frequency. A phase locked loop (PLL) demodulates the f.m. signal and a low pass filter is used to further reject the fundamental frequency and its harmonics.

3. CURRENT TRANSFORMER

The transmitter incorporates a CT to generate a current proportional to the line current. A burden is placed across the output of the CT to convert this current into a voltage suitable for connection to the VCO circuitry.

The design criteria for this CT are different from those of standard CTs. Firstly, there is negligible electrical insulation required between the CT and the current carrying conductor and secondly the VCO circuitry imposes a negligible volt-amp requirement. These relaxations in design result in a much smaller CT.

Equation 1 describes the volt-amp (VA) rating of the transformer.

$$VA = E_s I_s = E_s I_p / N_s \qquad (1)$$

where E_s = secondary voltage, I_s = secondary current
I_p = primary current, N_s = number of secondary turns

Equation 2 shows that the transformer core flux density B is directly related to the VA rating for a specified I_p. This normally imposes a minimum level of core flux density in standard CT design which conflicts with the requirement to minimise B in order to increase the accuracy of the CT.

$$B = \frac{1}{\omega A}\left[\frac{E_s}{N_s} + \frac{I_p R_n}{N_s^2}\right] \qquad (2)$$

where ω = line frequency ($100\,\pi$ rad s^{-1}),
A = transformer core c.s.a., R_n = secondary winding resistance

However in this hybrid design, since the VA requirement is negligible, E_s/N_s can be minimised. R_n and N_s^2 are directly proportional, the ratio of the two being determined by the size and geometry of the secondary winding. Increasing N_s does not therefore reduce $I_p R_n/N_s^2$. It is considered that an optimum value of N_s is that which makes E_s/N_s one tenth $I_p R_n/N_s^2$ since any further increase in N_s will result in only a small reduction in B.

4. PULSE FREQUENCY MODULATION (PFM)

Light is transmitted along an optical fibre from the transmitter to the receiver in the form of narrow frequency modulated pulses having a small mark space ratio (1:133).

The power contained in the PFM signal is proportional to the fundamental frequency and to the pulse width. The carrier fundamental frequency selected was 150 kHz to minimise phase error in the demodulated and filtered 50 Hz waveform. Reducing the pulse width requires the electronic circuits used to produce the PFM signal to have a wide bandwidth. An optimum choice of pulse width of 50 ns was made based on the availability of high speed CMOS devices.

The choice of fundamental frequency is directly related to power consumption and should be as low as possible. However, since the 50 Hz line frequency signal must be filtered from the PFM carrier in the final stage of the receiver, a low carrier frequency would result in excessive phase error in the output signal. The input signal to the low pass filter in the receiver consists of the 50 Hz signal and unwanted components with frequencies extending upwards from the fundamental frequency. The low pass filter used to remove these components introduces a phase lag in the 50 Hz signal which will be larger the nearer the filter cut-off frequency is to 50 Hz and the higher the order of the filter.

A phase lead network is used compensate, but compensation is not complete since the line current frequency is not constant at 50 Hz, but can vary between 49.5 and 50.5 Hz under normal operating conditions and between 47 and 51 Hz under fault conditions, BS 3938. It is therefore important to keep the phase lag to a minimum which is achieved by having a fundamental frequency much higher than the line frequency.

A carrier frequency of 150 kHz with a peak frequency deviation of +/- 5 kHz was used. The total phase error calculated does not exceed 3 minutes between 47 and 51 Hz.

5. OPTICAL ENERGISATION

Optical energisation of electronic circuits is not new. Typically square wave optical power is converted into square wave electrical power by a photovoltaic in the transmitter. The photovoltaic output voltage (typically 0.5 V peak) is amplified by an electronic circuit incorporating energy storage elements. The efficiency of such an optical energisation system is limited to 2-3%.

An alternative approach adopted in this system is to use an optical fibre bundle to divide the optical power to a series connected photovoltaic chain as shown in figure 2. An inexpensive tungsten halogen white light source is used to launch approximately 80 mW of optical power into a 2.5 mm diameter bundle. The bundle is randomly split into several tails each of

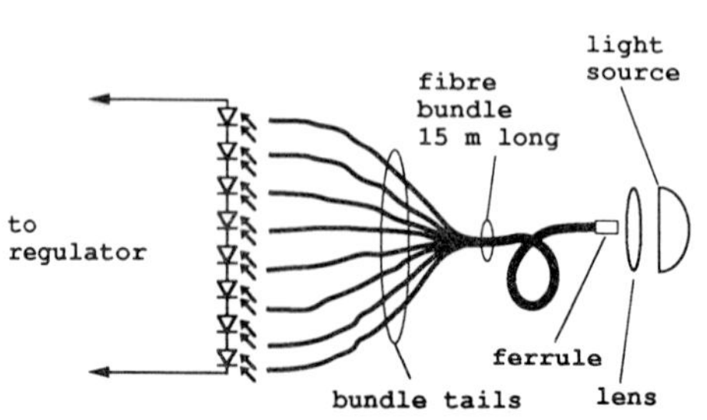

Figure 2 Optical energisation system

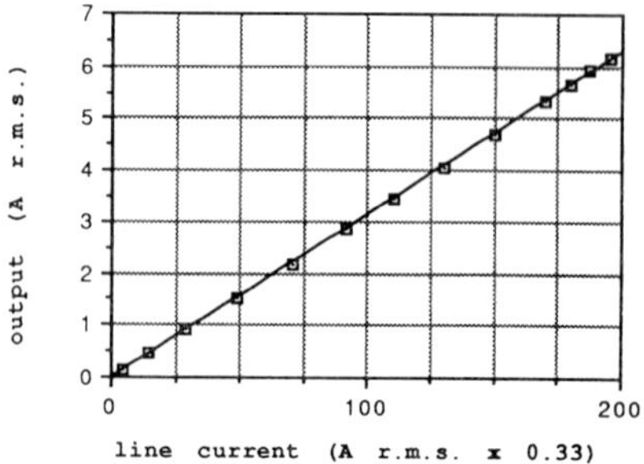

Figure 3 Transfer characteristic

which addresses a photovoltaic cell. By this method the optical power incident on each cell is equalised to better than 10%, an essential requirement for efficient operation of the chain. The system described maximises the efficiency and simplifies the optical energisation system since the optical input power is continuous and there are no voltage amplification circuits. The efficiency of the system was measured at approximately 7% at 24°C.

Voltage regulation of the output of the chain is necessary since this will vary with photovoltaic temperature and input optical power.

6. PERFORMANCE

The sensitivity of the system was set to 0.01 in order to mimic a 500:5 A ratio CT.

The measured signal to noise ratio of the system was approximately 54 dB with a bandwidth of 12 kHz limiting the resolution to 1 A r.m.s.

The linearity of the system was measured to be within the accuracy of the measurement system (2%). The characteristic measured at the V to I converter output is shown in figure 3. There was no measurable hysteresis in the characteristic to within the resolution of the measurement instrumentation (0.2%).

The accuracy of the system is determined by the temperature sensitivity of the transmitter which is approximately 400 ppm°C^{-1}.

The susceptibility of the hybrid transmitter to line voltage was below the resolution of the test instrumentation (0.1%) for line voltages of up to 11 kV r.m.s. For higher voltages the transmitter would require an anti-corona enclosure.

The phase error for the system was measured by comparing the output of the system with that of a reference CT whose phase error was specified to be less than 3 minutes. The measured phase error of the system was better than 20 minutes at one third full load current between 47 and 51 Hz. Sources of phase error include the receiver low pass filter, V to I converter and the transmitter CT.

The length of optical fibre used to connect the transmitter to the receiver was limited to 15 m due to optical attenuation in the fibre bundle. However it should eventually be possible to operate with optical fibres of approximately 1 km in length through the use of superior quality fibre bundles. The maximum length of the signal fibre is limited to several km by optical pulse dispersion.

The transmitter was subject to fault current and capacitor discharge tests by an electricity company. The fault current test involved passing 16 kA through a bar primary running through the centre of the transmitter for 3 seconds and then testing the transmitter for damage. The capacitor discharge test involved discharging capacitors through a bar primary running through the centre of the transmitter. The capacitor was discharged at 3 second intervals, recharging each time, for a total time of 15 minutes. This was done a total of three times with three different capacitors, 60 μF, 16 μF and 3.75 μF at voltages 6 kV and 32 kV respectively. The transmitter successfully passed both tests.

7. DISCUSSION

The optically coupled hybrid current measurement system described offers major advantages compared with conventional CTs. The most important is the reduction in cost and size due to the electrical isolation provided by the optical fibre. This advantage becomes pronounced with increased operating voltage. Other advantages include increase in the maximum allowed distance between the transmitter and receiver and immunity of the optical link to electromagnetic interference.

The electronic circuit used in the transmitter is simple, consumes little power and should prove extremely reliable.

The performance of the system in its present form in respect of accuracy, is consistent with limit of accuracy class 3 for metering CTs, BS 3938. The implementation of temperature compensation, and/or redesign of the transmitter to a higher specification would put the system into class 0.5 limit of accuracy for metering CTs.

8. REFERENCES

1. Sawa, T., Kurosawa, K., Kaminishi, T. and Yokata, T., 1990, IEEE Transactions on Power Delivery, 5, 884-891.

2. Berkebile, L.E., Nilsson, S. and Shan Sun, 1981, IEEE Transactions on Power Apparatus and Systems, PAS-100, 1498-1504.

3. Pilling, N.A, Holmes, R. and Jones, G.R. 1992, Proceedings Tenth International Conference on Gas Discharges and Their Applications, Swansea, 658-661.

4. BS 3938, 1973, Specification for current transformers, British Standards Institution.

9. ACKNOWLEDGEMENTS

The work was supported by Lucas Control Systems Products and the SERC. The South of Scotland Electricity Board is acknowledged for the use of their test facilities.

Bi-directional cladding power detection as a laser material processing monitor

D. Su, D.R. Hall and J.D.C. Jones
Physics Department, Heriot-Watt University, Edinburgh EH14 4AS, Scotland

ABSTRACT: We report a cladding power measurement technique for monitoring laser material processing and machining which uses a fibre optical beam delivery system. The method is based on a measurement of the power propagating in the cladding of the optical fibre in the direction from the work piece to the laser. The power reflected from the work piece and coupled into the cladding is shown to be a function of the position of the work piece relative to the beam waist formed by the output optics of the beam delivery system when the power is low and a function of laser-material interaction at elevated power levels.

1. INTRODUCTION

We have previously reported a technique for monitoring the power level guided in the cladding of a optical fibre[1]. The technique is capable of determining the efficiency with which the laser beam is launched into a fibre optical beam delivery system, the integrity of the system and the optical power reflected and scattered back into the fibre from the output end of the system.

In this paper we describe the development of a sensor based on cladding power detection applied to the monitoring and control of laser material processing. We have applied the technique in a fibre optic beam delivery system used with a high average power Nd:YAG laser, of a type typical of those used for material processing. The technique was illustrated in the detection of work piece position and in the real time monitoring of drilling and cutting of stainless steel. The technique involves no insertion loss, and is non-intrusive: it requires no optics at the work piece other than those already in place for beam delivery. The device consists of a main delivery fibre used for power beam transfer, a monitor fibre which is in optical contact with the main fibre and photodiodes connected to the monitor fibre to detect the optical signal in the monitor fibre. The principle of the bi-directional cladding power monitor is similar to that of a conventional fibre-directional coupler, but instead of coupling light from the core of the main fibre the cladding power monitor treats the main fibre as a multiguide device and couples the light guided in the cladding to the monitor fibre.

The laser beam launched into the main fibre from a laser source is normally retained in the fibre core. However, a fraction of the light scattered and reflected at the work piece is recaptured by the delivery fibre, and propagates back towards the source. A proportion of the reverse-propagating light is efficiently guided over distances of several metres within the cladding of the fibre. The reverse-propagating optical power is a strong function of the position of the work piece, and the state of the processing operation.

2. PRINCIPLE

When a laser beam from a fibre system is focused on a work piece the laser power is partly absorbed and partly reflected or scattered. The reflected beam is focused back into the fibre by the output optics in the opposite direction to the main laser beam. Depending on reflection conditions, such as the alignment and position of the work piece surface relative to the focused laser spot, a fraction of the reflected power is launched into the fibre cladding. During material processing, this reflected power carries information about laser machining process; however at power levels sufficiently low that there is no significant interaction between the laser beam and the work piece, the reflected power in the cladding depends mainly on the position of the work piece surface relative to the laser beam.

We first discuss using this technique to detect the focal position on a work piece at low power level. The optical power in the cladding is a function of the output intensity distribution of the fibre which depends on the beam launching conditions and the configuration of the fibre[2]. It is impractical to give a general model of the beam for a practical system; therefore, as an approximation we use the Gaussian Schell-model for the beam in the free space between the output of the fibre and the work piece, we thus consider that a beam waist exists coincident with the output end of the fibre, with radius w_0 and far-field divergence angle θ_0. The normalized power intensity is hence

$$I = \frac{2}{\pi w^2} \exp(-\frac{2r^2}{w^2}) \tag{1}$$

where w is the beam radius and is a function of the distance travelled by the beam and the properties of the relay optics; r is radial position for a coaxial system. Under the condition of normal incidence on the work piece surface, the power re-launched into the cladding is thus

$$p = \int_{r_1}^{r_2} I 2\pi r dr = \exp(-\frac{2r_1^{\ 2}}{w^2}) - \exp(-\frac{2r_2^{\ 2}}{w^2}) \tag{2}$$

where r_1 and r_2 are the fibre core and cladding radii respectively; w is the beam radius at the fibre end position after reflection from the work piece.

The beam radius is a function of the properties of the output optics and work piece position. A typical fibre output arrangement is shown in figure 1, in which lens F1 of focal length f_1 collimates the beam and lens F2 of focal length f_2 is used to image the near field profile of the fibre output onto the work piece, usually with de-magnification ($f_2 < f_1$). The distance between work piece and the ideal focus point (f_2 from lens F2) is x. In a practical system the distance between the lenses, d, is usually smaller than the optimum f_1+f_2 in order to save space.

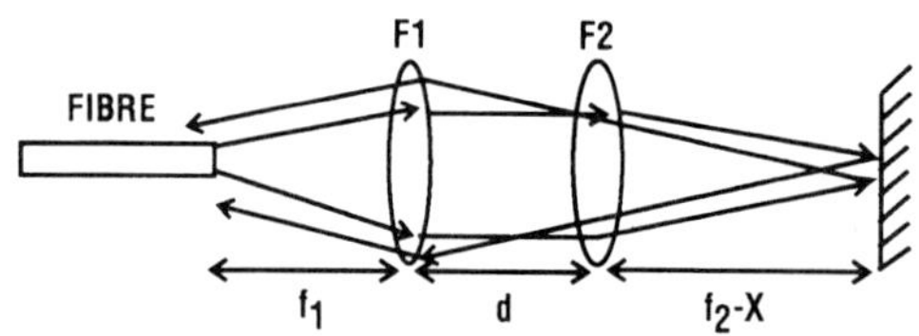

Fig.1 Output optics of a laser optical fibre beam delivery system.

In the above discussion we implied that the reflected beam can still be described by a Gaussian Schell-model, although this is a simplification of the real situation. However, we can generally divide the reflection from the work piece into two components: specular and scattered. In specular reflection the beam properties, e.g. its divergence and waist size, remain unchanged. On the other hand, scattered light behaves rather like a Lambertian source with its energy spreading over a larger angle than for a specular reflection, and its beam NA is only limited by the hard aperture of the output optics. The superposition of the two types of returned signal would normally result in a beam with different characteristics from those of a Gaussian Schell-model, as described by equation (1), thus presenting some difficulties in defining waist size and beam intensity. However, the scattered component is inefficiently re-coupled into the fibre because of its large numerical aperture, such that much of its energy will lie outside the aperture defined by the beam coupling optics. Therefore, for the illustrative purposes of the present analysis, we shall assume that the predominant fraction of the re-coupled light arises from the specularly reflected component. Hence we use the Gaussian Schell-model and the generalized ABCD-law[3,4] to calculate the beam radius w in equation (2). It is easy to show that the reflected beam radius at the fibre end for the configuration shown in figure 1 is

$$w^2 = w_0^2 \left\{ [1 - \frac{2}{f_2^2}(f_1 + f_2 - d)x]^2 + 4(\frac{f_1}{f_2})^4 (\frac{x}{z_0})^2 \right\} \quad (3)$$

where w_0 and θ_0 are the spot radius and divergence angle of the output beam from the fibre respectively and $z_0 = w_0/\theta_0$.

A fraction of the power reflected back into the cladding as a function of displacement x according to equations (2) and (3) can be detected and the output of the photodiode is proportional to the power propagating in the cladding in the direction opposite to the main beam.

For laser material process monitoring, the laser power is high and the interaction between a laser beam and matter is complicated, but in high power laser cutting and drilling of metals the process can be described as follows [5]: the first step is the heating of the surface due to laser energy absorption; when the temperature of the surface is high enough, melting, oxidation and plasma formation occurs. Oxidation increases the absorptivity of the laser energy, and the molten material is carried away by the assisting gas. During heating, the surface reflects a large fraction of the laser power depending on the type of the material, its surface finish and the temperature. Although the reflection is reduced after the surface is oxidized or melted, there is still some power reflected together with optical emission from the interaction between laser and material. The reflection is considerably reduced when the material is penetrated during drilling or cutting. The specular reflection is also very weak due to the deformation of the surface except at the beginning of a pulse, before the interaction has started. The main sources of the signal are non-specular reflection and light emission from the work piece which carries information about the machining process, and can be extracted and monitored without disturbing the power in the fibre core [1].

3. EXPERIMENTAL RESULTS

Experiments were carried out for two different applications: drilling and cutting of stainless steel plate of 2mm thickness; and work piece defocus detection at low power levels. In each case an industrial laser (Lumonics JK702) was used. The apparatus is shown in figure 2. The launched beam is monitored by photodiode 2 and the return signal from target T is monitored by photodiode 1 through a monitor fibre attached to the main beam delivery fibre. The input and return beam waveforms were sampled and recorded simultaneously. A main fibre of core diameter 1mm and a monitor fibre of core/cladding

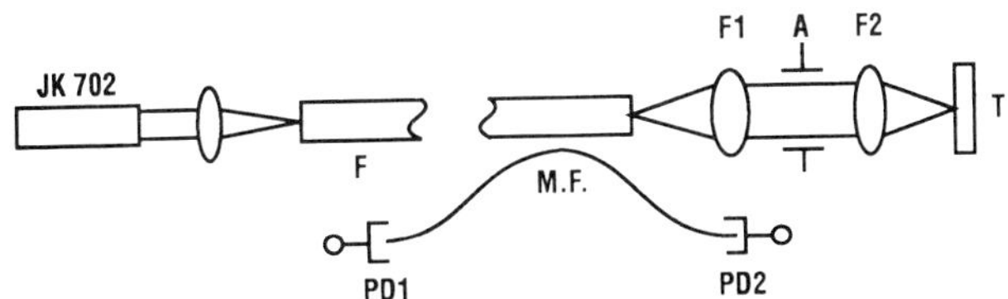

Fig.2 Experiment arrangement. JK-702: industrial Nd:YAG laser; F: main fibre; M.F.: monitor fibre; F1, F2: lenses of focal length f_1 and f_2; T: work piece; PD1, PD2: photodiodes; A: aperture.

diameter 50/125μm were used. In the work piece position sensing experiment the focal lengths of the lenses were f_1=125mm and f_2=100mm. The distance between the lenses (d) is 20mm. An aluminium block (T) was placed close to the focus of F2 and its axial distance was adjustable. An aperture of diameter 16mm was placed between the lenses to limit the scattered light whilst transmitting the main beam, which gives a NA of 0.064 for a collimated main beam. In the laser drilling and cutting experiments f_1=160mm and f_2=80mm.

Typical normalized results of the detector output as a function of the work piece axial displacement relative to the focus are given in figure 3, together with a theoretical curve, calculated according to equation (2) for the input parameters of w_0=0.4mm and θ_0=0.064.

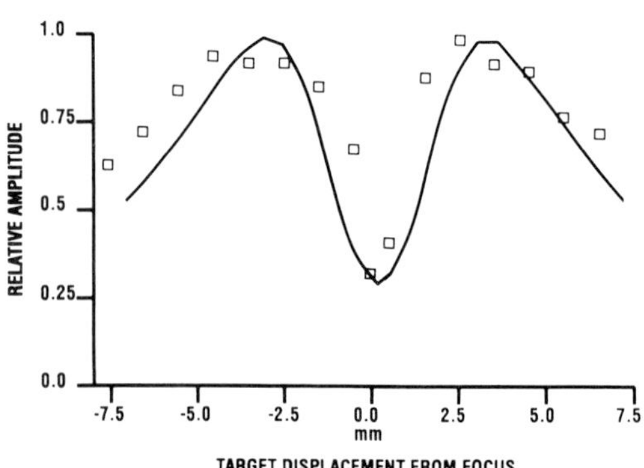

Fig.3 Output amplitude of PD1 versus work piece displacement with theoretical curve.

For laser drilling, the upper trace of figure 4 shows 6-pulse drilling of a 2mm thick stainless plate, and the lower trace shows 4-pulse drilling on the same sample. After 4 pulses the outlet opening size of the hole is only 0.1mm which caused no effect on the monitor signal; but after 6 pulses the outlet opening is 0.5mm and a drop in signal level of the sixth pulse is visible. A further pulse would develop the hole to its final size of about 0.58mm and the signal level then becomes stable at a constant amplitude due to cross-talk and other background reflections. Cross-talk is the small but inevitable and unwanted sensitivity of the return signal detector to light propagating in the forward direction in the fibre cladding.

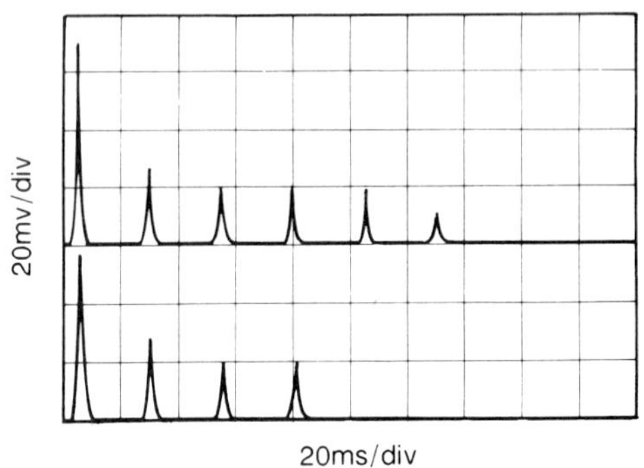

Fig.4 Monitor signal for laser drilling of stainless steel plate (upper and lower traces, 6 and 4 pulse respectively).

The process of cutting is different from that of drilling. During cutting, fresh material is fed continuously into the laser radiation area; therefore, unlike the case of drilling where penetration leads to no reflection signal, the material is not necessarily penetrated over the full extent of the laser beam. Air or oxygen assisted cutting can result in reduced surface reflection.

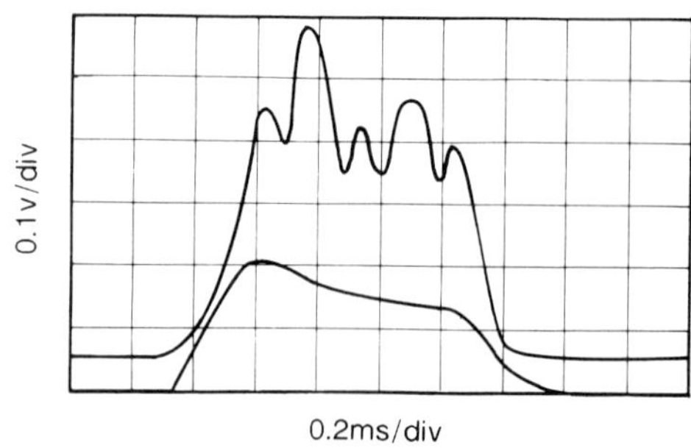

Fig.5 Monitor signal for argon assisted cutting (upper and lower traces 270 and 265mm/min feed-rates respectively).

Experiments were conducted on cutting of the 2mm thick stainless steel plate with argon as assist gas at different sample feeding speeds. High cutting speed results in failure to penetrate the sample. Trials were carried out using fixed laser pulse parameters and an average laser power of 210 W. Typical results are shown in figures 5. The upper trace of figure 5 shows a return pulse shape from an argon-assisted cut at a speed of 270mm/min and the lower trace at 265mm/min. The trace corresponding to higher feeding speed is the return signal representing a failed cut and the lower speed trace a successful one. The lower trace of the figure shows a relatively smooth pulse and bears similarities with the input pulse, with a high peak at the leading edge; conversely the upper trace has a relatively higher amplitude compared with its counterpart and shows random spikes superimposed on the pulse. We attribute those random spikes to the unstable process of light emission and material removal, although we do not have a complete explanation.

4. DISCUSSION AND CONCLUSION

When laser power is low the device can be used to locate the relative position of the work piece and laser focus. The detector output is a function of input spot size and divergence angle as well as the work piece displacement. The qualitative features of the figure 3 may be explained as follows. When the work piece surface is at the focus of the laser beam, most of the power is reflected into the fibre core and the fibre end is coincident with the reflected beam waist; the output is a minimum, and is a function of the spot and fibre sizes. As the work piece moves away from the focus, the reflected beam radius at the fibre end becomes larger and more light is launched into the cladding, hence the output of the detector increases; when the work piece is yet further away from the focus the decrease in intensity, due to beam radius increase at the fibre end position, dominates the monitor output. This process forms two peak outputs, besides the minimum value at the focus, as shown in figure 3.

For laser material processing we have shown that a high power laser beam delivery system with a cladding power monitor can be used to monitor metal drilling and cutting. The experiments successfully detected the difference between successful and unsuccessful machining processes. A drop in signal intensity is observed when a laser beam penetrates a metal sample; and different signal shapes and amplitudes distinguish a successful cut from an unsuccessful one. The application of the sensor described in this work would contribute to the automation and optimization of control parameters in laser material processing and hence increase reliability and productivity. The cladding power monitor is effective in real time monitoring of laser material processing; the principle can be extended easily to other applications such as laser medical applications, and possibly with the help of wavelength analysis of the return signal. The resolution of the work piece position measurement is appropriate for many material processing applications, and is achieved without intrusion or insertion loss.

ACKNOWLEDGEMENT

The authors thank Dr. C.L.M. Ireland of Lumonics Ltd (UK) for his encouragement and helpful discussions.

REFERENCES

[1] A.A.P. Boechat, D. Su and J.D.C. Jones "Bi-directional cladding power monitor for fibre-optic beam delivery systems", Measurement Science and Technology, 3(9), 897-901(1992).

[2] D. Su, A.A.P Boechat and J.D.C. Jones, "Beam delivery by large-core fibres: effects of launching conditions on near-field output profile", Applied Optics, 31(27), 5816-5821(1992).

[3] J. Turunen and A.F. Friberg, "Matrix representation of Gaussian Schell-model beams in optical system", Optics and Laser Technology, 18(5), 259-267(1986).
[4] M.W. Sasnett, "Propagation of multimode laser beams-The M^2 factor," Chap. 9 in The physics and technology of laser resonators, D.R. Hall and P.E. Jackson, Eds., pp.134-142, Adam Hilger, Bristol (1989).
[5] A.M. Prokhorov, V.I. Konov, I. Ursu and I.N. Mihailescu, Laser Heating of Metals, Adam Hilger, Bristol, 1990.

An optical fibre imaging system for the study of high speed motion

P.M. Weaver J.W. McBride

Departments of Mechanical and Electrical Engineering
University of Southampton
Highfield, Southampton, Hampshire, SO9 5NH, U.K.

ABSTRACT: This paper describes an imaging system employing an array of optical fibres and electronic light detection designed to study high speed motion. Light levels were detected electronically, and recorded digitally at high speed. The data were transferred to computer for permanent storage, analysis and image formation. Over 4000 images were captured at a maximum rate of 1 image per μs. The system has a number of potential applications, demonstrated in this paper by studies of the motion of the electric arc and the high speed contact mechanism in a miniature circuit breaker.

1. INTRODUCTION

The production of images of rapid events usually relies on photographic or video technology. However, such techniques are often difficult or impossible to implement under harsh conditions and in enclosed spaces. Where photographic or video techniques are used, the processing and interpretation of the recordings can be laborious and time-consuming. In addition the recording is too slow to provide detailed information on events occurring on a timescale of μs. High speed video is capable of a framing rate of the order of 1000 frames per second while high speed cine is capable of typically 10,000 frames per second. These limitations make conventional camera techniques less than ideal in areas such as switchgear research where the fast motion of electric arcs and contact mechanisms need to be studied. For these reasons a number of recent studies have employed electronic means to detect arc motion e.g. by virtue of its magnetic properties [1]. Recent advances in optical fibre technology have opened up new possibilities for studying arc motion by positioning an array of optical fibres in the arc chamber [2,3,4].

An electronic technique using optical fibre light guides is capable of overcoming many of the problems discussed above. Optical fibres permit access to enclosed spaces with a minimum of disturbance to the event being studied. Electronic detection and recording of the light can provide rapid response times of the order of microseconds and immediate access to data in electronic form for storage and analysis by computer. However, to achieve a spatial resolution comparable to camera techniques would require a prohibitive number of detectors and the associated data acquisition circuitry. A lower spatial resolution than could be obtained from a camera must therefore be accepted.

This paper describes the light detection by means of an array of optical fibres, the recording technique and a novel computational method used to construct an image of the arc from these data. Results are presented on the use of the system to study miniature circuit breaker arcs and contact mechanisms.

2. SIGNAL DETECTION AND DATA ACQUISITION

A schematic diagram of the electronics and instrumentation system is shown in Fig.1. An array of 47 optical fibres was positioned with optical access to the system under study, in this case the arc chamber of a miniature circuit breaker. Polymer fibres with a core diameter of 1mm were used. Attenuation in polymer fibre is higher than in glass - typically 200 dB km^{-1} at 665nm but over relatively short transmission distances required this is unlikely to impose serious limitations. The main advantages of polymer fibre lie in its ease of manipulation and an aperture comparable with the resolution required. Each optical fibre was recessed a distance **a** from the light source.

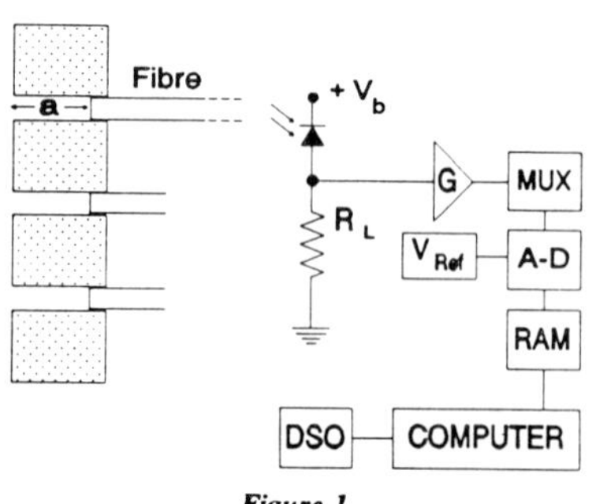

Figure 1
Instrumentation for optical fibre imaging system.

The light detector for each fibre consisted of a reverse biased photodiode ($\mathbf{V_b}$). Illumination of the photodiode caused a current to flow through the load resistor $\mathbf{R_L}$ to generate a voltage proportional to the light intensity on the fibre. The signal was passed through an amplifier of gain **G** (rise time 2.5μs) and an 8 way multiplexer and digitised by a 6 bit flash a-d converter. The value of the digital number was determined by a reference voltage $\mathbf{V_{Ref}}$. The digital information was recorded in RAM (32k*8bit). The timing of the multiplexer switching, a-d conversion and RAM write operations were synchronised so that one sample was made every clock cycle. A clock speed programmable up to 8MHz was used to provide a minimum recording time of 4.096ms. The system was connected to a p.c. computer via a 24 channel digital I/O card. The computer was used to set up the system and to initiate the data recording which could be synchronised with other instruments such as a digital storage oscilloscope (DSO). The transfer of data to the computer and storage on hard disc was subsequently effected under computer control.

The sensitivity of the light detection system needs to be adjusted for different applications. This can be accomplished by alteration of the following parameters:

(i) The fibre recess, **a**. The optical fibres are mounted in a tube that separates the front of the fibre from the light source. Increasing this recess reduces the light intensity on the fibre as well as reducing the effective aperture of the fibre [4].

(ii) The photodiode bias voltage, $\mathbf{V_b}$. The current through the photodiode is dependent on the bias voltage as well as the light intensity. A reduced bias voltage can be used to decrease the sensitivity of the apparatus

(iii) The load resistor, $\mathbf{R_L}$. The voltage input to the amplifier stage is the product of photocurrent and load resistance so a higher $\mathbf{R_L}$ gives a larger signal for a given light level.

(iv) The amplifier gain. The gain of the photodiode amplifier stage can be increased to make the apparatus more sensitive to low level light.

(v) The reference voltage for the a/d converter, $\mathbf{V_{Ref}}$. This can be reduced to increase the sensitivity of the device and allow the detection of lower intensity light.

Variation of some or all of these parameters allows the system to be optimised for a particular application. This paper describes two such applications - studies of arc motion and contact motion in a miniature circuit breaker.

3. IMAGE AND DATA ANALYSIS

To use the full potential of an image derived from intensity measurements at a limited number of points the image of the arc is presented as a set of contours each associated with a different intensity threshold. All the channels with an intensity greater than or equal to the threshold level were considered "on" and lie inside the contour; all those lower were "off". There are two aspects to the problem of constructing the contour:

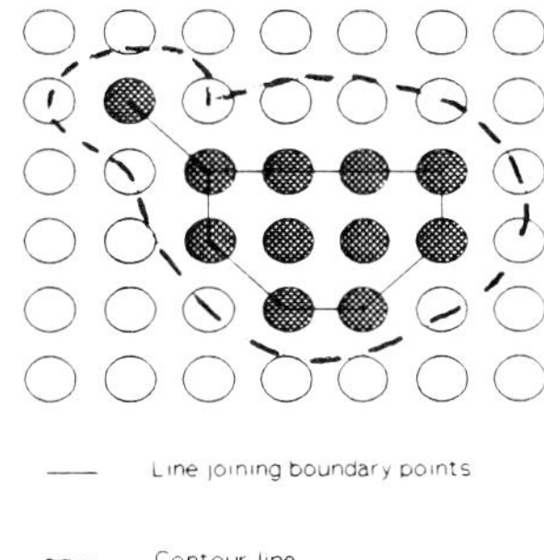

Figure 2
Sketch of a typical boundary with a contour line drawn around it.

(i) to identify the optical fibre channels that lie on the boundary of an "on" region. This produces a series of points which can be joined to make a line (Fig.2). However for a spur or a boundary consisting of only two points no area would be enclosed by a line joining the points and the image would not faithfully represent the spatial distribution of intensity.

(ii) a contour line must be drawn around the boundary points at a distance indicative of the area over which the optical fibre is sensitive or representative. Slightly reducing the distance between the boundary point and the contour line for contours of increasing intensity permits nesting of the contours.

The contour plotting technique was executed in software in a generalised form so that the program was capable of processing information from different arrangements of optical fibres to produce an animated image of the system being studied on the computer screen. The data could also be analysed for other information such as the centre of the intensity distribution.

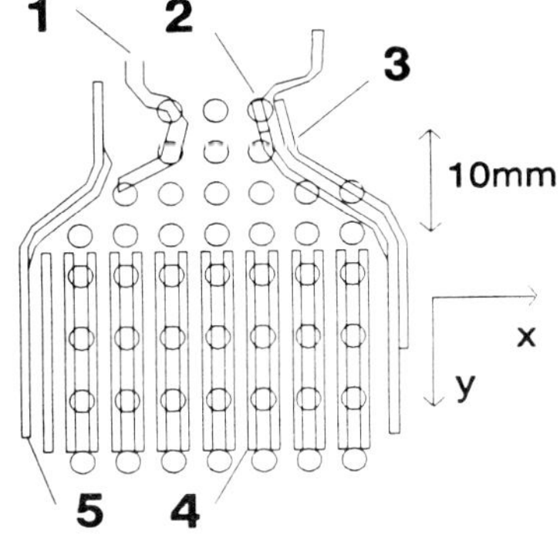

Figure 3
Schematic diagram of m.c.b. (see text for description).

4. RESULTS

4.1 Arc Motion

Miniature circuit breakers protect against potentially very large short circuit currents by rapidly breaking the circuit when an abnormal current is detected. A schematic diagram of the miniature circuit breaker used in the present study is shown in Fig.3 where (1) is the moving contact, (2) the fixed contact, (3) the fixed contact arc runner, (4) the arc splitter plates, (5) the moving contact arc runner. Circles show the optical fibre positions. When the contacts open an arc is drawn and the magnetic field generated by the current through the conductors propels the arc off the contacts into a stack of steel plates. The plates split the arc into a number of small arcs in series. This increases the arc voltage and helps reduce the current to zero in typically 3-4ms. The optical fibre imaging system is able to provide detailed information on the behaviour of the arc by detecting the emitted light. Images of the arc motion (peak current 3000A) are shown in Fig.4. The arc moves rapidly from the contact region into the splitter plates (Fig.4a,b). However, the contact gap cannot sustain the high voltage and the arc re-ignites in the contact region (Fig.4c). This sequence of events can occur many times during a single circuit breaking event. The re-ignition of the arc can cause damage in the contact area as well as reducing the performance of the circuit breaker. The imaging system can be used to study such phenomena such and to assess the effectiveness of design changes.

An arc trajectory was obtained from the light intensity distribution by calculating the centre of light intensity (x,y):

$$x = \frac{\sum I_i X_i}{\sum I_i}$$

$$y = \frac{\sum I_i Y_i}{\sum I_i}$$

The sums are over all of the fibres in the array where I_i is the light intensity at a given time for the fibre at position (X_i , Y_i). A typical arc trajectory is shown in Fig.5 with a simultaneously recorded voltage. Initially the arc is immobile or slowly moving (up to typically $20ms^{-1}$). As the arc approaches the splitter plates the voltage rises until a re-ignition occurs. The arc then moves once more towards the splitter plates but at a higher velocity (typically $100\text{-}300ms^{-1}$). This sequence of events is repeated many times (see Fig.4) before the circuit is broken. The trajectory information can be used to study how the arc voltage varies with position in the chamber and to obtain information on the arc velocity.

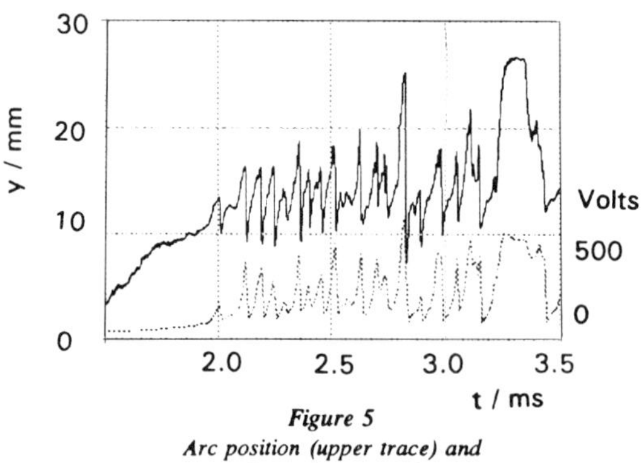

Figure 5
Arc position (upper trace) and voltage (lower trace).

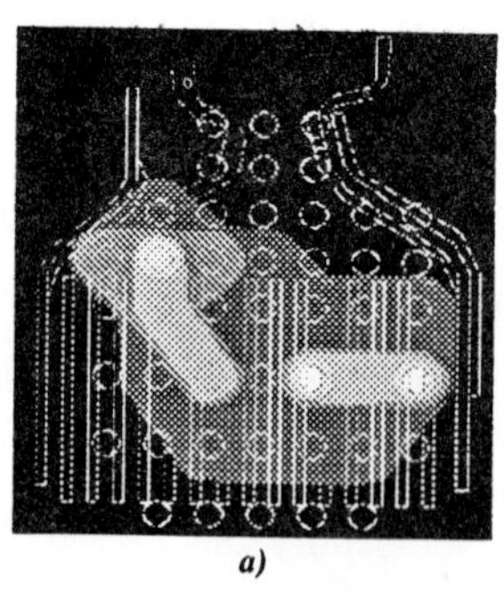

a)

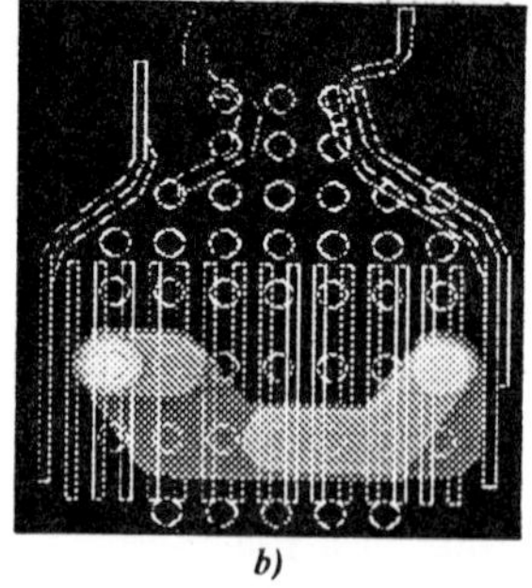

b)

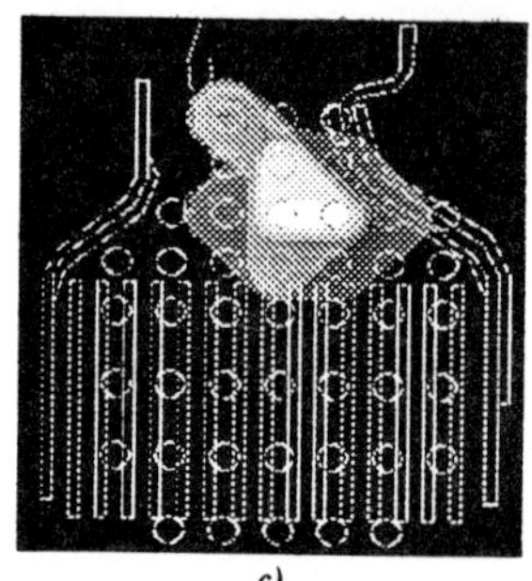

c)

Figure 4
Images of arc motion
a) 3063μs, 291V, 3063A
b) 3090μs, 485V, 2095A
c) 3210μs, 52V, 1676A

4.2 Contact Motion

To study the motion of the contacts the optical fibres were mounted in an array concentrated in the contact region. The array was backlit by a 150W projector bulb enclosed in a parabolic reflector. The contacts were short circuited so that, although the full short circuit current flowed through the contact actuator mechanism, no arc was formed. The light intensity was lower than that from a short circuit arc so the sensitivity of the data acquisition was increased by reducing the reference voltage for the a-d converter.

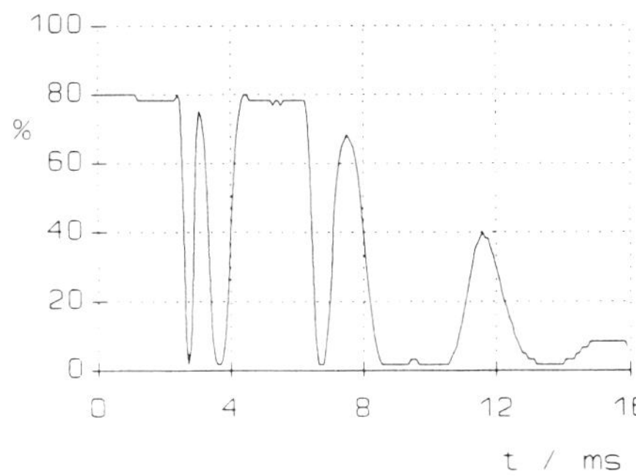

Figure 6
Light intensity signal (% full scale).

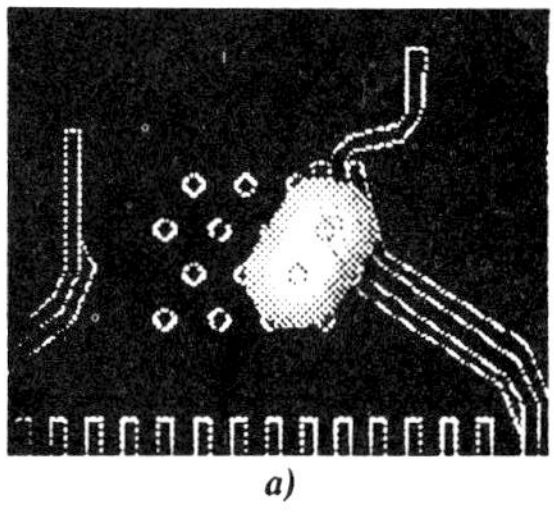
a)

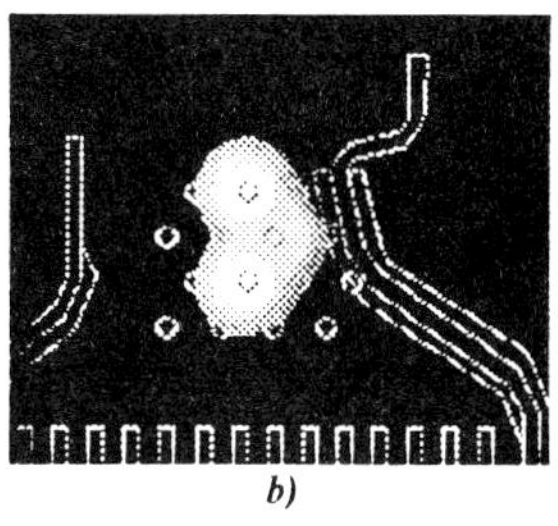
b)

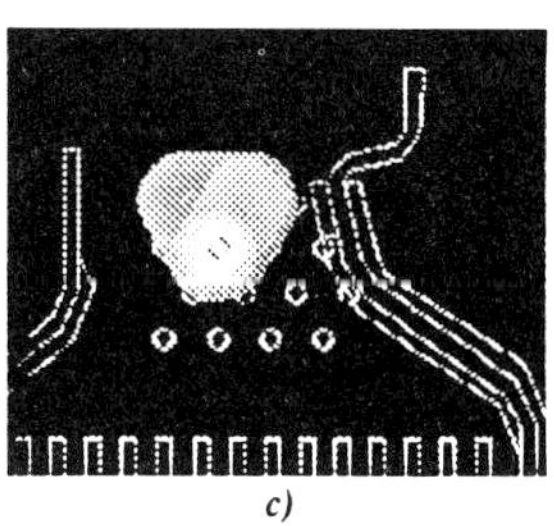
c)

Figure 7
Images of contact position at a) 1ms, b) 2.3ms, c) 15ms.

As the contact moved a shadow passed over the array. Each fibre detected variations in light intensity to give signals such as that shown in Fig.6. Initially the contacts were closed and the light intensity at this fibre position was a maximum. As the contact passed over the fibre the light level decreased then returned to maximum as the contact moved past. The contact rebounded against the arc runner and once more passed the optical fibre. Subsequent oscillations of the contact produced further variations in light intensity until the contact came to rest over the fibre.

To construct images of the contact motion the light intensity data were inverted and plotted as described above (Section 3). Typical results are shown in Fig.7. The resolution is too low to obtain detailed quantitative information from the images and further analysis of the data is required.

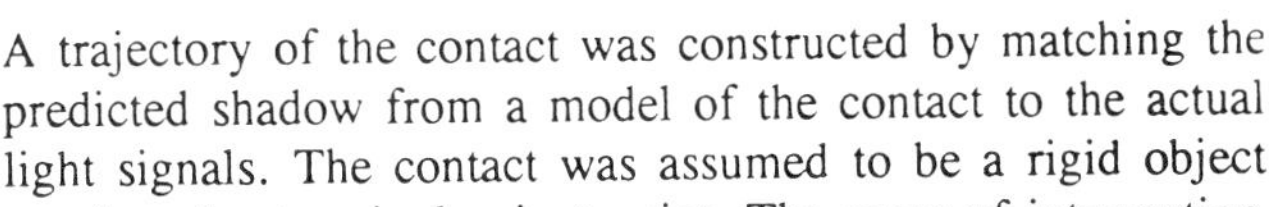

A trajectory of the contact was constructed by matching the predicted shadow from a model of the contact to the actual light signals. The contact was assumed to be a rigid object rotating about a single pivot point. The areas of intersection of the shadow and each fibre were calculated for positions across the arc chamber. At a given time the RMS deviation between the model and real intensity distributions was calculated for each model position. The model position that produced the lowest deviation was used as a measure of the contact position at that time.

The contact position was calculated at 100μs steps to produce a trajectory as shown in Fig.8 where the position $x = 0$ corresponds to the nominal open position of the moving contact. The contact opens and impacts with the arc runner at a small negative displacement. The impact occurs for a very short period of approximately 0.1 - 0.2 ms during which time energy is absorbed and the angular velocity at rebound is reduced. The maximum travel of the moving contact is 7mm. The velocity of the contacts opening is 7.13 ms^{-1} at the point of impact. After impact the contact rebounds and contact gap is reduced to 2.5mm at 3.5 ms after the start of the movement. At high a.c. currents ($>$10KA) the vibration of the moving contact after the impact period may have an important effect on the voltage characteristics and it may be advantageous to reduce this vibration.

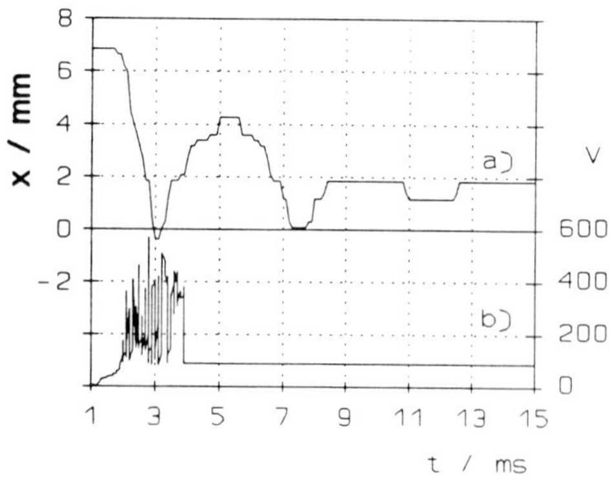

Figure 8
a) *Contact position.* ***b)*** *Arc voltage.*

5. CONCLUSIONS

An imaging system is described that uses an array of optical fibres and detecting electronics to observe high speed motion. Over 4000 images can be captured at a rate of 1 image per μs. The system has a number of potential applications.

Computer analysis of the data is used to re-construct animated images of the moving object and to obtain detailed trajectories of the motion.

The system is applied to the study of arc and contact motion in a miniature circuit breaker.

The analysis of the arc motion yields new information on the arc behaviour, its trajectory and velocity.

Detailed information on the contact dynamics is obtained from the optical study.

6. REFERENCES

[1] M. Mercier, A. Laurent, G. Velleaud, and F. Gary, "Study of the movement of an electric breaking arc at low voltage", J.Phys. D: Appl. Phys., 24, pp.681-684, (1991)

[2] M. Delaplace, "Video screen observation of arc behaviour inside a cut-off chamber", Revue Generale de l'Electricite, n.1, Jan. 1987, pp.26-32, (1987)

[3] J. Wassermann and W. Ziegler, "Quantitative recording of arc motion and structure through opaque walls employing optoelectronic sensors", J.Phys.E: Sci. Instrum., 21, pp.155-158, (1988)

[4] Weaver P.M., McBride J.W. "Arc motion in current limiting circuit breakers." Proc. 16th Int.Conf. on Electrical Contacts, Loughborough, pp.285-288 (1992)

The authors wish to acknowledge the financial support provided by S.E.R.C. and Crabtree Electrical Industries Limited and the technical contribution of Mr.M.L.Coombes.

Reduction of semiconductor laser noise, as applied to an optical microphone, using negative feedback

L S Karatzas, D A Keating, and M J Usher

Department of Cybernetics, School of Engineering and Information Sciences, University of Reading, PO Box 225, Whiteknights, Reading RG6 2AY, Berkshire, England.

ABSTRACT: A simple method using negative electronic current feedback in order to reduce the laser noise, when used in an interferometric application, is presented. The configuration of the experimental set-up and its application is shown. Noise spectrum plots under various conditions are examined. Conclusions are drawn on the advantages of negative feedback.

1. INTRODUCTION

The problem of laser noise, concerning applications like telecommunications and sensing [DAN80, PET81], is very important and widely researched. Laser noise can be classified to two types: amplitude and phase [DAN81, PET81, DAK86]. One of the most favourite techniques to achieve noise reduction is the use of negative electronic feedback. The feedback signal can be frequency [SAI85, OHT90] or current [VAN81, DAK86, YAM87, NEW89, KAN90, LEE90].

The laser diode is to be used in the interferometric detection of the displacement of a microphone membrane. The membrane will also be subject to negative force-feedback control, so the displacement will be restricted to few nanometers. Thus, the phase noise will be minimised [DAN81], and the largest noise contribution from the laser will be the amplitude noise. Current feedback will be used to minimise this contribution, and ultimately the limit posed in the system is the shot noise from the photodiode [DAN80, NEW89]. In the particular application for which the laser is intended [KAR91, KAR93, KEA91], the shot noise limit is calculated to be 1.39nA rms, in a bandwidth of 20kHz.

2. EXPERIMENTAL ARRANGEMENTS

2.1 Instrument characteristics

The laser diode used (Melles Griot, 06 DLL 907) is an index guided AlGaAs SHARP LT080MD model. It has a linearly polarised, collimated circular beam of 7.5mm diameter , and a high wavefront quality ($\lambda/4$ peak to peak). The wavelength of the radiation emitted is at 780nm, in a single transverse mode. The diode begins lasing after 47.9mA. The usual driving current is of the order of 56mA at 25°C, which corresponds to an output power of 3mW.

Using the Melles Griot 06 DLD 003 driver, temperature (stability ≤ 0.05°C for ambient drifts of ±10°C) and power stabilisation (stability of 0.2%) is possible, so mode-hopping is absent [DAN81, NEW89, LEE90]. The driver has an analog current modulation bandwidth of 200kHz, and power modulation bandwidth of 100kHz.

The wave analyser for the noise spectrum plots is a Hewlett-Packard HP3581A, the oscilloscope a HP54600A model, and the printer a HP Deskjet 500.

2.2 Experimental layout

Figure 1 shows the layout with no external feedback, and Figure 2 the arrangement when external feedback is applied to the laser.

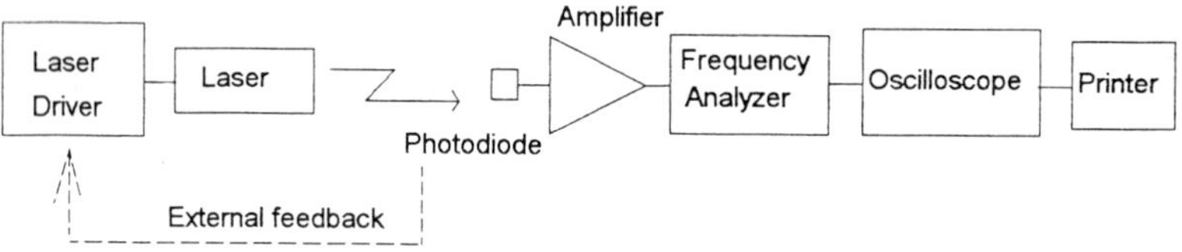

Figure 1. Experimental set-up.

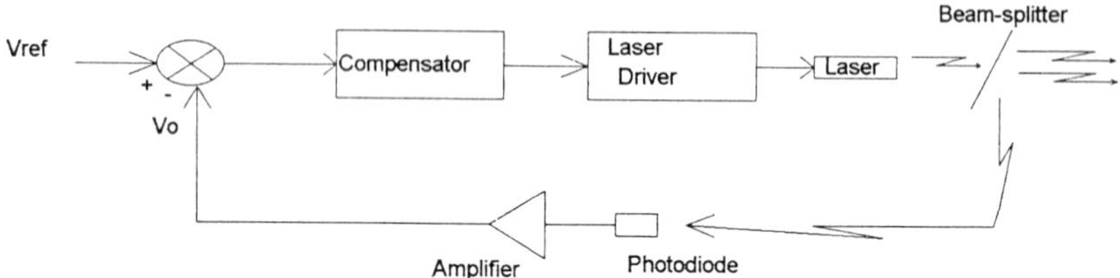

Figure 2. External electronic negative feedback applied to the laser.

The ratio by which the light is split, as shown in Figure 2, is explained in section 2.4. In this paper though, all the light from the laser is directed to the photodiode, giving a Vref of 7.5V. The laser driver modulation coefficient is 100mA/V. The photodiode is connected to the amplifier in a transimpedance configuration.

2.3 Compensator design

The laser driver-laser-photodiode-amplifier response (called the 'plant') can be described with the following transfer function:

$$tfp(j\omega) = \frac{Kp}{(1+j\omega t1)(1+j\omega t2)} \cdot \exp(-j\omega td) \qquad (1)$$

where $t1 = 2.7\times10^{-6}$s, $t2 = 0.7\times10^{-6}$s, $td = 1.5\times10^{-6}$s, and Kp is the plant gain.

This system would be unstable in closed-loop, so the gain had to be reduced. To improve the performance a compensator was designed, using the CODAS software (for control applications) for the control design, and the MathCAD software (for mathematical applications) for the numerical analysis. The resulting loop gain should be as flat and as high as possible, at the frequencies of interest (0Hz-20kHz). Adequate phase and gain margins are expected, and the lags should be introduced early, so their de-stabilising effects will be minimal at high frequencies. The compensator consists of two cascaded 'lag-lead' circuits, and its transfer function is:

$$\mathrm{tfc}(j\omega) = \mathrm{Kc} \cdot \frac{(1 + j\omega \mathrm{tb}) \cdot (1 + j\omega \mathrm{tc})}{((1 + j\omega \mathrm{ta})(1 + j\omega \mathrm{td})} \tag{2}$$

where ta = 1.59×10^{-3}s, tb = 1.59×10^{-5}s, tc = 5.3×10^{-6}s, td = 1.59×10^{-3}s, and Kc is the compensator gain.

In Figure 3 the compensator diagram is shown, where: ta = C1R2, tb = (C1R2R3)/(R2+R3), tc = (C2R5R6)/(R5+R6), td = C2R5, and Kc = ((R2+R3)/R1)×((R5+R6)/R4).

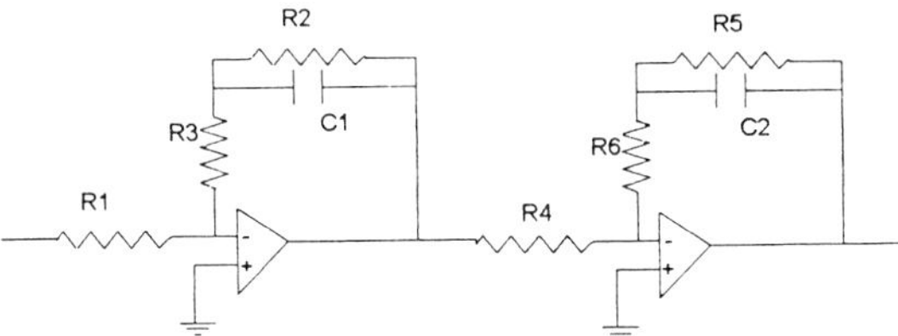

Figure 3. Compensator Diagram.

In Figure 4 the responses of the 'plant' + compensator (in ▫'s, open loop: -2dB at -180°) and of the stabilised plant (dashed) are shown.

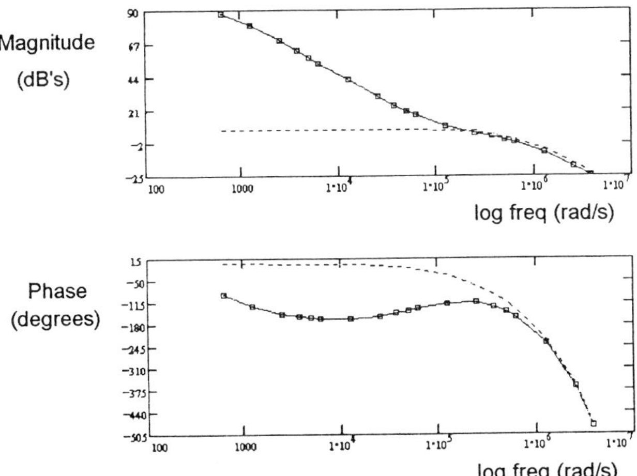

Figure 4. Stable frequency responses of plant+compensator, and plant alone.

It can be seen, that although this is not an optimal design, gain is increased dramatically, a minimum of ten times, compared to the response obtained by just reducing Kp.

2.4 Application of stabilised laser system

The original purpose for minimising the laser noise is discussed in [KAR91, KAR93, KEA91]. In short, the laser diode is to be used to provide the sensing of the movement of a microphone membrane (using a Fabry-Perot Interferometer, FPI). Furthermore, the microphone response is to be improved by using electrostatic force-feedback at the diaphragm (Figure 5, loop 2). Thus, in order to have minimal noise in the overall system, the laser source needs to be controlled (Figure 5, loop 1).The ratio by which light is split (using the polarising-beam-splitter PBS) is chosen so that the standing currents in the two photodiodes (OP & LCP) are equal.

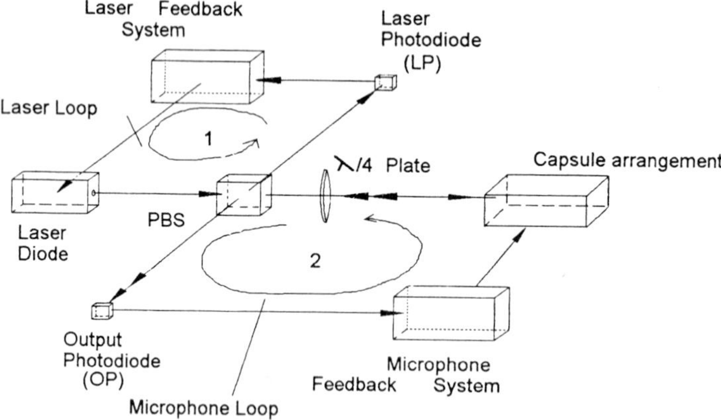

Figure 5. Layout of proposed system, showing the two control loops.

Because the path length difference in the FPI will be of the order of few nanometers, when the force-feedback is applied, the phase noise of the laser is minimised [DAN81], and the intensity noise will dominate. The resulting microphone will have a self-noise figure of -5.15dBA SPL and a dynamic range of 142.15dB.

3. NOISE RESULTS

In all the readings the laser driving current is 56mA at 25°C. Optical feedback was avoided with the PBS and quarter-wave plate (λ/4) arrangement (as shown in Figure 5, isolation of the order of 90dB was achieved), so further noise and instabilities were not introduced to the laser [DAK86]. The results were recorder from the wave analyser using the oscilloscope, and then plotted to the printer (as shown in Figure 1).

All measurements have been recorded in dBm/600ohms. Vertical scale: 20dB/div, and horizontal scale: 2kHz/div. The bandwidth was also kept constant at 100Hz. Input sensitivity was -10dB for all the readings.

Figure 6 shows the noise when the laser is on, at its normal operating point. After 2kHz the noise level is constant at -70 dBm. A more careful examination of the spectrum between 0Hz and 2kHz revealed some components of supply harmonics, which were eliminated when feedback was applied.

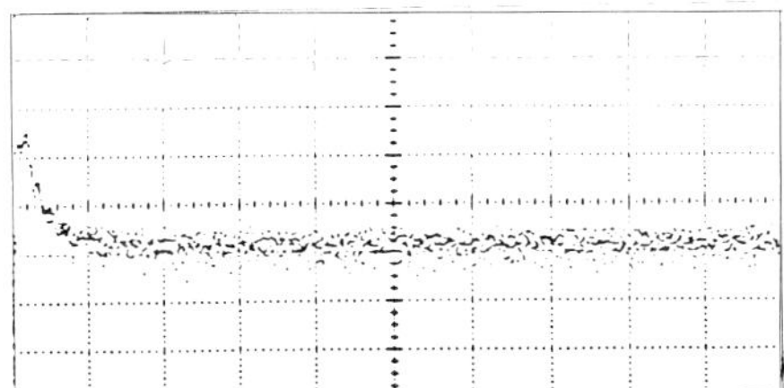

Figure 6. Natural spectrum of detected laser emission.

Figure 7 shows the laser noise with its internal feedback loop, with loop gain of 10. The noise is less than in Figure 6, at best by 15dBm, especially at low frequencies (≤6kHz). After that, the level tends to settle as before at -70dBm. The other two loop gain settings (x1, x2) did not improve noticeably the spectrum.

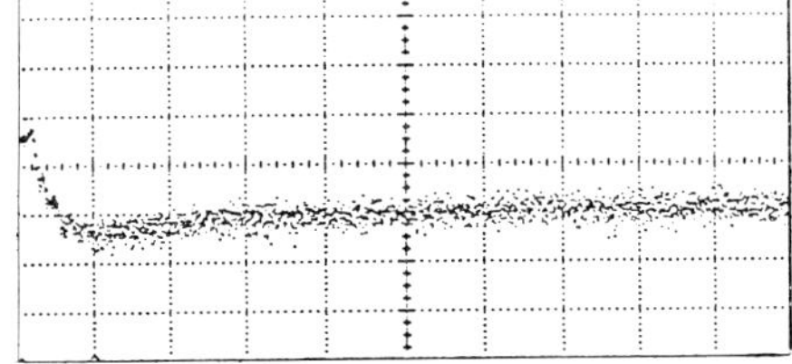

Figure 7. Laser noise with internal power control, loop gain of 10.

Figure 8 shows the effect on the noise when the external feedback is applied. An improvement of 35dBm is observed from the spectrum of Figure 6, at around 2kHz, and the noise level settles at -80dBm. Improvement of 20dBm from Figure 7 can also be seen. Throughout the 20kHz, the overall noise spectrum for the external feedback is better than the one obtained with the internal feedback. The noise spectrum with both feedback loops active (internal and external), is the same as in Figure 8.

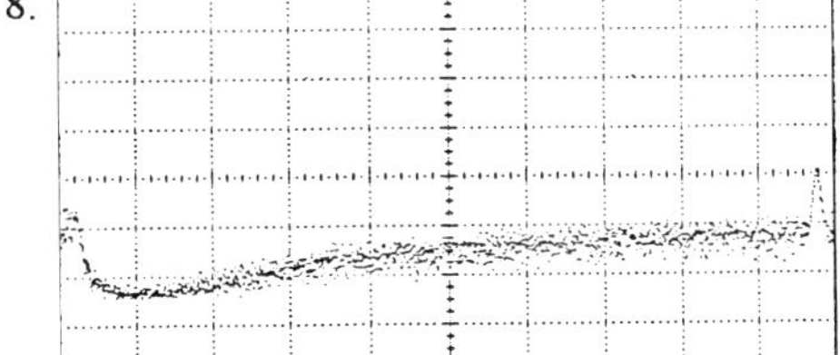

Figure 8. Laser noise with external feedback.

It has also been observed that the photodiode, when no input is present, does not contribute any detectable noise.

4. CONCLUSIONS

From the measurements, it can be concluded that external feedback is an effective method to minimise laser noise. The internal loop had an effect for a narrower range of frequencies than the external loop, which acted throughout the bandwidth of interest. It is unfortunate that the loop gain in the external loop was not as large as the one calculated from the design, but instabilities occurred, probably due to the internal current limit of the laser driver. Future work will include a better design of the loop, in order to reduce further the noise.

5. REFERENCES

[DAK86] Dakin J P and Withers P B, "The Reduction Of Semiconductor Laser Phase Noise For Sensor Applications", Optica Acta, Vol 33, No 4, 1986.

[DAN80] Dandridge A, Tveten A B, Miles R O, and Giallorenzi T G, "Laser Noise In Fiber-Optic Interferometer Systems", Appl Phys Lett, Vol 37, No 6, September 1980.

[DAN81] Dandridge A, Tveten A B, Miles R O, Jackson D A, and Giallorenzi T G, "Single-Mode Diode Laser Phase Noise", Appl Phys Lett, Vol 38, No 2, January 1981.

[KAN90] Kane T J, "Intensity Noise In Diode-Pumped Single-Frequency Nd:YAG Lasers And Its Control By Electronic Feedback", IEEE Photon Technol Lett, Vol 2, No 4, April 1990.

[KAR91] Karatzas L S, Keating D A, and Usher M J, "Development Of An Optical Microphone", Sensors: Technology, Systems and Applications, edited by KTV Grattan, Adam Hilger, 1991. (IOP, "Sensors and their Applications V", 22-25 September 1991, Edinburgh).

[KAR93] Karatzas L S, Keating D A, and Ucher M J, "Practical Optical Microphone", Proceedings of SERC/InstMC Symposium, March 1993. (SERC/InstMC, "Postgraduate Research in Control and Instrumentation", 30-31 March 1993, Nottingham).

[KEA91] Keating D A and Karatzas L S, "Optical Microphony", Audio Engineering Society (AES) Preprint, Session R-2, No 3153, October 1991. (AES 91st Convention, "Audio Fact and Fantasy - Reckoning with the Realities", 4-8 October 1991, New York).

[LEE90] Lee H S, Oh C H, Yang S H, and Chung N S, "Frequency Stabilization Of A Directly Modulated Semiconductor Laser", Rev Sci Instrum, Vol 61, No 9, September 1990.

[NEW89] Newson T P, Farahi F, Jones J D C, and Jackson D A, "Reduction Of Semiconductor Laser Diode Phase And Amplitude Noise In Interferometric Fiber Optic Sensors", Applied Optics, Vol 28, No 19, October 1989.

[OHT90] Ohtsu M, Murata M, and Kourogi M, "FM Noise Reduction And Subkilohertz Linewidth Of An AlGaAs Laser By Negative Electrical Feedback", IEEE J Quantum Electron, Vol 26, No 2, February 1990.

[PET81] Petermann K and Weidel E, "Semiconductor Laser Noise In An Interferometer System", IEEE J Quantum Electron, Vol QE-17, No 7, July 1981.

[SAI85] Saito S, Nilsson O, and Yamamoto Y, "FM Noise And Linewidth Reduction In A Semiconductor Laser By Means Of Negative Frequency Feedback Technique", Appl Phys Lett, Vol 46, No 1, January 1985.

[VAN81] Van De Grijp A, Koopman J C, Meuleman L J, Nicia A J A, Roza E, and Van Heuven J H C, "Novel Electro-Optical Feedback Technique For Noise And Distortion Reduction In High-Quality Analogue Optical Transmission Video Signals", Elec Lett, Vol 17, No 11, May 1981.

[YAM87] Yamada M, Nakaya N, and Funaki M, "Characteristics Of Mode-Hopping Noise And Its Suppresion With The Help Of Electric Negative Feedback In Semicoductor Lasers", IEEE J Quantum Electron, Vol QE-23, No 8, August 1987.

A simple fibre optic shaft monitor for measurement of rotational frequency

M. L. Everington and A. T. Augousti

School of Applied Physics, Kingston University, Kingston, Surrey KT1 2EE

Introduction

This paper describes the construction and characterisation of a simple fibre optic shaft monitor which utilises computationally intensive signal processing techniques to extract a measurement from the raw data. The system is based on the signal generated when an infra-red (IR) beam illuminating a shaft is reflected from the shaft, thereby creating a periodic signal. A range of processing techniques such as the use of Fast Fourier Transforms, auto correlation, and windowing are implemented on a PC and the relative efficiency of each technique is compared.

System Description

The overall system configuration is shown in Figure 1. Two separate optical fibres are jacketed together and polished down simultaneously to form a head which is aimed at the shaft being monitored. The two free ends of the optical fibres are connected to a control unit which houses the electronics. The output from an IR LED is coupled to one of the fibres thereby illuminating the shaft. Part of the reflected light is captured by the other fibre and transmitted back to the photodiode. The voltage signal produced by this is amplified and fed through a low pass filter.

The electronic signal from the control unit is connected to an A/D card residing in a PC. It is here that the processing to determine the fundamental frequency of the signal is carried out. The analogue signal is converted to a sequence of digital data which is stored, a reconstruction of a typical data sequence is shown in Figure 2. Then the algorithm for windowing, auto correlation, and Fast Fourier Transforms (FFTs) can be applied. These are implemented by software using C++ language which is quick enough to produce the result

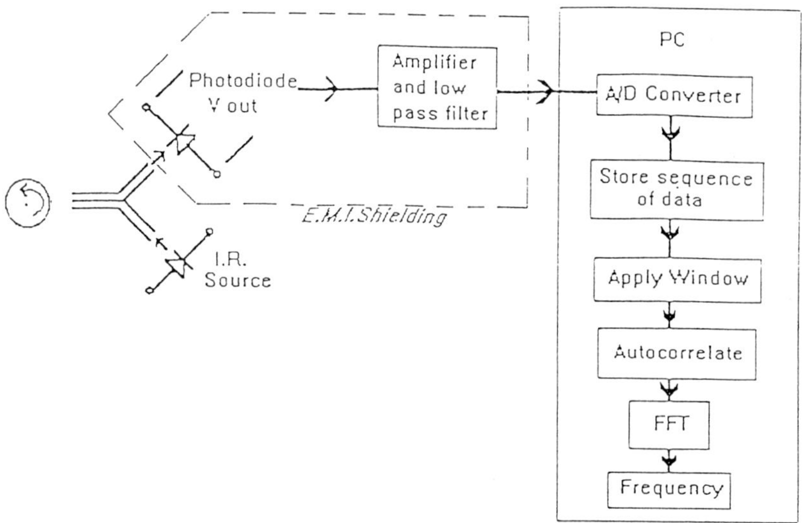

Figure 1. Optomechanical arrangement of system.

in under a second on sample lengths of 2048 points. The program endlessly captures another series of data and the processing is repeated.

The A/D converter (ADC) has a modest minimum settling time of 100 μs which in theory will produce a sampling rate of 10 kHz. However, since the ADC is software driven and the data values have to be stored for processing the maximum rates achieved are between 4 and 5 kHz which is adequate for this application.

The frequency of rotation is calculated for each series of data and printed on the screen. This is updated after each series of data is processed and the computer can also display graphically a trace of the waveform, the auto correlation function, and the frequency spectrum produced by the FFT algorithm.

Frequency Determination.

A good comparison of the various techniques used in the measurement of frequency is given by Cain and Yardim [1]. The two methods considered here are those of zero crossing counting and applying Fourier transforms.

The zero crossing frequency is extremely simple to calculate since it only involves counting the occurrence of successive samples having opposite algebraic signs. This is good for measuring pure sinusoids but problems are caused by higher order harmonics, DC offset in the A/D converter and in particular 50 Hz mains hum in the signal.

Fourier transforms are a useful method of representing time domain signals in the frequency domain and involve integrating a time domain signal to work out the energy density spectrum in the frequency domain. In the integration is performed numerically using a finite mesh. However, if a sequence is periodic it is known that it can be represented by a series of discrete summations known as the discrete Fourier transform (DFT). A theoretical justification of this is beyond the scope of this work but is covered in detail by Strum and Kirk [2] and by Rabiner and Schafer [3].

The calculation of the DFT requires a number of operations which increase with the square of the number of samples in the sequence being processed. This would prove time consuming on a PC if a reasonable sample length is to be used. However fast Fourier transforms (FFT) reduce the number of calculations required considerably by recognising that many of the calculations are repeated and redundant. The use of the FFT is the basis upon which the frequency determinations described have been based so far.

Techniques for Improving Signal Clarity

The use of correlation is accepted as a powerful method of overcoming noise and improving the clarity of periodic signals. Therefore the data sequences were auto correlated and the raw data was replaced with the auto correlation function (ACF) Although this result is an improvement of the signal clarity, the drawback with this process is that it increases the amount of computation required considerably.

The process of sampling unavoidably introduces discontinuities at the start and finish of any data series. The consequence of this are spectral leakage which makes the identification of closely situated frequencies difficult. The appropriate solution to this problem is windowing and a Hanning window which incorporates a raised cosine was applied to the data series. A 64 point version of a Hanning window is shown in Figure 5. This removes the effect of discontinuities by tapering the values of the end points towards zero. This process and the other types of windows commonly used have been investigated in detail by Harris [4].

Results and Discussion

The trace shown in Figure 2 is a graph of the raw data in a typical signal and it demonstrates a clear periodicity. However, it can be seen that higher order harmonics are superimposed on the signal. The auto correlation function of the raw data shown in Figure 3 demonstrates the effectiveness of the process in dealing with the problem. The harmonics have all but disappeared and the zero crossing technique could be used with confidence.

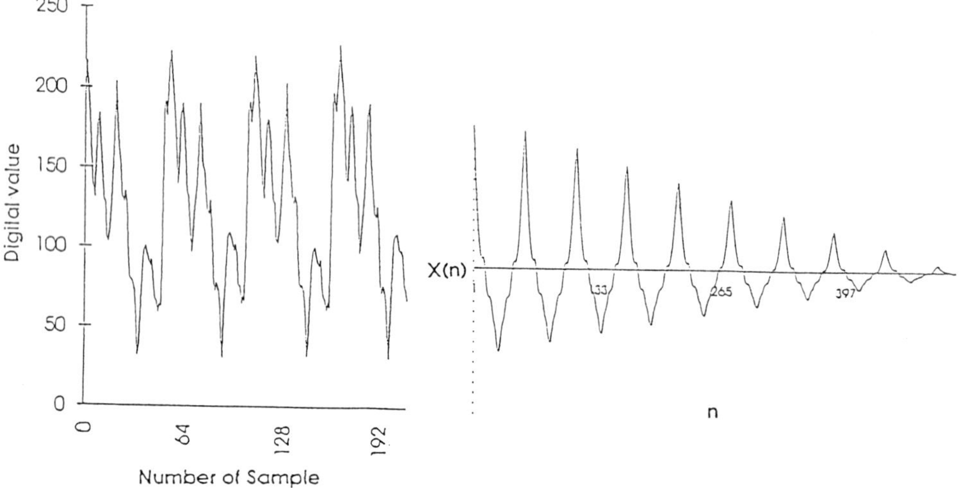

Fig. 2 Unprocessed signal from A/D converter.

Fig. 3 Auto correlation function of data in Fig. 2.

Figure 4 shows the frequency spectrum obtained by applying the FFT's to the raw data. Comparing this with Figure 6 shows the effect of windowing. It can be seen that the magnitude of the higher order harmonics are reduced and that the spectral resolution is improved even if these effects are unspectacular. In a case where the harmonics present are stronger this improvement could prove decisive in producing a reliable result.

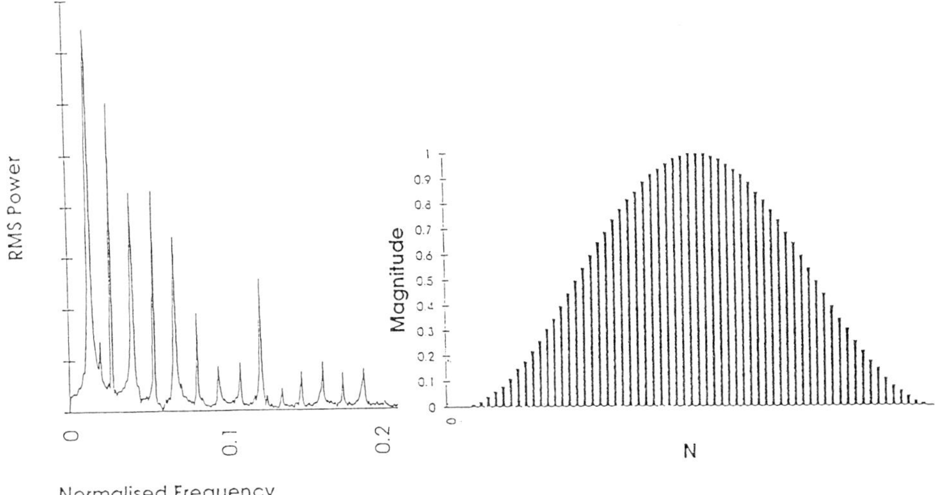

Fig. 4 FFT analysis on raw data.

Fig. 5 64 Point Hanning Window.

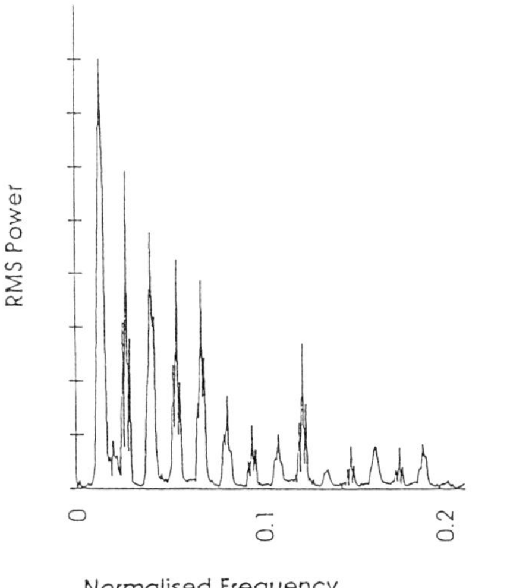

Fig. 6 FFT analysis of windowed data.

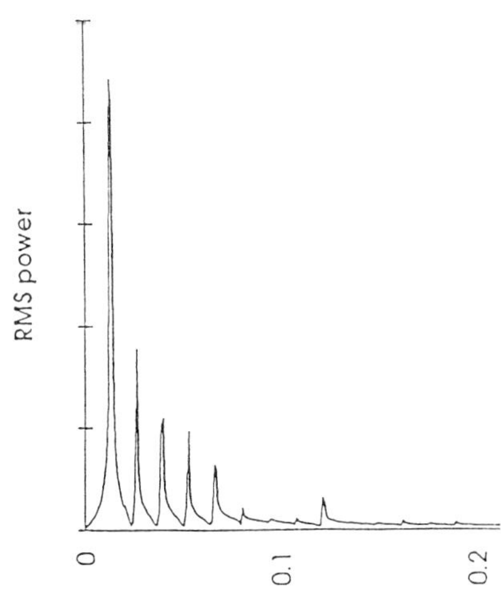

Fig. 7 FFT analysis of auto correlated data.

Figure 7 shows the effect of performing the FFT on the auto correlation function of the raw data. The improvements caused by the reduction of the higher order harmonics are clear and also the noise signals at higher frequencies have been almost completely cut out. The spectral spreading which is most pronounced on the upper side of each peak does not appear to interfere with the result since well defined peaks are obtained.

Conclusion

A simple fibre optic shaft monitor has been constructed, interfaced to a PC and its characteristics evaluated. The PC displays a graphical output in real time either in the time domain or the frequency domain. The response of the device depends on the level of resolution required, and response times of under a second using the FFT technique alone are achievable for sample sizes up to 1024 points, and approximately four to five seconds for both FFT and auto correlation for the same number of points. Although it was tested between approximately 100 rpm and 1000 rpm, the upper limit of the range is of the order of 100000 rpm, given the data acquisition rate achieved. The next objective of this work is to cross correlate signals from different positions along the shaft in order to obtain information about angular acceleration, torque and torsional stiffness.

References

1. **Cain G.D. and Yardim A.**, "Digital Estimation of Doppler Frequency in Waveband Signalling Applications" *Fifth International Conference on Digital Processing of Signals in Communications*

2. **Strum R.D. and Kirk D.E.** *First Principles of Discrete Systems and Signal Processing*, Massachusetts: Addison-Wesley 1988

3. **Rabiner L.R. and Schafer R.W.** *Digital Processing of Speech Signals*, New Jersey: Prentice-Hall, 1978

4. **Harris F.J.** "On the use of windows for Harmonic Analysis with the Discrete Fourier Transform" *Proc IEEE*, **66**, No. 1, pp. 51-83, Jan 1978

Optical fibre interferometric sensors using buffer guided light

F. M. Haran, J. D. C. Jones and J. S. Barton

Optoelectronics and Laser Engineering, Department of Physics, Heriot-Watt University, Riccarton, Edinburgh EH14 4AS.

A novel type of interferometric fibre optic sensor is described, based on a curved section of a buffered single mode optical fibre. Interference occurs between the core guided mode and the 'whispering gallery' mode guided within the fibre buffer. The fibre buffer coating thus serves as the sensing element. The technique is illustrated by its application to temperature measurement. A linear output is produced by using a two wavelength technique.

1. INTRODUCTION

An important class of fibre-optic sensors is based on the interaction of a measurand with the evanescent field of the guided wave. In a typical example, a portion of the cladding of the fibre is removed, and an overlay is deposited in the form of a thin film. A wave propagating through the overlay is thus modulated by the measurand. An advantage of such sensors is the ability to form the overlay from a material showing a predominant response to the desired measurand. In contrast, in conventional interferometric sensors the sensing element is generally single-mode optical fibre so that the measurand sensitivities are those of fused silica, with little control available to the designer. However, the construction of evanescent field sensors is laborious, requiring precise removal of the material, or reduction in the dimension of the guide, to access the field. Furthermore, the optical transfer function is complex, and is difficult to model analytically.

In this paper, we describe a technique which provides access to the evanescent field without removing cladding material: it is based on the introduction of a simple bend of suitable geometry into otherwise standard single mode optical fibre.

When an optical fibre is bent then loss occurs as the evanescent part of the guided field in the cladding in the outer part of the bend detaches itself from the guided wave at the radiation caustic, and propagates in a direction tangential to the curved fibre. Early theories assumed claddings of infinite extent [1] and provided a monotonic analytic expression for bend loss. However the cladding's finite extent leads to oscillations in bend loss as a function of the bend radius of curvature and wavelength, caused by a Whispering Gallery (WG) mode within the cladding, guided at the cladding/buffer interface [2] which recouples to the core guided mode. Buffered fibres also show these oscillations with a shorter period corresponding to a longer path for the WG mode reflected at the buffer/air interface [3]. These fibres have also been shown to have an oscillatory loss with temperature [4].

The establishment of this WG mode within the bent fibre is shown in figure 1. It can be seen from this figure that in effect the bend essentially creates a two beam Mach-Zehnder interferometer within a single fibre. The reference and signal arms are the core guided and WG modes respectively. Analogies with an evanescent field sensor may also be drawn, due to the bent fibre tapping off some of the core guided mode's evanescent field into the WG mode. This way of accessing the evanescent field is non-intrusive and the amount of power tapped from the core can be controlled more accurately and easily by changing either the bend radius or length to suit a particular application. The sensing elements in these types of sensors is the buffer coating, which can be formed from a material chosen to suit a particular measurand.

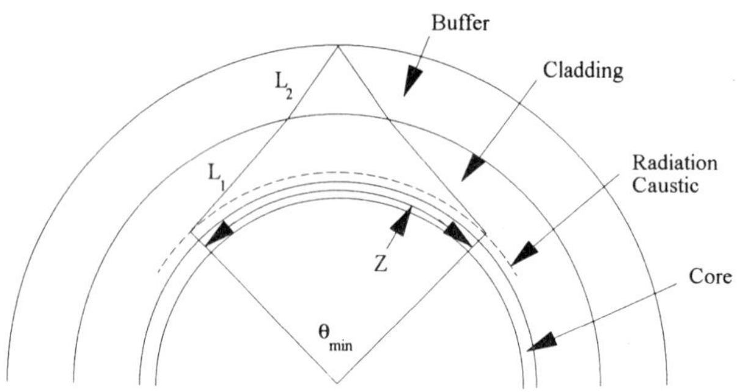

Figure 1: Bend loss model.

As an illustrative example, we have demonstrated the concept for the application of temperature measurement. In this case, the standard buffer coating is an appropriate sensing material: its thermo-optic coefficient is large and well characterised.

In common with all interferometric sensors, a means of demodulating the optical signal to generate an output linear with the measurand is required. We demonstrate a suitable technique based on dual wavelength illumination.

2. THEORY

When a bend is formed in a single-mode optical fibre, electromagnetic radiation is ejected from the core at the radiation caustic, setting up WG modes which propagate in the cladding [2] or the buffer [3,4], as shown in figure 1. The WG modes may recombine coherently with the core guided mode, once again at the radiation caustic, such that the phase difference between the WG mode and the core guided mode is:

$$\phi_o = \frac{2\pi l}{\lambda} - \beta Z + \Psi \tag{1}$$

where Z is the physical path length of the core guided mode as shown in figure 1, Ψ is a phase constant, and l corresponds to the optical path length of the WG mode:

$$l = 2(L_1 n_{cl} + L_2 n_b) \tag{2}$$

where L_1 and L_2 are the geometrical path lengths in the cladding and buffer respectively as illustrated in figure 1, n_{cl} and n_b are the refractive indices of the cladding and buffer respectively.

The thermal sensitivity of an optical path is given by:

$$\frac{1}{L}\frac{\partial\phi}{\partial T}=\frac{2\pi}{\lambda}(n\alpha+\beta) \tag{3}$$

Where α is the thermal expansion coefficient, β is the thermo-optic coefficient, L is the geometrical length, $\partial\phi/\partial T$ is the temperature coefficient of optical phase, and λ is the free space wavelength.

The path lengths Z and L ($L = L_1 + L_2$) will experience temperature-induced phase changes $\Delta\phi_{co}$ and $\Delta\phi_{WG}$ respectively, with the relative phase change, $\Delta\phi$ given by:

$$\Delta\phi=\frac{2\pi\Delta T}{\lambda}\left[2L_1\left(n_{cl}\alpha_{cl}+\beta_{cl}\right)+2L_2\left(n_b\alpha_b+\beta_b\right)-Z\left(n_{co}\alpha_{co}+\beta_{co}\right)\right] \tag{4}$$

Where ΔT is the temperature change, L_1 and L_2 are the physical path lengths shown in figure 1. The subscripts co, cl and b represent core, cladding and buffer respectively.

It can also be seen from figure 1 that there will be a minimum bend angle required to allow recoupling between the WG and core guided modes. This minimum angle, θ_{min}, can be calculated from the geometry and the optical properties of the fibre.

The relationship between the optical power transmitted by the bend, P, and the differential phase, $\Delta\phi$, is a function of the arc length of the bend. With reference to figure 1, it may be seen that for short bends the WG mode does not return to the radiation caustic, and the power ejected by the bend is lost. For longer bends, the WG mode interferes with the core guided mode, such that:

$$P\approx P_o\left[1+V\cos\left(\Delta\phi+\phi_o\right)\right] \tag{5}$$

where V is the visibility of interference, and P_o and ϕ_o are constants. If the bend length is increased further, then higher order interference is possible. However, for the experiment described in this paper, bend lengths were chosen to satisfy the assumptions of equation (5).

Equation (5) indicates that the sensitivity fades at the points $(\Delta\phi + \phi_o) = N\pi$, where N is an integer. Hence, to linearise the sensor output, a two wavelength interrogation scheme was used. The sensor is alternately illuminated with wavelengths λ_1 and λ_2, producing power transmission P_1 and P_2 respectively, and chosen such that $\phi_o(\lambda_1) - \phi_o(\lambda_2) = \pi/2$. Hence,

$$\Delta\phi=\tan^{-1}\left(\frac{\hat{P}_1-\hat{P}_2}{\hat{P}_1+\hat{P}_2}\right) \tag{6}$$

where $\hat{P}_i = P_i / P_{oi}$.

In practice, the phase offset $\phi_o(\lambda_1)$ - $\phi_o(\lambda_2)$ is not exactly $\pi/2$. Furthermore, $\Delta\phi(\lambda_1)$ - $\Delta\phi(\lambda_2)$ = $f(T)$, because the refractive indices and thermo-optic coefficients are wavelength dependent, especially those of the buffer coating. Hence, equation (6) is refined taking into account the parameters of the practical sensor.

3. EXPERIMENTAL

A standard telecommunications fibre was subject to a controlled bend in the experimental arrangement shown in figure 2. The fibre was bent in a 4 mm radius with a bend angle of 45° around a brass mandrel which incorporated a Peltier cooler for temperature control. Temperature-induced bend loss oscillations were investigated by stepping the Peltier cooler in 1°C increments from 12°C to 40°C. The power transmitted through the fibre was measured at each increment using an InGaAs photodetector with lock-in detection. Four minutes was allowed between each reading to allow the buffer to come to thermal equilibrium. Temperature oscillations were taken at two wavelengths, λ_1 = 1300 nm and λ_2 = 1302.5 nm to obtain a linearised output.

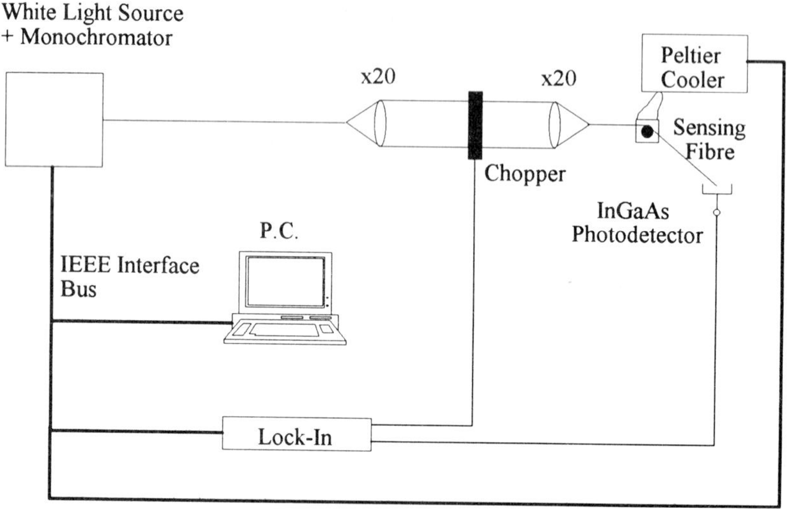

Figure 2: Experimental arrangement.

4. RESULTS

Figure 3 shows the temperature induced bend loss oscillations obtained at the two wavelengths λ_1 and λ_2. It can be seen from this figure that the periodicity of the oscillations are slightly different, as indicated in section 2. The effect is exacerbated because for the buffer material the terms $n\alpha$ and β are of similar magnitude but of opposite sign. Hence, a small dispersive effect in β produces a strong wavelength dependence in $\partial\phi/\partial T$. Furthermore, the phase offset between the two data sets was not exactly $\pi/2$. The processing algorithm compensated for the non-quadrature phase offset, but not the temperature dependence of $\Delta\phi$. Figure 4 shows the phase recovered by the processing algorithm plotted against the temperature measured by a thermistor embedded in the processing mandrel, indicating a sensitivity of 0.37 rad K^{-1}.

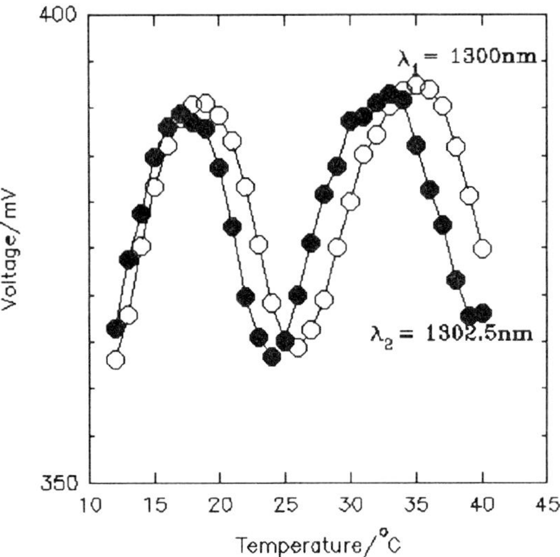

Figure 3: Temperature induced bend loss oscillations obtained at λ_1 and λ_2.

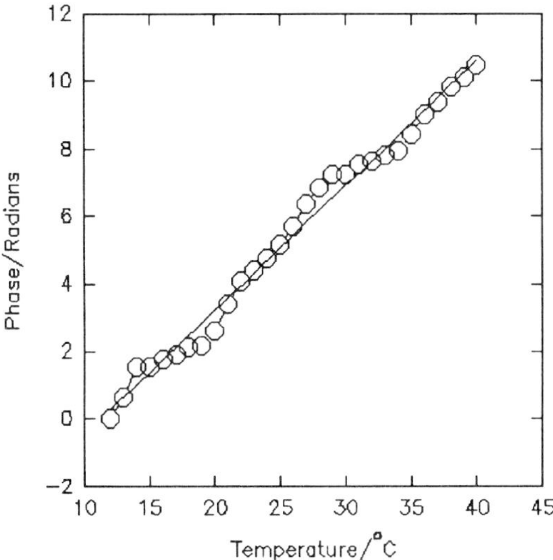

Figure 4: Linearised output obtained from the results shown in figure 3.

5. DISCUSSION

The temperature range of the sensor is set by the properties of the buffer coating. For the acrylate used in the present experiments that range is -50 to 85°C. However, caution must be exercised when using the sensor below the dew point; condensation at the buffer/air interface will modify the WG mode. The linearity of the sensor was demonstrated to ±0.92°C, with the error arising from the assumption of the temperature dependence of $\Delta\phi$; the resolution is <0.1°C.

Temperature measurement was intended only as a simple illustration: the WG mode is analogous to both evanescent field sensors and the Mach-Zehnder interferometric sensors, and is appropriate for similar applications, although much easier to construct.

6. CONCLUSIONS

This paper has demonstrated for the possibility of a new type of simple single-mode fibre-optic sensor involving a simple bend which uses the buffer coating as the sensing element. These sensors have similarities with both evanescent field sensors and Mach-Zehnder interferometric sensors, but require no special fibre components, or mechanical processing of the fibre material. A practical temperature sensor was constructed which used a standard single-mode fibre with dual wavelength passive demodulation.

7. ACKNOWLEDGEMENTS.

The authors thank Colin Paton, Peter Wilson and Dave Stockton of British Telecom (BT) for many helpful discussions and Prof. Kazuo Ono of Ehime University, Japan for theoretical contributions. This research was partially supported by BT and the Science and Engineering Research Council (UK), whom FMH thanks for the provision of a research studentship.

8. REFERENCES.

[1] D. Marcuse, 'Curvature loss formula for optical fibers,' J. Opt. Soc. Am. **66**, 216, (1976).

[2] A. J. Harris and P. F. Castle, 'Bend loss measurements on high numerical aperature single-mode fibers as a function of wavelength and bend radius,' J. Lightwave Technol. **LT-4**, 34, (1986).

[3] R. Morgan, J. S. Barton, P. G. Harper and J. D. C. Jones, 'Wavelength dependence of bend loss in monomode optical fibers: effect of the fiber buffer coating,' Opt. Letts. **15**, 947, (1990).

[4] R. Morgan, J. S. Barton, P. G. Harper and J. D. C. Jones, 'Temperature dependence of bend loss in monomode optical fibres,' Elec. Letts. **26**, 937, (1990).

A comparison of the matrix and star topologies for optical sensor networks

J C Walker *, R Holmes and G R Jones

Department of Electrical Engineering and Electronics, Liverpool University, Brownlow Hill, PO Box 147, Liverpool L69 3BX
* now at Merseyside Fire and Civil Defence Authority, Mulberry House, Canning Place, Liverpool L1 8JB

ABSTRACT: Over the past few years there has been considerable interest in the networking of optical fibre sensors. To date much of the work has been concentrated on the multiplexing technique used, while the choice of topology has been a secondary issue, in part determined by the choice of multiplexing technique. This paper considers two topologies, star and bus, both of which have been used in conjunction with a Spatial Light Modulator (SLM), and compares in detail the relative performance of these topologies.

1. INTRODUCTION

A SLM comprises a number of pixels which can be independently switched. By uniformly illuminating one side of the SLM and positioning optical fibres in front of the pixels on the other side it is possible to modulate the light launched into each of the optical fibres.
Time Division Multiplexing (TDM) is the simplest form of multiplexing that can be employed with an SLM, and it was for this reason that this multiplexing technique was used to test both the matrix and star topologies: It is also possible to use Frequency Division Multiplexing and Code Division Multiplexing with both topologies. In TDM the pixels are opened and closed in a strict sequence. By allowing only one pixel to be open at a time, the light incident on the receiver is a series of pulses, each pulse of light corresponding to a single sensor. The relative timing positions of the pulses are determined by the software which allows identification of the pulse for an individual sensor. The response of a sensor is evaluated by measuring the light intensity with all the pixels closed, the dark level, and then measuring the pulse height with the relevant pixel open, the light level. The dark level is subtracted from the light level, the difference being representative of the modulation state of the sensor.
A non zero dark level occurs due to the modulation wavelength range of the SLM device being smaller than the spectral range of the white light source used to illuminate the SLM. An infra red filter is used to reduce this unmodulated light, but it is important that the residual dark level is totally removed from the sensor signal, since the dark level is in effect a source of crosstalk.

2. SLM STAR TOPOLOGY

An eight sensor star network is shown in figure 1. A fibre is butted up to each activated pixel so that light is transmitted through the fibre to the sensor. The measured average optical power from these fibres at the point of coupling to the sensors was found to be -22 dBm @ 850 nm. Light from all sensors is coupled into a single fibre bundle and is incident on the detector.

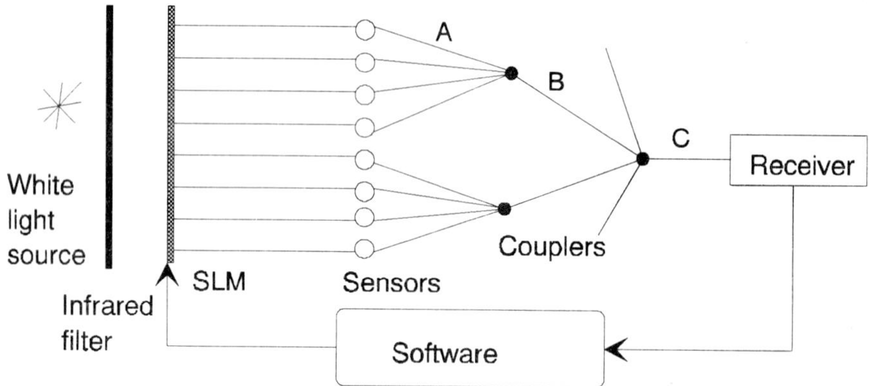

Figure 1. The Star topology

The advantage of the star toplogy is that, it uses a single source and detector, the optical losses are equalised, thus reducing the dynamic range requirement of the receiver, and it minimises the total number of fibre bundle couplers required in the network. The main disadvantage of this topology is that for a N sensor network, N output fibres, from the source, are required.

3. SLM MATRIX TOPOLOGY

A three by three SLM sensor matrix network is shown schematically in figure 2. The main tail of a SMA terminated fibre bundle, which was used as a 4 into 1 coupler, was butted up to each pixel. The advantage of butting a fibre bundle to the SLM rather than a single fibre connected to a fibre bundle, is that more light can be coupled into the fibre bundle in this manner. In tests it was found that an average of -18 dBm @ 850 nm of optical power was emitted by each of the secondary tails of the fibre bundle at the point of coupling to the sensors, this compares to an average of -22 dBm @ 850 nm of optical power which is emitted by each single fibre before connection to the sensors in the star network.
Three of the secondary tails in the fibre bundle coupler were attached to photoelastic strain sensors, the fourth tail remained unconnected. The output of each of the sensors was connected to a secondary tail of a four way fibre bundle coupler, in the same manner as in the star network. Three fibre bundle couplers were used, each one was attached to three sensors while the main tail of each bundle was connected to a transimpedance optical receiver.

There are three advantages of using fibre bundle couplers in this way to implement the matrix topology;
i) ideally optical losses for all paths through the network are equal, which ensures that the optical power from all sensors at the receiver are equalised, as is the case for the star toplogy.
ii) the number of fibre bundle couplers following the sensors, in a matrix topolopy is less than in a star topology using a SLM; this reduces the optical attenuation and should lead to an improvement in SNR.
iii) crosstalk is minimised, due to the presence of fewer possible optical paths for back reflected modulated light.

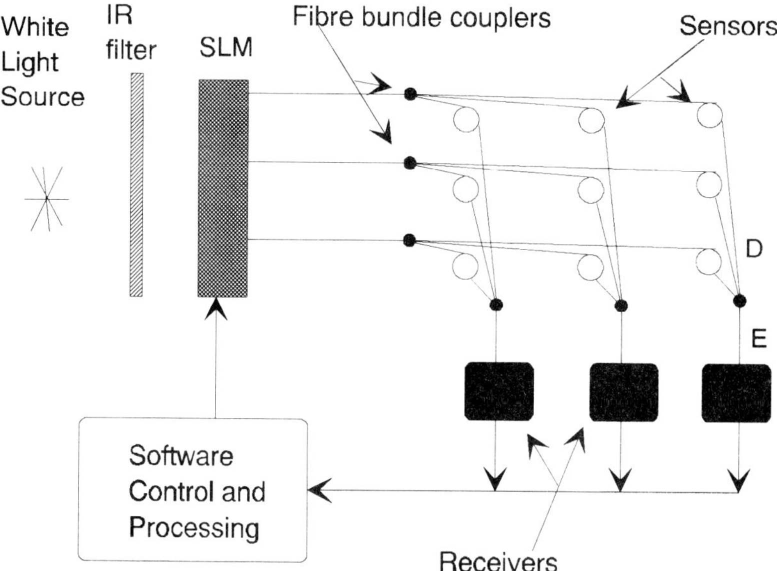

Figure2 A three by three SLM Matrix network

4. ADVANTAGES OF THE SLM MATRIX TOPLOGY COMPARED WITH THE SLM STAR TOPLOGY

In certain circumstances where space is at a premium, it is desirable to have a number of sensors connected to a single fibre lead; a fibre lead being an optical fibre connection between the source or receiver and the rest of the sensor network. In a SLM N sensor star network, there are N+1 fibre leads; N fibre leads connected to the SLM and one fibre lead connected to the receiver. In a SLM N sensor matrix network, there are $2*N^{1/2}$ fibre leads, half are connected to the SLM and half are connected to the receivers. Thus in a SLM star sensor network, the number of sensors per fibre lead is

$$N/(N+1) \tag{1}$$

while the number of sensors per fibre lead in a SLM sensor matrix network is

$$N/(2*N^{1/2}) \tag{2}$$

It can be seen that the matrix network has an increasing sensor to fibre lead ratio while the ratio for the star network tends towards unity with increasing numbers of sensors in the network. If the sensors per fibre lead ratio is the dominant consideration then the matrix network has an obvious advantage compared with the star network.

A further advantage of the matrix network is that the duty cycle time is shorter. For a N sensor matrix network which uses TDM, the duty cycle time is

$$2*t*N^{1/2} \qquad (3)$$

where t is the TDM time slot.

Equation 3 follows since each receiver is connected to $N^{1/2}$ sensors and the duty cycle has to address only $N^{1/2}$ sensors. A N sensor star network using TDM has a duty cycle time of

$$2*t*N \qquad (4)$$

It can be seen by comparing equations 3 and 4, that a N sensor matrix topology has a duty cycle time which is smaller than that of a N sensor star topology by a factor of $N^{1/2}$.

Finally, the SLM matrix topology can potentially have higher SNRs than a SLM star topology. This is due to there being one fewer fibre bundle coupler in a matrix toplogy, between a sensor and the receiver, than in a star topology network with the same number of sensors, since each receiver has to be coupled to only $N^{1/2}$ sensors rather than N sensors.

It should be remembered that the size of the matrix topology will be limited by the availability of large coupling ratio fibre bundle couplers.

5. Disadvantages of the SLM matrix toplogy compared with the SLM star topology

A SLM star topology requires only one receiver for N sensors, whereas in a SLM square matrix topology, $N^{1/2}$ receivers are required. However, the receivers can be electronically multiplexed after the first stage of detection, and so the cost, space, and weight of the receivers can be minimised.

A more restrictive constraint is the physical toplogy of the matrix topology. While a matrix topology is ideal for supporting sensors that are relatively close together in a confined space, it is not ideally suited to networking sensors which are positioned over a large site. Examples of these two situations are monitoring strain on an aircraft wing for which the matrix approach would be ideal and monitoring strain along a pipe which would be better suited to the SLM star topology. In conclusion it is the application that determines whether the matrix network is the optimum topology.

6. SNR and crosstalk results for the Matrix and Star Topologies

A nine sensor matrix topology and an eight sensor star topology were implemented. The average SNR was measured by recording the SNR for all the sensors in the both topologies while they were in a steady state. The average SNR for the matrix topology was 66 dB within a receiver of bandwidth 60 Hz (Walker 1992), while the average SNR in a 60 Hz bandwidth for an eight sensor star topology was 65 dB (Walker 1991). The crosstalk ratio was found by measuring the mean signal for each sensor, then optically removing at least one of the sensors from each topology and measuring the mean signal for each of the remaining sensors. The difference in the mean signals between the two sets of readings was due to crosstalk effects. The average crosstalk ratio was found to be -68 dB for the matrix topology while for the star topology the crosstalk levels were below -70 dB.

7. DISCUSSION

There is a significant reduction in the duty cycle time when a matrix topology is used. In this case the duty cycle time of the three by three matrix topology was 44% of the duty cycle time for the eight sensor star topology. The theoretical improvement in SNR for a matrix topology was expected to be larger than 1 dB because of the one fewer fibre bundle coupler used after the sensors in the matrix topology compared to the star topology, thus the optical attenuation will be lower in the matrix topology. When the dominant noise source is shot noise, then the expected SNR improvement for the matrix toplogy compared with the star toplogy is equal to the difference in dB in the incident optical power on the receiver for the two topologies. The reason for this is that the photocurrent is proportional to the optical power and the SNR is proportional to the photocurrent when the dominant noise is shot noise. Thus the improvement in the SNR between the two systems is equal to the ratio of the incident optical powers and so the improvement in the SNR is

$$10\log_{10}(P1/P2) = 10\log_{10}P1 - 10\log_{10}P2 \quad (5)$$

where P1and P2 is the incident optical power on the receiver for the matrix and star topologies respectively.

The matrix toplogy has a 4 dB higher input optical power than the star toplogy, section 3, and for a nine sensor array one fewer fibre bundle coupler in the optical path after each sensor. The optical loss has been measured between points A and B in the star topology as being -5 dB, while the attenuation between points B and C is -13 dB, figure 1. Between points A and B the output fibre from the sensor is coupled into the secondary tail of the first fibre bundle coupler; the core diameter of the secondary tail of the fibre bundle coupler is larger than that of the sensor output fibre thus there is reasonable coupling efficiency. However, between points B and C, figure 1, the main tail of the first fibre bundle coupler is connected to a secondary tail of the second fibre bundle coupler. The core diameter of the main tail is larger than that of a secondary tail, thus in this case the coupling efficiency is reduced and the attenuation is increased. In the matrix topology the optical attenuation between points D and E, figure 2, is -5 dB due to the connection of the sensor output fibre to a secondary tail in the fibre bundle coupler. Since there are no further fibre bundle couplers in the optical path, the optical attenuation following a sensor in a matrix topology is less by 13 dB compared with a star topology. Thus the expected SNR improvement is 17 dB. However, as is the case in the star topology, the dominant noise source in the matrix topology was external. This limits the SNR, because the noise power is independent of both the incident optical power on the detector and the receiver gain so that decreasing the optical attenuation in the network has a limited effect on the SNR. Hence only a marginal improvement in the SNR for the matrix toplogy was observed. It should be noted that because both the matrix and star networks used the TDM multiplexing system, there was only a small difference in optical dynamic range of the two topologies due to only one sensor transmitting at any one time. If the external noise source could be removed or attenuated then it is expected that the matrix topology would have a superior SNR to that of the star topology.

The crosstalk levels for the matrix topology were larger than those estimated for the star topology. A source of additional crosstalk in the matrix topology is the presence of back reflection optical paths. These cause modulated light from one sensor to be transmitted through another sensor so causing crosstalk, but this has been calculated to be of the order of -130 dB. However, measured electrical crosstalk between the receivers in the matrix topology is -66 +/- 2 dB, which agrees well with the crosstalk measured in the tests. Further, this indicates that the main source of crosstalk was receiver based. Therefore the effects of crosstalk can be reduced by improving the receiver design.

9. Conclusion

In the tests the SNR and the crosstalk performance of the SLM matrix and star topologies were broadly similar. It is anticipated that with an improved receiver design the SNR performance of the matrix toplogy could be significantly superior to that of the star toplogy for a specified number of sensors. The main advantages that the SLM matrix topology has compared with the SLM star topology are that one fewer fibre bundle coupler is required between the sensor and the receiver, thus reducing the attenuation, the response time of the network is shorter and the topology can support a greater number of sensors for a given SNR.

10. References

Walker J. C.,Holmes R. and Jones G. R., 1991, "Multiplexing optical sensors using a Spatial Light Modulator", Elec. Letts., Vol. 27 No. 22, pp 2022-2023.

Walker, J. C., Holmes, R., and Jones, G. R., 1992, "A Nine element optical sensor matrix using a spatial light modulator", Elec. Letts., Vol 28, No 17, pp 1627-1628

11. ACKNOWLEDGEMENT

The authors are grateful to Lucas Control Systems Products for their support and assistance.

A neural network based optical wavefront sensor

A.F. Armitage, R. McKeating & A.M. Smillie
Dept. of Electrical, Electronic & Computer Engineering,
Napier University, 219 Colinton Road, Edinburgh, EH14 1DJ

C.M. Humphries & E.Atad
Royal Observatory, Blackford Hill, Edinburgh, EH9 3HJ

ABSTRACT: Preliminary results are presented from a novel optical wavefront sensor. An array of samples from the wavefront is captured using a ccd camera and a Hartmann-like mask. A centroiding operation determines the displacement of the image spots. An artificial neural network system then interpolates the distorted wavefront, which is described via a Zernike polynomial. The outputs of the neural network are found to agree with the original wavefront to a high degree of accuracy.

1. INTRODUCTION

Optical astronomy from ground based sites is limited in many respects by distortions to the optical wavefronts arriving at the detector system. Light from distant stars arrives at the Earth's atmosphere in the form of a spherical wavefront of such a large radius that it is effectively a perfect plane wavefront. Passage through the atmosphere and any imperfect optical system distorts this wavefront. For some years, active optics have been used to reduce some of the slower or static distortions. Autoguiding systems with a typical bandwidth of 0.1 Hz can help to reduce the effects of slow variations, such as thermal or gravity induced distortions of the telescope structure. However, these active systems cannot cope with the rapid (approximately 10-100 Hz) variations produced by atmospheric turbulence.
Recently, much work[1,2] has been done on adaptive optics for optical astronomy. This implies being able to measure and correct for atmospheric turbulence with a bandwidth of about 100 Hz in the infra-red, and at greater rates in the visible.

2. WAVEFRONT DISTORTION AND MEASUREMENT

Temperature variations in the atmosphere cause refractive index variations at scales from tens of centimetres to hundreds of metres. This results in a complex phase aberration of any optical wavefront passing through the atmosphere. The real part of the aberration (known as 'seeing') corresponds to a phase distortion of the wavefront, while the imaginary part corresponds to intensity fluctuations (scintillation). Over the surface of a large mirror, the phase distortion can lead to a variation of up to 10 μm in the 'troughs' and 'valleys' of a wavefront. Adaptive optics systems attempt to measure and correct these phase variations. The scintillation term can not easily be corrected without reducing the light intensity; this is inappropriate for astronomical applications. An adaptive system that can correct for atmospheric fluctuations will also be able to correct minor distortions in the telescope optics.

The atmospheric seeing depends on the site, and on the prevailing weather conditions. The power spectrum in the variations can be described[3] approximately by a Kolmogorov power spectrum of turbulence. The distorted wavefront typically has a complicated shape that can be modelled by a set of orthogonal polynomials. These are generally defined using polar coordinates, since telescopes have circular apertures. For a telescope with a central obscuration, polynomials that are orthogonal over an annular pupil are properly required, but for obscuration ratios less than about 0.3 it is found that Zernike circle polynomials[4,5] are generally adequate. These have the advantage that the low order terms correspond to the classical (Seidel) aberrations such as coma, spherical aberration, and astigmatism.

Measurement of the wavefront distortions can be made by a variety of sensors. Most of these measure the slope of the wavefront, as optical frequencies are too high for direct measurement of the phase. For example, shearing interferometry is well suited to precise measurements of static aberrations that arise from imperfections of the optical system. An alternative is the classical Hartmann test, in which a mask projects an array of point images whose deviation from their reference positions provides a measure of the wavefront distortion. In astronomical applications light levels are generally low, so the Shack-Hartmann variant may be preferred. Here, the mask of pin-holes is replaced by an array of lenslets; these have a superior light gathering power and improved sensitivity since the individual images are in focus.

Recently, artificial neural network techniques have been applied[6,7,8]. Phase distortions are deduced from a pair of images of a reference star. One of the images is in focus, and the other is deliberately out of focus. Using the star Vega as the reference star, good agreement was reached between the measurement of phase distortion from the neural network system and from measurements made by a conventional Shack-Hartmann sensor.

3. SYSTEM DESCRIPTION

In our application to the adaptive compensation of atmospheric seeing fluctuations at an infrared astronomical telescope, the incoming wavefront is sampled at visible wavelengths by a Hartmann mask located at a re-imaged pupil of the telescope (Figure 1a). A lens behind the mask forms a defocused image matched to the dimensions of the CCD detector such that the individual sub-aperture images corresponding to the 4 x 4 array of holes in the mask each have a diameter of about five pixels. For a Shack-Hartmann array of lenslets (Figure 1b), the entire pupil produces a similar array of sub-aperture images. In either case, the image centroids are determined by taking intensity weighted moments along x and y co-ordinate directions of the CCD output.

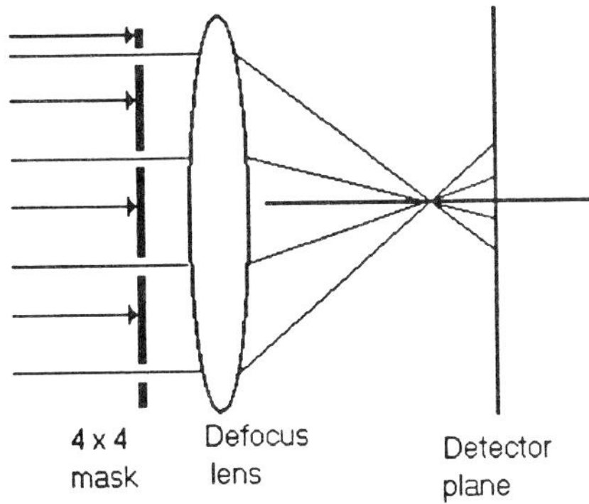

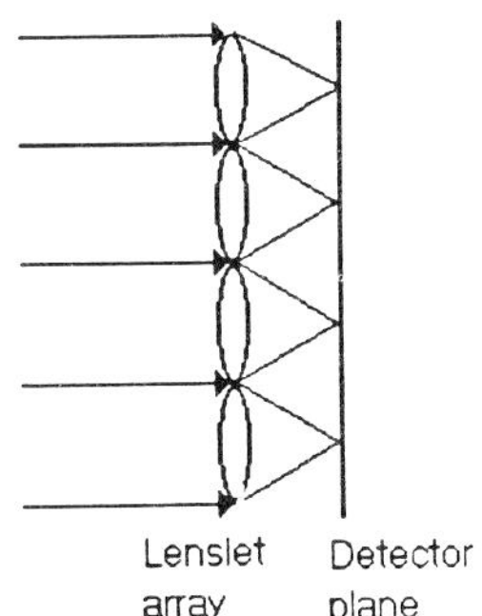

Fig. 1. (a) Modified Hartmann mask. (b) Shack-Hartmann array

4. IMAGE DATA REDUCTION

The main tasks to be performed by this part of the wavefront sensor may be summarized as follows.

1. Acquire a reference wavefront from a point source.
2. Determine image formation on the sampled wavefront (ie, locate the "spots" or sub- images).
3. Find the centre of each sub-image.
4. Repeat 1 to 3 for the wavefront being measured.
5. Determine the deviation vectors, ie ,the difference in magnitude and direction between the reference and measured centres for each sub-image in the reference image and the corresponding sub-images in the arbitrary image. These are the raw tilt data.

With regard to the image formation function a set theory approach was adopted. This relies upon forming classes from the image data such that sub-images may be formed. For each sub-image, the pixels within each sub-image must be greater than the background/foreground value or threshold value T. With reference to Fig 1 we are in the fortunate position of being able to employ simple thresholding in the cases of a Shack-Hartmann lenslet arrangement and of a Hartmann mask. In the first case a background/foreground distinction is formed by the change of irradiance due to the lenslets and in the second the distinction is formed by the fact that light can only pass only through the holes that form the mask. The sub-images are identified by searching for contiguity in both x and y directions. In implementing this algorithm we may make use of two consequences of the fact that we raster scan memory. First we form image subsets by testing for contiguity in the x axis for each y location. Having reached the last location in memory we may perform a search back towards the origin to test for contiguity in the y axis. Once subsets in the x axis have been formed, we need only establish that a single member of one subset may form a union with another subset since, if one member forms a union, all will do likewise (from the definition of class union). The algorithm implementation is shown in Fig 2.

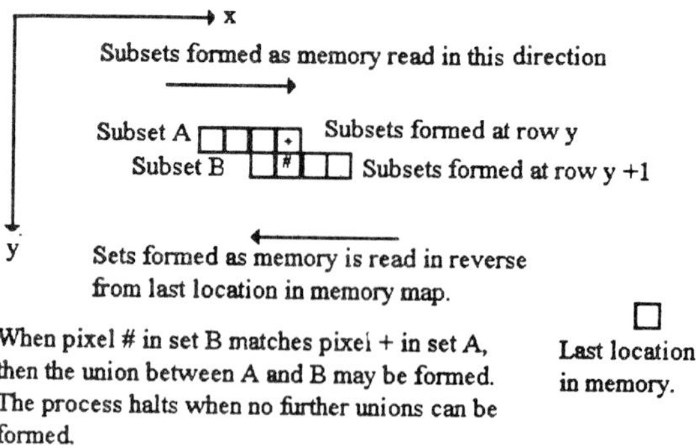

Fig. 2. Formation of sub-images

An advantage of the algorithm is that it is guaranteed to halt. This is a consequence of the "two pass" nature of forming the sub-images. Memory (or store) is read up to some defined location, then read in reverse back to the start of the search point thus giving a defined end point, irrespective of whether or not any data satisfying the given description were found. It may appear somewhat wasteful to have to suffer the time penalty of reading data memory twice. However the bulk of the search is performed in the upward pass. Thus any data found partially satisfying the desciption have been identified. These basic data may then be assigned to a smaller store and the downward path search may be performed on this reduced data set, thus incurring a minimal time penalty.

A further consideration is one of image noise ie, "rogue" pixels that are greater than threshold, but that are not members of any sub-image class. The most obvious method might be to employ a median or low pass filter. This approach would also alter the profile of the genuine sub-images, a process that would affect the process of finding the centres of each sub-image. The approach adopted has been dubbed a "tiling" filter since it involves finding the cross section of each sub-image as a set of tiles (Fig 3). We may then define a spatial threshold that just specifies the cross section of a valid image. Any other image smaller than this threshold may be rejected. To ensure that the correct sub-images are used between each frame (ie, sub-image 1 is the same between any two frames) a simple form of scene labelling is employed. The labels correspond to a label for each lenslet or hole being used in the sensor. Thus each sub-image is consistently labeled between frames. The final stage is determination of the centroids. This is achieved by taking moments about the centre of the bounding rectangle of each sub-image. For the prototype system, the number of sub-images is simply equal to the number of holes in the 4 x 4 mask.

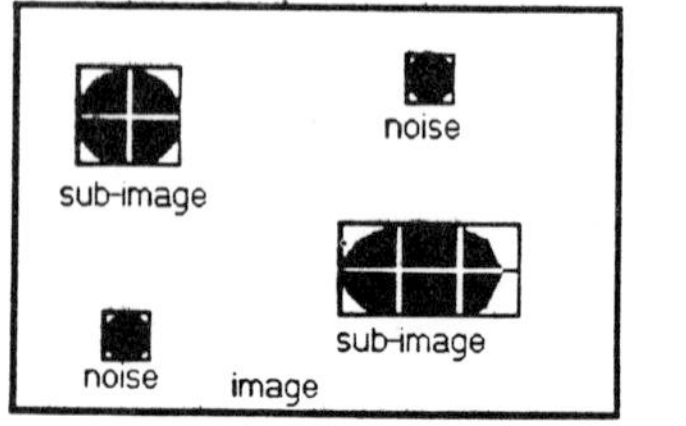

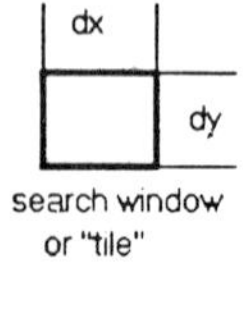

Fig. 3. The tiling filter for thresholding.

5. NEURAL NETWORK STRUCTURE AND TRAINING

The neural network used was a simple two layer perceptron[9] consisting of an input buffer, one hidden layer, and an output layer. Each layer was fully connected to the one before, giving two layers of weighted connections. The input buffer took the 32 x and y deviations from the 16 sub-images and presented them to the network. The hidden layer consisted of 16 processing elements (PEs), each of which performed a hyperbolic tangent transfer function of their inputs. The output layer was made up of 8 PEs, each of which passed their input data through a sigmoid transfer function. The outputs produced were in the form of Zernike coefficients, with each one representing a certain type of aberration e.g. x-tilt, astigmatism or spherical aberration.

Supervised learning using the back propagation algorithm[10] was the training method applied to the problem. Sets of inputs and their correct, or desired, outputs were presented to the network. The connection weights were initially produced randomly and the inputs were fed forward through the network to produce a set of outputs. Comparison of these outputs with the desired outputs produced a local error (e_k) at each output unit k. The aim was to reduce these errors by adapting the connection weights. The change in the weight connecting PE_j in the hidden layer with PE_k in the output layer was given by:

$$\Delta Wk_j = l_{coef}\, e_k x_j + m\Delta Wk_j$$

where x_j is the output from the hidden layer PE, l_{coef} represents the learning coefficient, and m is a momentum term.

The value of l_{coef} affects the size of the weight change and therefore the rate of learning. The momentum term feeds through a portion of the previous weight change for that connection which reinforces general trends while cancelling out oscillatory behaviour.

The development and training of the data was done using the NWorks software package[11]. Once the weights had been calculated, the network was implemented in the C programming language.

The training data were produced using a Fortran program written by Dr. G. Catalan which randomly generated a set of Zernike coefficients, and then calculated the corresponding x and y deviations. Three files, each containing 1000 examples of input/output data, were produced. Each one was presented to the network four times giving a total of 12000 examples. The RMS error, the most accurate measure of network performance, was reduced to approximately 0.012 over an output range of 0.6, giving a 2% error.

6. TESTING AND RESULTS

The network was tested using new data, generated in the same way as the test data, but not used as part of the training process.

One such set of data is shown below. The desired outputs (the Zernike coefficients) and actual outputs shown here are typical of the results produced.

DESIRED	-2.302	-0.043	-4.070	2.924	2.728	-0.605	0.857	-2.025
ACTUAL	-2.401	-0.065	-4.129	3.087	2.846	-0.721	0.887	-2.279

The coefficients are in units of wavelength. It can be seen that there is very good agreement between the desired and actual outputs, considering that the 4x4 mask is providing a limited sample of the incoming wavefront. With new data, the system performs well, though at the time of writing only simulated data generated by computer has been used. Some of the expected benefits of the neural network approach, such as resistance to noise, are only expected to become apparent when real data from a telescope are used.

7. CONCLUSIONS

The encouraging results achieved so far have demonstrated the feasibility of the method. There are two main limitations with the current system. The first is the use of the Hartmann mask, which means that much of the light is obscured. This will limit the number of stars that can be used for measuring the atmospheric aberration. Work on a Shack-Hartmann lenslet array should overcome this problem. The second limitation is that the speed of the system is not adequate for true adaptive control. The neural network simulation, running on a '486 pc, generates the coefficients in under one second. However, the initial centroiding operation takes several seconds to run. Alternative sensing methods and/or more powerful computing hardware will be required to reduce the time needed to calculate the Zernike coefficients from the measurement of the aberrated wavefront. The next stage will be to connect up a deformable mirror to achieve a complete closed-loop system. At that stage it may be more efficient to train the network to generate actuator control signals directly, rather than the Zernike coefficients that are currently produced.

8. REFERENCES

1.Beckers J.M. & Merkle F., Astrophysics and Space Science **160**:345-351, 1989.
2.Merkle F., Physics World, **4**:33-38, 1991.
3. Kolmogorov A., in *Turbulence: Classic Papers on Statistical Theory*, Friedlander S.K. and Topper L., eds., Interscience, New York, 1961.
4. Zernike F., Physica, **1**:689, 1934.
5. Born M. and Wolf E., *Principles of optics*, sixth ed., Pergamon, Oxford, 1983.
6. Angel J., Wizinowich P., Lloyd-Hart M. & Sandler D., Nature, **348**:221-224, 1990.
7.Sandler D.G., Barrett T.K., Palmer D.A., Fugate R.Q. & Wild W.J., Nature, **351**:300-302, 1991.
8. Lloyd-Hart et al., The Astrophysical Journal, **390**:L41-L44, 1992.
9. Lippmann R.P., IEEE Acoustics, Speech, and Signal Processing Magazine, April: 4-22, 1987.
10. Rumelhart D.E., & McClelland J.L. eds. *Parallel Distributed Processing*, Vol 1, MIT Press, 1986
11. Copyright NeuralWare, Inc. Pittsburgh, PA., 1991.

Profile measurement of optically rough surfaces by fibre optic interferometry

D.P.Hand, T.A.Carolan, J.S.Barton, J.D.C.Jones
Department of Physics, Heriot-Watt University, Riccarton, Edinburgh EH14 4AS

ABSTRACT: A rugged interferometric fibre optic instrument for non-contact profiling of optically rough surfaces has been developed, designed to characterise machined metal surfaces without removal from the machine tool. The sensor is a robust and compact Fizeau interferometer, in which one reflection is derived from the test surface. Performance is demonstrated with face-milled steel surfaces, with steep local gradients and local variations in reflectivity of a factor of 100,000. The measured horizontal resolution is 7μm, and the noise-limited vertical resolution 0.3nm/√Hz.

1. INTRODUCTION

We present an interferometric fibre optic instrument for profile measurement of optically rough surfaces, incorporating a remote, compact, non-contact optical probe which is sufficiently rugged to allow operation in realistic engineering environments. The instrument has been developed to measure profiles of machined metal surfaces without removing them from the machine tool, in order to gain information about the state of tool wear [1]. It complements a previously reported fibre interferometric sensor used for acoustic emission measurement during the machining process [2,3] (also providing tool wear information [4]), with the aim of constructing an automated tool wear monitoring system. The machined metal surfaces of interest typically have features with peak heights in the range of 1-10μm and widths from 10μm-1mm, and overall scan lengths of up to a few cm are required.

To minimise sensitivity to ambient conditions, given the requirement of machine shop operation, the sensor is based around a small Fizeau interferometer, only a few mm long and 1mm in diameter, formed by the cavity between a fibre end and the test surface and incorporating a small focusing lens. This is connected to a remote optical transceiver via a single mode fibre optic cable. Signal processing is based on a 4 phase-step algorithm [5], achieved by current modulation of the laser diode source, providing immunity to the extreme variations in surface reflectivity (a factor of 100,000 is not uncommon), found with machined surfaces.

Alternative surface profilers are in existence, based both on mechanical and optical techniques, but are unsuitable for this application. Mechanical profilers, which operate by traversing a diamond stylus across the surface, are delicate and require quiet, clean environments, precluding their use in a machine shop. Additionally, the diamond stylus can cause undesirable scratches on the surface. The few currently available non-contact optical profilers, meanwhile, whether based on full-field [6], or scanning techniques [7] are only suitable for profiling optically smooth surfaces, and would be unable to cope with the steep gradients and widely varying reflectivities of typical machined surfaces.

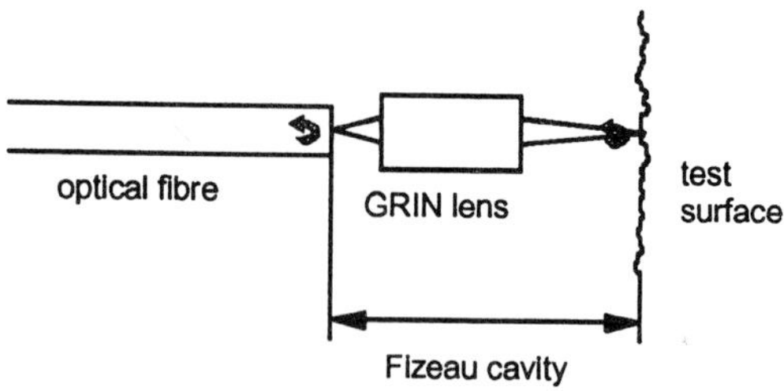

Figure 1: Fizeau interferometer probe

2. INSTRUMENT DESIGN

The Fizeau interferometer probe which forms the basis of the system is illustrated in figure 1. It consists of the cavity between the cleaved fibre end and the test surface, with an intermediate graded index (GRIN) lens focusing onto the surface. The size of the focused spot on the surface is set in the fabrication stage by the separation between the fibre end-face and the lens. For the specific lens used, the feasible spot diameter range is 7μm to 1mm. Light is brought in through the fibre, and its partial reflection at the cleaved end acts as the reference, interfering with the signal beam returning from the test surface. Light returning through the fibre is therefore modulated according to:

$$I = I_o\left\{1+V\cos\left(\frac{4\pi\nu D}{c}\right)\right\} \tag{1}$$

where ν is the light frequency, c the speed of light, and D the optical length of the Fizeau cavity. V is a visibility term, and depends on the reflectivities of both the cleaved fibre end and the test surface.

A passive phase-stepping technique is used to recover D. A minimum of three phase-steps are required, for the three variables I_o, V and D; however four steps are used to allow the possibility of error checking. Phase-stepping is achieved by modulating ν, the optical frequency of the laser diode source, produced by modulating the diode injection current. The amplitude of each step is adjusted to be equivalent to a phase change in the interferometer of $\pi/2$ rads. Hence, D is given by:

$$D = \frac{c}{4\pi\nu}\tan^{-1}\left(\frac{I_1-I_3}{I_2-I_4}\right) \tag{2}$$

where I_n is the light intensity returned from the nth phase step, for which the induced phase shift is $n\pi/2$; the I_n are normalised by the launched laser power.

This equation returns a value of D corresponding to an optical phase in the range $-\pi/2$ to $\pi/2$. In general, the probe to surface distance can be written as $D_m = D + (2m-1)c/8\nu$ where the integer m labels the fringe order. The profile of the surface may therefore be determined unambiguously if the probe/lens arrangement is scanned parallel to the surface and measurements of D are made sufficiently close together that for sequential values of D_m, the optical phase change is less than π. In addition, each set of four phase-step measurements must be sufficiently closely spaced to ensure that all are effectively of the same point on the surface.

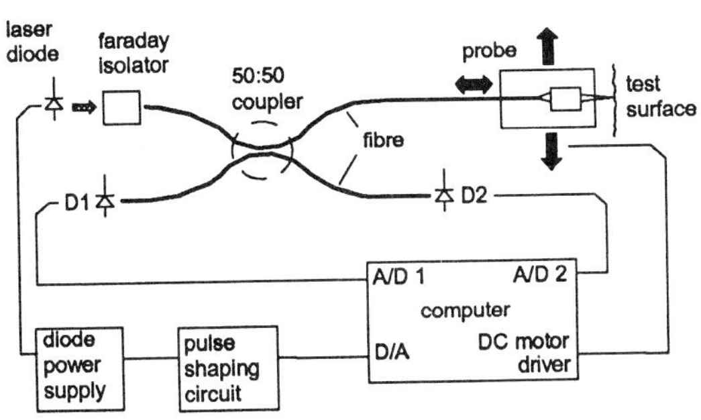

Figure 2 : Schematic diagram of optical surface profile instrument

A schematic diagram of the complete instrument is shown in figure 2. The probe is connected to the rest of the system by a single mode fibre, which has the important feature of being insensitive to any external perturbations, such as bending, vibrations or temperature fluctuations, since the Fizeau is unaffected by the phase and polarisation of the input light. In addition to measuring the return signal from the probe at photodiode D1, the input intensity is also monitored at D2. This is vital, since the injection current modulation necessary for phase-stepping also causes amplitude modulation, which must be compensated before applying equation (2). Signals detected at D1 are therefore simply divided by those at D2, which also cancels any errors arising from changes in optical launch efficiency. A PC containing an A/D and D/A board controls the entire instrument, providing the modulation signal for the laser diode whilst simultaneously recording and analysing the data, in addition to controlling the dc motorised stage which scans the probe across the test surface.

3. APPLICATION

In order to verify the operation of the optical instrument, a calibration piece for a scanning electron microscope, consisting of a series of lines 51μm apart, was used as a test surface. The resultant profile is shown in figure 3, demonstrating a vertical resolution of 10nm, and horizontal resolution of 7μm (limited by the spot size). The normalised vertical resolution corresponding to the interferometer noise floor was $0.3\text{nm}/\sqrt{\text{Hz}}$. To test operation with realistic machined metal surfaces, finishing cuts were made on EN24 annealed steel blocks by face-milling with a Wadkins CNC milling machine. Scanning these with the optical system, surface profiles such as that shown in figure 4(a) were obtained. The characteristic machining marks are clearly visible, with the fundamental periodicity corresponding to the rotation

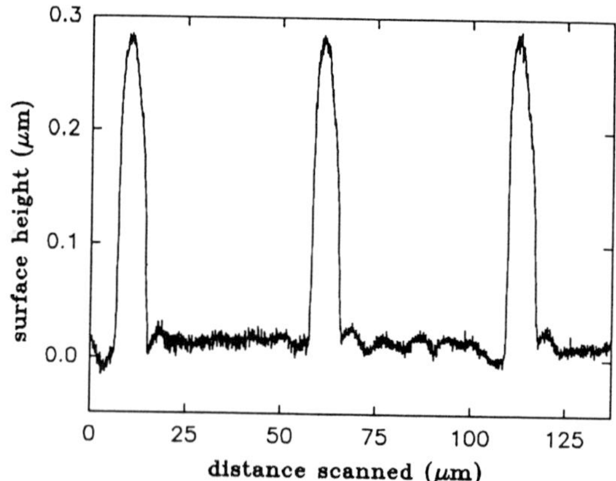

Figure 3 : Profile of SEM calibration piece obtained with optical instrument

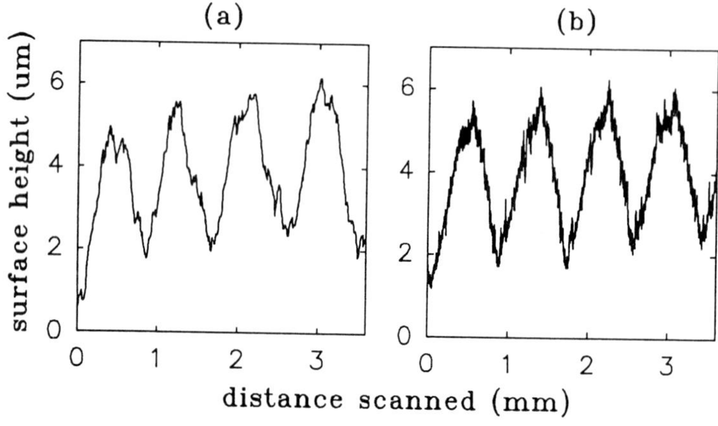

Figure 4 : Sample surface profile of face-milled steel block obtained (a) with optical instrument and (b) with Talysurf

frequency of the milling tool holder. A profile of a similar part of the surface generated with a commercial mechanical stylus profiler, the Rank Taylor Hobson Talysurf 5, is shown in figure 4(b) for comparison. The profiles do not exactly match because it is not feasible to register the test piece within the limits of their horizontal resolutions, but the similarities are clear. These results demonstrate the instrument's capability to cope with steep surface gradients and large variations in returned power. The visibility was observed to vary over the range from 0.02-0.64 (corresponding to a reflectivity range of 4×10^{-6} to 0.46).

4. CONCLUSIONS

A rugged non-contact interferometric fibre optic sensor has been developed for measuring surface profiles of the optically rough surfaces produced during metal machining processes. It has a vertical resolution of 10nm and horizontal of 7μm (dependent on spot-size). Profiles of both an SEM calibration piece and a typical machined surface have been successfully measured, verifying the operation of the instrument. The main focus of the work is to provide a surface finish sensor for inter-operational in-situ use on machine tools, but other possible applications exist, in particular profiling of soft materials, or those with highly variable reflectivities.

Acknowledgement: This work was partially supported by the ACME Directorate of the SERC, UK.

REFERENCES

[1] J.Raja, D.J.Whitehouse: Int. J. Prod. Res. **22** (1984) 453.
[2] R.McBride, T.A.Carolan, J.S.Barton, W.K.D.Borthwick, J.D.C.Jones: Meas. Sci. Tech., in press (1993).
[3] D.P.Hand, T.A.Carolan, J.S.Barton, J.D.C.Jones: Opt. Comm., in press (1993).
[4] E.Kannatey-Asibu, Jr., D.A.Dornfeld: Wear **76** (1982) 247.
[5] P.Hariharan: Appl. Opt. **28** (1989) 27.
[6] B.Bhushan, J.C.Wyant, C.L.Koliopoulos: Appl. Opt. **24** (1985) 1489.
[7] C.J.R.Sheppard, H.J.Matthews: J. Mod. Opt. **35** (1988) 145.

A self-referenced fibre optic temperature sensor using the temperature dependent threshold voltage of twisted nematic liquid crystals

J. Mason and A.T. Augousti

School of Applied Physics, Kingston University, Kingston, Surrey KT1 2EE

Introduction

The authors have previously reported the use of liquid crystals as transducing elements in novel configurations [1-3]. The use of liquid crystals in these devices relies on the temperature dependence of their alignment and relaxation times when activated in an LCD, as opposed to their more conventional usage when it is the temperature dependence of their spectral reflection characteristics that is utilised [4-6]. The liquid crystals in these latter instances have been directly thermotropic, and have not been activated in an electrical fashion. This paper reports a sensor which uses the temperature dependence of the activation voltage of the liquid crystals in an LCD. The device is self-referenced in that two time-multiplexed signals are produced when the LCD is alternately activated and inactivated. The output signal is a ratio of these two signals, and remarkably is almost independent of intensity over an order of magnitude. The output signal diminishes by nearly 80% over a temperature range of under 60°C. The division of the two signals is performed using a look-up table, and this is compared with a manual calculation.

Experimental Design

Figure 1 shows the variations of intensity of reflected light from an LCD with increasing applied voltage, for a series of different temperatures. It is evident from this graph that there exists a range of voltages over which the reflected intensity depends on temperature for a given voltage applied to the LCD. The isotherms in Figure 1 could be considered to form a surface in a 3-D space in which temperature is plotted perpendicularly to the two axes shown. In this case one can usefully characterize the surface by taking slices parallel to the three perpendicular planes, and indeed Figure 1 is a superposition of four slices. Figure 2 shows a slice which is parallel to the applied voltage-temperature plane, with a fixed value of

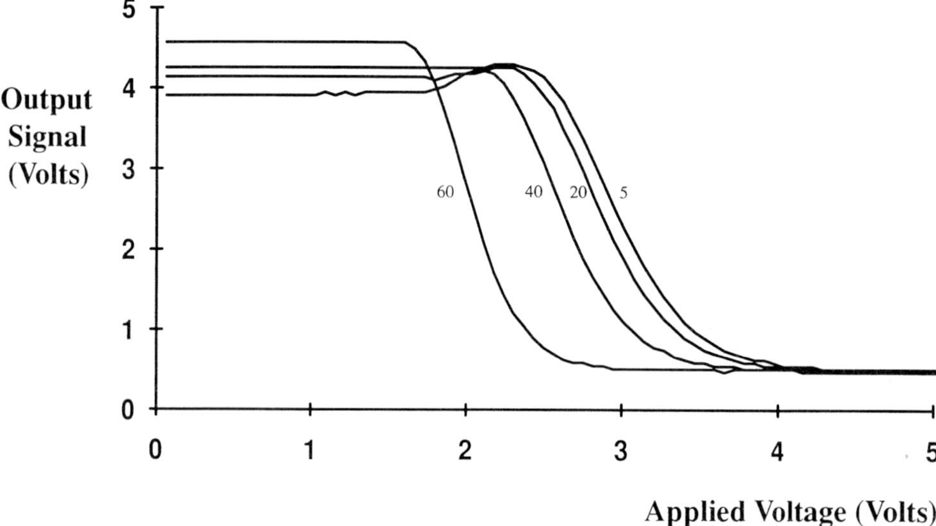

Figure 1. Intensity versus voltage for a family of different isotherms.

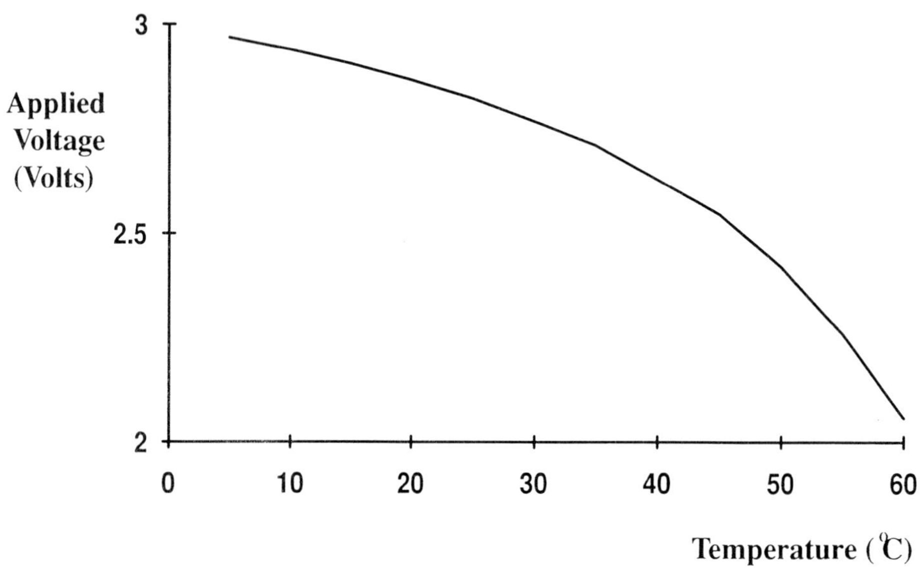

Figure 2. Applied voltage versus temperature for a given intensity.

intensity giving 2.46V of output voltage. Figure 3 shows two slices taken parallel to the output signal-temperature plane at two applied voltages of 0V and 2.5V. These are most useful for our purposes since they indicate that one signal is temperature-dependent (and relatively linear) whilst the other is relatively independent of temperature and would therefore serve as a useful reference level.

A system was constructed therefore to address the LCD, to detect the reflected intensity when the LCD was activated and inactivated and to automatically divide one signal by the other and display the output. A block diagram of this system is shown in Figure 4. A segment of an LCD is continuously illuminated by a red LED (HLMP 8150) using an optical fibre, and the light that is reflected from the back face of the segment is transmitted via a second optical fibre to a photodiode with a built-in inverting amplifier, which produces what has been termed the output signal. Due to the brightness of the illuminating LED, the reflected intensity is high and the signal must therefore be attenuated before entering the ADC (by a factor of approximately 3), which is performed using an inverting amplifier, which additionally makes the signal positive once more.

In order to obtain the ratio of the two signals, a control circuit was constructed which coordinated the addressing of an EPROM via the two data latches with the alternate activation and inactivation of the LCD. The activation state was achieved by energising alternately at high frequency two photovoltaic devices (in fact, matched LEDs operating as detectors) whose operation has been described elsewhere [1,2]. The signals representing the activated and inactivated states are alternately latched into the two 8-bit latches and provide the upper and lower bytes of a 2-byte address in the EPROM. The EPROM was programmed in such a way that each address contained the ratio of the upper and lower bytes of the address, and this value was output to the ADC which displayed the ratio on a DVM. This value was recorded over a specific temperature and the ratio was compared with one that was calculated manually using data that had been collected earlier. These results are displayed in Figure 5.

Discussion

The ratio signal which was obtained after some digital signal processing is relatively linear in its dependence on temperature, and varies by a factor of 5 as the temperature decreases.

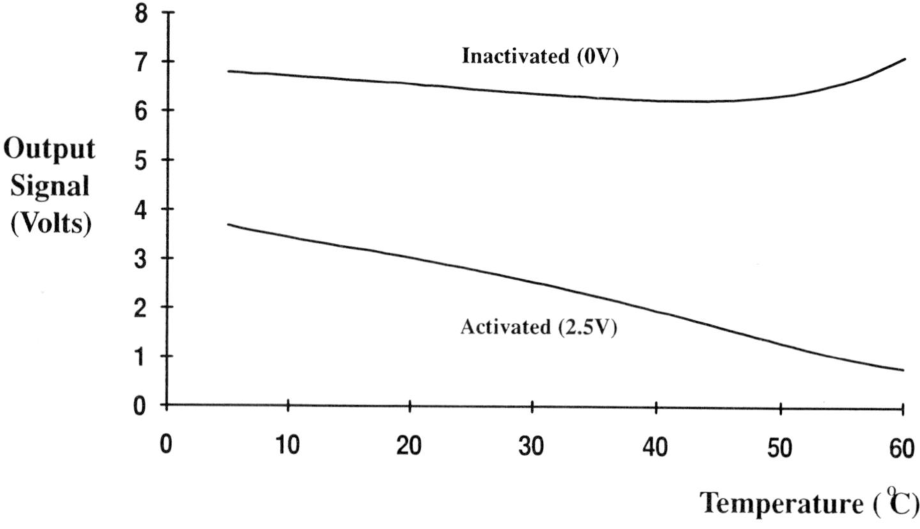

Figure 3. Temperature response of the LCD in activated and inactivated states.

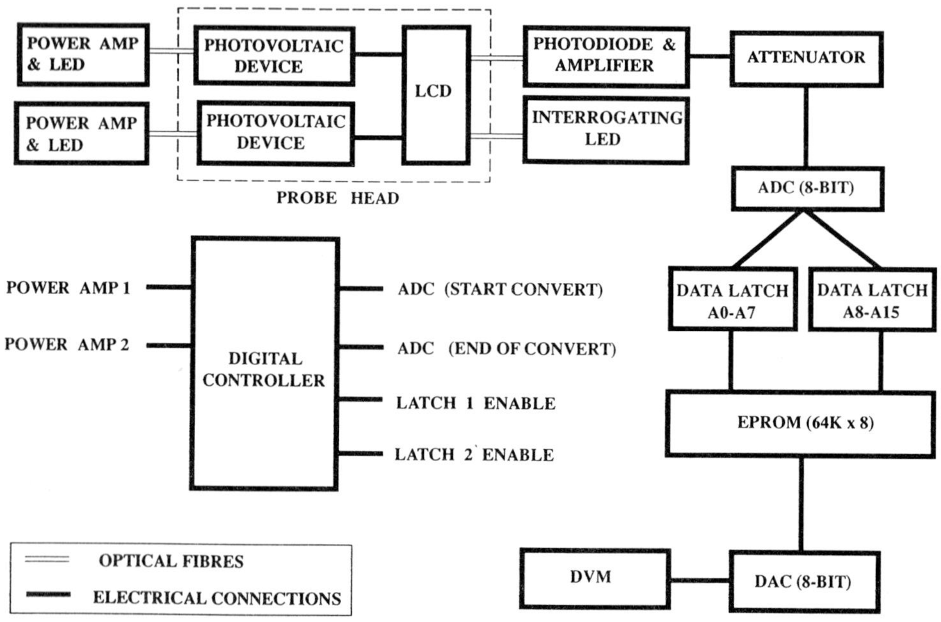

Figure 4. Block diagram of optical arrangement and electronic processing

The hysteris of the system is negligible on the scale of Figure 5, and the response time of the system is of the order of one second. The discrepancy with the manual observations is of most interest, and is attributed to the regime in which the manual readings were taken. In this situation, the LCD was activated and inactivated manually, with irregular intervals between activation. This appears to produce a slight variation in the reflected intensity observed. Whilst this suggests that the calibration curve (Figure 5) may depend on the frequency of activation, this is not perceived to be a problem, since the clocks which are used to control the system are highly accurate.

The most remarkable aspect of the system is the degree of immunity to intensity variation provided by the reference channel. The ratio of the two multiplexed signals was observed to be constant to less than 1% when the intensity of the continuous illuminating LED was varied by an over one order of magnitude.

Conclusion

A fibre optic temperature sensor has been constructed which has a probe head which is optically powered and addressed. It produces an output which diminishes almost linearly by over 80% over a temperature range of nearly 60°C. It has a response time of approximately one second and is effectively intensity-independent.

References

1. **Mason J and Augousti A T** special issue of *Optical Engineering* **31**(8) 1992 p1663-6

2. **Mason J and Augousti A T** presented at the *14th Symposium on Photonic Measurements*, Sopron, Hungary, 1-3 June 1992

3. **Augousti A T, Mason J and Koenders M A** Proceedings of the *International Conference on Electronic Measurement and Instruments*, 20-22 October 1992, Tianjin, China pp337-42

4. **Augousti A, Mason J and Grattan K T V** *Sensors - Technology, Systems and Applications* Adam Hilger 1991 p339-45 (Conf. Proceedings of Sensors and Their Applications V, Edinburgh, 22-25 Sept. 1991)

5. **Mason J and Augousti A T** Proceedings of the *2nd Applied Optics and Optoelectronics Conference*, 14-17 September 1992, Leeds, UK pp233-4

6. **Coles H J, Bone E L, Bowdler E R and Gleeson H F** *Proc SPIE - Int. Soc. Opt Eng.* (USA) 1990 **949** pp185-90

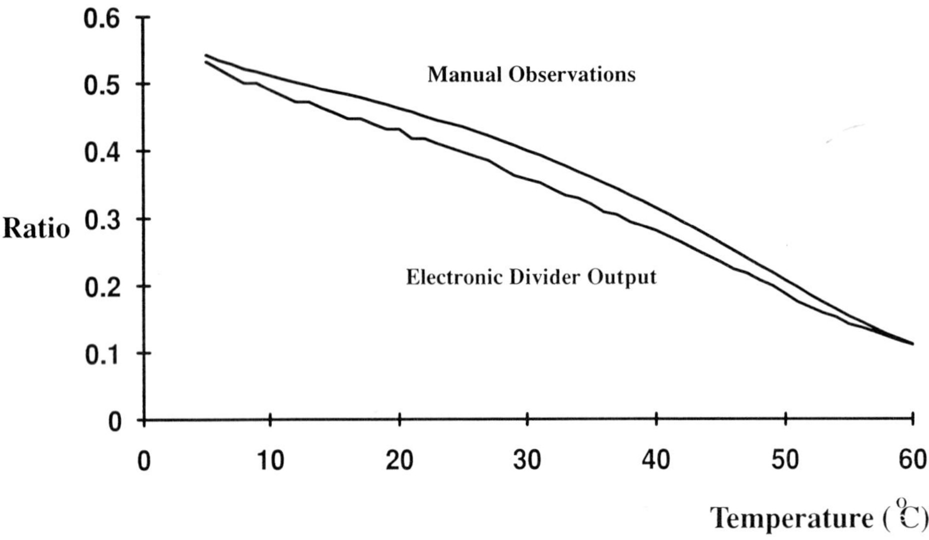

Figure 5. Ratio signal versus temperature.

A HiBi fibre polarization modulation scheme for ellipsometric measurements

R. Chitaree, K. Weir, A.W. Palmer & K.T.V. Grattan

Measurement and Instrumentation Centre,
Department of Electrical, Electronic,
and Information Engineering,
City University,
Northampton Square,
London, EC1V OHB, England.

ABSTRACT: A HiBi fibre polarisation modulation scheme is introduced for hi-speed ellipsometric measurements. It has been demonstrated that the simple system can be used to determine the optical constants of a non-absorbing material with an accuracy of the order 2%.

1. INTRODUCTION

A polarisation modulation technique which can be used to generate a rotating plane polarised light beam has many potential real-time applications in polarimetric and ellipsometric measurement systems. Polarisation modulation ellipsometric techniques offer particular advantages over static systems, in that no mechanical adjustment of the optical components is required and the high rate of polarisation modulation provides the potential for high speed measurement.
A technique to achieve this using free-space bulk optical components and fibre optics has previously been described[1]. Such techniques have been investigated and their experimental performance[2] compared in terms of the degree of polarisation and ellipticity. These results have demonstrated that a higher quality output was achieved using the fibre optic scheme. Such a scheme has been configured with a variation in the intensity output of the beam of $< 1\%$ in the required polarisation during a 2π polarisation change with less than 1% of its energy in the orthogonal axis. This technique has now been applied to the measurement of refractive index of a non-absorbing sample using the ellipsometric technique.

2. THEORETICAL BACKGROUND

The state of the electric field, E_i, of the rotating plane polarised light can be written as a vector , using Jones notation[3], as

$$E_i = \begin{bmatrix} E_0\cos(2\pi f_m t) \\ E_0\sin(2\pi f_m t) \end{bmatrix} \qquad \text{..........(1)}$$

where f_{m} = frequency modulation, E_0 = a constant and t = time.

By following the change in the characteristics of the light passing through the optical arrangement described in the figure 1, the alternation in polarisation of the light beam can be written in terms of the product of Jones matrices as,

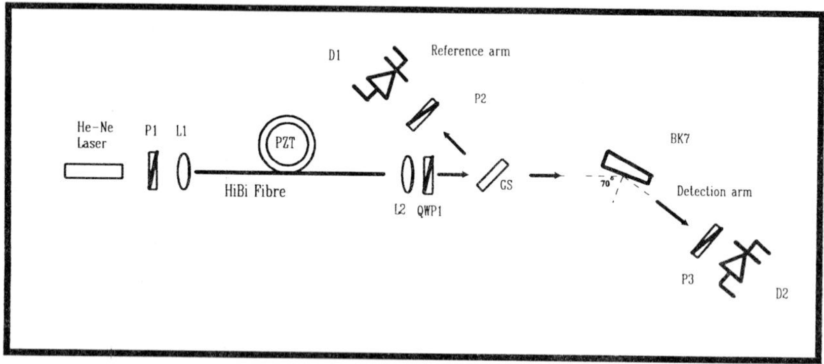

Figure 1: HiBi fiber Polarization Modulation arrangement for Ellipsometric measurements. Ps, Polarising Prisms; Ls, Lenses; QWP, Quarter Wave Plate; GS, Glass Slide; D, Detector.

$$E_f = \underbrace{\begin{bmatrix} \cos^2\theta & \cos\theta\sin\theta \\ \cos\theta\sin\theta & \sin^2\theta \end{bmatrix}}_{\text{polariser}} \underbrace{\begin{bmatrix} r_p & 0 \\ 0 & r_s e^{i\Delta} \end{bmatrix}}_{\text{sample}} \underbrace{\begin{bmatrix} E_0\cos(2\pi f_m t) \\ E_0\sin(2\pi f_m t) \end{bmatrix}}_{\text{rotating light}} \qquad \text{..........(2)}$$

where E_f describes the electric field of the detected light, r_p, r_s = the Fresnel reflection coefficients for incident polarisations p, parallel to and s, perpendicular to, the plane of incidence, Δ = the phase difference introduced by a sample, and θ = the orientation of a polariser with respect to the plane of incidence.

Consequently, the intensity of the light emerging is given by,

$$I_f = E_f \cdot E_f^* \qquad \text{..........(3)}$$

and

$$I_f = I_0\{[\frac{R_p\cos^2\theta + R_s\sin^2\theta}{2}] + [\frac{R_p\cos^2\theta - R_s\sin^2\theta}{2}]\cos(4\pi f_m t) + 2\sqrt{R_p R_s}\cos\theta\sin\theta\cos(2\pi f_m t)\sin(2\pi f_m t)\cos\Delta\} \quad \text{..........(4)}$$

where $R_p = r_p^2$, $R_s = r_s^2$.

In the particular case of interest, the polariser is orientated at an angle of $\frac{\pi}{4}$ to the plane of incidence; therefore, eq.(4) may be simplified to:

$$I_f = \frac{I_0}{2}\{(\frac{R_p + R_s}{2}) + (\frac{R_p - R_s}{2})\cos(4\pi f_m t) + \sqrt{R_p R_s}\sin(4\pi f_m t)\cos\Delta\} \quad \text{.....(5)}$$

Since the light used in this experiment is rotating polarised light in which its azimuth depends upon the modulation signal, two particular orientations, the perpendicular and parallel directions, can easily be identified from a reference signal.

When the light is modulated to be parallel to an incident plane, the intensity detected is given by,

$$I_{parallel} = \frac{I_0 R_p}{2} \quad \text{..........(6).}$$

Similarly, when it becomes perpendicular to the incident plane, the intensity is written as,

$$I_{perpendicular} = \frac{I_0 R_s}{2} \quad \text{...........(7).}$$

Thus there is sufficient information to determine the refractive index of a sample from the change in polarisation of light reflected from it. The equation[4] used to determine the refractive index of a sample, is as follows,

$$\frac{N_1}{N_0} = \sin\phi_0[1 + (\tfrac{1-\rho}{1+\rho})^2\tan^2\phi_0]^{\frac{1}{2}} \quad \text{..........(8)}$$

where $\rho = \frac{r_p}{r_s} = \left(\frac{R_p}{R_s}\right)^{\frac{1}{2}} e^{i\Delta}$; the ratio of the complex Fresnel reflection coefficients,

where ϕ_0 = an incident angle, N_0 = the complex index of refraction of medium 0 air),N_1= the complex index of refraction of medium 1(the sample).

For a non-absorbing sample, $\Delta = 0^{o}$ or 180°. Thus, it is possible to determine the refractive index of such a sample, N_1, if the values of angle of incidence and N_o are given.

3. EXPERIMENTAL DETAILS

The polarisation modulation scheme employed is an interferometric technique produced by winding a length(10m) of HiBi fibre (York HB-600) around a cylindrical PZT (fig.1). This was driven sinusoidally to stretch the fibre and modulate its birefringence. Light from a HeNe laser(λ = 632.8 nm) was polarised at 45° to the eigenaxes of the fibre and launched into the fibre by means of a x10 objective lens. The output from the fibre was then collimated (via a further objective lens) and passed through a quarter-wave plate with its optical axis at 45° to the axes of the fibre. This transforms the output from each eigenmode into circularly polarised light of opposite handedness which combine to give a linear state. The relative modulation of the path length then modulates the azimuth of the linear state. The reference signal was provided by sampling a small fraction of the beam (via a glass slide at an incidence angle of 3°) through a polarising prism onto a photodiode. The main beam was incident on the sample and the reflected signal detected via a polariser (at 45° to the axes) and photodiode. The detected and reference signals were then transferred to computer for analysis.

4. EXPERIMENTAL RESULTS

The intensity of the light beam received after reflection from an optical wedge (BK7 which is used as a standard sample), inclined at 70°, was measured. A wedge shape is used to separate the front and back reflections from one another in order to avoid unwanted interference. The wedge, BK7, has no absorption at the wavelength (λ = 632.8 nm.) used.

The measurement approach is summarised in Fig. 2. From these measurement ρ is obtained i.e.,

$$\rho = \left(\frac{I_0 R_p}{I_0 R_s}\right)^{\frac{1}{2}} = \left(\frac{6.45}{49.48}\right)^{\frac{1}{2}} \qquad \text{..........(9).}$$

Substitution into equation (8) gives the refractive index of BK7 as 1.531 with a standard deviation of 0.014 from this mean value. This is compared to the known refractive index of the sample, given as 1.519 while the index of medium 0(air), N_o is given as 1.0, and the incident angle is 70°.

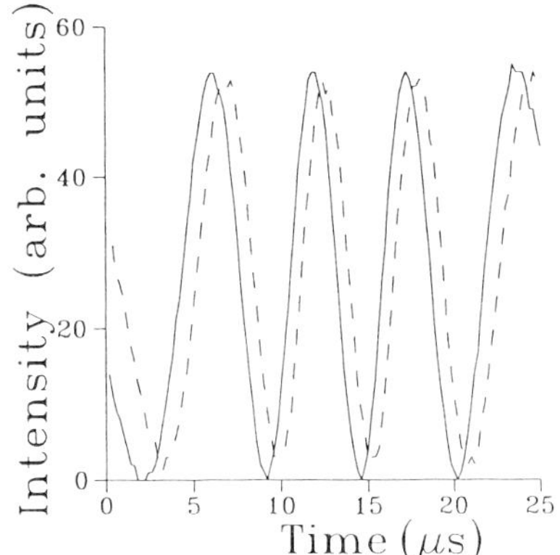

Figure 2: Output signals at detection arm
Solid line – signal from sample,
Broken line – reference signal in which peaks correspond to perpendicular states and troughs correspond to parallel states of polarisation.

Sources which may contribute to the measurement error are the accurate determination of the incidence angle, the orientation of the polarisation components and the digitisation process. Work is continuing to reduce the systematic error from these sources.

5. CONCLUSION

In conclusion, it has been demonstrated that the technique described, using simple procedures, allows the reliable calculation of the refractive index of a non-absorbing sample.
Furthermore, when the azimuth of polarisation is at 45° to the plane of incidence, the detected beam intensity, I_{45}, together with $I_{parallel}$ and $I_{perpendicular}$ allows the value of the phase shift, Δ, to be calculated from eq. (5).
From this calculation the parameter, ρ, can be determined accurately. This allows the calculation of the complex refractive index, permitting measurement on absorbing materials and thin films. This is an area of continuing current work, representing the potential for the use of this technique.

6. ACKNOWLEDGEMENTS

The authors are pleased to acknowledge support from the Science and Engineering Research Council(SERC). R C acknowledges support from the Thai Government by way of a studentship.

REFERENCES

1. R P Tatam, J D C Jones and D A Jackson, *Optical Polarisation state control scheme using fibre optics or Bragg cells,* J. Phys. E: Sci. Instrum. 19, p. 711-717, 1986.
2. R Chitaree, K Weir, A W Palmer & K T V Grattan, in *Polarisation Modulation in Free Space and Fibre Optics,* Conference on Applied Optics and Opto-Electronics , 14-17 September 1992 , Leeds, UK., Proceedings published by Institute of Physics, p.220-222, 1992.
3. R Clark Jones, *A New Calculus for the Treatment of Optical Systems,* J. Opt. Soc. Am. 31, p. 488-493, 1941.
4. R M A Azzam and N M Bashara, *Ellipsometry and Polarised Light,* North Holland, Amsterdam, 1977.

Optical sensors for carbon dioxide based on ionophores

S. Ozawa[1], W. Simon[2]

1 On leave from Central Research Laboratory, Hitachi, Ltd., Kokubunji, Tokyo 185, Japan

2 Department of Organic Chemistry, Swiss Federal Institute of Technology (ETH), Universitätstrasse 16, CH-8092 Zürich, Switzerland (deceased on November 17, 1992)

ABSTRACT: Optical sensors (optodes) for Carbon Dioxide (CO_2) based on ionophores (lipophilic ion-complexing reagents) have been developed. The sensor is essentially a plastic membrane of about 3 μm thickness, which incorporates a chromoionophore (lipophilic pH-indicator dye), a lipophilic ionic compound, and optionally an ionophore for (bi)carbonate. In the presence of humidity, it alters its UV-visible spectrum upon exposure to gaseous CO_2 of different concentrations. The ionophore enhances the sensitivity towards CO_2.

1. INTRODUCTION

To quickly determine the concentration of CO_2 in various media, a number of CO_2 sensors have been developed. They can be roughly classified into electrodes and optodes, which employ electrical and optical signal as the output, respectively. The Severinghaus CO_2 electrode [1,2] is a representative of the former type.

An optode for CO_2 was first realized by Lübbers and Opitz [3]. The recognition of CO_2 was carried out by means of an internal buffer solution, the same as for the Severinghaus type electrode, but a pH indicator dye was used instead of the pH electrode to report the pH change induced. Since then, several varieties of this type of CO_2 optode have been described [4-9]. However, the use of the buffer solution makes the sensor complex, unstable, and difficult to handle, so it is preferable to avoid it. An optode for CO_2 without such buffer solution was described in a recent report [10].

Independently, we have developed optodes that selectively respond to NH_3, which performed perfectly without any buffer solution [11, 12]. They were based on a NH_4^+ ion-selective optode membrane [13], which was designed to selectively detect NH_4^+ ion in aqueous solutions by employing ionophores (lipophilic ion-complexing reagents). Successively, similar gas optodes for CO_2 were developed [14], and their formation, function, and advantages are summarized in this report.

2. EXPERIMENTAL

Reagents. Chromoionophore: 3-octadecanoyl-7-hydroxycoumarin (C828) was obtained from Lambda Probes and Diagnostics, Austria. Ionophore: 4-(dodecylsulfonyl)-1-trifluoroacetylbenzene (ETH 6019) was synthesized according to [15]. Lipophilic cationic site: methyltridodecylammonium chloride (MTDDACl); plasticizer: bis(2-ethylhexyl) sebacate (DOS); matrix: poly(vinyl chloride) (PVC), high molecular weight, were obtained from Fluka AG, Switzerland. Note that all the reagents have nonpolar functional groups which provide high lipophilicity, *i. e.* affinity towards organic solvents rather than water.

Optode preparation. The optode essentially comprised an optode membrane, and optionally a gas-permeable membrane. The optode membrane consisted of plasticized PVC (DOS:PVC = 2:1 by weight) and 50 mmol kg^{-1} each of the chromoionophore and MTDDACl. Optionally, 100 mmol kg^{-1} of the ionophore was also employed. The optode membrane was prepared by first dissolving the reagents in tetrahydrofuran and then spin-coating onto a quartz plate. The resulting membrane of about 3 μm thickness was conditioned with an appropriate buffer solution. This was used as the optode for gas phase samples. For determination of gases dissolved in aqueous solution samples, the entire surface of the membrane was directly covered by a gas-permeable membrane made of Poly(tetrafluoroethylene).

Apparatus and Procedure. The optode was placed in a flow-through cell, and its output was evaluated in the transmission mode by placing the cell inside a UV-VIS spectrophotometer [11]. Gas phase samples were prepared by diluting the standard gas with N_2 [12], and aqueous samples were prepared by adding excessive H_2SO_4 to aqueous solutions of the sodium salts of the respective acid.

3. RESULTS AND DISCUSSION

Response mechanisms. The active components in the CO_2 optode membrane phase (m) react with CO_2 in the sample phase (s) in the presence of excess humidity as follows:

$$\nu L^{-}(m) + p\overline{L}(m) + H_2O(s) + CO_2(s) \leftrightarrow \nu LH(m) + p\overline{L}\cdot\cdot H_{2-\nu}CO_3^{\nu-}(m) \quad \ldots(1)$$

where LH and $\overline{L}$ stand for the chromoionophore and the optional neutral ionophore. The electroneutrality in the membrane phase is ensured by lipophilic cationic sites Q^+, such as quaternary ammonium ions. ν is the degree of deprotonation of the carbonic acid, and p is the coordination number of the ionophore $\overline{L}$. According to a general derivation [16], the response function of the optode can be derived similarly as follows:

$$K_{eq} = \frac{(1-\alpha)^{\nu}(C_Q - \alpha C_L)}{C_G \nu \alpha^{\nu}\{C_{\overline{L}} - \frac{p}{\nu}(C_Q - \alpha C_L)\}^p} \quad \ldots(2)$$

where α is the ratio of free to total chromoionophore concentration in the optode membrane. C_L, $C_{\overline{L}}$, and C_Q are the respective total concentrations of the membrane components, and C_G is the partial pressure of the CO_2 gas. Keq is the overall equilibrium constant for the reaction (1). The value of α is related to the output of the optode membrane, *i. e.* absorbance A by

$$A = A_1 \alpha + A_0 (1-\alpha) \quad \ldots(3)$$

where A_0 and A_1 are the respective limiting absorbance values for $\alpha=0$ and $\alpha=1$.

Spectral response of the optode. The optode membrane based on C828, ETH 6019, and MTDDA+ as the chromoionophore, ionophore, and lipophilic cationic compound showed spectral change as shown in Fig. 1 when exposed to different concentrations of humidified CO_2 gas. The absorbance band at 424 nm is due to the unprotonated form of C828, and the intensity decreases about 35 % when increasing the concentration of humidified CO_2 by 100 volume %. Although this is not a very large change, differences could be easily distinguished by conventional optics. The band at 252 nm due to uncomplexed form of ETH 6019 is almost invisible, which indicates that the ionophore is mostly complexed probably with either H_2O or (bi)carbonate.

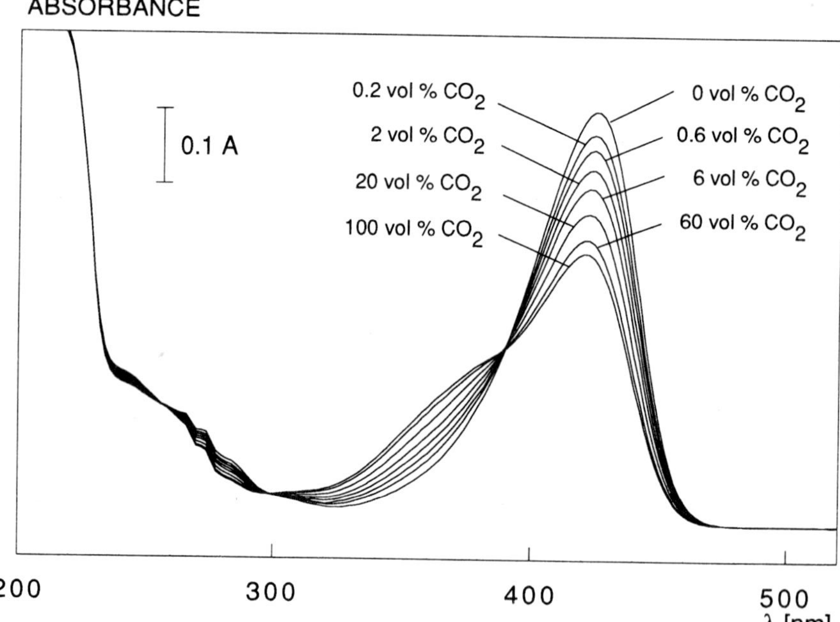

Fig.1 Absorption spectra of the gas optode based on C828, ETH 6019, and MTDDACl. Samples were humidified CO_2 gas diluted with humidified N_2 gas.

As is shown in Fig. 1, the change in the spectra are gradual, and perfectly converging isosbestic points are achieved. These facts suggest that in the present optode membrane, the chromoionophore is involved in a single straightforward response reaction, such as proposed in equation (1). We consider that this is primarily accomplished by the use of highly lipophilic reagents for the membrane materials, which reduces the side effects *e. g.* caused by the dissolution of the reagents by water.

Normalized output of the optodes. A normalized absorbance, α, defined by equation (3) at the wavelength of maximal absorption can be used to describe the response function of the optodes. Fig. 2 shows plots of thus obtained α value against the logarithm of the partial pressure of CO_2, with the optode described above as well as a blank prepared without the ionophore. Theoretical response functions are shown with solid lines, which were calculated according to equation (2) with appropriate Keq values and by assuming $\nu=2$, p=2 and $\nu=1$, p=0 for the optodes with and without the ionophore, respectively. It can be seen that the theoretical curves fit fairly well to the experimental plots, so that the response mechanism described above is supported as the first approximation. One possible reason for the small but significant deviation of the theoretical curve could be that the number of coordination of the ionophore (p) and the degree of deprotonation of the carbonic acid (ν) may not be uniform as is assumed.

It is also apparent that the optode with the ionophore shows a response at lower CO_2 concentrations than the one without the ionophore. This is a proof that the ionophore contributes to enhancing the response reaction with CO_2.

The second effect of the ionophore, *i. e.* to induce the selectivity towards CO_2, has not yet been observed. (Optodes with and without the ionophore both showed response to aqueous CH_3COOH at concentrations about 1.5 orders of magnitude lower than to aqueous CO_2). This is in contrast with the remarkably

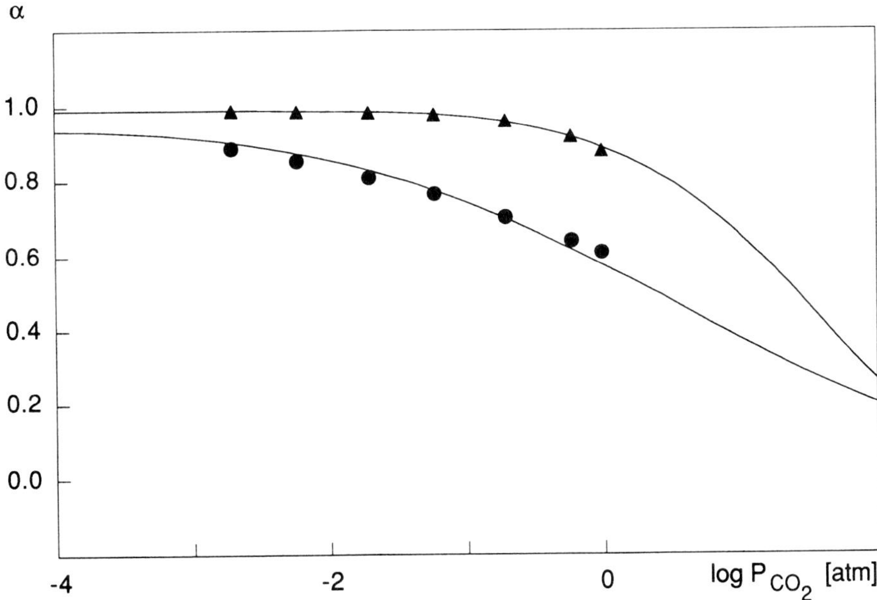

Fig.2 Normalized absorbance of the CO_2 optodes with (●) and without (▲) the ionophore ETH 6019. C828 and $MTDDA^+$ were used as the chromoionophore and the cationic site, respectively.

high selectivity obtained with the ionophore-based NH_3 gas-selective optodes [11]. Presumably, a CO_2 optode with higher selectivity may also be realized if a highly selective ionophore for (bi)carbonate ion could be obtained.

4. ACKNOWLEDGEMENTS

We would like to thank Dr. E. Pretsch and Dr. D. Wegmann at ETH for helpful discussions. S. Ozawa gratefully acknowledges support by the management of Hitachi Ltd. for his stay at ETH, in particular by H. Miyagi and Dr. Y. Watanabe.

5. REFERENCES

[1] R. W. Stow, R. F. Baer, B. F. Randall; Arch. Phys. Med. Rehabil. 38, 646 (1957).

[2] J. W. Severinghaus, A. F. Bradley; J. Appl. Physiol. 13, 515 (1958).

[3] D. W. Lübbers, N. Opitz; Z. Naturforsch. 30c, 532 (1975).

[4] J. L. Gehrich, D. W. Lübbers, N. Opitz, D. R. Hansmann, W. W. Miller, J. K. Tusa, M. Yafuso; IEEE Trans. Biomed. Eng. 33, 117 (1986).

[5] T. Hirschfeld, F. Miller, S. Thomas, H. Miller, F. Milanovich, R. W. Gaver; J. Lightwave Technol. 5, 1027 (1987).

[6] K. Goswami, J. A. Kennedy, D. K. Dandge, S. M. Klainer; Proc. SPIE-Int. Soc. Opt. Eng. 1172, 225 (1989).

[7] Z. Zhujun, W. R. Seitz; Anal. Chim. Acta 160, 305 (1984).

[8] C. Munkholm, D. R. Walt; Talanta 35, 109 (1988).

[9] O. S. Wolfbeis, L. J. Weis, M. J. P. Leiner, W. E. Ziegler; Anal. Chem. 60, 2028 (1988).

[10] A. Mills, Qing Chang, N. McMurray; Anal. Chem. 64, 1383 (1992).

[11] S. Ozawa, P. C. Hauser, K. Seiler, S. S. S. Tan, W. E. Morf, W. Simon; Anal. Chem. 63, 640 (1991).

[12] S. J. West, S. Ozawa, K. Seiler, S. S. S. Tan, W. Simon; Anal. Chem. 64, 533 (1992).

[13] K. Seiler, W. E. Morf, B. Rusterholz, W. Simon; Anal. Sci. 5, 557 (1989).

[14] S. Ozawa; Dissertation ETH Zürich No. 9647 (1992).

[15] Ch. Behringer, B. Lehmann, J.-P. Haug, K. Seiler, W. E. Morf, K. Hartmann, W. Simon; Anal. Chim. Acta 233, 41 (1990).

[16] W. E. Morf, K. Seiler, P. R. Sφrensen, W. Simon; Ion-Selective Electrodes, Vol. 5; E. Pungor (Ed.), Akadémiai Kiadó, Budapest, 1989, page 141-152.

A polarimetric optical fibre sensor immune to temperature variations for detecting vibrations

F. Pigeon, S. Pelissier, A. Mure-Ravaud, H. Gagnaire and C. Veillas

Laboratoire Traitement du Signal et Instrumentation U R.A. CNRS 842
Faculté des Sciences
23, rue du Docteur Paul Michelon, 42023 Saint-Etienne cedex 2, France

ABSTRACT : A study of temperature effects on a polarimetric vibration sensor using a telecommunication grade monomode fibre is presented. Insensitivity to temperature effects is obtained by a feedback system compensating for the low frequency linear birefringence variations.

1. INTRODUCTION

In polarimetric optical fibre sensors, the phenomenon to be detected modifies the birefringence of the fibre and then the polarization of the light. Many authors have described such sensors using high birefringence fibres [1]. In sensors applications, these fibres are very sensitive to temperature effects. A temperature insensitive fibre optic displacement sensor has been proposed by Imaï et al [2]. In their method, the fibre transducer consists of a pair of coils of polarization maintaining fibres spliced with their polarization axes at right angles to each other. In a previous paper [3], we have shown the possibility of using a telecommunication grade monomode fibre to design a polarimetric sensor to detect mechanical vibrations. Such fibres are low cost and have a smaller sensitivity to temperature than high birefringence fibres. Unfortunately, the method of Imaï for thermal compensation cannot be used because of the lack of fixed polarization axes. This paper deals with the study of temperature effects on characteristics of the sensor and shows the method to eliminate these effects.

The transducer consists of a coil of optical fibre. It detects variations in the linear birefringence of the fibre when vibrations induce variations in the radius of curvature of the coil. This sensor needs a bias point adjustment which is obtained by a compression of an auxiliary coil [3] : for good working conditions, a circularly polarized light at the input end of the fibre must be linearly polarized at the output end. The temperature variations induce modifications of the linear birefringence and therefore a displacement of the bias point. These variations can be compensated by modifying the compression of the auxiliary coil. Discrimination between linear birefringence variations due to vibrations or to temperature is possible thanks to the difference of frequency of the two phenomena : low frequencies for temperature and higher frequencies for vibrations. A feedback system acting on the auxiliary coil allows to keep constant the low frequency component of the signal. The sensor is so immune to temperature variations.

2. EXPERIMENTAL SET-UP

The experimental set-up is shown on figure 1. Linearly polarized light from the He-Ne laser is circularly polarized by a quarter wave plate and then is focused onto the fibre end using a microscope objective. A few meters of single-mode optical fibre (34 ST 1101 from EOTec Corporation) is used. The polarization state at the end of the fibre is determined with a Wollaston prism and two photodiodes. The sensing element is a coil (C_S) of this fibre with a single turn of 26 mm diameter. The mechanical vibrations are applied on a point of the coil diametrically opposite to the fixation point and so modify the radius of curvature of the fibre.

The auxiliary coil C_f is realized with the same fibre. It is a solenoidal winding of nine turns each of 26 mm diameter. The radius of curvature can be varied by compressing C_f with the help of an actuator. A major part of the fibre is put in a temperature regulated heat chamber.

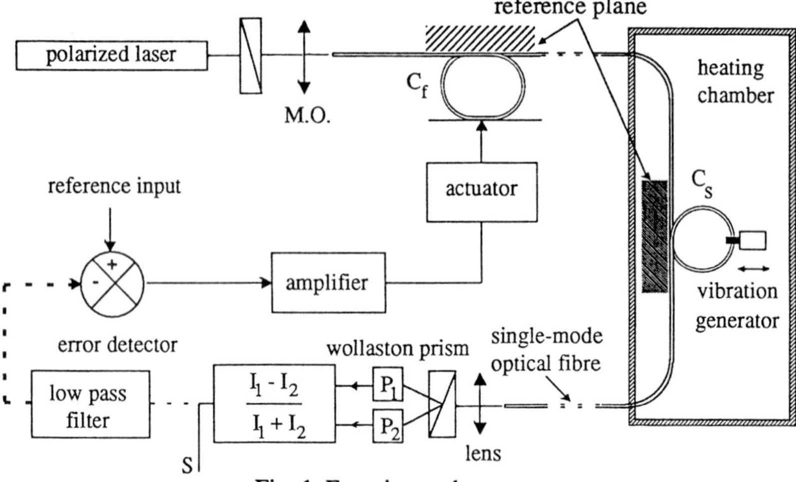

Fig. 1. Experimental set-up

3. TEMPERATURE INFLUENCE

In the absence of vibrations, we have studied the variations of the parameters of birefringence of the total length of the fibre for various compressions of C_f. These parameters can be chosen to be the orientation of the Wollaston prism θ_a corresponding to maximum signal intensity and the phase difference ϕ_0 between the two orthogonal linearly polarized modes. The value of $|\sin \phi_0|$ is deduced from the measurement of the detected signal for two orthogonal positions of the Wollaston prism [3]. The values of ϕ_0 are plotted versus the compresion of C_f for two temperature values (figure 2).

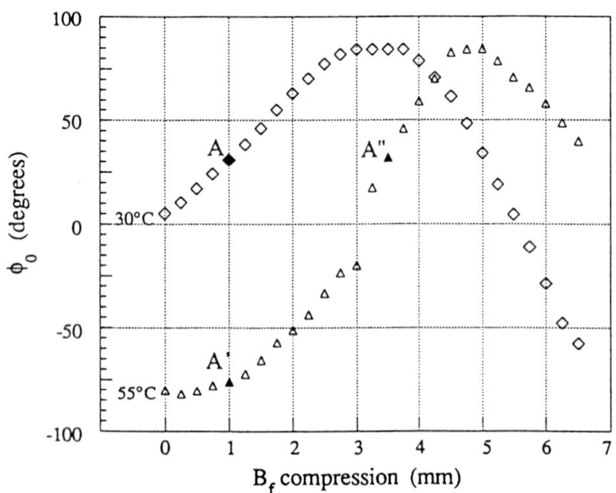

Fig. 2. Variation of ϕ_0 versus the compression of the coil

It can be seen that the variation of ϕ_0 with compression is not linear. This result is explained by the existence of both circular and linear birefringences. Therefore an initial bias point adjustement is needed. The best linearity and dynamic range of the sensor are obtained for a bias point in the middle of the linear part of the curve. In that case, the responsivity is 1 V/mm. The maximum output signal without distortion is 10 V. The minimum amplitude of the vibrations that can be detected is as low as 10 µm. The frequency range lies from dc to 2 kHz and depends on the coil diameter [4], [5]. From figure 2, it can be seen that the curve is shifted to the right as temperature increases.

Figure 3 shows the measured values of θ_a versus compression. Variations of temperature also induce a shift of the curve.

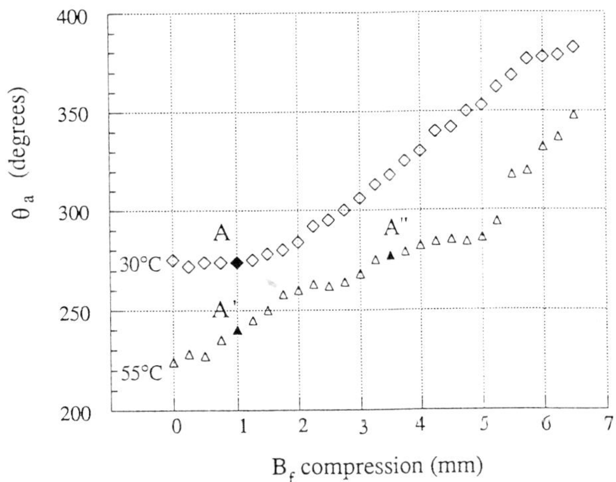

Fig. 3. Angular position of the Wollaston prism versus compression of the coil.

For 30°C , the bias point is represented by point A on the curves (figures 2 and 3). This is obtained by a 2 mm compression of C_f. A vibration of 1 mm amplitude and 60 Hz frequency is applied on C_s and the temperature is varied from 30 to 55°C. The amplitude of detected signal increases and distortion appears. This is due to the displacement of the the bias point from A to A'. The relative variation of the signal is plotted on figure 4. Its maximal value is about 50%.

4. INSENSITIVITY TO TEMPERATURE VARIATIONS

In order to keep constant the responsivity and the dynamic range of the sensor, the bias point must remain located in the linear part of the curves shown in Figures 2 and 3. So, as the temperature varies from 30°C to 55°C, the compression of C_f has to be increased from 2 mm to 7mm. By this way the bias point remains on an horizontal line from A to A". This is automatically done by using a feedback represented by the dashed line on Figure 1. This feedback system is built up in a simple manner. The low frequency component of the signal is compared to a reference. The sign of the difference drives the step by step motor of the compression actuator. A little hysteresis is introduced for stability purpose. Figure 4 shows the relative variation of the signal versus temperature when the feedback system is working. It is less than 10%. This result shows that this feedback greatly improves the performances of the sensor.

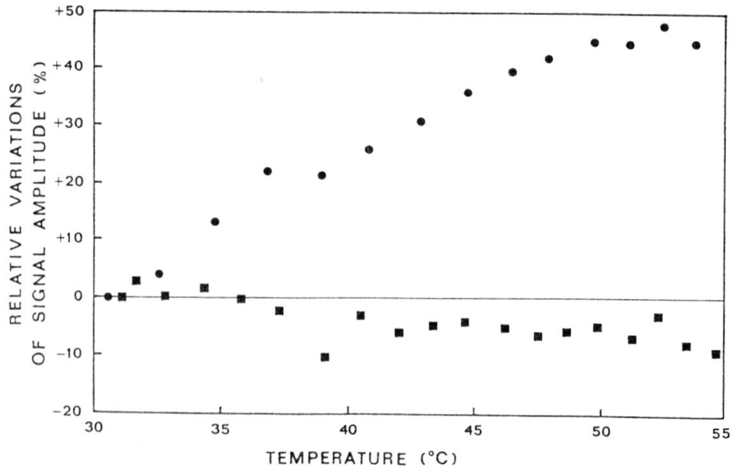

Fig. 4. Signal amplitude versus temperature
● without feedback
■ with feedback

5. CONCLUSION

We have studied the effects of temperature on the characteristics of birefringence of a telecommunication grade monomode fibre used as a sensor. A feedback system allows to keep constant these characteristics. This method gives a good insensitivity of the sensor to temperature variations. This system is usable for any length of fibre and any temperarure range by adjusting the number of turns of the auxiliary coil. A He-Ne laser has been used for convenience but a 1.3 μm source could be used with an appropriate fibre.

REFERENCES

[1] W. Eickhoff : 'Temperature sensing by mode-mode interference in birefringent optical fibres', Opt. Lett.,1981, **6**, pp 204.

[2] Y. Imai, M. A. Rodorigues and K. Iizuka : 'Temperature insensitive fibre coil sensor for altimeters', Applied optics, 1990, **29**, pp 975-978.

[3] F. Pigeon, A. Mure-Ravaud, C. Veillas and H. Gagnaire : 'Utilisation d'une fibre optique unimodale standard en capteur polarimétrique.Application à la détection de vibration mécanique', J. phys. III, 1991, **1**, pp 1323-1335.

[4] F. Pigeon, S. Pelissier, A. Mure-Ravaud, H. Gagnaire and C. Veillas : 'Optical fibre Young modulus measurement using an optical method', Electronics Letters, 1992, **28**, pp 1034-1035.

[5] F. Pigeon, A. Mure-Ravaud, C. Veillas and H. Gagnaire : 'Measurement of mechanical shocks with a monomode optical fibre sensor', Sensors and their applications V, Edinburgh, 22-25 september 1991.

Linear and scalable optical sensing technique for displacement

D. Kalymnios
University of North London, School of Electronic and Communications Engineering and Applied Physics
Holloway Road, London N7 8DB, U.K.

ABSTRACT : A novel optical method for measuring displacement is described. It is suitable for both opto-electronic and fibre-optic sensors and has a linear response and a scalable range. Preliminary results with an opto-electronic sensor are given.

1. INTRODUCTION

Displacement measurements are required in many industrial applications. The range of displacement could be in the submicron region for some applications or extend to many millimetres in others. In all of these cases either mechanical, capacitive, magnetic or optical techniques may be usable. Optical techniques are mostly preferable because of the non-contact operation and the accurate and fast response they offer. In addition, optical techniques that can also use fibres, are particularly desirable in electrically noisy environments (1,2).
When considering the many reported fibre-optic displacement sensor techniques, with millimetres measuring range, one can arbitrarily but conveniently classify them in either short-range (3-6), 10mm or less, or long-range, (7,11) a few tens of millimetres or more. A range of 100 mm has also been reported (12). In most, multimode glass fibres are used and the signals are coupled to the fibres using larger collection area optics, such as GRIN rods or other lenses.
In many applications the cost of a glass fibre optic sensor system may be considered prohibitive. This may be more true in cases where the processing electronics need only be a few metres away from the sensor head or in cases where electrical interference is not that severe to necessarily warrant the use of fibres. In such circumstances cost saving could be achieved if one could use Plastic-Optical-Fibres (POF) or well screened optoelectronic sensors. POF are currently gaining popularity for many short-distance link applications and many types of other sensors (13).
The proposed method for measuring displacement, is particularly suited for the development of cheaper sensor systems as mentioned above. The method has a linear response over a defined operational range and is applicable to both 'optoelectronic versions' and 'fibre-optic versions' of displacement sensors. Another feature of the method is in the operational range which is scalable. With POF of large core diameter (1mm) it is envisaged that additional signal collection optics will not be necessary for the shorter operational ranges, with reflective targets. Evidence for the validity of the proposed method, has been obtained using an optoelectronic version of the sensor.

2. PRINCIPLE OF OPERATION

The method is based on the inverse square law principle. In the arrangement shown in Fig.1, the photodetectors voltage signals are, for reasons of simplified analysis, 'approximated' to be of the form $S_1 = K_1P_0/(2x-x_0)^2$ and $S_2 = K_2P_0/(2x+x_0)^2$ where $K_{1,2}$ is a generalised constant for each detector and P_0 is the radiant intensity (w/sr) at the detector positions. The constant is of form $K_i = r_iA_it_iG(R_i)$, where r is the detector responsivity, A its active area, t a transmission parameter due to optics or a length of fibre, and $G(R_i)$ is the detector gain. It is accepted that K_1 and K_2 will in general be different because any of the three parameters r, A, t, could be different. One can, however, have fixed $G(R_1)$ and vary $G(R_2)$ so that the voltage signals S_1 and S_2 could become the same when both detectors have the same light radiant intensity incident on them.. This is in fact the detector calibration procedure and is performed prior to arranging them as in Fig 1. An important requirement for the sensor is for the detectors to be situated as close to the optic axis of the sensor as possible (i.e. minimum y). Only then can the radiant intensity of the emitter be assumed to be the same for both detectors.
This requirement could essentially be satisfied with emitters which have a fairly constant and symmetrical polar distribution around the normal. On the basis of the above assumptions one can now process the two signals in the following way.

Considering the ratio of the sum and difference of the signals one can easily show that:

$$S_R = \frac{S_1 + S_2}{S_1 - S_2} = \frac{x}{x_0} + \frac{x_0}{4x} \qquad (1)$$

since x = R + x_0, and for R = kx_0, k = integer
Then:

$$S_R = (k + 1) + \frac{1}{4(k + 1)} \qquad (2)$$

In a similar way it can easily be shown that in the case of diffuse (Lambertian) point targets (to be realised with a collimated beam),

$$S_D = \frac{x}{2x_0} + \frac{x_0}{2x} \qquad (3)$$

and:

$$S_D = \frac{1}{2}[(k + 1) + \frac{1}{(k + 1)}] \qquad (4)$$

Both S_R and S_D are plotted in Fig 2 for various k, and clearly these are quite linear. The slope of the graphs are close to 1 and 0.5 respectively, and the second term contributions in equation (2) and (4) representing the deviation from linearity decrease with increasing k. The exact form of S_R computed in the ranges $4x_0$ to $16x_0$ is also a straight line but with a slope 0.7% higher than the approximated one shown in Fig 2. Defining a percentage deviation from linearity as:

$$\alpha_{L,R} = \frac{(dS/dR)_{R_2} - (dS/dR)_{R_1}}{(dS/dR)_{R_1}} \times 100\% \qquad (5)$$

then it can be shown that,

$$\alpha_{L,R} \simeq \frac{100}{4(k_0+1)^2}\% \quad and \quad \alpha_{L,D} \simeq \frac{100}{(k_0+1)^2}\% \qquad (6)$$

It is interesting to note that linearity of response for the sensor is suggested to be wholly determined by the choice of k_0 i.e. a minimum range in multiples of x_0. As an example for a minimum range of $4x_0$ i.e.k_0 = 4, in the case of the reflective target a linearity of 1% can be achieved. Similar performance with a diffuse target would only be achieved at a minimum range of $9x_0$. It is not immediately evident from the above approximate analysis at what actual value of x_0 the predicted sensor performance will no longer be valid. The resolution of the method can also be obtained to a good approximation from equation (2) and (4) by neglecting the second term.

$$dR_R \simeq x_0 dS_R \qquad (7)$$

and

$$dR_D \simeq 2x_0 dS_D \qquad (8)$$

Obviously resolution will be ultimately determined by the ability of the processing electronics to discern changes in $S_{R,D}$. This in turn will depend on the signal to noise ratios present. On a conservative estimate that changes in $S_{R,D}$ of 1% are discernible, then resolutions of a few tens of microns seem feasible. Alternatively, the percentage resolution at a range of $10x_0$ could be about 0.1%.

Sensor performance stability should not be affected by emitter power variations or similar temperature effects in the detectors. The detected signal $S_{R,D}$ is believed to be largely unaffected by these changes. The same will not be true, however, if in the case of using fibres these undergo transmission changes by differing amounts for whatever reason.

3. EXPERIMENTAL RESULTS AND DISCUSSION

An optoelectronic version of the proposed sensor was built and tested for the validity of the method. This consisted of two PIN photodiodes of active area 7.5 mm^2 and an infrared emitter of total output power of 1.5mW and $\pm 50^0$ half-intensity polar points. Detectors to emitter separation y, was 7mm. The detected signals were the voltages across the load resistors R1 and R2 (variable) of the respective detectors. Calibration of the detectors had been performed prior to assembly of the sensor and in the manner already outlined. Target displacements with a resolution of $1\mu m$ could be made manually. The quantity S_R was calculated from the measured signals S_1 and S_2 for two targets, a front surface aluminium mirror and an aluminium dull surface plate. The results are shown in fig 1, where S_R is plotted versus multiples of x_0, k and with x_0 set at 5 mm.

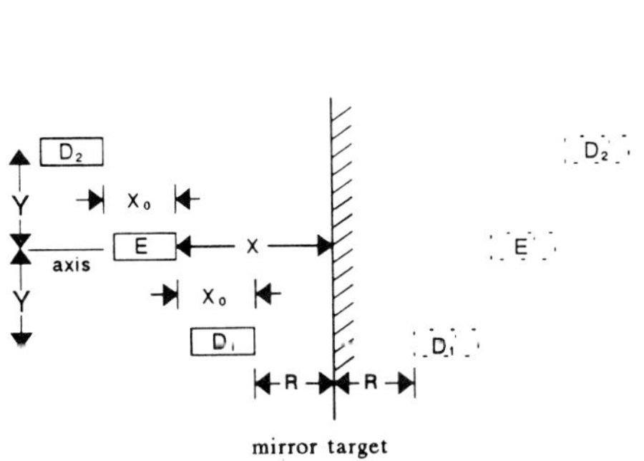

Fig.1. Emitter (E) and detectors (D) arrangement in the sensor head

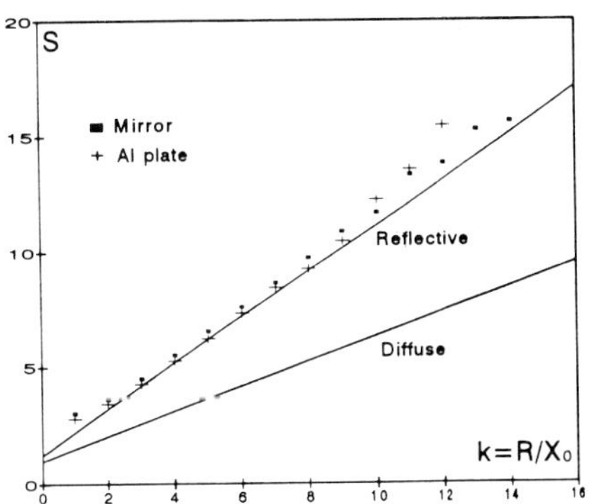

Fig.2. The theoretical response (solid lines) of the sensor arrangement and initial experimental results for two targets.

The results indicate a fairly linear response between $3x_0$ and $9x_0$ for both targets. The deviation from linearity as defined by equation (5) and for a range interval of $4x_0$ i.e. k = 4 - 5 and k = 8 - 9 , is found to be 16% for the mirror and 27% for the aluminium plate. The 'average slope' i.e. a change in S_R per mm over the range $3x_0$ to $9x_0$ differs from the exact theoretical by 9.2% and 6% respectively for the above targets. Signal to noise ratio at the range of $10x_0$ was over 40dB. Beyond the range $9x_0$ the experimental S_R points begin to deviate more significantly from the theoretical response. With an error in S_1 and S_2 not greater than 1% the experimental value of S_R for this range should agree with the theoretical to within 4%. All the above deviations from theory gave rise to suspicions about the polar distribution of the source, since the detectors were calibrated under identical signal conditions. Indeed it was established that the polar distribution of the source was not symmetrical. Between -10^0 and $+10^0$ from the normal it exhibited a linear increase in radiant intensity by about 4.4%. At the range of $10x_0$ therefore, an operational angle of $\pm 8^0$, the signals are in error to about 3.5%. This should give rise to an error in S_R of about 14%. For a symmetrical polar distribution of the source, this type of error should not arise. In a developed sensor S_R would normally be represented by an analogue voltage. As long as a linear response can be achieved, relative deviations from the true $S_{R,D}$ response should not be important, for a particular target. Different linear responses for different targets (i.e. partially reflective and partially diffuse) are anticipated with the present signal processing technique. This however will need to be established experimentally with various targets.

4. CONCLUSIONS

A novel method for realising both optoelectronic and fibre optic sensors has been described which is based on the inverse square law for radiation. The main features of the proposed sensors are those of linear response and scalability. These features have been illustrated using an approximate analytical method which does not differ significantly from the exact calculations. Initial experimental results obtained with an electronic version of the sensor, for two targets, suggest that the features born out by the analysis are valid. Optimisation of the performance of the sensor is thought to depend on the use of appropriate emitters that closely approximate the emission characteristics assumed in the analysis.

ACKNOWLEDGMENTS

The author wishes to thank his colleague I.W.Rogers for many useful discussions, and N. Ioannides for contributing the experimental results, part of his final-year BEng project.

REFERENCES

1. B. Culshaw : 'Optical Systems and Sensors for Measurement and Control'. J. Phys. E:Sci Instrum, vol 16. 1983 pp978-986.
2. B.E.Jones : 'Optical Fibre Sensors and Systems for Industry'. J. Phys. E:Sci Instrum, vol 18. 1985 pp 770-781.
3. A.M. Scheggi et al : 'Optical Fibre Displacement Sensor'. SPIE, vol 798 Fiber Optic Sensors II(1987) pp72-74.
4. M. Brenci et al : 'Fibre Optic Position Sensor Array'. Optical Fibre Sensors 1988, Technical Digest Series, vol 2 (Optical Society of America, Washington DC 1988) pp141-144.
5. E. Bois et al : 'Loss Compensated Remote Fiber Optic Displacement Sensor for Industrial Applications'. Proc. EFOC/LAN 88, June 29-July 1 1988, pp246-250.
6. S.D. Cusworth and J.M. Senior : 'A Reflective Optical Sensing Technique employing a GRIN Rod Lens'. J. Phys. E:Sci Instrum 20 (1987) pp102-103.
7. C.A. Wade et al : 'Optical Fibre Displacement Sensor based on Electrical Subcarrier Interferometry using a Mach-Zender configuration'. SPIE, vol 586, Fibre Optic Sensors (1985) pp223-229.
8. G.I. Frank et al : 'General Purpose Position Sensor'. SPIE, vol 586, Fibre Optic Sensors (1985) pp238-243.
9. S. Ramakrishan et al : 'Line Loss Independent Fibreoptic Displacement Sensor with Electrical Subcarrier Phase Encoding'. Optical Fibre Sensors 1988. Technical Digest Series, vol 2 pp133-136.
10. H. Hosokawa et al : 'Integrated Optic Microdisplacement Sensor using a Y Junction and a Polarization Maintaining Fibre'. Optical Fibre Sensors 1988. Technical Digest Series, vol 2 pp137-140.
11. G. Murtaza et al : 'Intensity Modulated Optical Fibre Sensor System for measuring up to 20mm of linear displacement'. IEE Colloquium, 'Fibre-optic Sensor Technology', May 1992. Digest no. 1992/128.
12. K. Iwamoto and I Kamata : 'Liquid-level Sensor with Optical Fibers'. Appl. Optics, vol 31, no. 1 1992, pp51-54.
13. Conference Proc : 'Plastic Optical Fibres and Applications'. Paris, June 22-23 1992. ISBN 3-90584-05-8.

Twisted optical fiber liquid level sensor

A.Suhadolnik, J.Možina
Faculty of Mechanical Engineering, University of Ljubljana, Aškrčeva 6, 61000 Ljubljana, Slovenia

ABSTRACT: A new optical fiber liquid level sensor was made. The basic mechanism involved is light transmission between the transmitting and the receiving fiber, which are stripped and twisted together. In this contribution the sensor structure and liquid level measurements are described.

1. INTRODUCTION

Several optical fiber sensors have been developed for the liquid level measurements. A single point liquid level sensor based on the refractive index change has been described [1]. An optical fiber with a prismatic or U-shaped tip was immersed into a reservoir and the effect of total refraction change caused by filling the reservoir was observed [2, 3]. The continuous monitoring of liquid level is possible by bending the optical fiber which is connected to the float [4]. The differential light absorption in the absorptive medium [5, 6] and the optical radar system [7] are also used for liquid level determination. Another effect to determine the level position is achieved with a fluorescent fiber and an external light [8]. A digital optical fiber liquid level sensor with reflectors between two parallel fibers [9] and a continuous sensor based on the light transition between two parallel fibers have also been developed [10].

The working range of those sensors is limited by the properties of the measured medium. The absorption and the index of refraction of the liquid influence the amount of the light passing from the transmitting to the receiving part of the sensor. The wetting effect is another important parameter for all immersion sensors. On the other hand the fluorescent optical fiber sensors suffer because of the external disturbing light.

In this paper the optical fiber sensor for the continuous liquid level measuring is presented. The sensor consists of two twisted fibers which are stripped on the whole sensing region. In addition the basic principle of the operation and the measuring examples with this optical fiber sensor are shown.

2. PRINCIPLES OF OPERATION

The liquid level sensor is composed of two tightly twisted multimode step index profile optical fibers (Fig. 1).

The plastic fiber, 1 mm in diameter, is made of the PMMA (polymethyl methacrylate) material. Both fibers claddings are stripped in the sensor region in order to achieve the direct surface contact with the surroundings medium. The input light, which is launched into the fiber, passes into the sensing region of the transmitting fiber. Since the cladding is removed, the light can get out of the transmitting fiber into the liquid medium. Only a minor part of the light reaches the receiving fiber and travels as guiding modes of this fiber to the photodetector. The amount of the light power passing between both fibers per

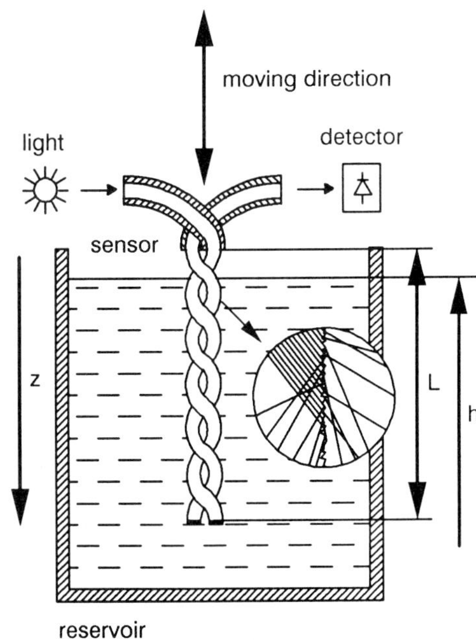

Fig. 1. Twisted optical fiber liquid level sensor

unit length depends on the length of the sensor immersed in the liquid. The liquid level can be determined with the light power measured at the output end of the receiving fiber. For better sensitivity, the transmitting fiber surface must be uniformly damaged. When the fiber surface is rough, more light exits the transmitting fiber as it is shown in the circle of Fig. 1.

With the increasing fiber surface roughness the sensitivity of the sensor is enlarged. The light power radiation from the transmitting fiber per unit length is expressed by the equation:

$$\frac{dP_1(z)}{dz} = P_0\beta(z)\ e^{-\beta(z)z}, \tag{1}$$

where P_0 is the light launched into the input end of the transmitting fiber, P_1 is the light power radiated from the transmitting fiber and depends on the longitudinal coordinate z, and $\beta(z)$ the light attenuation coefficient in the transmitting fiber. The optical power P_2, which is collected in the receiving fiber, is equal:

$$P_2 = \int_0^L \gamma(z)dP_1(z), \tag{2}$$

where $\gamma(z)$ is the transition coefficient and L the effective sensor length. The coefficient $\beta(z)$ depends on the fiber surface roughness of the transmitting fiber. The transition coefficient $\gamma(z)$ depends on the light absorption in the liquid, the distance between both fibers, the number of receiving fibers, the fibers geometry, the wetting effect and the refractive index of the liquid. With combining both equations, and taking into account the different transition coefficients between immersed and unimmersed parts of the fibers, the output optical power from the receiving fiber $P_2(h)$ is given by:

$$P_2(h) = \int_0^{L-h} \gamma_g(z)P_0\beta(z)e^{-\beta(z)z}dz + \int_{L-h}^{L} \gamma_l(z)P_0\beta(z)e^{-\beta(z)z}dz, \tag{3}$$

where $\gamma_g(z)$ and $\gamma_l(z)$ represent the transition coefficients of the immersed and the unimmersed part of the sensor. Variable h denotes the liquid level position in the reservoir. The attenuation coefficient $\beta(z)$ and the transition coefficients $\gamma_g(z)$ and $\gamma_l(z)$ can be determined experimentally. The theoretical curves are presented in Fig. 2a for the radiated power $P_1(z)$ and for different constant attenuation coefficients $\beta(z) = \text{const.}$ at sensor length $L = 50\text{mm}$.

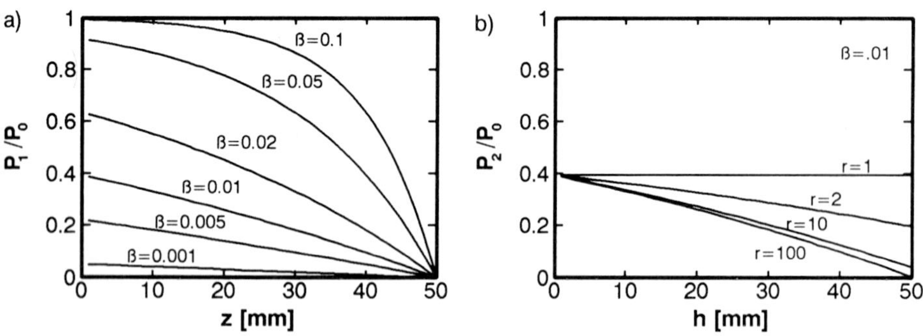

Fig. 2. Theoretical response a) with the attenuation coefficient β as the parameter b) with the transition coefficient ratio r as the parameter

In Fig. 2b, the theoretical curves of the optical power $P_2(h)$ on the output end of the receiving fiber are shown at different transition coefficient ratios r. The ratio r is defined by $r = \gamma_g/\gamma_l$. It is assumed in the theoretical model that all the optical power from the transmitting fiber is transferred to the receiving fiber, but in fact only a part of the radiated light reaches the receiving fiber.

In our experimental arrangement the rising of the liquid level is simulated by moving the twisted fiber sensor with the stepper motor. A suitable light source and an optical power meter with Si photodiode are used for signal monitoring. Both the stepper motor and the optical power meter are under computer control. The sensor tip is protected with a plastic cover to prevent the light reflectance from the bottom of the reservoir. The measuring system is additionally protected from the unwanted external light. The water is used as the testing liquid in the level position measurements.

3. MEASUREMENTS AND RESULTS

The most important parameter of the sensor is the uniform surface roughness on the radiative sensor surface. The sensor sensitivity depends on the surface roughness and is proportional to the damage level. In Fig. 3 the sensor responses are shown with the increasing surface roughness from the curve a through c. The active sensor length is 35 mm. If the preparation of the sensor surface is uniform, the sensor response is linear (Fig. 4).

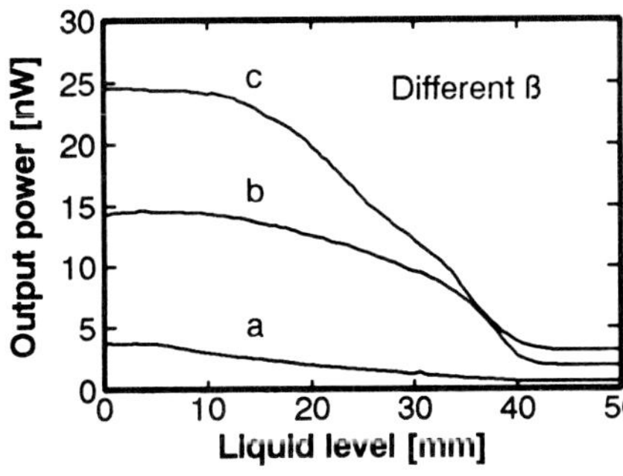

Fig. 3. Sensor responses with the increasing surface roughness (a - low, b - medium, c - high)

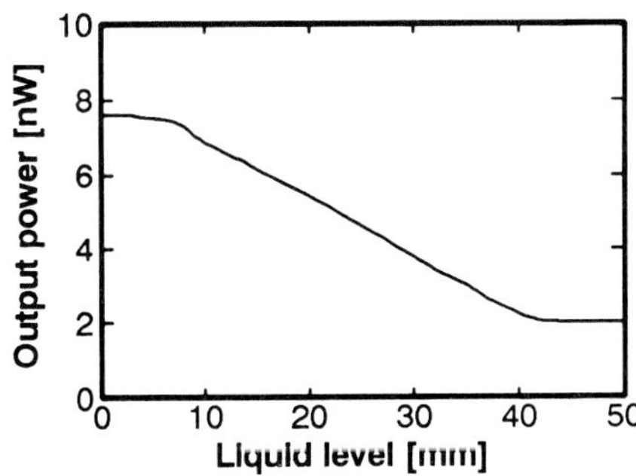

Fig. 4. Linear sensor response with the change of the liquid level

The wetting effect is observed during the measurements. When the sensor is pulled out of the liquid, a thin film of the liquid is adsorbed on the sensor surface. The film has a higher index of refraction than the air and acts as an additional cladding around the sensor (Fig. 5) [11].

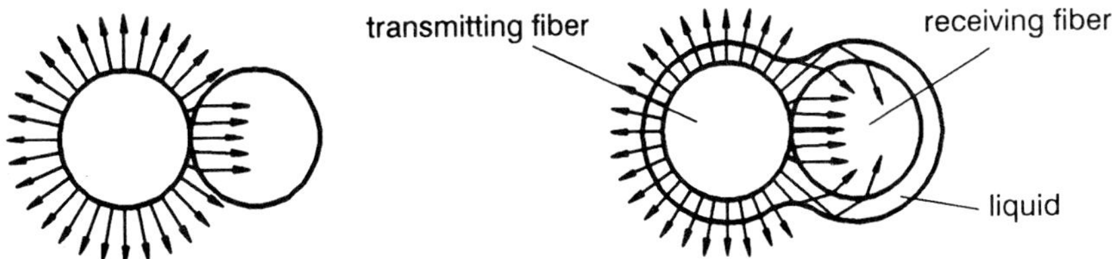

Fig. 5. Thin liquid film influence on the light transmission between the fibers

After the first immersion of the sensor tip in the liquid the light is guided more effectively to the receiving fiber and the signal rises significantly due to the wetting effect (Fig. 6). Further immersions of the sensor into the liquid medium create a small hysteresis effect (curve b, c in the Fig. 6). With a proper choice of the number of the fibers turns, the thin film becomes stable with time, if the time delay between immersing and rising of the sensor tip is not so long that the fibers would dry. Fig. 7 shows the time stability of the thin film in hours after rising the sensor tip out of the liquid medium. With the growing number of turns the transition of the light between the transmitting and the receiving fiber is increased (Fig. 8). A larger number of turns increases the sensor response, but the linearity of the sensor suffers.

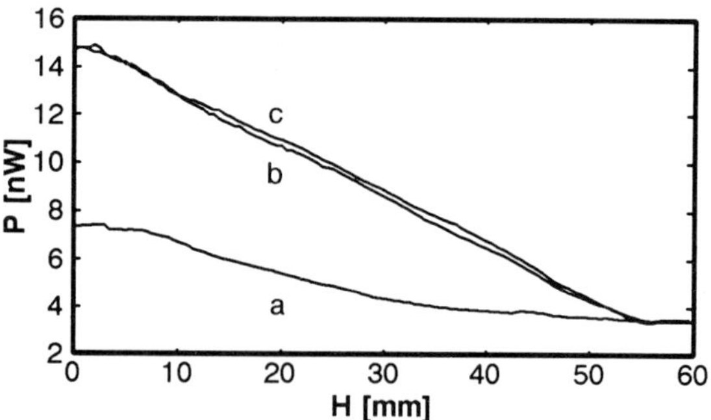

Fig. 6. Wetting effect a - first sensor immersion in the liquid, b - rising sensor out of the liquid, c - second immersion in the liquid

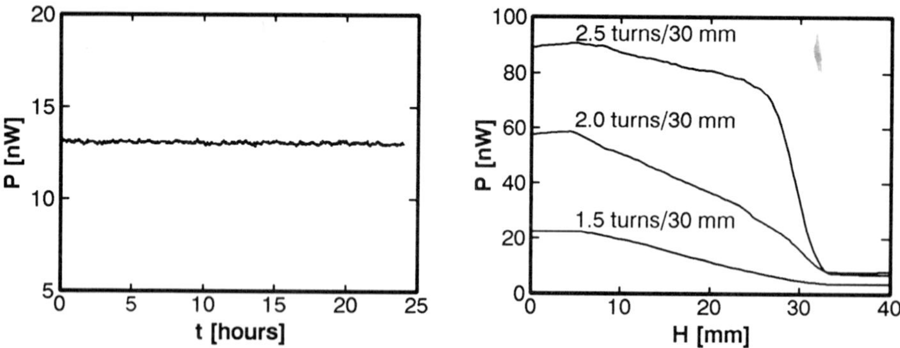

Fig. 7. Time stability of the light power after rising the sensor out of the liquid and drying it

Fig. 8. Sensor response for different numbers of turns

4. CONCLUSIONS

The twisted fiber optic continuous liquid level sensor was developed. Suitable sensitivity for liquid level sensing and signal stability after rising the sensor out of the liquid medium was achieved. In this work the influence of the wetting effect and the number of turns were also discussed.

5. ACKNOWLEDGEMENT

The authors wish to thank Marko Črnigoj for his help by experimental work.

6. REFERENCES

1. J.E.Geake; Journal of Scientific Instruments, Vol. 31, p-260, 1954
2. J.Niewisch; Siemens Forsch.- u. Entwickl.-Ber. Bd. 15, Nr. 3, p-115, 1986
3. K.Spenner *et al*; I'st Conf. on Optical Fiber Sensors, London, IEE 221, p-96, 1983
4. K.Spenner; Technisches Messen, Vol. 51, No .9, p-329, 1984
5. J.P.Dakin, M.G.Holliday; I'st Conference on Optical Fiber Sensors, London, IEE 221, p-91, 1983
6. C.P.Yakymyshyn, C.R. Pollock; Journal of Lightwave Technology, Vol. LT-5, No.7, p-941, 1987
7. D.A.Jackson; I'st Conference on Optical Fiber Sensors, London, IEE 221, p-100, 1983
8. A.T.Augousti, J.Mason, K.T.V.Grattan; Rev. Sci. Instrum., Vol.61, No.12, p-3854, 1990
9. J.A.Morris, C.R.Pollock; Journal of Lightwave Technology, Vol. LT-5, No.7, p-920, 1987
10. S. Ramakrishnan, R.T.Kersten; II'nd Conf. on OFS, Stuttgart, VDE-Verlag, Berlin, p-105, 1984
11. E.Smela, J.J.Santiago-Aviles; Sensors and Actuators, 13, p-117, 1988

Finite element analysis of the stress distribution in 'bow-tie' optical fibres with sensor application

Yueai Liu, B.M.A. Rahman, K.T.V. Grattan
City University
Department of Electrical, Electronic, and Information Engineering
Northampton Square, London EC1V 0HB, UK

ABSTRACT: A detailed formulation of the use of the finite element method, applied to the analysis of the stress distribution in optical fibres, is given. Results obtained show the potential of the method in fibre characterization, for sensor use.

1. INTRODUCTION

Single optical fibres which can maintain a state of polarization over a long distance of propagation are reliable in coherent optical communication, fibre optical sensing systems and many other application fields[1][2][3]. However, the commonly used single mode fibres with a complete circular symmetry cannot effectively maintain a state of polarization, as they are very sensitive to a number of handling factors. It is well known that the photoelastic effect is widely used for analysing the stress distribution in materials. Therefore, it becomes possible to maintain the state of polarization in an optical fibre by maintaining a certain stress in it. In this paper, the stress induced by temperature and external forces in optical fibres will be analysed; the finite element analysis formulation used to achieve this will be given. An understanding of these machanisms is very useful in the design of fiber sensors and communication systems using such effects.

2. STRESSES AND STRAINS IN OPTICAL FIBRES

Generally speaking, optical fibres are produced in an atmosphere at very high temperature. Therefore, when the fibres are cooled to room temperature, stresses develop if the materials used have different thermal expansion coefficients. Further, external forces can be applied to produce deliberate stresses in the fibre itself. In the fabrication of a non-symmetrical fibre, such as a bow-tie fibre, the change of temperature during the cooling stage produces stress, which in turn introduces a nonsymmetrical refractive index change due to the elesto-optic eefect.

The structure of an optical fibre is always long in one dimension and very limited in the other two transverse dimensions. As a result of this special feature, the strain status of optical fibres is believed to be such that the strain component along the optical fiber, $\epsilon_z \equiv 0$. Therefore, the stress-strain relationships is found to be as follows[4].

$$\sigma_x = \frac{E}{(1-\mu)(1-2v)}((1-\mu)\epsilon_x + \mu\epsilon_y) - \frac{\alpha E \Delta T}{(1-2\mu)} \quad (1)$$

$$\sigma_y = \frac{E}{(1-\mu)(1-2v)}(\mu\epsilon_x + (1-\mu)\epsilon - y) - \frac{\alpha E \Delta T}{(1-2\mu)} \quad (2)$$

$$\sigma_z = \mu(\sigma_x + \sigma_y) - E\alpha\Delta T \quad (3)$$

$$\tau_{xy} = \frac{E}{2(1+\mu)}\gamma_{xy} \quad (4)$$

where ϵ_x, ϵ_y, and ϵ_z are the three axial strain components, γ_{xy} the shear strain component, while σ_x, σ_y, and σ_z are the three axial stress components, τ_{xy} the shear stress component.

E is the Young's modulus, μ is the Poisson's ratio, α is the thermal expansion coefficient, and ΔT is the temperature change.

These form the basic foundation for the finite–element stress analysis in optical fibres discussed in this work.

3. FINITE ELEMENT METHOD ANALYSIS

To study the distribution of stresses in an optical fibre, the finite element approach is a highly suitable method[4][5] to be applied due to its flexibility and power. For this purpose, the equilibrium equations need to be established through minimizing the total energy (P_t) of the fibre system,

$$P_t = \frac{1}{2} l \int_s [\boldsymbol{\sigma}]^T [\epsilon\ ] ds - l \int_L [\boldsymbol{\delta}\ ]^T \mathbf{F} dl \tag{5}$$

where l is a length of the optical fibre, $\boldsymbol{\sigma}$ and ϵ are the stress and strain vectors respectively, $\boldsymbol{\delta}$ the displacement vector at any point, and $\mathbf{F}$ the external forces applied. The first integration takes place over the cross section of the fibre, and the second integration along the boundary of the fibre section. The first integration represents the internal strain energy, while the second indicates the work contribution of the applied loads.

In the finite element method, this total potential energy is expressed as the summation of the element total potential energy p_t^e, the partial deviation of which to the nodal displacements is found to be:

$$\frac{\partial P_t^e}{\partial \boldsymbol{\delta}^e} = \mathbf{K}^e \boldsymbol{\delta}^e - \mathbf{F}^e \tag{6}$$

where

$$\mathbf{F}^e = l \int_{L_e} [\mathbf{N}]^T \mathbf{F} dl \tag{7}$$

are the equivalent nodal forces for the element.

$$\mathbf{K}^e = \frac{1}{2} l \int_{se} [\mathbf{B}]^T \mathbf{D} \mathbf{B} ds \tag{8}$$

is termed as the element stiffness matrix. The element displacement is considered as being linearly related to the nodal displacement.

$$\boldsymbol{\delta}\ = \mathbf{N} \boldsymbol{\delta}^e \tag{9}$$

where $\mathbf{N}$ is the set of interpolation functions termed the shape functions. The strains within the element can also be expressed in terms of the element nodal displacements:

$$\epsilon\ = \mathbf{B} \boldsymbol{\delta}^e \tag{10}$$

where $\mathbf{B}$ is the strain matrix generally composed of derivatives of the shape functions. Finally, the stresses may be related to the strain by use of an elasticity matrix $\mathbf{D}$, as follows

$$\boldsymbol{\sigma} = \mathbf{D} \epsilon \tag{11}$$

The summation of the terms in Eq.(6) over all the elements, when equal to zero, results in a system of equilibrium equations for the complete continuum. These equations are then solved by any standard linear equation solving technique to yield the nodal displacements.

Considering the plain–strain problem of optical fibres results in[1]

$$\boldsymbol{\sigma} = \begin{pmatrix} \sigma_x \\ \sigma_y \\ \tau_{xy} \end{pmatrix}, \tag{12}$$

$$\boldsymbol{\epsilon} = \begin{pmatrix} \epsilon_x - (1+\mu)\alpha\Delta T \\ \epsilon_y - (1+\mu)\alpha\Delta T \\ \gamma_{xy} \end{pmatrix} \tag{13}$$

and

$$\mathbf{D} = \frac{E}{(1+\mu)(1-2\mu)} \begin{pmatrix} (1-\mu) & \mu & 0 \\ \mu & (1-\mu) & 0 \\ 0 & 0 & (1-2\mu)/2 \end{pmatrix} \tag{14}$$

If triangular elements are applied, the strain matrix $\mathbf{B}$ can be obtained as,

$$\mathbf{B} = \frac{1}{2A} \begin{pmatrix} y_j - y_k & 0 & y_k - y_i & 0 & y_i - y_j & 0 \\ 0 & x_k - x_j & 0 & x_i - x_k & 0 & x_j - x_i \\ x_k - x_j & y_j - y_k & x_i - x_k & y_k - y_i & x_j - x_i & y_i - y_j \end{pmatrix} \tag{15}$$

Where x_i, x_j, x_k and y_i, y_j, y_k are the x and y coordinates of the three vertices of the triangular element.

For a material under stress, its optical property can be represented by an index ellipsoid, the principal axes of which coincide with the principal axes of stress at the point. Allowing n_x, n_y and n_x to be the principal refractive indices for waves vibrating parallel to the principal stress σ_x, σ_y, and σ_z respectively at any point and n_0 be the index of refraction for the unstressed material. Then, the following equations express the relationship between the principal refractive indices and the principal stresses[6]:

$$n_x - n_0 = C_1\sigma_x + C_2(\sigma_y + \sigma_z) \tag{16}$$
$$n_y - n_0 = C_1\sigma_y + C_2(\sigma_x + \sigma_z) \tag{17}$$
$$n_z - n_0 = C_1\sigma_z + C_2(\sigma_y + \sigma_x) \tag{18}$$

where C_1, C_2 are the optoelectronic coefficients of the material. Hence, the principal refractive indices of the fiber materials can be obtained from Eqs (16), (17), (18) after the principal stresses in the optical fiber are obtained from the result of the above finite element analysis.

4. COMPUTER SIMULATION RESULTS

Using the established formulation, a FORTRAN program has been developed to analyse the stress distribution in several optical fibres. Stresses in optical fibres with different structures are considered in the work, representing different optical fibers used in a number of different application areas[1][2].

[1]Since the strain component in the z axis is zero, the stress component in this direction has no contribution to the total potential energy of the element. Therefore, it is not considered here.

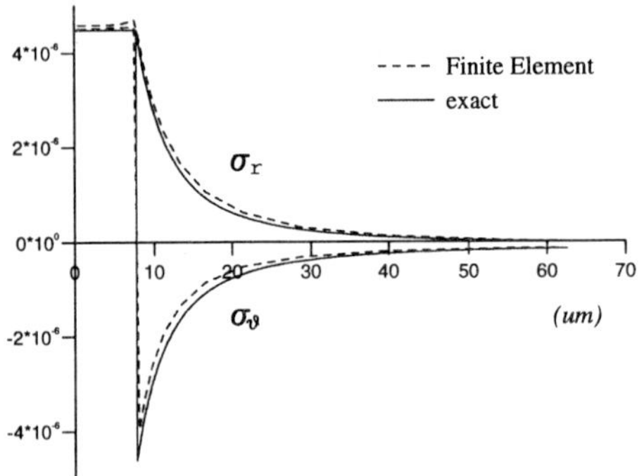

Figure 1: Comparison of the thermal stress distributions in a circular fibre calculated by the finite element method with their corresponding analytical solutions.

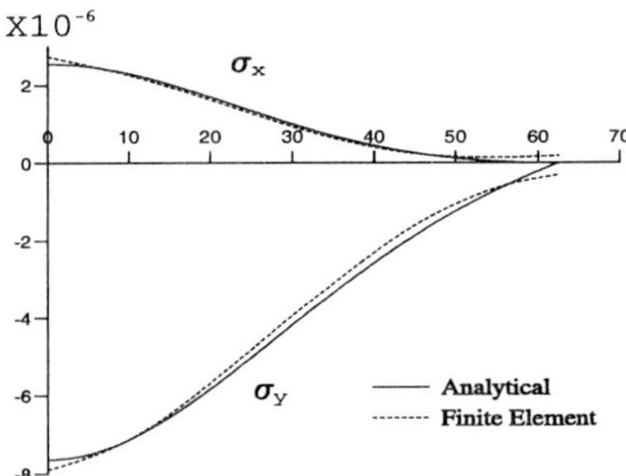

Figure 2: Comparison of the stress distributions along the x axis calculated by the finite element method with their corresponding analytical solutions.

First of all, circular fibres under thermal stress and load, respectively, are analysed. Since the analytical stress distributions of fibres under thermal stress and load are obtainable, they can be used to test the correctness and effectiveness of the developed program.

For a circular fibre under thermal stress, the stress distributions are given as follows[4].

$$\sigma_r = \begin{cases} \frac{(b^2-a^2)}{2b^2}\frac{(\alpha_2-\alpha_1)}{(1-\nu)}E\Delta T & (0 \leq r \leq a) \\ \frac{a^2}{2b^2}\frac{(\alpha_2-\alpha_1)}{(1-\nu)}E\Delta T(\frac{b^2}{r^2}-1) & (a \leq r \leq b) \end{cases} \tag{19}$$

$$\sigma_\theta = \begin{cases} \frac{(b^2-a^2}{2b^2)}\frac{(\alpha_2-\alpha_1)}{(1-\nu)}E\Delta T & (0 \leq r \leq a) \\ -\frac{a^2}{2b^2}\frac{(\alpha_2-\alpha_1)}{(1-\nu)}E\Delta T(\frac{b^2}{r^2}+1) & (a \leq r \leq b) \end{cases} \tag{20}$$

where a and b are the core and cladding radii of the fibre respectively. The Young's modulus and Poisson's ratio for both the core and the cladding are considered to be similar, α_1 and α_2 are the thermal expansion coefficients of the core and cladding respectively. These stress distributions are shown by the solid lines in Figure 1, when $a = 7.8\,\mu m$, $b = 62.5\,\mu m$, $E = 0.00783\,kg/\mu m^2$, $\nu = 0.186$, $\alpha_1 = 14.85 \times 10^{-7}\,^0C^{-1}$, $\alpha_2 = 5.4 \times 10^{-7}\,^0C^{-1}$, $\Delta T = -1000.0\,^0C$. The broken lines in Figure 1 show the finite element solution of the stress distributions in the same optical fibre. From Figure 1, it can be seen clearly that the finite element solution agrees very well with the analytical solution.

The computer calculation of the stress distributions of the same circular fibre under load also gives a numerical solution which is very close to the analytical one. The analytical stress distributions along the x axis, when the load is applied along the y axis, are given as follows[4].

$$\sigma_x = \frac{W_0}{\pi b}\left[1 - \frac{4b^2x^2}{(b^2+x^2)^2}\right] \tag{21}$$

$$\sigma_y = -\frac{W_0}{\pi b}\left[\frac{4b^2x^2}{(b^2+x^2)^2} - 1\right] \tag{22}$$

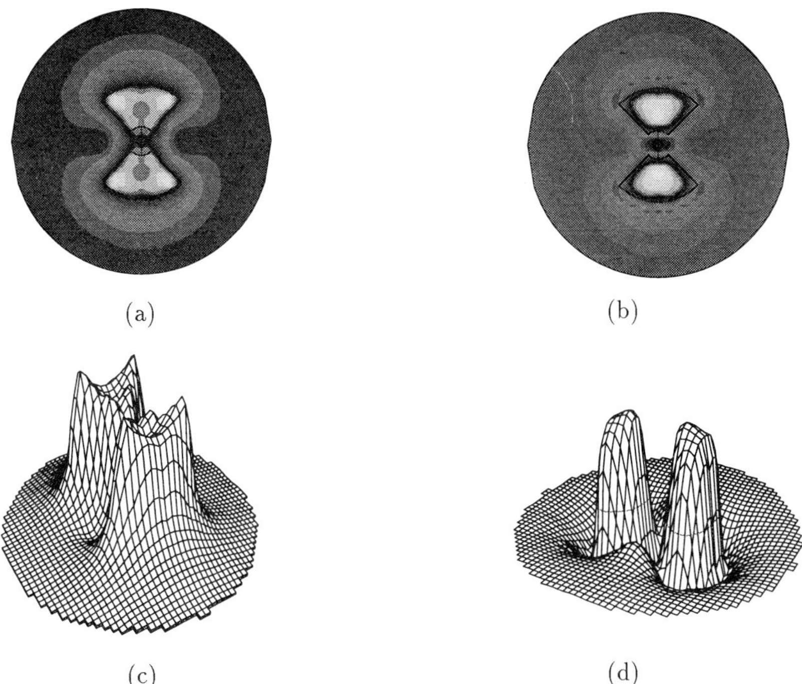

Figure 3: The stress distributions in a bow–tie optical fiber, where (a) and (c) are the contour map and the 3–D display of σ_x, and (b) and (d) are the contour map and 3–D display of σ_y.

These stress distributions are presented by the solid lines in Figure 2. The broken lines in Figure 2 presents the numerical results of the finite element. The load applied is $W_0 = 0.0005\ kg/\mu m$.

With the established program, the stress distributions in many different types of fibres may be calculated. Figure 3 shows the results of the calculated thermal stress distributions in a bow–tie type optical fibre, where , Figure 3(a) illustrates the contour map of the x component of the stress distribution, Figure 3(c) shows the corresponding 3–D display; Figure 3(b) illustrates the contour map of the y component of the stress distribution, and Figure 3(d) the 3-D display of the y component of the stress distribution.

The refractive index distributions can then further calculated and are shown in Figure 4. It can be seen that n_x, n_y, and n_z are obviously all different from each other as the consequence of the existence of the stress in the bow–tie optical fibre. The different values of n_x, n_y, and n_z cause different light polirisations to experience different values of refractive indices and inturn this will result in different propagation constants of the two orthogonal modes, and as a result, the power transfer between them will be reduced. Thus the fibre can maintain the input state of polarisation to a high degree.

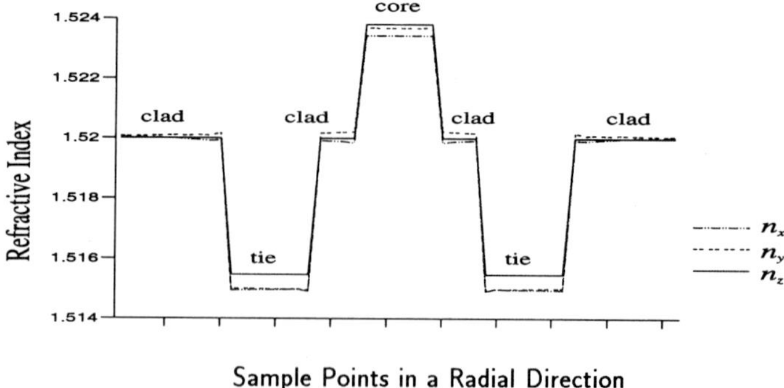

Figure 4: The refractive indices distributions in a stressed bow–tie fibre.

5. CONCLUDING REMARKS

The finite element analysis of stress distributions in optical fibres has been carried out. Refractive indices changes in the fibre are calculated from the result of the finite element analysis of the stress. These results are now being used to investigate the optical behavior of optical fibres under stress, using the vector **H** field finite element method[7]. As a result, the understanding of optical sensor devices using such fibres can be better, and system designs can be optimized ahead of construction.

6. ACKNOWLEDGEMENT

The research project is funded by SERC, the Science and Engineering Research Council, whose support is gratefully acknowledged.

References

[1] J. Noda, K. Okamoto, Y. Sasaki, "Polarization–Maintaining Fibers and Their Applications", *J. Lightwave Technology*, **LT–4**, pp.1071–1089(1986).

[2] Y. Yamamoto, T. Kimura, "Coherent Optical Fiber Transmission Systems", *IEEE J. Quantum Electron.*, **QE–17**, pp.919–935(1981).

[3] R. Griffiths, "Recent and Current Developments in Distributed Fiber Optics Sensing for Structural Monitoring", *Proc. SPIE*, **985**, pp.69–76(1988)

[4] K. Okamoto, T. Hosaka, T. Edahiro, "Stress Analysis of Optical Fibers by a Finite Element Method", *IEEE J. of Quantum Electronics*, **QE–17**, pp.2123–2129(1981).

[5] E. Hinton and D. R. J. Owen, *Finite Element Programming*, (Academic Press, Inc., (London) Ltd.), (1977).

[6] K. Tajima, M. Ohashi, Y. Sasaki, "A New Single–Polarization Optical Fiber ", *J. Lightwave technology*, **7**, pp.1499–1503(1989).

[7] B.M.A. Rahman, F.A. Fernandez, & J.B. Davies, "Review of Finite Element Methods for Microwave and Optical Waveguides", *Proc. IEEE*, **79**, pp.1442–1448(1991)

Analysis of surface plasmons in evanescent wave fibre-optic sensors using the finite element method

B.M.A. Rahman, P.A. Buah, K.T.V. Grattan
City University
Department of Electrical, Electronic, and Information Engineering
Nothamptom Square,London EC1V 0HB
Tel: +44-71-477-8123 Telefax: +44-71-477-8568

ABSTRACT: The propagation characteristics of the optical guided modes in multilayer metal-clad planar optical waveguides have been investigated using the finite element method, employing an accurate magnetic field formulation. Results for symmetric and asymetric 3-layer and 6-layer structures in the nanometer range, of resonance, field profiles and supermodes, have been presented. The dependence of the modal field characteristics on the metal thickness has been obtained.

1. INTRODUCTION

The recent prediction of long-range surface-plasmon modes on thin metal films has stimulated theoretical interest in this topic, as a result of the diverse applications it has found, which includes the enhancement of second harmonic generation, its use for fibre polarisers and the determination of the thickness and optical constants of thin metal films, with application to advanced optoelectrochemical sensors.

1.1 Background

Surface plasmon modes which are attenuated electromagnetic modes supported by either a single metal dielectric interface, or more composite structures like a thin metal film surrounded by semi-infinite dielectrics and vice versa, are inherently TM-polarised. At the wavelength of operation, the real part of the metal-dielectric permittivity must be negative. Some of the surface-active media (media having complex dielectric constants with negative real parts) that have been used so far are gold, silver, aluminium, indium, iron, nickel, and tungsten [1]. In this work the coupling between fibre modes and metal modes has been analysed using an accurate finite element approach. The structure considered is a multilayered coupled metal-cladded guide and an optical fibre with an oil buffer layer. This structure has a lot of application and has been studied both experimentally and numerically[1,2]. However due to the very small size of the oil buffer layer, the determination of the resonance condition by the variation of the oil thickness is complicated. Results for this type of problem, utilising the advantages of the finite element method[3], are presented. The finite element method has become a powerful tool throughout engineering, valued for its flexibility and versatility, being used in the analysis of complicated structural, thermal, fluid flow, semiconductor and electromagnetic problems. As optoelectronic devices continue to evolve and their dimensions shrink from the micrometer to the nanometer scale, computer simulation based on the finite element method is playing a more and more important part in the analysis, design and optimization of such advanced optical devices, this being more cost-efficient than empirical design. Also, as the range of guided wave devices becomes more intricate, the need for analysis becomes more demanding, and this may be met with advanced computer techniques.

To apply the finite element method [3], the guide cross section is first suitably divided into a patchwork of triangular subregions called elements. Each element can have a different value of permeability μ, permittivity ε, nonlinear coefficient n_2, loss, gain or anisotropy. An accurate vector magnetic field formulation [3] has been developed which is "exact-in-the-limit" for the characterization of a wide range of optical waveguides, including optical fibres, planar optical and microwave waveguides, semiconductor lasers, and so on. The finite element formulation using a full $\mathbf{H}$ vector field, where each

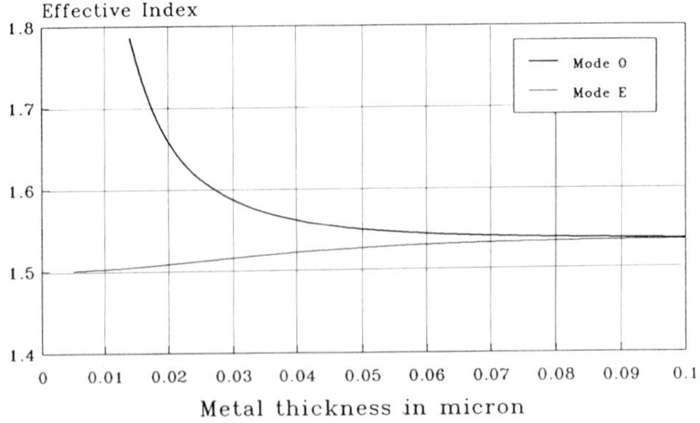

Figure 1: Variation of the effective index with metal thickness for a 3-layer symmetrical planar waveguide

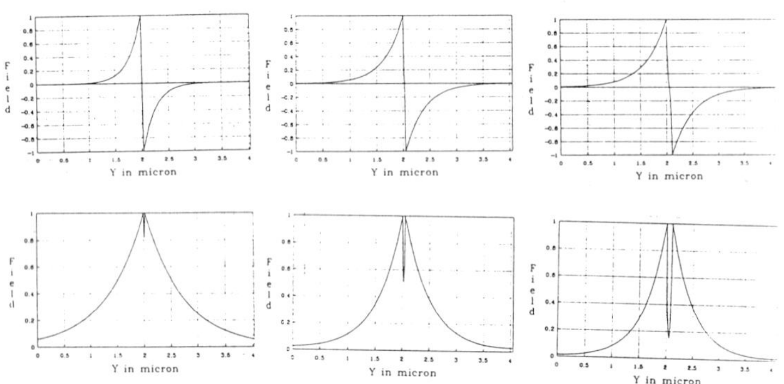

Figure 2: Variation of the field along the transverse axis

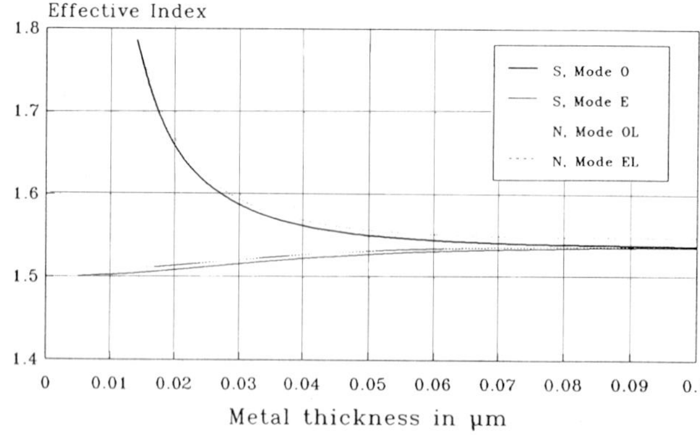

Figure 3: Variation of the effective index with metal layer thickness for symmetrical and asymmetrical waveguides.

field component, H_x, H_y, and H_z, is separately represented by a function which is continuous over the whole transverse plane of the waveguide device, is more accurate than the alternative $\mathbf{E}$ field formulation, as in the former case the field continuity is automatically satisfied across the dielectric interfaces.

2. RESULTS AND ANALYSIS

2.1 Analysis of symmetrical metal wavegiudes

In the first example a symmetrical structure, where a thin aluminum layer with thickness, h microns, is bound on both sides by glass with refractive index 1.5, is analysed. The refractive index of the central metal layer is considered to have a value j6.841 (imaginary) for infrared light of 0.850 micron in wavelength. For such a structure the eigenfield of the first mode is anti-symmetric (odd) and that of the second mode is symmetric (even) and field decays exponentially in the two cladding layers as well as inside the central metal layer, where its relative dielectric constant is negative (ϵ_m=-46.8). The effective indices (β/k_0) of both the guided modes are higher than the refractive indices of the two identical cladding layers ($n_c = n_s = 1.50$). For a single metal/dielectric interface, it is possible for an optical wave to propagate as a surface mode with the field decaying exponentially in both the metal and dielectric regions. The antisymmetrical and symmetrical modes in three layer structures can be considered as the first (odd) and second (even) coupled supermodes of the two individual modes at the two isolated metal/dielectric interfaces. Figure 1 shows the variation of the effective indices of the two coupled supermodes with the thickness of the central metal layer, h. When this thickness of the central layer increases, the separation of the two interfaces also increases and the eigenvalues of the both supermodes converge to that of the single metal/dielectric surface mode. Figure 2 shows the eigenvectors of the first (odd) and second (even) supermodes for metal of thicknesses 0.025, 0.05 and 0.1 microns. As the metal thickness increases, the first supermode (odd-mode) decays slowly in the adjacent cladding layers, whereas for the second supermode (even-mode), the field decays faster in the cladding layers.

2.2 Analysis of asymmetrical metal waveguides

Next a non-symmetrical three-layer structure has been analysed. For this nonsymmetrical structure the central metal layer is bounded by two different dielectrics, when the refractive indices of the top cladding and lower substrate regions are 1.51 and 1.50 respectively. When the central metal thickness is large, two surface modes uncouple to two distinct modes of the two isolated metal/cladding (n=1.51) and metal/substrate (n=1.50) respectively. Figure 3 shows the variation of the eigenvalues of the two supermodes for a non-symmetrical structure in comparison with that of the earlier symmetrical structure. Figure 4 shows the field variation for the two supermodes for metal thicknesses 0.025, 0.05, and 0.1 microns respectively. For smaller values of metal thickness, the odd- and even-like modes of the non-symmetrical structure are very similar to the odd and even modes of the symmetrical structure. Slight asymmetry in the field variation can be observed, but for higher metal thicknesses this asymmetry is much more pronounced for the same dielectric ratio (here 1.50/1.51). This is due to the coupling of two nonsynchronous surface modes, where only a small amount of power transfer is possible between these isolated modes, when they are not phase matched.

For the symmetrical structure, equal power is carried by the top and lower claddings for both symmetrical and anti-symmetrical modes. However, the power fraction for the even supermode increases with the metal thickness as, besides an increase in the metal thickness, the field in the two outer claddings also decreases faster. Figure 5 shows the variation of the power fractions P_c, P_s, and P_m, the power carried by the top cladding, the lower substrate and the central metal region respectively with the metal thicknesses for the non-symmetrical structures. For the odd-like first supermode, the field maximum is at the top cladding/metal interface and the cladding power fraction, P_c increases

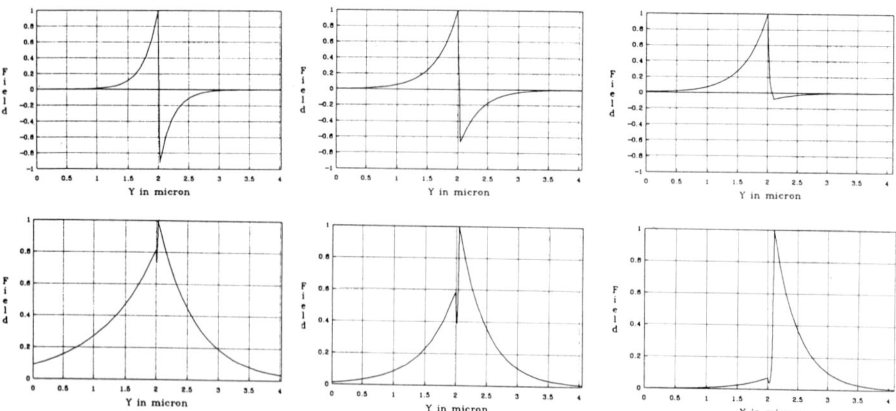

Figure 4: Variation of the field along the transverse axis

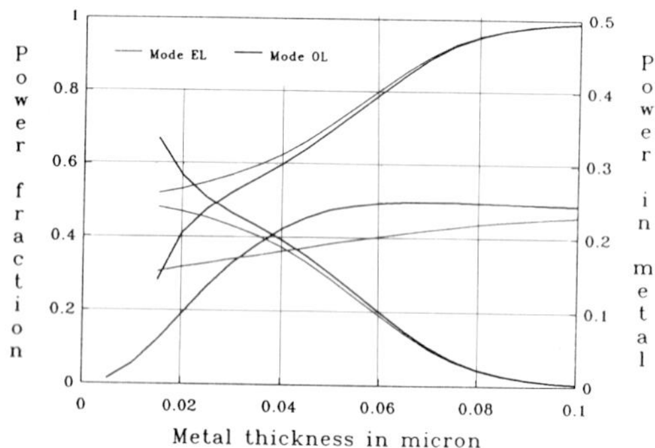

Figure 5: Variation of the power confinement ratio in three layers for the even-like and odd like modes for an asymmetrical structure.

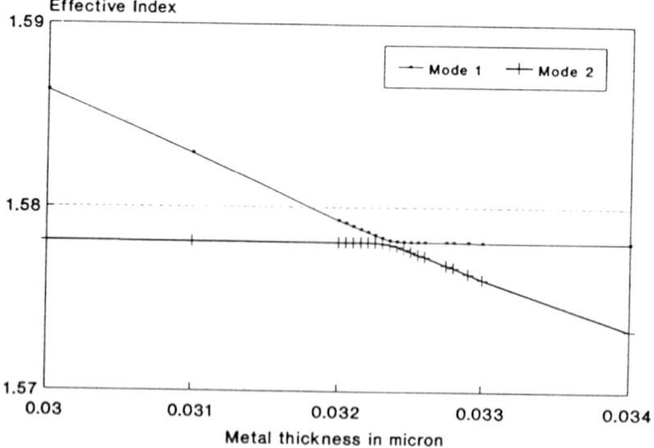

Figure 6: Variation of the effective index with the metal layer thickness showing phase matching at h=.03235μm.

with the metal thicknesses. Similarly, P_s decreases but P_m increases slowly with h which will in turn increase the total loss of this mode. For the even-like second supermode the field maximum is at the metal/substrate interface and P_s increases monotonically with the metal thickness, h. P_c decreases and P_m increases initially and settles to a constant value with metal thickness variation. It can be observed for the even-like second mode that although the field maximum is at the metal/substrate interface(see Fig 4d) for a smaller metal thickness, more power is carried by the top cladding region as there the field decays more slowly.

2.3 Simulation of polished metal-clad optical fibre

A multi-layer structure representing a optical fibre with variable metal thickness coupled to the earlier described non-symmetrical metal structure has been been simulated. This type of structure has been widely considered for polished fibre based polarisation sensitive devices [1,2]. We have considered coupling to non-symmetrical metal modes by varying the metal thickness to achieve optimum mode coupling to the fibre mode. This type of structure was simulated with a six-layer planar formation. The odd-like or even-like modal effective index can be varied by changing the metal thickness or by changing the refractive index of the matching oil layer. Figure 6 shows the change of the effective indices of the odd-like metal mode with metal thickness. The effective index of the fiber mode is unchanged with the metal thickness and when this is 0.03235 micron, these two modes intersect each other. For a metal thickness below 0.03235 micron, the fibre mode is the second supermode (here the term supermode means coupled mode between the fibre mode and the odd-like mode, which is itself a coupled surface mode, of the metal structure). For a metal thickness above 0.0324 micron the fibre mode is the first supermode. Only when the metal thickness is about 0.03235 micron, two modes (the fibre mode and the metal mode) are phase matched and only in this condition is power transfer between the modes possible. It is important to notice that if the metal thickness is changed only by 0.001 microns there will not be any appreciable power transfer between the two modes. The tolerance can be changed by reducing the polished cladding thickness and the phase matching region can be changed by altering the index of the matching oil. Figure 7 and 8 show the composite coupled modes at the phase-matching condition when the fibre and the odd-like metal modes are coupled and when the even-like metal mode is coupled to the fiber mode respectively.

3. SUMMARY

The effective index of the fibre mode lies between the value of the guide and core refractive indices. The effective indices of the plasmon modes lie above the refractive index of the bounding layers and for the first antisymmetrical mode it decreases sharply with the metal thickness whereas for the second symmetrical mode it increases slowly with the metal thickness. The finite element method is capable of modelling multilayer composite structures incorporating fibre and metal layer to locate phase matching condition. This efficient design procedure can be used to study the effect of device structures such as metal thickness, refractive index and thickness of the oil layer etc. and to optimise the design of such polarising devices. Thus it can be seen that the main advantage shown of the finite element method is its ability to handle complex structures, besides this, it can introduce anisotropy, nonlinearity, loss/gain or the effect of strain if necessary, for more advanced analysis. Further since the method handles the complete coupled structure (supermodes) it preserves the orthogonality condition even in strong coupling conditions unlike the coupled mode method.

References:

[1] Michael N. Zervas, IEEE Photonics Techn. Lett., vol2, no.4, p. 253, April 1990.
[2] W. Johnstone, G. Stewart, T Hart,and B. Culshaw, J. Lightwave Technol., LT-8, p538,1990.
[3] B.M.A. Rahman, F.A. Fernandez, and J.B. Davies, Proc. IEEE, vol.79, p.1442, 1991.

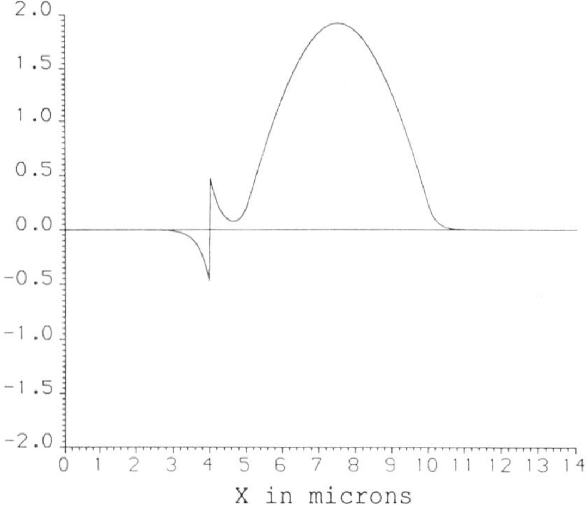

Figure 7: Variation of the field along the transverse direction showing coupling between fibre mode and odd-like plasmon mode.

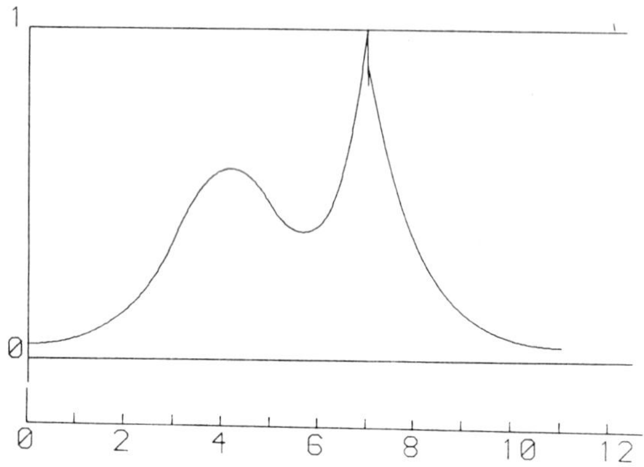

Figure 8: Variation of the field along the transverse direction showing coupling between fibre mode and even-like plasmon mode.

Design and characterization of optical directional couplers using the finite element method

T. Wongcharoen, B.M.A. Rahman and K.T.V. Grattan
City University
Department of Electrical, Electronic and Information Engineering
Northampton Square, London EC1V 0HB

ABSTRACT: A discussion of coupling length variation with fabrication tolerances and asymmetry for directional couplers used as elements of optical sensors is presented, using the finite element method, employing an accurate vector H-field formulation. The effect of the tapered guide sections, leading to the directional couplers, on the coupling length of the device is also presented.

1. INTRODUCTION

Directional couplers are among the most important optical elements used in the fabrication of various optical devices such as power dividers, power couplers, optical switches, modulators, wavelength filters, polarization splitters or transformers etc. in fibre optical and integrated optical circuits, for optical communication and sensing systems.

In order to understand their operating characteristics and achieve an optimization of these directional coupler based devices, it is important to be able to predict the coupling properties between the guides. Since the design, modification, development and optimization of such devices can be both tedious and expensive, a clear need exists for an accurate and versatile design method to ensure high and consistent quality in use.

Over the years, there has been considerable interest in the theoretical analysis of such devices. Many simple methods have been developed and used, such as the method of Marcatili [1], the coupled mode theory [2], and the effective index method [3] for such analysis. However, in these simple methods the waveguide cross-section is very much restricted to simpler cross-sections. The Beam Propagation method [4] has been used to find the power transfer for axially variant coupled waveguides but with a restriction on the small index differences and it provides only scalar solutions.The finite difference method [5] and scalar [6] and vector [7] finite element method have also been used to characterise such devices.

2. THE FINITE ELEMENT METHOD

In the finite element method the waveguiding region is subdivided into a patchwork of a finite number of subregions called elements. Each element can have different shapes and sizes and different refractive indices, and using many such triangles a complex waveguiding cross-section can be accurately modelled. The vector **H**-field formulation has been extensively used for the bench-mark solution of a wide range of practical optical waveguide problems [8]. This formulation is "exact-in-the-limit" and particularly suitable for the optical waveguides where the magnetic field is naturally continuous across the dielectric interfaces. Applied to optical directional coupler problems [7], this method permits an accurate computation of the coupling length between two completely arbitrary shaped optical guides. In this procedure, since the eigenvalues (here the propagation constants, β) and the eigenvectors (here the vector magnetic fields, $\mathbf{H}(x,y)$) of the complete directional coupler structure rather than that of the two individual guides are calculated, the resultant supermodes are always orthogonal to each other, even when they are strongly coupled.

3. APPLICATION OF THE METHODS

To show the usefulness of this method, some results for coupled channel waveguides are presented. Such results can also be obtained for different types of waveguides, including optical fibres and semiconductor rib waveguides, with if necessary, graded or anisotropic or nonlinear materials. For this example, it is considered that the core, substrate and top cladding refractive indices are 2.3, 2.29 and 1.0 respectively, typical of the channel waveguide used. The width and depth of the waveguide are W and D (microns) respectively. The operating optical wavelength is 0.850 µm.

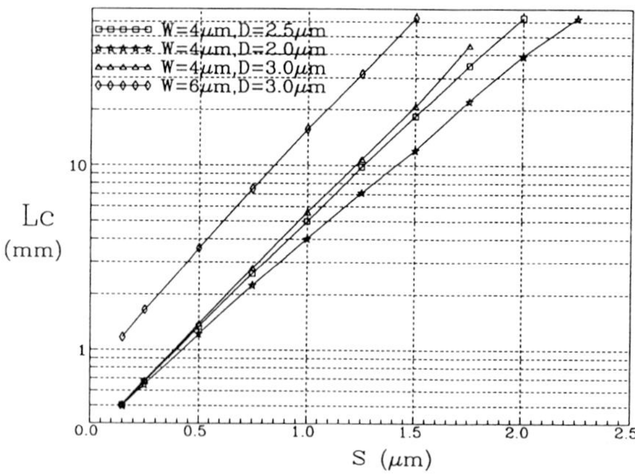

Figure 1 Variation of the coupling length with the separation distance.

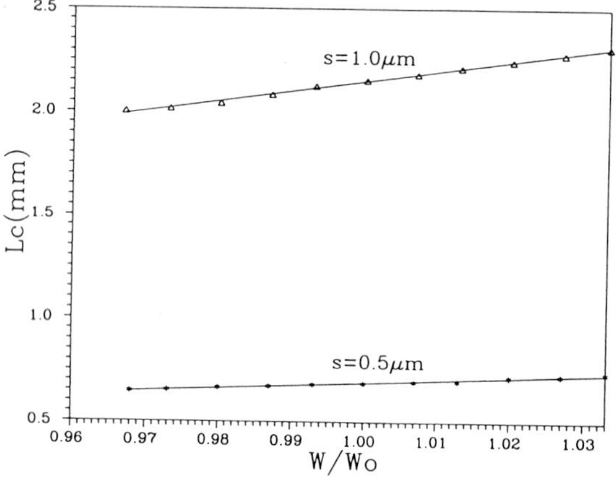

Figure 2 Variation of the coupling length with W/W_o.

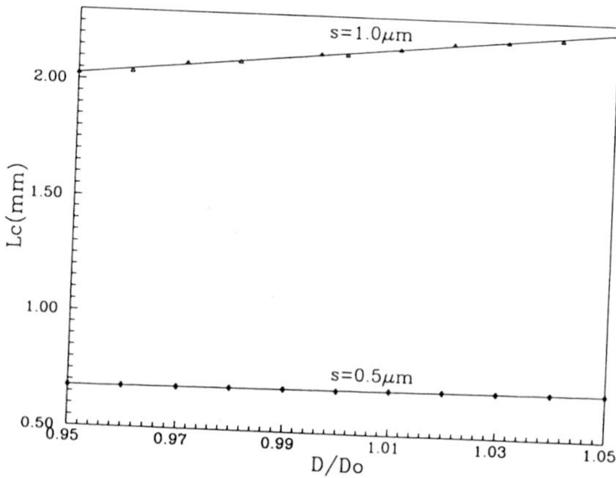

Figure 3 Variation of the coupling length with D/D_o.

In a directional coupler, two optical waveguides are placed in parallel, at close proximity. The interaction of the overlapping evanescent fields of the guided modes in the individual guides causes power exchange between the coupled guides. This power exchange can be controlled by adjusting the synchronization and the coupling coefficient between the two guides. Complete energy transfer can occur between the two coupled identical waveguides if their length is equal to the coupling length, L_c, for a given separation between the guides, $2s$. The coupling length, L_c, may be defined as the length required to achieve 180 degree relative phase shift between two normal modes propagating along the directional coupler, i.e.

$$Lc = \frac{\pi}{\Delta\beta} = \frac{\pi}{\beta_e - \beta_o} \tag{1}$$

where β_e and β_o are the propagation constants of the even and odd supermodes of the coupled guides.

4. RESULTS

Figure 1 shows the variation of the coupling length, L_c, with the half separation distance, s, for four different waveguide cross-sections. It can be seen from this figure that L_c becomes larger with increasing s and follows an exponential law so the L_c vs s plot is linear on a log-linear scale. To reduce the device length, a benefit in practical applications where space may be limited, it is important to have a shorter value of L_c which in turn requires strong coupling between the guides. Unlike the finite element method, many simple methods of analysis which are based on the perturbation of the normal modes of the two individual guides, fail to preserve the orthogonality of the supermodes and power conservation criteria in strong coupling situations.

Due to fabrication tolerances, it is not always possible precisely to control the waveguide dimensions W and D or the refractive indices in different regions. It is important to see the effect of such a variation on the coupling lengths. Figure 2 shows this variation of the coupling length with the change in width fraction (W/W_o) for two different separation distances. When the half-separation distance, $s = 0.5$ μm, the central coupling length required is 0.688 mm, compared to 2.13 mm when $s = 1.0$ μm. Although $\Delta Lc/\Delta W$ is smaller for smaller separation distances, its effect on the power transfer will be detrimental as $\Delta L_c/L_c$ becomes higher, which introduces significant cross-talk. Similarly figure 3 shows the variation of the coupling length with the tolerance of the guide depth (D/D_o).

However, the effect of structural asymmetry due to the fabrication procedure is much more severe in its effect on their performance than the symmetrical structural deviations. This is because the two modes in the two isolated nonidentical guides will have different propagation constants, β, and will not be phase matched to transfer power effectively between the guides. In the next example coupling between two nonidentical guides with widths and depths W_1, W_2, D_1 and D_2 respectively is considered. Figure 4 shows the variation of L_c with W_2 for different separation distances when W_1, D_1, D_2 are 2.6, 2.5, and 3.0 microns respectively. The coupling length shows its peak when W_2 is such that two isolated waveguides are phase matched. When the two guides are not phase matched, although the coupling length is smaller, the power transfer between two coupled modes will be very small as their propagation constants are different. Although such unintentional loss of phase matching in couplers designed to be symmetrical may not be desirable, specially designed asymmetrical couplers have excellent wavelength-dependent coupling ratios and find their application in the construction of wavelength division multi/demultiplexing systems.

In optical circuits, it is essential to have bent or curved guiding sections to connect the directional coupler region to other optical devices such as optical fibres. The separation between the guides is increased outside the coupling region either by introducing a sloped straight or a curved S section, as shown in figure 5(a) and 5(b). These bent or curved sections will introduce an offset or radiation loss. It is desirable that power transfer should be only in the actual directional coupler region and there should not be any power transfer between the two connecting guides, but in fact, there will always be some transfer of power in this region.

If the angle θ is smaller, to obtain a certain final separation, a longer transition length, B, is required making the total device length longer. These tapered transition sections leading towards and away from the coupler region also introduce additional power transfer between the guides which can be written

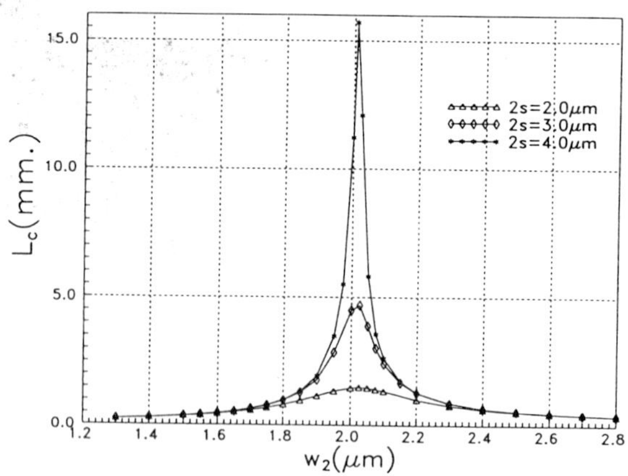

Figure 4 Variation of the coupling length with the second guide width, W_2.

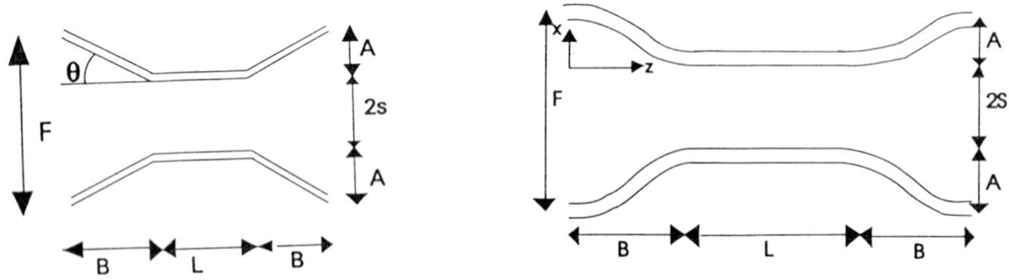

Figure 5 Input and output sections of directional couplers (a) straight line approach (b) curved *S* shaped approach.

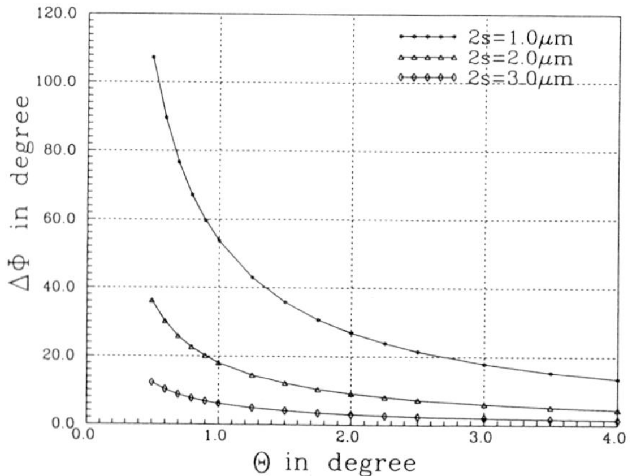

Figure 6 Variation of the additional phase shift, ΔΦ with the slope angle θ.

as an additional phase shift, $\Delta\Phi$, from the two transition regions. Figure 6 shows the variation of $\Delta\Phi$ with angle θ for different separation distances, with F is fixed at 50 μm. When the angle θ is increased, for a given F, the required horizontal distance B will be smaller and the additional $\Delta\Phi$ will also be reduced, but it will introduce more loss due to sharp transitions. It can be observed that $\Delta\Phi$ reduces with θ, and also for higher separation distances.

In the design of a directional coupler, its effective length, L_e, is usually chosen so that

$$\Delta\beta L_e = n\pi \tag{2}$$

where n is an odd integer. If the additional phase shift $\Delta\Phi$ is not taken into account, the effective length of the directional coupler is not exactly equal to an odd number times the coupling length, and so the performance of the directional coupler will be degraded. and the power output from port 1, P_1, is not quite zero. This cross-power ratio, when the effect of $\Delta\Phi$, is neglected may be given by:

$$P_1/P_0 = \sin^2(\Delta\Phi/2) \tag{3}$$

where P_0 is the total power. However, as this additional phase shift can be accurately calculated for different designs, if the length of the parallel section (L_c') is shortened appropriately so that $L_e = L_c$, only then may 100 percent power transfer be possible. Figure 8 shows the length of the parallel section for a straight line approach for different angles θ and separation distances. Similarly, figure 9 shows the actual coupling length for the use of the curved section approach.

5. CONCLUSIONS

Accurate characterization of symmetrical and asymmetrical directional couplers with approach guides has been demonstrated using the powerful finite element method. This computer aided modelling ability can be utilised to design and optimize a wide range of directional coupler based optical sensor devices.

6. REFERENCES

[1] E.A.J. Marcatili, "Dielectric rectangular waveguide and directional coupler for integrated optics", *Bell Syst. Tech. J.*, **48**, pp.2071-2102, 1969.

[2] A. Yariv, "Coupled-mode theory for guided wave optics", *IEEE J. Quantum Electron.*, **QE-9**, pp.919-933, 1973.

[3] H.C. Cheng and R.V. Ramaswamy, "Symmetrical directional coupler as wavelength multiplexer-demultiplexer: Theory and experiment", *IEEE J. Quantum Electron.*, **QE-27**, pp.567-574, 1991.

[4] M.D. Feit, J.A. Fleck Jr., and L. McCaughan, "Comparison of calculated and measured performance of diffused channel-waveguide couplers,", *J. Opt. Soc. Amer.*, **73**, pp.1296-1304, 1983.

[5] N. Schultz, K. Bierwirth, F. Arndtand U.Koster, "Rigorous finite-difference analysis of couple channel waveguides with arbitrary varying index profiles", *J. Lightwave Tech.*, **LT-9**, pp.1244-1253, 1991.

[6] L. Bersiner, U. Hempelmann and E. Strake, "Numerical analysis of passive integrated-optical polarisation splitters: comparison of finite-element method and beam-propagation method results", *J. Opt. Soc. Amer.* B, **8**, pp.422-433, 1991.

[7] B.M.A. Rahman and J.B. Davies, "Finite-element solution of integrated optical waveguides", *J. Lightwave Tech.*, **LT-2**, pp.682-688, 1984.

[8] B.M.A. Rahman, F.A. Fernandez and J.B. Davies, "Review of finite element methods for microwave and optical waveguides", *Proc. IEEE*, **79**, pp.1442-1448, 1991.

[9] D. Marcuse, "Directional couplers made of nonidentical asymmetrical slabs, Part I, Synchronous coupler", *J. Lightwave Tech.*, **LT-5**, pp.113, 1987.

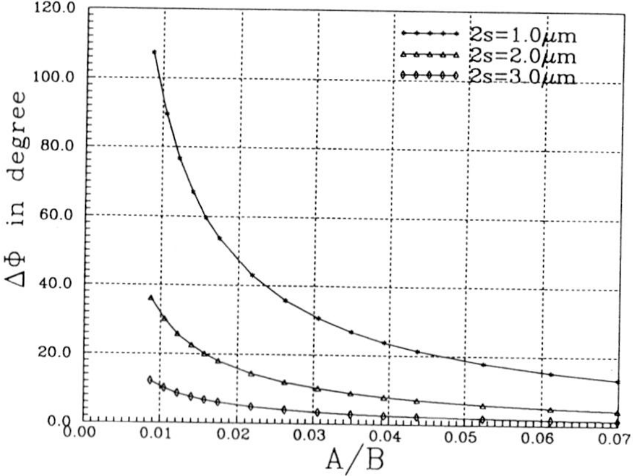

Figure 7 Variation of the additional phase with *A/B* ratio.

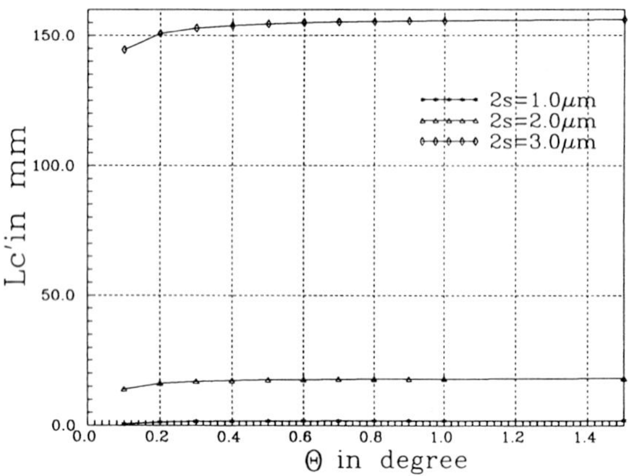

Figure 8 Effective coupling length variation for straight line approach.

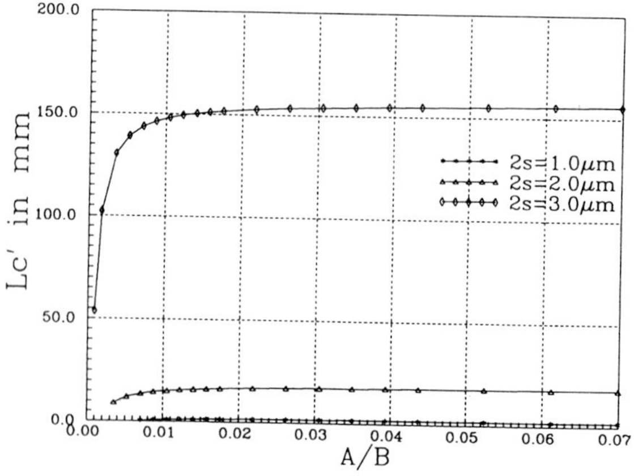

Figure 9 Effective coupling length variation for curved *S* section approach.

Temperature measurement using a resonant hybrid optoelectronic system

A.T. Augousti and J. Mason

School of Applied Physics, Kingston University, Kingston, Surrey KT1 2EE, UK

Introduction

Liquid crystals (LCs) have found wide use as transducing elements in the measurement of a range of physical parameters, primarily in the measurement of temperature and static electric and magnetic fields [1-3]. The physical property most commonly employed in this mode of operation is the dependence of the spectral reflectivity of the LC materials on the measurand in question , particularly in the measurement of temperature [4,5]. Thus thermochromic LCs are most often used either in their raw state or in microencapsulated form [4]. It is rather more unusual to find liquid crystal devices employed directly as transducers. This paper describes the development of a temperature measurement system based on such a device, namely a Liquid Crystal Display (LCD). The advantages of this system as configured are primarily that the system has an output directly in the frequency domain, the probe head is fully optically powered, and the resonant frequency of the system is strongly intensity-independent. The remainder of this paper describes the optoelectronic configuration of the system and the characteristics of this system.

Optoelectronic system description

The system setup is as shown in Figure 1. The main features of the system are based on a similar configuration described elsewhere [6], although this system differs in a number of important respects. The system consists of a LCD segment addressed by an unmodulated red LED via an optical fibre. The light reflected by the LCD segment forms the signal which is used as input to the electronic system controller. The signal is electronically differentiated, and following some signal processing involving a Phase Locked Loop (PLL) the resulting signal is converted into a unipolar square wave with 50% duty cycle (see Figure 2). The positive value of the square wave is a TTL level, namely 5V. The frequency of the square wave is halved, and by use of a NOT gate, the complement of the wave is also generated. Each of these signals modulates the current to one of two super bright red LEDs respectively, which each address separate photovoltaic devices configured in such a way as to provide a voltage of alternating polarity across the LCD segment. The photovoltaic devices chosen are in fact pairs of red LEDs hardwired in series, acting in reverse to their normal function. These were chosen following extensive investigation into a suitable choice for such devices, since they are optically well matched to the sources, provide a sufficiently high voltage to switch

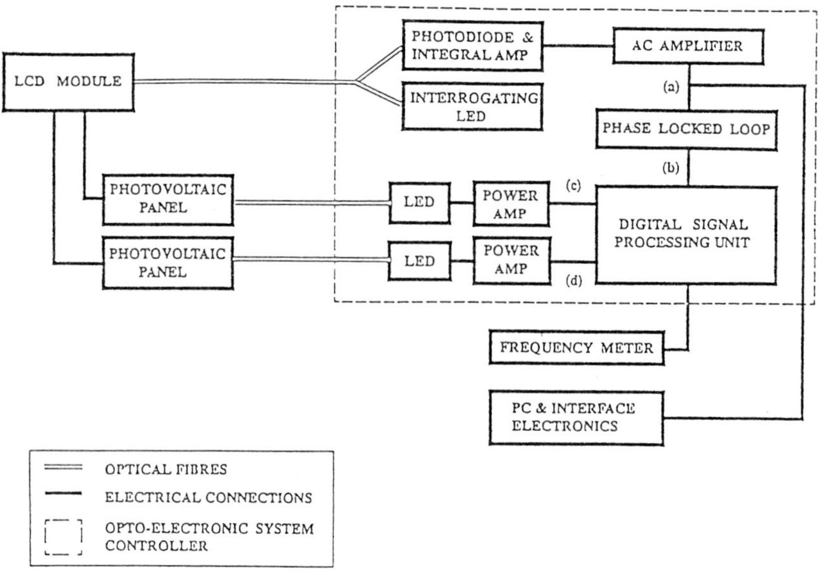

Figure 1. Block diagram showing the optoelectronic system configuration.

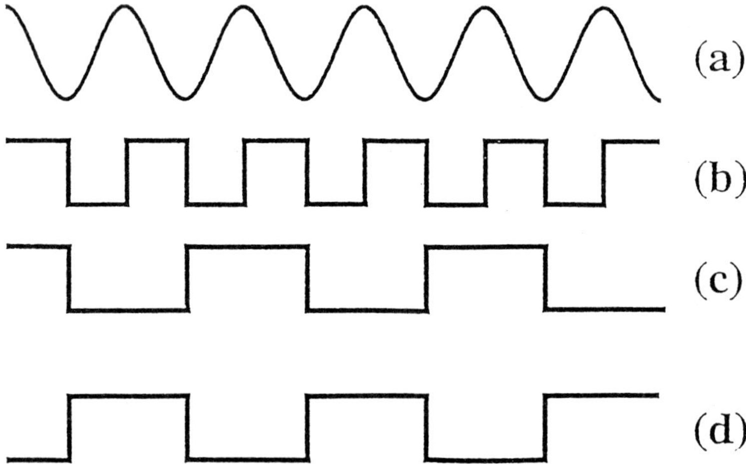

Figure 2. Timing diagram showing the operation of the system. The symbols a-d represent the waveform at various parts of the circuit as indicated in Figure 1.

the LCD segment when two LEDs are cascaded, and have a sufficiently high forward bias voltage so that they do not form a heavy load when the other LED pair is switched on. Furthermore, the LEDs saturate to their maximum voltage output at relatively low levels of illumination and in effect act as stabilised voltage sources. The low current which they source in relation to a photodiode is not a hindrance, since the LCD sinks much lower values than are available.

The operation of the circuit will not be clear from the description of the system above. It can be summarised as follows. At any one time, one or other of the LEDs will be switched on. Assume that the display is darkening. The reflected signal will therefore be diminishing, and the derivative tending towards zero (from below). The arrangement of the PLL and the digital signal processing unit is such as to generate an ON state when this value is reached. This switches on the other red LED which optically powers the alternate photovoltaic device, and the voltage across the LCD is reversed. The LCD rapidly begins to brighten, passes through a maximum of reflected intensity, and the derivative signal becomes negative. As the darkening rate gradually decreases, the cycle repeats itself, with the other powering LED being switched ON, and forcing the LCD to lighten and then darken once more. The system is intrinsically unstable in both states, and reverts to the other eventually. This particular configuration differs from others previously described elsewhere [6], and dispels any question of the system finding some static position of equilibrium.

The oscillation of this system was found to be remarkably stable. The oscillatory frequency depends upon the alignment time of the liquid crystals to an applied electric field. Once more this configuration is superior to the previous one in that the duty cycle is highly symmetrical since only alignment times are involved, whereas in the previous configuration both alignment and relaxation times were important. A typical output showing the reflected signal is shown in Figure 3 (data captured directly via a PC). Figure 4 shows the variation of this resonant frequency with temperature.

Discussion

A simple mathematical model has been developed, described more fully elsewhere [7], which reproduces the behaviour observed experimentally. The system can be represented by a single nonlinear second-order differential equation

$$\ddot{\phi} = \alpha V_0 sign(\dot{\phi}) \sin\phi + \beta \cos\phi - \gamma \dot{\phi} \tag{1}$$

where ϕ represents the angle between the mean director for the liquid crystal system and the

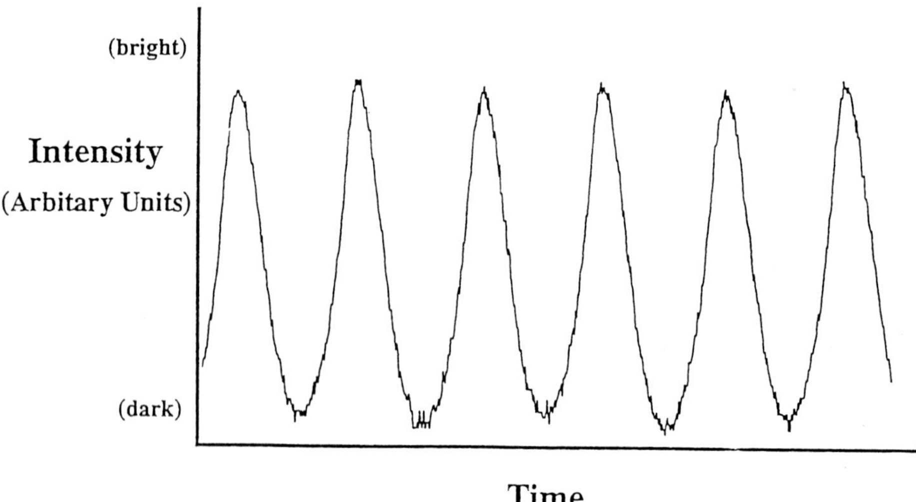

Figure 3. A typical output showing the variation in reflected intensity with time at room temperature.

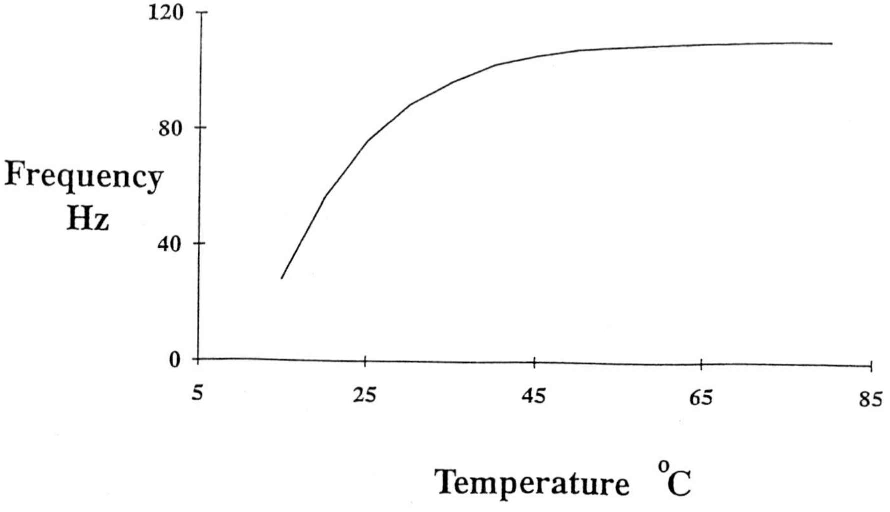

Figure 4. Graph of the resonant frequency versus temperature.

line joining the two plates of the display, and α, β, and γ are coefficients which represent alignment, restoring and damping forces respectively.

By varying the value of the viscosity parameter, one may obtain the resonant frequency (in normalised units), as a function of the parameter γ. This is displayed in Figure 5. In fact, the plot is one of frequency (in arbitrary units) against the logarithm of the inverse of viscosity. The reason for this choice is that according to many sources, for instance [8], the viscosity depends on the temperature in a negative exponential fashion. Thus this figure corresponds to the earlier experimental plot of frequency versus temperature, and the two show good agreement. One can see a substantial variation in the resonant frequency with viscosity. However, the lower values of the viscosity form a boundary for this graph, since as the viscosity decreases, the amplitude of the swings grows until motoring occurs, i.e. when the liquid crystals begin to perform complete revolutions. This is a weakness of this model, since energy is always put into the system by the aligning term, with only the viscous term acting to dissipate it. It is inevitable therefore that the swings will eventually exceed the 'elastic limit' for any given value of the elastic restoring torque constant, and motoring will occur. This places a limit on the range of values of the viscosity for which sensible results for the frequency can be obtained.

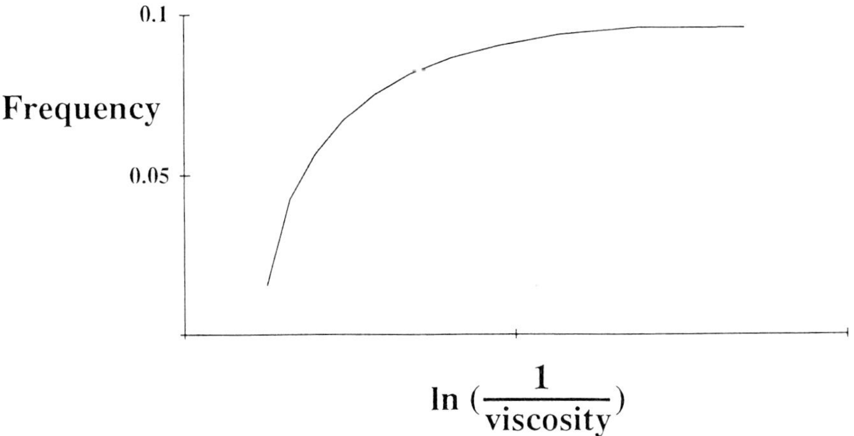

Figure 5. Graph of the resonant frequency versus ln(1/γ) obtained from the model for values of α and β of 0.8 and 0.4. The graph is limited at low values of viscosity, since the amplitude of the swing then exceeds $\pm\pi/2$, and the model breaks down, but as one might expect, the frequency appears to approach a fixed value as the viscosity tends to zero.

Conclusion

We have built and demonstrated a self oscillating system based on the alignment response times of liquid crystals. The optical power required to energise the probe head is small, and the probe head can certainly be miniaturised into a compact form. The probe head is sensitive to temperature through the dependence of the LC viscosity on temperature. The output of the system is in a desirable digital form, and is largely intensity independent. The oscillation frequency varies by approximately a factor of 3 as the temperature changes from room temperature to 80°C.

A mathematical model based on simple physical principles has been proposed which contains the relevant features of the system. It was solved numerically, and exhibited the characteristics which were observed experimentally, namely stable oscillation, and the strong dependence of the frequency on temperature.

This work is part of a continuing project, and further areas which will be investigated include

(a) miniaturisation of the probe head into a compact form

(b) refinement of the mathematical model by using more realistic functional forms for the aligning, restoring and dissipative terms

(c) inclusion of real values for the parameter used in Equation 1.

When the two latter objectives have been achieved, it may well be possible to employ this relatively simple and inexpensive system to rapidly characterise samples of LC formulations, in particular establishing the value of the bulk viscosity. More advanced perturbation techniques may be employed using the error derived from observed and predicted intensity waveforms to obtain information about the functional form of the forces involved, and their absolute values.

References

1. **Augousti A T, Grattan K T V and Palmer A** *J.Phys.E: Sci. Instrum.* 1988 **21** p817-9
2. **De Rossi D, Bennasi A, L'Abbate A and Dario P** *J. Biomed. Engng.* 1980 **2** pp257-61
3. **Livingston G K** *Radiat. Environ. Biophys.* 1980 **17** pp233-43
4. **Augousti A T, Mason J and Grattan K T V** *Sensors: Technology, Systems and Applications* ed K T V Grattan (Bristol: Adam Hilger) 1991 pp339-45
5. **Coles H J, Bone E L, Bowdler E R and Gleeson H F** *Proc SPIE - Int. Soc. Opt Eng.* (USA) 1990 **949** pp185-90
6. **Mason J and Augousti A T** *Opt. Eng.* **31**(8) 1992 1663-6
7. **Augousti A T, Mason J and Koenders M A** Proceedings of the *International Conference on Electronic Measurement and Instruments,* 20-22 October 1992, Tianjin, China pp337-42
8. **Sage I** *Thermotropic Liquid Crystals* ed G W Gray (Chichester: John Wiley and Sons) 1987 pp64-98

A flap angle control based on fibre optical measurements

P V P Yupapin, K Weir*, K T V Grattan*and S Kusamran

Optoelectronic Instrumentation Group, Department of Applied Physics, Faculty of Science, King Mongkut's Institute of Technology Ladkrabang (**KMITL**), Ladkrabang, Bangkok 10520, **Thailand**.

*Measurement and Instrumentation Centre, Department of Electrical, Electronic and Information Engineering, **City University**, Northampton Square, London EC1V 0HB, **U.K.**

Abstract:

An investigation of the use of a fibre optic sensor configured as a flap angle control device is presented. A highly birefringent fibre (Hi-Bi) incorporating a laser diode as a light source is employed and demonstrated for the measurement of the relationship between the coupled light power and flap angle. The force-induced coupling of power between the two eigenmodes of the light propagating in the fibre allows this angle to be measured and controlled. Results obtained have shown that this device may have potential in industrial or aerospace applications. Theoretical aspects of the sensor system in this arrangement are discussed.

1. Introduction

Fibre optic sensors are recognized as attractive devices for measurement and control, offering a number of advantages, e.g. small size, light weight, and they can be used in high voltage, electrically noisy, high temperature environments. Such devices have been the subject of significant development because of these advantages.

In this work, a study is made of the use of a Hi-Bi fibre in which force-induced cross-coupling of optical power within the fibre is allowed to occur [1], in response to an externally applied force. The angle of the force may be adjusted and controlled with respect to the fast axis of the sensing fibre. To achieve this, light from a multimode laser diode is linearly polarized and launched in one of the two eigenmodes of the D-shaped Hi-Bi fibre employed [2]. The power coupled into the second eigenmode provides information on the applied force angle of the coupling device. The device susceptibility to temperature variation is also investigated and discussed.

2. Operating Principle

When linearly polarized light is launched into one of the two eigenmodes of a Hi-Bi fibre, it normally remains in the same polarization state. The output intensity with the polarizer crossed will be a minimum, while it will be a maximum in a parallel position. When a force is applied to the fibre, a certain amount of optical energy may be transferred between the two polarization modes. The device to achieve this coupling consists of a pair of wedges each of width w, and the fibre is sandwiched between them. If the angle of the applied force is adjusted at the coupling point by rotating the sensing fibre, as shown schematically in Figure 1, then the coupling ratio, k, at that point can be related to the force angle, and this is given by [1]

$$k=G[\sin 2\varphi \sin(\pi Bw/\lambda)]^2 f^2 \tag{1}$$

Here G is a constant, φ is the applied force angle between the direction of applied force and the fast axis of the fibre, B is the fibre birefringence, λ is the light source wavelength, w is the width of the coupling region (i.e. the coupling tooth width), and f is the force per unit length (i.e. F/w). From equation (1), the maximum coupling term can be obtained when the applied force angle is given by

$$\varphi = (2n+1)\pi/4 \tag{2}$$

and the width of the coupling teeth,

$$w = (2n+1)L_B/2, \tag{3}$$

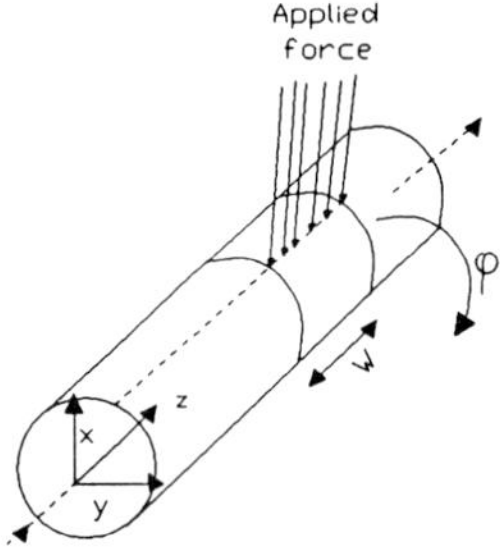

Figure 1 Schematic of the coupling point unit using optical fibre.

where n=0,1,2....., and L_B is the fibre beat length. This means that the first maximum of output intensity can be detected when $\varphi = 45^0$ and $w = L_B/2$. The other parameters in equation (1) can be arranged to be constant, and so the coupling ratio, and thus the coupled intensity will vary with the applied force angle, φ. Thus, the required measurand may be obtained.

3. Experimental Arrangement

A fibre optic sensor using this principle and its application has been demonstrated. The system was used to measure coupled power between the two eigenmodes of light propagating in the sensing fibre used.

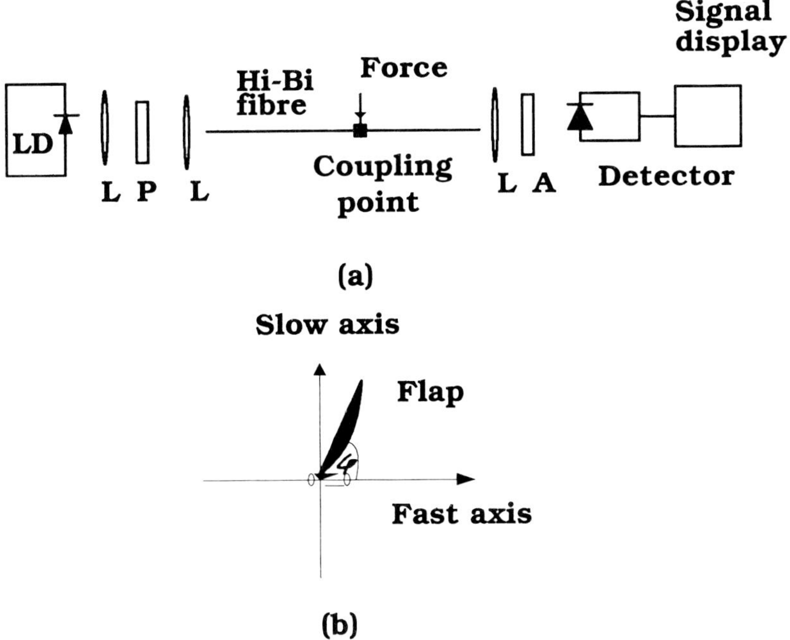

Figure 2 (a) Illustration of the experimental arrangement: LD Laser diode, Ls Lenses, P Polarizer, A Output Polarizer, (b) A schematic sensing unit.

The measurement of coupled power and the applied force relationship could be observed by placing a detector directly beyond the output polarizer in the sensor system as shown in Figure 2(a). The light source used was a multimode laser diode manufactured by Sharp (type LT023MD) [3] with an output of 3 mW, operating at a wavelength of 780 nm. The output light, which was collimated by a x10 objective lens with NA = 0.25, passed through a polarizer (P). It was then focussed into the Hi-Bi fibre via a x10 objective lens. The fibre used was manufactured by the Andrew Corporation [2] with a length of 1.5 m, an operating wavelength of 780 nm, a birefringence of $1.5x10^{-4}$, and a calculated beat length of 5.5 mm [1]. A length of the fibre constituted the sensing unit where the force was applied. The light output from the fibre was then collimated and passed through a polarizer (A) to allow selection of the light polarization emerging from the fibre. Light was launched via one of the two modes into the fibre through control of the polarizer P. The polarizer A was aligned in the crossed position throughout the experimental measurements.

The angle of applied force (with respect to fibre axes) could be controlled by rotating the fibre in the sensing unit, as illustrated in Figure 2(b). A constant force was applied on the coupling point. When the force angle was adjusted, the amount of cross-coupling of power from one mode to the other could be detected and read from the signal display device. The fibre was generally rotated through 2π to investigate its coupling efficiency and also to determine experimentally the relationship between the coupled power and angle. The coupled power can be affected by parameters such as the magnitude of the force, its angle with respect to the fibre fast axis, φ, and the coupling tooth width, w, as discussed earlier (see equation (1)). The sensing unit was also placed in the temperature controlled environment system, and the sensor temperature dependency tested.

4. Results and Discussion

The results reported here were obtained from the D-series Hi-Bi fibre, and a rotating sensing device (as shown in Figure 2(b)) could be fixed to the fibre. The widths of the coupling teeth used were 0.5 mm and 2.5 mm, and a constant force was applied to the sensing unit (i.e. the coupling point). The applied force angle was varied for each set of measurements. The same procedures were repeated for each of the tooth widths tested in the sensor system.

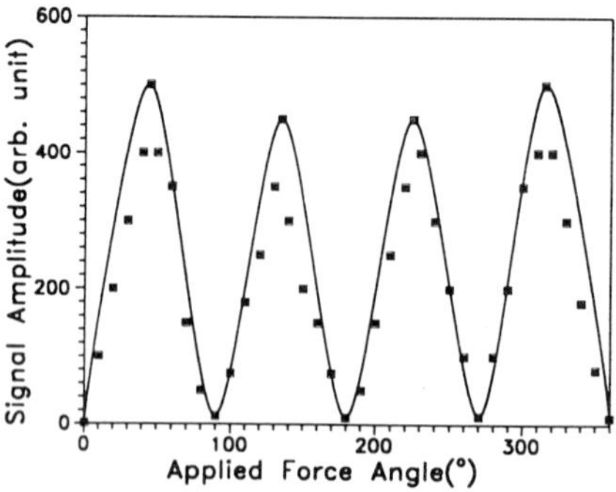

Figure 3 Plot of signal amplitude against applied force angles using D-Series fibre at room temperature (~20^0C) .

Figure 3 shows the relation between applied force angle and output amplitude, using a coupling tooth width of 2.5 mm and a constant force of 2 N. The force angle was rotated through 2π radians (360^0). These results show that at $\varphi = 45^0$, i.e. applied force angles of 45^0, 135^0, 225^0 and 315^0, then a maximum coupling of power could be obtained. Thus, the force angle could be applied to the sensing unit with a limited angle of $\pi/4$ radians (45^0) before the result becomes ambiguous due to the periodicity of the signal. In this test, an uncertainty in rotation angle of 1^0 is assigned to the system.

Figure 4 shows a plot of normalized coupled intensity against applied force with two difference tooth widths. The narrower tooth width of 0.5 mm produced a bigger effect, but in practice, its smaller dimension gave rise to problems in that the fibre could be easily damaged or broken. In normal applications, this effect may be reduced by using the wider tooth width.

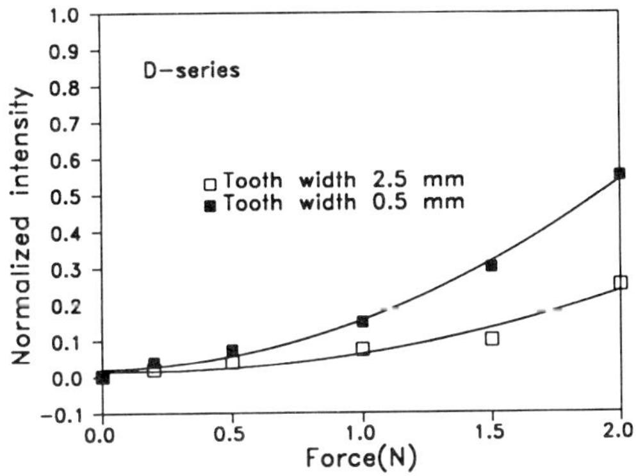

Figure 4 Plot of normalized intensity against applied force using D-series fibre at applied force angle 45^0.

A sensing unit with a length of fibre of 60 cm was placed in a temperature controlled environment system to investigate and test the sensing device temperature characteristic. This was studied from room temperature (~20 ^{0}C) up to 100 ^{0}C with a constant force of 2 N, and tooth width of 2.5 mm. Figure 5 shows that a change in output intensity may result from the temperature variations.

In engineering such a sensor scheme great care would be required; a constant force must be applied independent of angle, and the fibre must not twist as the flap rotates. In addition some compensation scheme would be required to overcome thermal effects. These further aspects of system design are now being considered.

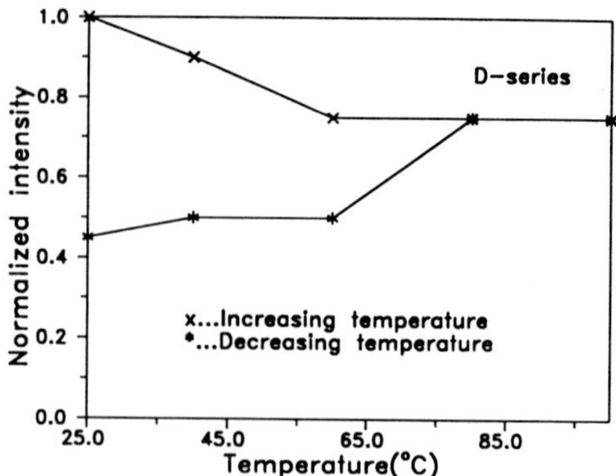

Figure 5 Plot of normalized intensity against the change in temperature using D-Series Hi-Bi fibre with a constant force of 2 N, tooth width of 2.5 mm at applied force angle of 45^0.

6. Conclusion

A fibre optic sensors using D-Series Hi-Bi fibre as a sensing element for measuring angle of applied force has been presented. The device may be used to measure the force-angle-induced coupling power of the fibre employed. By mean of this method, the output intensity and thus the flap angle caused by the applied force were evaluated quantitatively. This investigation has shown the potential of the use such a scheme for industrial applications.

References

[1] YUPAPIN, P. V. P. WEIR, K. GRATTAN, K.T.V. and PALMER, A. W. (1993). A study of polarization maintaining fibre characteristics using an interferometric technique for sensor applications, submitted to **IEEE** Journal of Lightwave Technology.

[2] Polarization maintaining optical fibres, The Andrew Corporation, Bulletin 1295D, Available from Andrew Antenna, Weybridge, Surrey, U.K.

[3] SHARP laser diodes, Laser diode user's manual, Available from Access Pacific Ltd., Kymbrook School House, Kimbolton Road, Keysoe, Bedford MK44 2HH, U.K.

Section E

FLOW MEASUREMENT AND PROCESS TOMOGRAPHY

Sensors for multiphase flow measurement

M.L.Sanderson

Department of Fluid Engineering and Instrumentation, Cranfield Institute of Technology, Cranfield, BEDFORD MK43 0AL, England

ABSTRACT : This paper examines some of the work currently being undertaken within the Department of Fluid Engineering and Instrumentation at Cranfield Institute of Technology into the development of sensors for multiphase flowmeasurement. The methods include those based on static charge and NMR. Analysis methods based on time series analysis from impedance and ultrasonic sensors for flow regime identification using chaos theory are described together with work on image processing from B-scan ultrasonic images of multiphase flow.

1.INTRODUCTION

Multiphase flowmeasurement remains one of the most significant outstanding problems in flow measurement. The ability to measure, to an acceptable accuracy, the mass flow rate of oil, water and gas being produced from an oil well without separating the flows would have significant benefits to the oil industry. Within the chemical and food industries many of the flows which are required to be measured are multiphase with often solid, liquid, and gas components. An ability to measure these flows more accurately would give rise to improved process efficiency and product yield. Within the water industry the treatment of waste requires the measurement of multiphase flows which from an environmental point require improved measurement.

Sensors in a multiphase flow system may be required for the following purposes :-

- Measurement of the volume/mass flowrates of the individual components of the multiphase flow.
- Identification of the fraction of each of the components in the multiphase flow.
- Flow regime identification
- Flow imaging

The methods which have been developed in the oil industry for the measurement of the flowrates have generally depended on creating a homogeneous flow pattern by the use of a mixer and then measuring the velocity of the mixture using a flowmeter such as a venturi, turbine, or correlation flowmeter.(1),(2),(3) Fraction measurement has usually been undertaken using a combination of impedance and radiation measurements.(4) The most common method of flow regime identification in the laboratory has been by visual inspection. Flow imaging by tomographic methods is now being developed.(5)

The Department of Fluid Engineering and Instrumentation (DFEI) at Cranfield Institute of Technology has a considerable research programme into multiphase flow involving the development of novel sensing and signal processing schemes to address some of the above problems. Some of these will be described below.

2. NOVEL SENSORS

Work is currently being undertaken on the development of sensors using ultrasonic,(6), electromagnetic (7), electrostatic (8) and NMR (9) techniques.

2.1 A static charge flowmeter for measuring oil/gas flows

Flowing dielectric fluids have associated with them imbedded static charges as a consequence of the charge layer on the wall being entrained into the bulk of the fluid by turbulent eddies or by excess charge developed in the fluid created by pumping. The steady state theory of charging in turbulent pipe flow has been studied by Abedian and Sonin (10) amongst others. Al-Rabeh and Hemp (11),(12) have demonstrated that it is possible with a simple electrode structure to measure the flow of a single phase fluid using the imbedded static charge and the 'spatial filter' concept. It is also possible to measure the individual flowrates of oil and gas in a two-phase mixture using the technique in a wide variety of flow regimes including bubbly, plug, wavy, slug, and wavy slug.

The overall system is shown in figure 1. The sensor comprises a static charge flowmeter consisting of eight copper ring electrodes around a tube of dielectric material and a capacitance transducer consisting of two helical electrodes of the same pitch 180° out of phase with each other. Each helix is exactly two turns long. This is a configuration which has previously been shown to provide void fraction measurements independent of the phase distribution (13). The whole sensor is screened by a grounded cylindrical metal case with end plates.

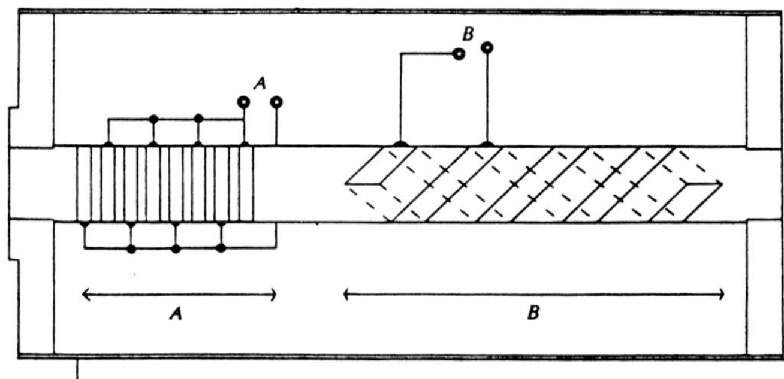

A : Static Charge Electrodes B : Void Fraction Electrodes

Figure 1 : Static Charge Flowmeter

The static charge flowmeter generates a spectrum as shown in figure 2 which is a function of the gas and liquid flowrates. A representative frequency from this spectrum is obtained. This may be either the peak frequency or a measure of the overall bandwidth. It has been shown that the most consistent result for a particular oil is the point at which the noise has fallen to a fixed level. In horizontal pipework this frequency correlates most closely with the total flowrate, Q_t. The capacitance between the helical sensors is related to the void fraction of the mixture. The measurements are then combined to provide the measurement of the liquid flowrate, Q_l, the gas flowrate, Q_g, and the void fraction, under the assumption of no slip between the phases.

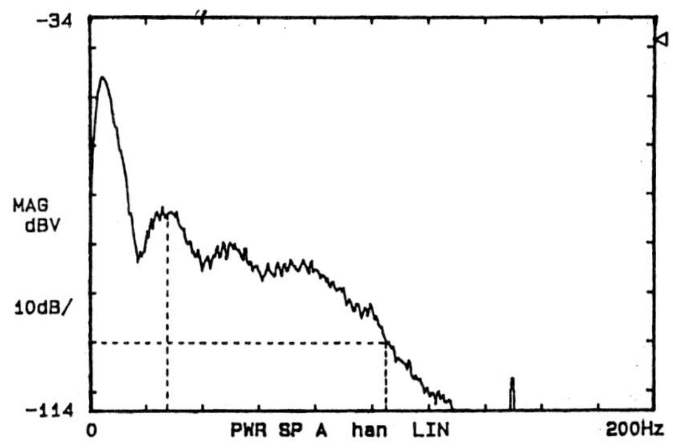

Figure 2 : Static Charge Spectrum

The technique has been experimentally validated on a 2 cm diameter flow rig using BP 180 as the dielectric fluid and nitrogen as the gas in horizontal pipe flow. Figure 3 shows the results obtained from the static charge sensor for a series of total flowrates and liquid flowrates. Figure 4 shows the output of the capacitance sensor as a function of void fraction. The experimental results show that the static charge element of the meter provides an accuracy of approximately ±7% of rate from 0.14l/s to 0.56l/s. The void fraction system has an accuracy of 0.01 for void fractions between 0.05 to 0.62.

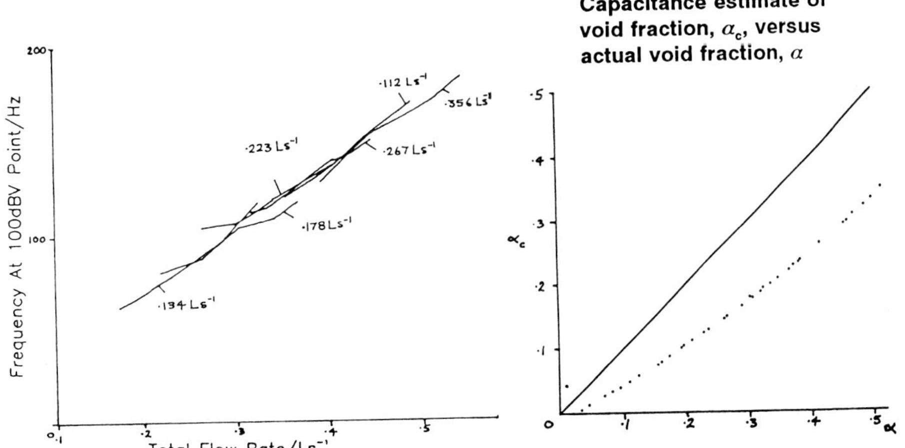

Figure 3 : Calibration of Static Charge Flowmeter

Figure 4 : Void Fraction Calibration

The experiments have demonstrated that it is possible to extend the static charge flowmeter principle from use on single phase liquids to two-phase oil-gas mixtures and still maintain a reasonably acceptable degree of overall flowrate and void fraction measurement.

2.2 Pulsed H-NMR for in two phase flow

Nuclear Magnetic Resonance for hydrogen nuclei (H-NMR) is a method already in use for the oil industry. The current research is to investigate the potential of H-NMR for well head monitoring.

There is an abundance of H-nuclei in all three phases (oil,water and gas) which makes NMR promising for this task. The NMR-signal of oil and water is different in terms of the spin-spin relaxation times (T_2 - relaxation). This can be used to distinguish between individual phases. the signal decay also depends on the efflux of excited material during the measurement time which gives a measure of the flow.

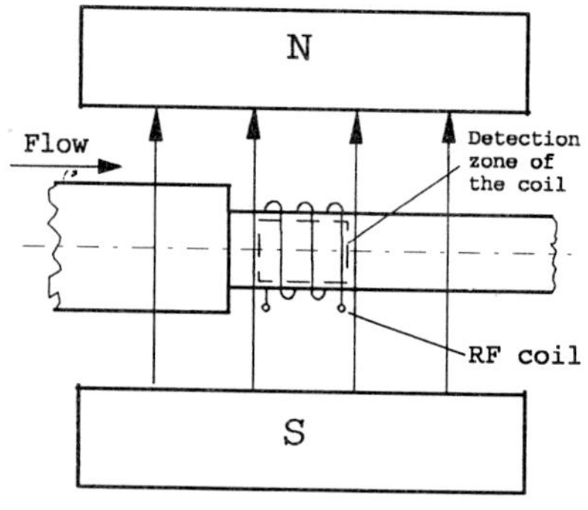

Figure 5: Modified Spectrometer

Figure 5 shows a modified laboratory spectrometer which has been used for these experiments. An initial radio frequency magnetic field excites the hydrogen nuclei within the measurement zone. This induces a high frequency voltage in a rf coil which surrounds the excited material. The initial amplitude of this signal is proportional to the amount of H-nuclei of all components of the flow.

Since the T_2 relaxation time of oil is relatively short compared to water a, biexponential signal decay is obtained as shown in figure 6. After the signal due to the oil has decayed then the observed signal is due only to the water signal and allows a calculation of the initial signal of the water component by back extrapolation. In this way the composition of the oil/water phase can be determined. The efflux of excited material during the observation time causes an additional loss of signal. This loss is a measure of the flowrate. Increasing the flowrates of the components results in a decrease of the corresponding T_2 values of the bi-exponential signal curve.

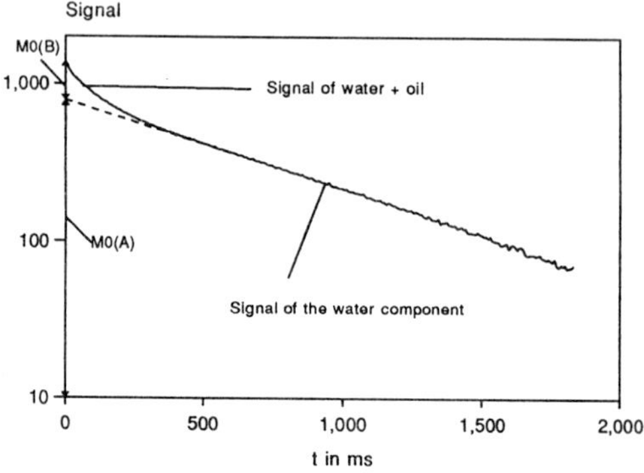

Figure 6: Biexponential Decay

The above method is potentially insensitive to phase slip. In addition it appears not to be affected by wax deposits on the tube walls. Since the actual flow sensor is a simple coil wound around a non-metallic tube the method is also non-intrusive and highly robust against abrasive fluid components, such as sand. Figures 7 and 8 show the measured and calculated flowrates of oil and water for a 50-50 oil water mixture obtained using this method. Further work is now being undertaken to examine the use of T_1 (spin-lattice relaxation) to distinguish between the flow of oil and that of water.

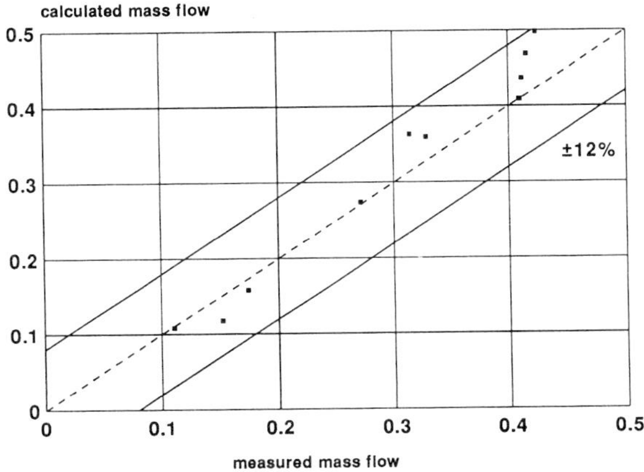

Figure 7: Oil Measurement

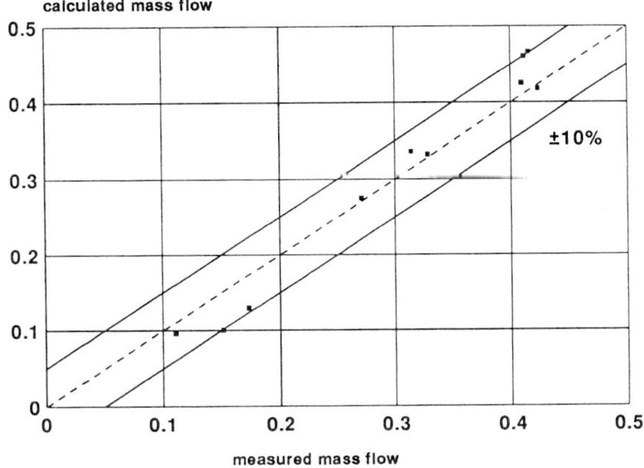

Figure 8: Water Measurement

3. FLOW REGIME IDENTIFICATION (14)

All existing flowmeters are to some extent flow regime dependent. These flow regimes have been characterized for both horizontal and vertical flows under laboratory conditions. The usual method for identifying the regime being visual inspection. The generation of these various flow regimes such as stratified, plug, slug, bubbly, annular flow regimes exist is real systems has been shown to be configuration dependent. Visual inspection of the flow is generally not practical in many systems from the point of view of safety since these systems operate at high pressures and there is generally considerable difficulty in keeping optical windows clean. Corrosion in multiphase systems is flow regime dependent. It would therefore be desirable to have sensors which would identify the flow regime. In a multiphase flowmeasurement system their output could be used to correct the output of a conventional flowmeter. In the area of the corrosion they could be used to predict the life time of the pipeline or particular piece of equipment. Work is currently being undertaken within DFEI to identify the flow regime by taking the signal from a single transducer and undertaking time series analyses using ideas from chaos theory to determine the fractal dimension of the signal from the transducer.

Vertical oil water flows have been investigated using both resistive and ultrasonic sensors. The resistance as a function of time was monitored and the amplitude of the ultrasonic beam scattered through 150° from a 1 MHz continuous wave ultrasonic beam incident across the diameter of the pipe. The measurements were made with a water superficial velocity of 0.4 m/s and the oil hold up 0.05. Figures 9 and 10 show the signals from the resistive and ultrasonic experiments respectively.

The instantaneous state of a dynamical system can be represented as a point in a multi-dimensional phase space. The trajectories of all physically realizable systems trend towards a finite attractor whose shape and complexity are determined by the underlying dynamics. If the system is chaotic the trajectory has a sensitive dependence on initial conditions, although the gross shape of the attractor is independent of the initial conditions. The trajectory of a chaotic system never repeats itself and therefore the attractor by definition must be fractal.

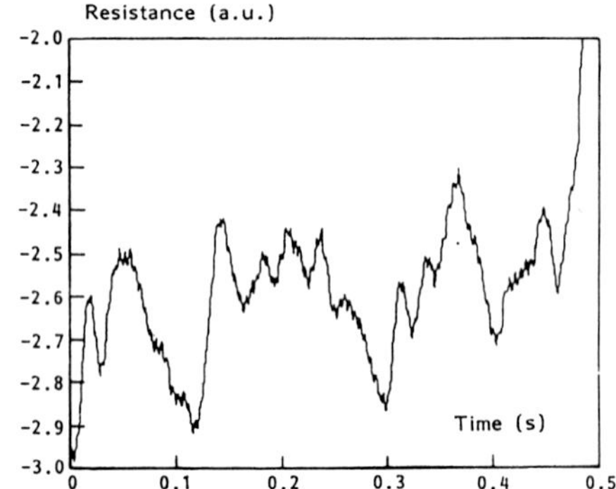

Figure 9: Resistive Data

In order to construct an attractor for a dynamical system, a knowledge of all of the independent variables governing the behaviour is required. It has been shown however that an attractor can be reconstructed from the time series from a single transducer. The attractor is reconstructed by taking the experimental time series $X_0(t)$ and creating other series of data from it by using delays i.e.

$$X_1(t) = X_0(t + \tau)$$
$$X_2(t) = X_0(t + 2\tau)$$

$$X_n(t) = X_0(t + n\tau)$$

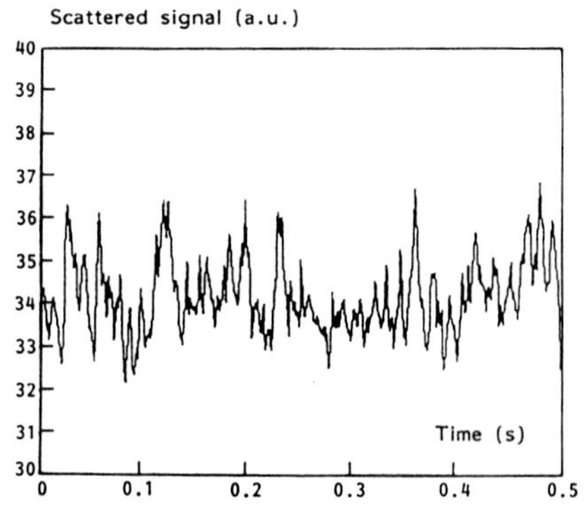

Figure 10: Ultrasonic Data

Figures 11 shows the attractor in $X(t),X(t + \tau)$ space; and figure 12 in $X(t),X(t + 7\tau)$ space. It is clear that the trajectory is less knotted as the delay is increased.

The fractal dimension of the attractor can be obtained by the method proposed in (15). A correlation is defined as :-

$$C(\varepsilon) = \lim_{N \to \infty} \lim_{\varepsilon \to 0} \frac{1}{N^2} \sum_{\substack{i,j \\ i \neq j}}^{N} \theta(\varepsilon - |R_i - R_j|)$$

where N is the number of data points and θ is the Heaviside step function that tests whether R_j lies within the hypersphere of side 2ϵ from R_i. R_i and R_j are points on the reconstructed attractor in the M-dimensional space.

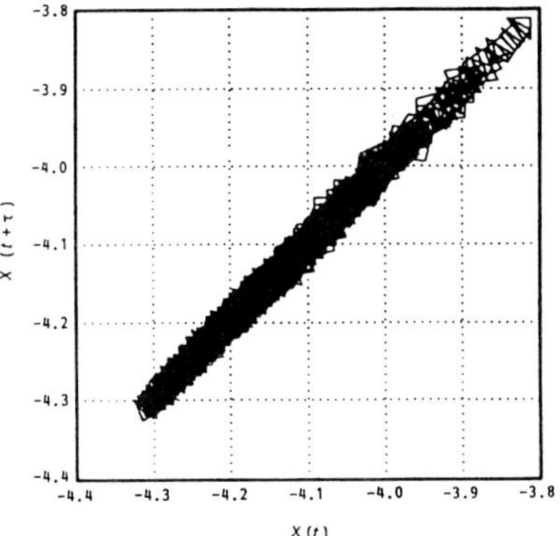

Figure 11: X(t),X(t+τ) space

For an attractor with a fractal dimension D

$$C(\varepsilon)_{\varepsilon \to 0} \propto \varepsilon^D$$

Thus a plot of $\log_{10}C(\epsilon)$ against $\log_{10}\epsilon$ will have slope of D, provided that the space has a dimension greater than or equal to D and the attractor is fully expanded. This requires an embedding dimension, M, greater than 2D + 1.

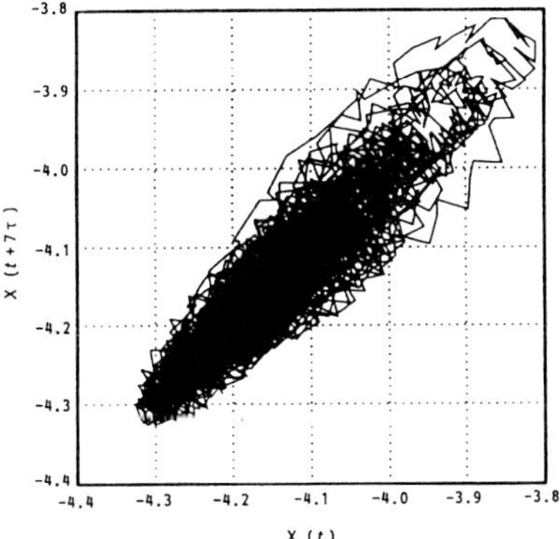

Figure 12: X(t),X(t+7τ) space

Applying these methods to the resistive and ultrasonic data and plotting the dimension of the attractor against the embedding dimension produces the result shown in figure 13 which shows that for both sets of data the dimension of the attractor asymptotes to 5 for M> 12. It can be seen that the fractal dimension measured is the same for both transducers and therefore it is reasonable to assume that they are both measuring the dynamics of the flow.

It would therefore be reasonable to assume that if the fractal dimension of the flow gives a measure of the complexity of the behaviour and a guide to number of parameters governing the dynamics of the flow that different flow regimes would have different fractal dimensions.

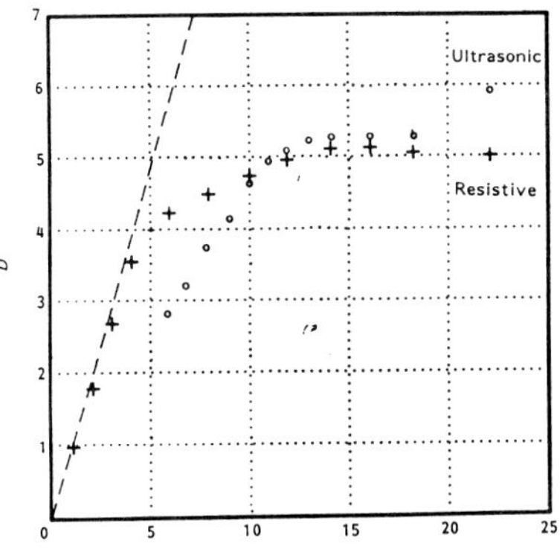

Figure 13: Fractal Dimension

The present algorithm requires massive computational power. However it may be possible to determine the fractal dimension in real time using either analogue or digital circuits.

4. FLOW IMAGING

There is increasing interest in visualising multiphase flows using electromagnetic field imaging (resistive,capacitive or inductive) methods or ultrasonic methods. The aim of this work is generally to provide a reconstruction algorithm which generates anthropomorphic images i.e. images similar to those obtained by the human eye if the pipe was provided with a viewing section. However in many cases the flow patterns interact with the original field in such a way as to as to make this reconstruction of flow field difficult. One particular example of this in a two phase liquid/gas system is the effect caused by multiple scattering of ultrasound from individual gas bubbles. Under these circumstances it will not be possible to obtain a reconstruction which identifies individual bubbles. However an image can still be obtained from which information can be extracted.

Work is being undertaken (16) to use ultrasound to detect the gas phase in a two phase flow and attempt to infer the characteristics of the flow using suitable image processing techniques. The system uses a commercial B-scan which is normally used for medical diagnosis. The image from this B-scan is either stored on a video recorder or passed to a frame grabber interfaced to a 386/33 microprocessor and a real time image processing system.

Figure 14 shows the speckle pattern which is obtained on the output of the B-scan for a series of gas void fractions and flowrates. The histogram of the pattern is sensitive to the gas void fraction as shown in figure 15.

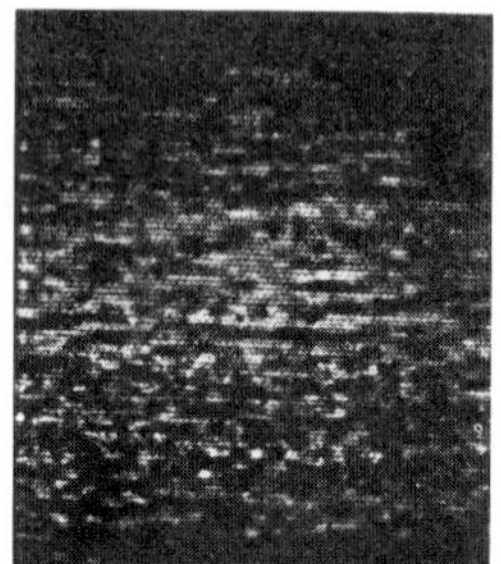

Figure 14: Speckle Patterns at Various Void Fractions and Flowrates

Figure 15: Statistics of Speckle Pattern at Various Void Fractions and Flowrates

The coherent nature of the image allows the use of methods available for coherent images in general and in particular those used in the study of Radar images. The statistics of a coherent image are related to the Gamma distribution.

$$\phi(x;\alpha,\beta) = \frac{\beta^{\alpha}x^{\alpha-1}\exp(-\beta x)}{\Gamma(\alpha)}$$

for $x \geq 0$, $\alpha \geq 0$, and $\beta > 0$. Given a density function a gamma distribution is fitted by performing a logarithmic least squares fit, i.e. compute α and β for which

$$\sum (\ln\hat{\varphi}_i - \ln\varphi_i)^2$$

is a minimum. α and β are then related to the void fraction.

Examination of the 2D spectrum of the speckle pattern shows that the spectrum is white across the flow and is bandlimited along the flow. As the flow rate increases the speckle pattern in the direction of the flow becomes streaky. The average width of the spectrum along the flow is inversely proportional to the flow velocity.

Preliminary studies have been made using a rig which provides control over the liquid flowrate and gas void fraction. The system has been used to study horizontal two-phase flows in pipes of 75mm internal diameter at velocities up to 3 m/s. Comparisons have been made between the statistics of the speckle pattern obtained using the system and the gas void fraction present in the flow. Comparisons have also been made between the 2D spectrum of the patterns and the flowrate.

5. CONCLUSIONS

The problem of multiphase flow measurement remains as yet generally unsolved. The likely solution however is likely to require a combination of novel sensing technologies; existing technologies with intelligence built in concerning the flow regime in which the sensor is operating; and some means of identifying the flow regime. The application of image processing techniques is likely to lead to additional information being generated from the multiphase flow from which phase fractions and velocities can be determined. The work described above offers some contribution to those advances.

6. ACKNOWLEDGEMENTS

I am extremely grateful to my colleagues and students in the department for allowing me to draw upon their work for this paper.

7. REFERENCES

1. Hammer E.A. and Nordvedt J.E., "The application of a venturi meter to multiphase flow measurement", Sensors: Technology, Systems and Applications, ed K.T.V.Grattan, pp 233-238, Adam Hilger, Bristol.
2. Hall A. and Shaw C., "Field experience of a two-phase flow measurement", North Sea Flow Metering Workshop, NEL, Sept 1988.
3. Kinghorn F.C.,"Challenging areas in flow measurement", Measurement and Control, **21**, pp 229-237, 1988.
4.Dykesteen E. and Frantzen K.H., "Multiphase fraction meter developed and field tested", Oil and Gas Journal, pp 50-55, Feb 1991.

5. Thorn R., Huang S.M., Xie C.G., Salkeld J.A., Hunt A. and Beck M.S.,"Flow imaging for multi-component flow measurement", FMI,**1**, No5, pp 259-268, 1990.
6. Oddie G.M., The Characterisation of multicomponent (liquid) flows using scattered ultrasound, PhD Thesis, CIT, 1992.
7. Kraaft R.G., Multiphase Flow in Electromagnetic Flowmeters, PhD Thesis, CIT, 1993.
8. Ellis S.M., Hemp J. and Sanderson M.L. "A static charge flowmeter for measuring two-phase flows" to be presented at IMEKO, South Korea, Oct.1993.
9. Bayer A., Carr-Brion K. and Sanderson M.L., "Pulsed H-NMR for Simultaneous Measurement of Flow Rates and Concentrations in Two Phase Flow (oil/water)", PSTI Technical Forum, Aberdeen, May 1992.
10. Abedian B. and Sonin A.A., "The Theory of Electric Charging in Turbulent Pipe Flow", Journal of Fluid Mechanics, **120**, pp 199-217, 1982.
11. Al-Rabeh R.H. and Hemp J., UK Patent No 2063482
12. Al-Rabeh R.H. and Hemp J., "A new method for measuring the flowrate of insulating fluids", Int. Conf. on Advances in Flow Measurement Techniques, Warwick, September 1981.
13. Sami M., Abouelwafa A., and Kendall E.J., "The use of capacitance sensors for phase percentage determination in multiphase pipelines", IEEE Trans Instrum. and Meas. **IM-29**, No.1, March 1980.
14. Oddie G.M.," On the detection of a low-dimensional attractor in a disperse two-component (oil-water) flow in a vertical pipe, FMI,**2**, No4, pp 225-231, 1991.
15. Grassberger P. and Procaccia I., "Dimensions and entropy of strange attractors from a fluctuating dynamics approach", Physica, **13D**, pp 34-45, 1984.
16. Blackledge J.M., "Analysis of two-phase flows using ultrasonic scanning" Process Industry Journal, pp 27-29, May 1992.

Review of process tomography image reconstruction methods

C. G. Xie

Department of Electrical Engineering & Electronics, Process Tomography Unit, UMIST, Manchester M60 1QD

Abstract - Tomography techniques based on different physical principles have been used for various scientific and industrial applications. Many image reconstruction methods have been developed. This paper reviews, categorises and compares different algorithms in a systematic way. The information compiled also indicates the application areas of process tomography systems and the limitations of different imaging modalities.

1. INTRODUCTION

The use of tomography techniques for scientific and industrial applications is becoming widespread and numerous image reconstruction methods have been developed. The techniques used cover almost all the energy spectra ranging from high frequencies to low frequencies, and thus range from high resolution to low resolution. For each technique, different operating modes and reconstruction algorithms are used. For example, among nucleonic techniques there are tomographic methods based on transmission, emission or scattering principles, and the image reconstruction algorithms are totally different. The intention of this paper is to review various tomography techniques, the associated image reconstruction methods and their typical applications. The main results are tabulated in Tables 1-5. This review covers nucleonic, optical, microwave, magnetic resonance (MR), acoustic and electrical techniques. Schematic diagrams of each technique are also included for clarity.

2. CATEGORISATION OF TOMOGRAPHY TECHNIQUES & RELATED IMAGE RECONSTRUCTION METHODS

Three major tomography techniques can be identified from Tables 1-5, they are transmission, diffraction and electrical tomography (scattering may be considered as a diffraction phenomenon). Other tomography techniques include emission, reflection, time-of-flight, interferometric and magnetic resonance. Transmission, diffraction and electrical tomography techniques are now described in some detail.

Transmission Tomography

Some radiation sources, such as X-rays and γ-rays, are highly penetrative because of their extremely short wavelengths compared with the size of a practical object. Rays emitted from these sources are able to pass through an object without being bent but with their radiation intensity attenuated. The degree of attenuation is mainly dependent on the density of the material. Tomographic systems can be constructed by using these sources and the appropriate detectors mounted outside the object of interests. They are generally called *transmission tomography* systems. Most high-resolution, nucleonic imaging systems are based on this principle (Table 1), such as commercially available X-ray CT scanners. The use of transmission tomography in optical systems (mainly laser-based) and ultrasonic systems has also been found (Tables 2 and 4). Care should be taken, however, when non-nucleonic sources are used. Ultrasonic transmission-tomography systems, for example, may produce erroneous results since reflection and/or diffraction effects may dominate, depending on the properties of the object being interrogated and on the frequency of the ultrasonic waves. Many transmission tomography algorithms have been well developed for medical CT scanners; the detailed descriptions of these algorithms can be found in the books by Herman [1] and Natterer [2]. Typical algorithms are Fourier inversion, convolutional backprojection and ART. Nucleonic transmission-tomography systems have been mainly used, mostly in laboratories; its resolution is very high (at mm scale) but the responses of the systems are relatively slow.

Diffraction Tomography

Rays and waves interact with materials in very different ways due to the fact that they have very different frequency spectra. Unlike X-rays, for example, microwaves have a wavelength comparable to the size of the objects, thus refraction and diffraction that lead to 'ray' bending cannot be neglected. Reconstruction algorithms described for transmission tomography systems, where the probing rays are straight lines, become inapplicable. Different reconstruction algorithms based on wave diffraction principles should be used. The aim of diffraction

Table 1 - Tomographic techniques: nucleonic methods

Technique	Principle schematic	Reconstruction	Application
Nucleonic Transmission Photon (X-ray & γ-ray) Neutron	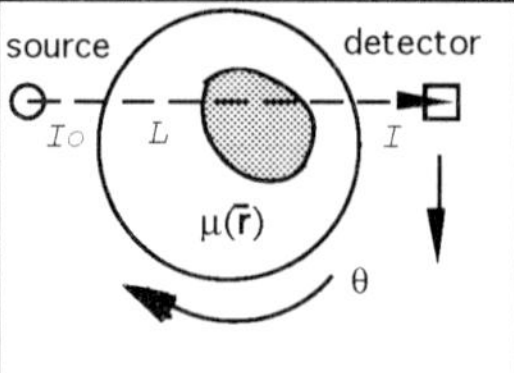 source, detector, Io, L, I, μ(r̄), θ	Unknown: attenuation $\mu(\vec{r})$. *Direct method* [1-2]: Fourier inversion; Filtered back-projection. *Iterative method* [1-2]: Algebraic reconstruction technique (ART)	Multiphase flow imaging [3]. Mixing study [4]. Fluidised or packed bed imaging [5]. NDT & E [6].
Nucleonic Emission Positron emission tomography (PET) Single photon emission CT (SPECT)	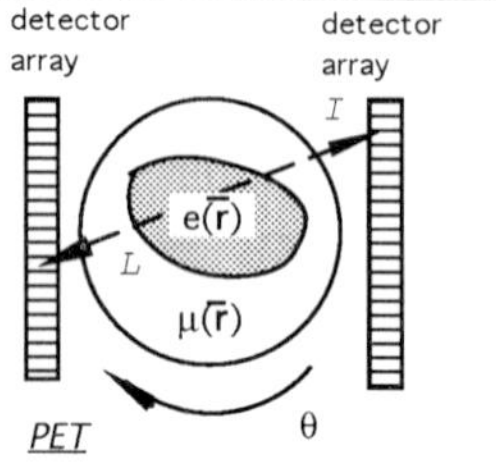 detector array, detector array, I, e(r̄), L, μ(r̄), PET, θ	Unknown: emission $e(\vec{r})$. For PET: Direct method - Fourier inversion [2]. For SPECT: Iterative methods based on statistical approach [7]. Direct methods using matrix pseudoinversion [8].	NDT & E in nuclear industry [9]. Particulate flow imaging [10].
Nucleonic Scattering Neutron scattering X- or γ-ray Compton scattering	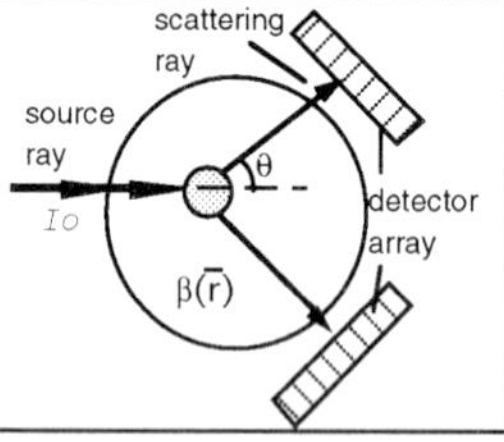 scattering ray, source ray, Io, θ, detector array, β(r̄)	Unknown: voidage $\beta(\vec{r})$. ART based: Solve matrix equation based on regularised least-squares inversion (direct-problem solved at every iteration) [11].	Gas-liquid flow imaging [11]. NDT & E [12].

Table 2 - Tomographic techniques: optical methods

Technique	Principle schematic	Reconstruction	Application
Optical Transmission	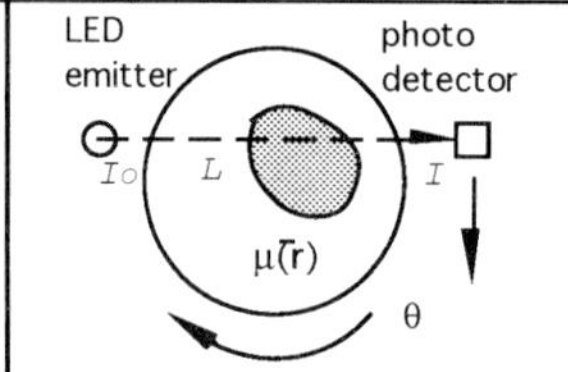 LED emitter, photo detector, Io, L, I, μ(r̄), θ	Unknown: optical attenuation $\mu(\vec{r})$. Similar to nucleonic transmission [13].	Flow study and combustion diagnostics [13-15].
Optical Emission (infrared)	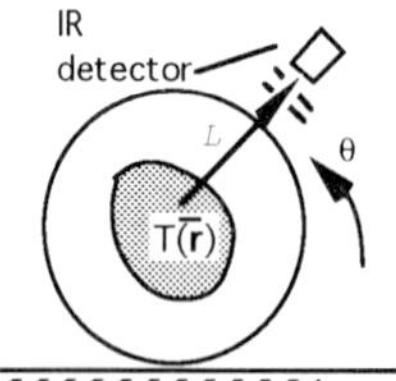 IR detector, L, θ, T(r̄)	Unknown: temperature $T(\vec{r})$. Filtered backprojection [16-17].	Temperature imaging [16,18]. Plasma diagnostics [17].
Optical Interferometric	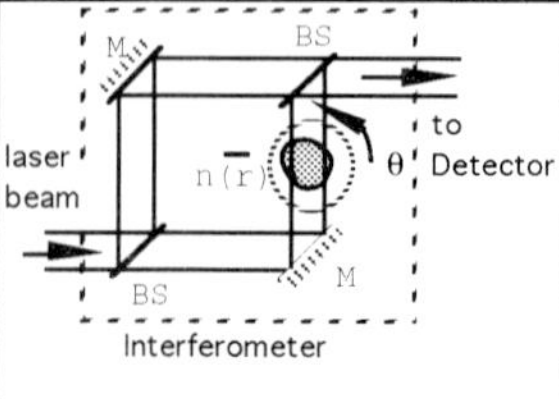 M, BS, to Detector, laser beam, n(r̄), θ, BS, M, Interferometer	Unknown: refractive index $n(\vec{r})$. ART [19,22]. Series expansion [21].	Temperature imaging [19]. Mixing study [20,22]. Flow imaging [21].

tomography is to reconstruct the properties (dielectric or acoustic) of a scatterer irradiated with waves (electromagnetic or acoustic) from the measurements of the scattered fields taken at multiple angles around the scatter. Reconstruction algorithms based on the well-known *Diffraction Slice Theorem* were developed for microwave [23] and ultrasonic [42-44] tomography systems. Such algorithms are generally based on Born's or Rytov's approximations when solving the related wave equations so that the scattered field within the scatterer being imaged can be neglected. These algorithms are suitable for weakly scattering objects; they usually fail when applied to strong scatterers. For strongly scattering objects, reconstruction algorithms based on moment method and matrix pseudoinversion were developed [24]. Techniques based on iterative methods, such as simulated annealing, were also developed [25]; they produced images of improved quality, but are often computationally time consuming. Diffraction tomography systems are of moderate resolution and have found applications mainly in non-destructive testing and evaluation.

Electrical Tomography

Excitation sources (voltage or current) for use with electrical capacitance, electrical resistance and electrical inductance tomography systems are generally of low frequencies (say below 5 MHz). Therefore, unlike microwave systems which usually operate in the GHz range and thus are described by the wave equations, electrical tomography systems are described by equations governing the *electrostatic field* - usually by the Poisson's equations. It is well known that, for electrostatic fields, when electric flux (or current) lines are encountered by an interface of different permittivities (or conductivities), the flux (or current) lines will deflect, i.e. bend. Therefore image reconstruction algorithms described for straight-ray transmission tomography are not applicable. Since propagation of electromagnetic waves and the distribution of electrostatic field lines are governed by different differential equations, reconstruction algorithms for diffraction tomography may not be suitable. Therefore different image reconstruction algorithms have been developed for electrical tomography. Owing to the promising potential of electrical tomography for routine industrial as well as biomedical applications, over the last decade tremendous efforts have been made in developing image reconstruction algorithms. A diversity of algorithms, both non-iterative and iterative, have been emerging. Backprojection algorithm is a typical qualitative, non-iterative algorithm [46,51]. A typical quantitative, iterative algorithm is the one based on the modified Newton-Raphson method [54-56]. Electrical tomography systems are of relatively low resolution, however, they have found widespread applications in various industrial applications due to their low cost and fast response.

3. SUMMARY

- There are three major tomography techniques: - transmission, diffraction and electrical tomography. They are based on, respectively, straight-ray propagation (e.g. X-rays), wave propagation (e.g. microwaves) and mostly conservation field (e.g. capacitance imaging based on electrostatic field).
- Image reconstruction algorithms can be broadly classified in two categories: transform (direct) methods (e.g. Fourier inversion) and iterative methods (e.g. ART).
- Problems associated with transmission tomography are mostly linear. Related algorithms are well developed for medical CT scanners, with convolutional backprojection algorithm being the most popular. For industrial applications, limited data problems are often common, therefore, ART based algorithms incorporating some prior knowledge are often used.
- Problems associated with diffraction tomography are inherently nonlinear. Some algorithms are developed with or without Born approximation. Born approximation is invoked for objects exhibiting weak scattering and the related algorithms are mostly based on Fourier inversion. For strong scatterers, algorithms based on moment methods and matrix pseudoinversion are usually used.
- Problems associated with electrical tomography are also inherently nonlinear. Many algorithms have been emerging, with backprojection methods (mostly non-iterative) and Newton-Raphson methods (iterative) being the popular ones.

REFERENCES

1. Herman G T, *Image reconstruction from* projections, Academic Press, New York (1980)
2. Natterer F, *The mathematics of computerised tomography,* John Wiley & Sons, Chichester (1986)
3. Vinegar H J and Wellington S L, *Rev. Sci. Instrum.* **58**, 96 (1987)
4. Wang S Y et al., *Appl. Opt.* **24**, 4021 (1985)
5. Hosseini-Ashrafi M E, Tüzün U and MacCuaig N, in *Tomographic Techniques for Process Design and Operation*, edited by M S Beck et al. CEC Computational Mechanics Publications (1993)
6. Rapaport M S and Gayer A, *NDT & E Int.* **24**, 141 (1991)

Table 3 - Tomographic techniques: microwave & magnetic resonance methods

Technique	Principle schematic	Reconstruction	Application
Microwave Diffraction	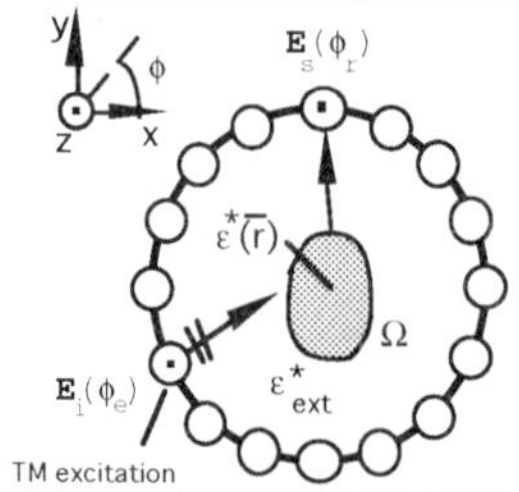	Unknown: permittivity $\varepsilon^*(\vec{r})$. *With Born approximaton:* Fourier inversion [23]. *No Born approximation:* Moment method and pseudoinversion [24]. Simulated annealing [25]. Modified Newton-Kantorovich method [26].	Fast on-line control of conveyed products; NDT of reinforced concrete structures [27].
Magnetic resonance	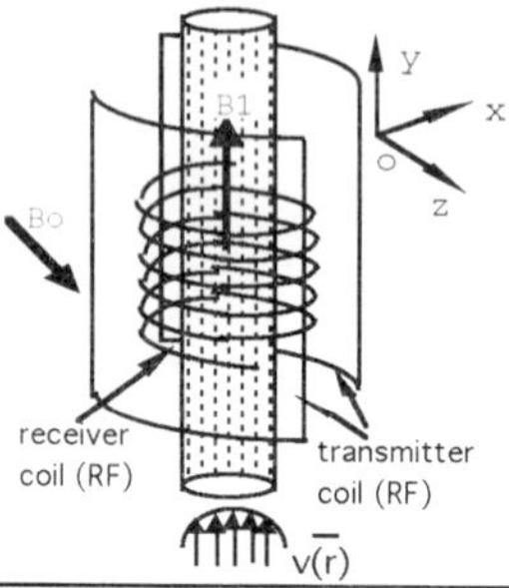	Unknown: velocity $v(\vec{r})$ and/or concentration. Fourier inversion [28-29].	Flow velocity imaging [28-30]. Imaging study of concentrated solid suspensions [31].

Table 4 - Tomographic techniques: acoustic methods

Technique	Principle schematic	Reconstruction	Application
Acoustic Transmission	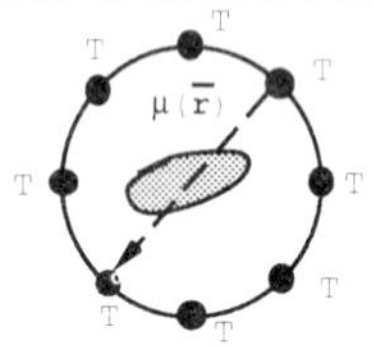	Unknown: acoustic attenuation $\mu(\vec{r})$. Similar to nucleonic transmission [32].	Bubbly flow imaging [32].
Acoustic Reflection	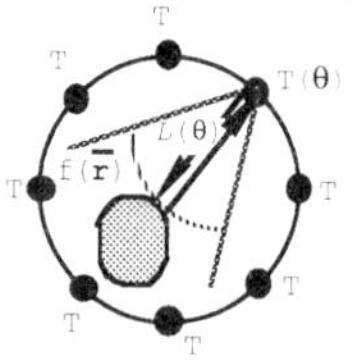	Unknown: reflection interface $f(\vec{r})$. Backprojection [33].	Gas/liquid flow study [33].
Acoustic Time-of-flight	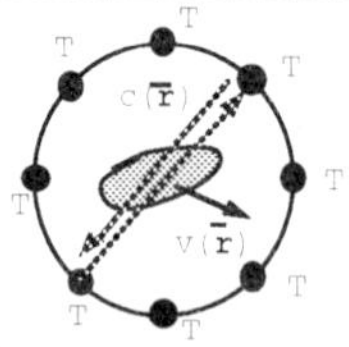	Unknown: sound and flow velocities $c(\vec{r})$ and $v(\vec{r})$ Backprojection [34]. Series expansion [35]. ART [36-37]. Transform methods (Fourier inversion or filtered backprojection) [39-41].	Flow void imaging [34]. Furnace temperature imaging [35-36]. Flow velocity imaging: scalar [37-38], vectorial [39-40]. NDE of solids and composites [41].
Acoustic Diffraction	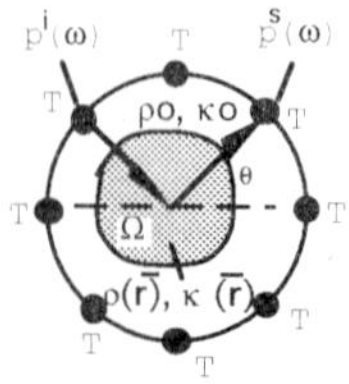	Unknown: compressibility and density $\kappa(\vec{r})$ and $\rho(\vec{r})$. *With Born approximation:* Fourier inversion [42-44].	Fluid study [42-43]. NDT & E [45].

7. Green P J, *IEEE Trans. Med. Imaging* **9**, 84 (1990)
8. Smith M F et al., *IEEE Trans. Med. Imaging* **11**,165 (1992)
9. Phillips J R et al., *Electrical Power Research Institute Report* EPRI-NP 1952 (1981)
10. Parker D J et al., in *Tomographic Techniques for Process Design and Operation*, edited by M S Beck et al. CEC Computational Mechanics Publications (1993)
11. Hussein E M A and Meneley D A, *Int. J. Multiphase Flow* **12**, 1 (1986)
12. Babot D, Berodias G and Peix G, *NDT & E Int.* **24**, 247 (1991)
13. Beiting E J, *Appl. Opt.* **31**, 1328 (1992)
14. Santoro R J et al., *Int. J. Heat & Mass Transfer* **24**, 1139 (1981)
15. Goulard R and Emmerman P J, in *Inverse Problems in Optics*, H. P. Baltes ed., Vol. 20 of Topics in Current Physics, Springer-Verlag, New York (1980)
16. Uchiyama H, Nakajima M and Yuta S, *Appl. Opt.* **24**, 4111 (1985)
17. Hino M et al., *Appl. Opt.* **26**, 4742 (1987)
18. Hall R J and Bonczyk P A, *Appl. Opt.* **29**, 4590 (1990)
19. Hertz H M, *Opt. Commun.* **54**, 131 (1985)
20. Snyder R and Hesselink L, *Opt. Lett.* **13**, 351 (1988)
21. Cha S, *Opt. Eng.* **27**, 557 (1988)
22. Mewes D, Fellhölter A and Renz R, in *Tomographic Techniques for Process Design and Operation*, edited by M S Beck et al. CEC Computational Mechanics Publications (1993)
23. Broquetas A et al., *IEEE Trans. Microwave Theory Tech.* **39**, 836 (1991)
24. Caorsi S, Gragnani G L and Pastorino M, *IEEE Trans. Microwave Theory Tech.* **39**, 845 (1991)
25. Garnero L et al., *IEEE Trans. Microwave Theory Tech.* **39**, 1801 (1991)
26. Takenaka T, Harada H and Tanaka M, *Microwave and Optical Technology Letters* **5**, 94 (1992)
27. Bolomey J C and Pichot C, *Int. J. Imaging System Technol.* **2**, 144 (1990).
28. Kose K et al., *J. Physical Soc. Japan* **54**, 81 (1985)
29. Caprihan A and Fukushima E, *Physics Reports* (Review Section of Physics Letters) **198**, 195 (1990)
30. Tyszka M, Hawkes R C and Hall L D, *Flow Meas. Instrum.* **2**, 131 (1991)
31. Sinton S W and Chow A W, *J. Rheology* **35**, 735 (1991)
32. Wolf J, *Part. Part. Syst. Charact.* **5** 170 (1988)
33. Weigand F and Hoyle B S, *IEEE Trans. Ultrasonics, Ferroelectrics and Frequency Control* **36**, 652 (1989)
34. Gai H et al., *Proc. 3rd Int. Conf. Image Processing Applications* Warwick U.K. 18-20th July, pp.237 (1989)
35. Norton S J, Testardi L R and Wadley H N G, *J. Res. Nat. Bureau Standards* **89**, 65 (1984)
36. Schwarz A, in *Tomographic Techniques for Process Design and Operation*, edited by M S Beck et al. CEC Computational Mechanics Publications (1993)
37. Sato T and Shiraki M, *J. Acoust. Soc. Am.* **76**, 1427 (1984)
38. Ko D S, DeFerrari H A and Malanotte-Rizzoli P, *J. Geophys. Res.* **94**, 6197 (1989)
39. Norton S J, *Geophys. Journal* **97**, 161 (1989)
40. Braun H and Hauck A, *IEEE Trans. Signal Processing* **39**, 464 (1991)
41. Rose J L and Ditri J J, *Proc. Ultrasonic Symposium*, 4-7th Dec. Honolulu, Hawaii Vol.2 pp.991-5 (1990)
42. Duchêne B, Lesselier D and Tabbara W, *IEEE Trans. Ultrasonics, Ferroelectrics and Frequency Control* **35**, 437 (1988)
43. Blackledge J M et al., *J. Phys. D: Appl. Phys.* **20**, 1 (1987)
44. Pourjavid S and Tretiak O, *IEEE Trans. Ultrasonics, Ferroelectrics and Frequency Control* **38**, 74 (1991)
45. Capineri L et al., *Ultrasonics*, **30**, 275 (1992)
46. Xie C G et al., *IEE Proc.G* **139**, 89 (1992)
47. Xie C G et al, in *Tomographic Techniques for Process Design and Operation*, edited by M S Beck et al. CEC Computational Mechanics Publications (1993)
48. Isaksen Ø and Nordtvedt J E, *ibid.*
49. Mckee S L et al., *Proc. IChemE Research Event*, pp.726, 6-7 January, Birmingham (Publisher: IChemE) (1993)
50. Fasching G E and Smith N S, *Rev. Sci. Instrum.* **62**, 2243 (1991)
51. Barber D C, *Clin. Phys. Physiol. Meas.* **11**, Suppl.A, 45 (1990)
52. Kotre C J, *Clin. Phys. Physiol. Meas.* **11**, Suppl.A, 275 (1990)
53. Yorkey T J, Webster J G and Tompkins, *IEEE Trans. Biomed. Eng.* **34**, 898 (1987)
54. Yorkey T J, Webster J G and Tompkins, *Proc. Annu. Int. Conf. IEEE Engineering in Med. Biology Society* **8** pp.339 (1986)
55. Hua P et al., *IEEE Trans. Medical Imaging* **10**, 621 (1991)
56. Abdullah M Z, Quick S V and Dickin F J, in *Tomographic Techniques for Process Design and Operation*, edited by M S Beck et al. CEC Computational Mechanics Publications (1993)
57. Ilyas O M et al. *ibid.*
58. Lopes E P and Lopes E P, *IEE Proc.-F*, **139**, 27 (1992)
59. Daily W et al., *Journal Water Resources Research* **28**, 1429 (1992)
60. Yu Z Z et al., *Proc. Int. Conf. Electronic Meas. Instrum.*, 20-22 October, Tianjin, China. pp.486 (1992)
61. Purvis W R et al., IEE Proc.A, **140**, 135 (1993)

Table 5 - Tomographic techniques: electrical methods

Technique	Principle schematic	Reconstruction	Application
Electrical Capacitance	$Q_\theta(\phi)$; $V(\theta)=V$; $\varepsilon(\bar{r})$; electrodes	Unknown: permittivity $\varepsilon(\vec{r})$. Backprojection [46-47]. Iterative method based on optimisation technique [48].	Mainly for processes involving insulating materials. Gas/oil, gas/solids two-phase flow imaging [47,49]. Fluidised bed imaging [50].
Electrical Resistance	$V_\theta(\phi+\alpha)$; $V_\theta(\phi)$; $i(\theta+\alpha)$; $\sigma(\bar{r})$; $i(\theta)$; electrodes	Unknown: conductivity $\sigma(\vec{r})$. Backprojection between equi-potential lines [51] and based on sensitivity coefficient [52]. Perturbation method [53]. Newton-Raphson method [54-56].	Mainly for processes involving conducting media. Hydrocyclone imaging [56]. Mixing study [57]. Geophysical prospecting, environmental monitoring [58-59].
Electrical Inductance	rotating parallel excitation field **B**; B; $\mu(\bar{r})$ $\sigma(\bar{r})$; detecting coils	Unknown: permeability $\mu(\vec{r})$ and conductivity $\sigma(\vec{r})$. Backprojection [60].	A new technique suitable for, e.g. metallurgical and/or mineral processes [60].
Electrical Induced current impedance	magnetic driving coil; voltage sensing electrodes; $\varepsilon(\bar{r})$, $\sigma(\bar{r})$; induced current; $\bar{j}$	Unknown: permittivity $\varepsilon(\vec{r})$ and conductivity $\sigma(\vec{r})$. Backprojection [61].	A new technique [61].

Electromagnetic tomography (EMT): a new process imaging system

Z Z Yu, A J Peyton, M S Beck, L A Xu+
Department of Electrical Engineering and Electronics, UMIST, Manchester, UK, M60 1QD.
+ Department of Electrical Engineering and Automation, Tianjin University, China.

ABSTRACT: An electrical tomographic system which is inductive in nature is presented. The system operates by interrogating the object space with a variety of magnetic field strength patterns and measuring the resultant flux densities on the periphery. Distributions ferromagnetic and/or electrically conductive material can be detected. The paper gives an overview of the imaging method, a description of the EMT system and reconstruction algorithm, some simulation results and sample images.

1. INTRODUCTION

Electrical tomographic techniques are proving to be successful at providing low cost, non-intrusive systems for imaging the internal distribution for medical and process applications. At present, resistance and capacitance sensing methods are well established [1-3]. For example, electrical capacitance tomography (ECT) has been used for a variety of process applications problems such as multi-component flow measurements in oil pipelines, imaging of pneumatic conveyors and observing combustion phenomena, whilst electrical impedance tomography (EIT) has been used to measure tracer migration in soils, and to monitor the performance of hydrocyclones.

Magnetic methods employing the measurement of electrical inductance however have up to now received relatively little attention although we have completed a feasibility study [4]. Magnetic sensors have been considered previously for crack detection with the use of eddy current tomography [5]. For medical applications, new developments have included a magnetic imaging system employing single excitation and detection coils [6] and an induced currents impedance imaging system where currents are magnetically induced within the object medium, with detection by voltage sensing electrodes [7]. This paper describes a novel process imaging system based on electromagnetic tomography.

2. THE PRINCIPLE OF EMT SYSTEM

The underlying principle of the EMT system may be summarised as follows. The object space is energised by an AC sinusoidal magnetic field created by a number of excitation coils. Material within the space causes the spatial distribution of the magnetic field $B(x,y)$ to be distorted, with the field distortion being dependant mainly on the material properties (electrical conductivity $\sigma(x,y)$, magnetic permeability $\mu(x,y)$, size, shape and position) and on the frequency, ω, of the applied magnetic field. According to [8], it has

$$\nabla \times [(\sigma(x,y) + j\omega\varepsilon_0)^{-1}[\nabla \times (\mu(x,y)^{-1} \cdot B(x,y))]] = -j\omega B(x,y) \qquad (1)$$

and

$$\nabla \cdot B(x,y) = 0 \qquad (2)$$

where $B(x,y)$ is represented by a complex phasor. These expressions assume materials of linear property (i.e. $\sigma(x,y)$ and $\mu(x,y)$ are independent of $B(x,y)$), neglect free charges and take $\varepsilon_r = 1$ (air) over the object space.

The tomographic imaging system excites the object space with a number of field strength patterns and then measures the radial field components for each pattern around the outside of the space. Enough excitation patterns are used to ensure that sufficient data can be obtained to enable an image to be reconstructed. Clearly, an accurate image reconstruction would require a non-linear or iterative reconstruction algorithm which solves the complicated equations (1) and (2). However, for demonstration purposes, a simple reconstruction algorithm of the type below was used in this study.

$$I = R \cdot M \tag{3}$$

where I is the image vector, M the normalised measured signal vector and R the transformation matrix.

The system uses one set of coils to interrogate the object space with a variety of field strength patterns and another set of detection coils to measure the radial component of flux densities on the periphery of the space. The measured signals are fed into a host computer which reconstructs images of the material distribution.

3. SYSTEM DESCRIPTION

The experimental system is shown in figure 1, it consists of three main sections, namely the primary sensor, the sensor electronics and the image reconstruction computer.

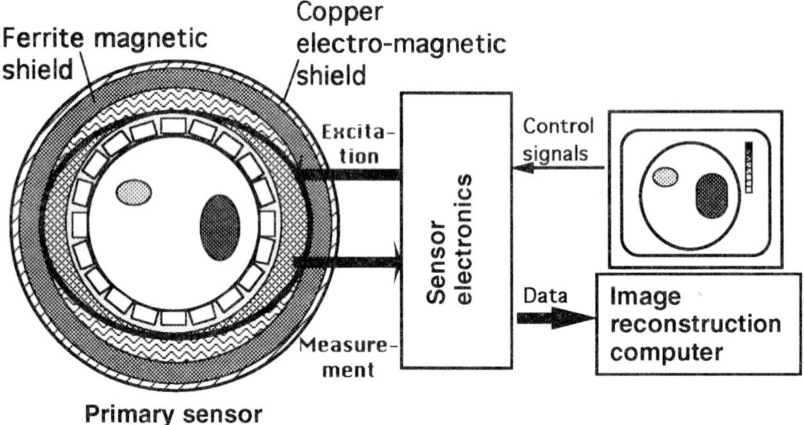

Figure 1. Block diagram of the experimental system

3.1 Primary sensor.

The primary sensor is composed of four assemblies, namely: excitation coils, detection coils, magnetic confinement shield and external electromagnetic shield, and has a 75 mm internal diameter. The excitation coils consist of two pairs of orthogonal (X-Y) windings. Each winding surrounds the object space and has an appropriate turns distribution to ensure that a parallel magnetic field is created within the object space when the object space is empty [9]. The applied field has a frequency of 500 kHz and the direction of the field can be controlled over 360° by the relative magnitudes of the excitation currents in the X and Y windings. The currents are set by the host computer and a large number of field projections are possible. At present only five equally spaced field projections (0°, 72°, 144°, 216° and 288°) are used. There are 21 individual detection coils, which are equally positioned on the circumference, enabling 21 concurrent measurements of the peripheral field to be taken for each particular field projection. Consequently, a total of 105 independent measurements are obtained for image reconstruction. The ferrite magnetic confinement shield both concentrates the field inside the image space and prevents interference from external objects. The electromagnetic shield further confines the flux and improves electromagnetic compatibility.

3.2 Sensor electronics.
The sensor electronics has two main functions, to control the excitation coils and to condition the outputs from the detector coils. The excitation circuitry contains a 500 kHz sine wave generator, two 12-bit DACs, two analogue multipliers which control the magnitude and phase (0° or 180°) of the excitation sine waves, and two voltage-to-current output stages. The detector circuitry having 21 channels, each channel is composed of an input buffer and an amplitude detector. The channels are multiplexed and subsequently digitised with a 12-bit ADC.

3.3 Reconstruction computer
The image reconstruction computer controls the measurement procedure, acquires the object data and implements the image reconstruction algorithm.

4. SIMULATION
An important feature of this EMT sensor is the provision of a parallel magnetic field across the object space when the space is empty. This parallel field is achieved by the choice of turns distribution used for the excitation windings. To verify the design of these windings and the sensitivity of the array, two

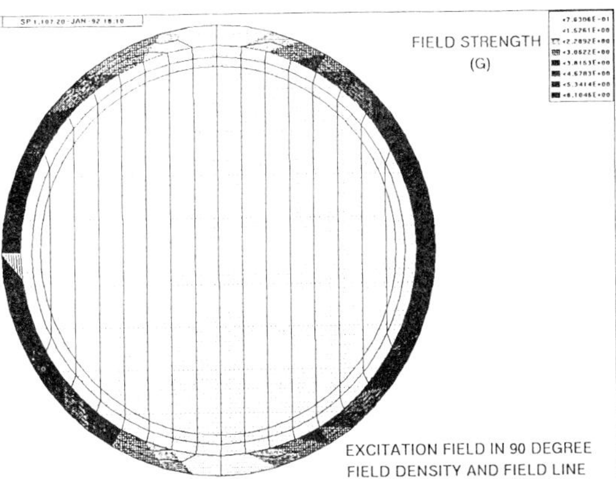

Fig. 2. The excitation field

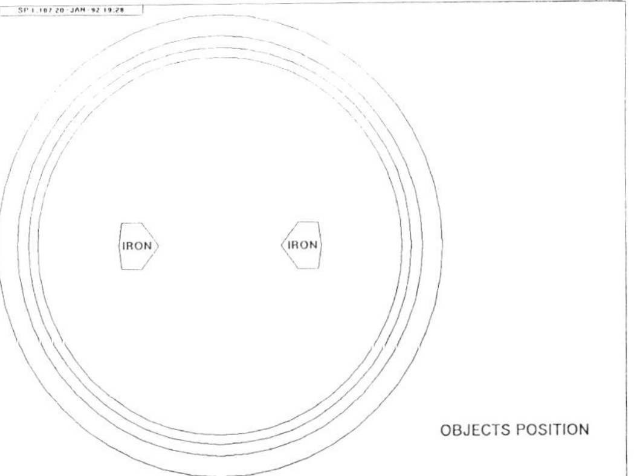

Fig. 3. Position of two conductive bars.

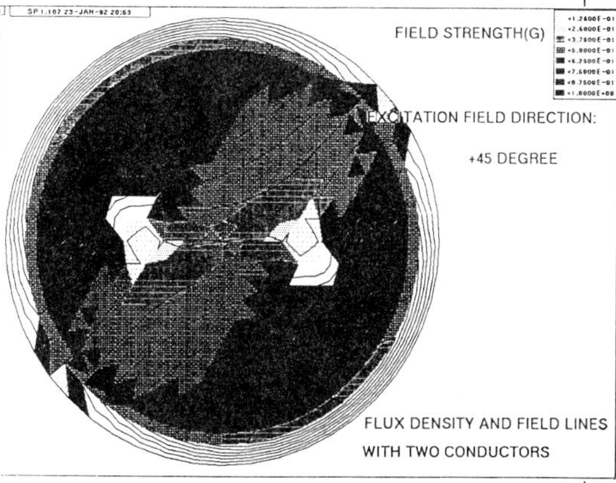

Fig. 4. The distorted field

dimensional computer simulation was performed using a commercial finite element electromagnetic software package, SLIM. The predicted flux density distributions on the boundary of the object space were consistent in each case with the actual values measured by the detection coils.

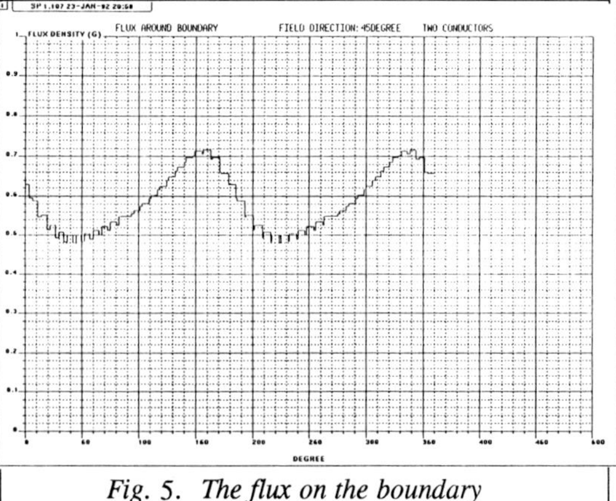

Fig. 5. The flux on the boundary

Figure 2 shows the simulated field lines and flux density (by the dot concentration) of the excitation field when the sensor is empty and illustrates the parallel magnetic field created by the current flowing in the excitation windings. When objects are present the field is distorted, for example, two conductive bars in the position as shown in figure 3, will distort the 45° parallel field as shown in figure 4. Figure 5 shows the resultant flux density on the boundary from 0° to 360°, which is equivalent to the measured signals for this projection.

5. IMAGE RECONSTRUCTION ALGORITHM

The derivation of the linear reconstruction algorithm [10] for this EMT system was performed in two parts. Firstly the forward problem was solved by an experimental method as follows. The image area was evenly divided into 144 small squares (12 x 12) which correspond to the pixels in the reconstructed image. The squares were assigned Cartesian co-ordinates. The output of the detection coils were measured for each of the five field projections with every pixel occupied in turn by a copper sample bar. These measurements describe the response of the system to each individual pixel. The data was normalised according to:

$$s(p,c,x,y) = \frac{M(p,c,x,y) - M_E(p,c)}{M_F(p,c) - M_E(p,c)} \qquad (4)$$

where M(p,c,x,y) is the value measured by detection coil, c, in the projection, p, with the (x,y) pixel occupied by the sample bar. M_E (p,c) and M_F (p,c,) are the outputs of detection coil, c, for projection, p, with the object space empty and completely filled with copper respectively.

To create the image reconstruction algorithm, the response data was organised into a transformation matrix, which converts the 105 sensor measurements into an image. The reconstruction image I(x,y) is represented by pixel grey levels obtained by the following expression.

$$I(x,y) = k \cdot \exp\{[\sum_{p=1}^{5}\sum_{c=1}^{21} M(p,c)\cdot S(p,c,x,y)] / [\sum_{p=1}^{5}\sum_{c=1}^{21} S(p,c,x,y)]\} \qquad (5)$$

where *M(p,c)* is the normalised signals from the 21 detection coils for the 5 field projection. The exponential function helps to enhance grey level contrast and a suitable scale factor k is required to display the image on a screen.

6. RESULTS

Figure 6 and 7 show images for a number of conductivity distributions. Figure 6 shows the image of a single copper bar, beneath the image are the measurement profile by 105 normalised measurements, and at the side a 3D graphical representation of the image. Figure 7 shows images of other distributions. Clearly, the images obtained show the correct characteristics of the actual contents of the object space. Similar images were obtained when ferrite rods were inserted into the object space. As conductive material tends to excludes the flux whereas ferrite material concentrate flux, the former produces positive pixel values while the later produces negative values.

It takes 200ms to capture 105 data and approximately 6 seconds for image reconstruction on a 386 PC .

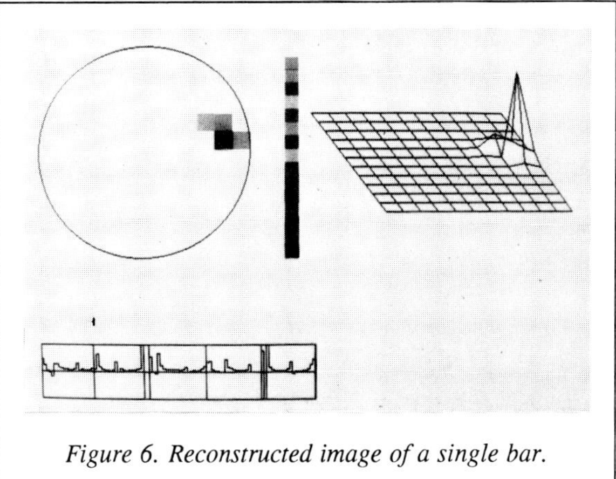

Figure 6. Reconstructed image of a single bar.

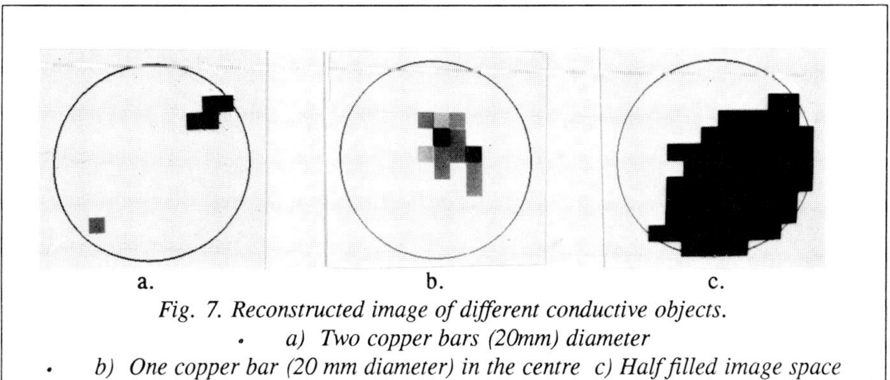

a. b. c.

Fig. 7. Reconstructed image of different conductive objects.

a) Two copper bars (20mm) diameter

b) One copper bar (20 mm diameter) in the centre c) Half filled image space

7. CONCLUSIONS

The experimental system described in this paper has clearly demonstrated the feasibility of EMT. Some important features of this system are:

i. non-contacting
ii. provides information on the distribution of conductivity and/or ferromagnetic material
iii. operates with a relatively high excitation frequency and so has the potential for high speed operation
iv. the sensor employs separate coils for excitation and detection purposes.

The separate excitation coils allow considerable flexibility in the operation of the sensor. For example, the number of independent samples taken for each frame is not dependant on the number of detection channels and can be increased by simply using more field projections. This may help to increase image quality. Alternatively, the excitation coils could be designed for specific flux patterns (such as the parallel field described here), to excite preferentially particular regions within the object space, or possibly to reduce 3D effects.

The images obtained from the system are on the whole representative of the real object distributions. However, their quality is restricted. A significant software limitation of the system is with the simple linear reconstruction algorithm which does not accurately take into account the considerable variations in the magnetic flux patterns caused by the presence of conducting or ferromagnetic objects (the soft field effect). Consequently, image fidelity especially in the central region is poor as in the results earlier. It is expected therefore that iterative or non-linear reconstruction algorithms will eventually become an important part of future EMT systems. A major hardware limit of the system is considered to be the data acquisition section which measures the magnitude of the AC peripheral flux densities. In fact $B(x,y)$ has real and imaginary components related to the conductivity, permeability distributions and to the excitation frequency. It is expected that future EMT systems will require phase sensitive acquisition schemes which can acquire these two independent components.

REFERENCES

1. Dickin FJ, Hoyle BS, Hunt A, Huang SM, Ilyas O, Lenn C, Waterfall RC, Williams RA, Xie CG, and Beck MS. "Tomographic imaging of industrial process equipment: techniques and applications". IEE Proceedings-G. Vol.139. No. 1. Feb. 1992, pp 72-82.
2. Huang SM, Xie CG, Thorn R, Snowden D, Beck MS, "Design of sensor electronics for electrical capacitance tomography". IEE Proceeding-G, Vol. 139, N0.1, Feb. 1992, pp. 83-88.
3. Xie CG, Huang SM, Hoyle BS, Thorn R, Lenn C, Snowden D and Beck MS. "Electrical capacitance tomography for flow imaging: system model for development of image reconstruction algorithms and design of primary sensors". IEE Proceeding-G, Vol.139, No.1, Feb.1992,pp.89-98.
4. Beck MS, Campogrande E, Morris M, Williams RA and Waterfall RC (eds) "European concerted action on process tomography", conference proceeding, CEC Brite-Euram, Karlsruhe Germany, March 1993.
5. Harrision DJ, "Eddy current tomography" IOP short meeting on tomography and scatter imaging 1988,pp.91-98.
6. S Al-Zeibak and Saunders NH. "Feasibility study of *In Vivo* electromagnetic imaging.", Phys.Med.Biol., Vol.38, 1993, pp. 151-60.
7. Tozer RC, Simpson JC, Freeston IL, and Mathias JM. "Non-contact induced current impedance imaging". Electronics letters, 9 April 1992, Vol.28 No.8, pp.773-4.
8. Hayt WH. "Engineering Electromagnetics, 4th edition, McGraw-Hill, 1981.
9. Yu ZZ., Conway WF., Dickin FJ., Xie CG., Beck MS., and Xu LA, 'Field design for electromagnetic tomography system', Proceeding. 1992 Int. Conf. on Electronic Measurement & Instruments, Tianjin, PR China, 20-22 Oct 1992.
10. HERMAN, G. T.,: 'Image reconstruction from projections' academic press, 1980.

Principles of tomographic characterization of a two-particle type dispersion

E. O. Etuke‡, D. A. Hall§, R. C. Waterfall‡, M. S. Beck‡, R. A. Williams†, and T. Dyakowski†.
§ *Department of Material Science,* † *Department of Chemical Engineering,* ‡ *Department of Electrical Engineering and Electronics, University of Manchester Institute of Science and Technology, Manchester M60 1QD, UK.*

Abstract - Process engineers would wish to follow the progression of mixing of different particle types as they disperse throughout a process vessel. This paper examines the principles for discriminating between two types of particles in a dispersion. The measurement technique uses multiple frequencies to excite the suspension and exploits the frequency dependent impedance of at least one of the dispersed phases. The technique, though initially tested for a single pair of electrodes, can be extended to tomographic imaging of particulate dispersions by utilizing multiple electrodes, eg. 12 electrodes.

1. INTRODUCTION

Electrically, material identification revolves around studies of permittivity and conductivity. One approach seeks to exploit the fact that different materials have different dielectric relaxation times and frequencies. *Dielectric relaxation frequency* is the frequency at which the polarization of the dipoles in a given dielectric fails to follow an applied alternating electric field. Many dielectric relaxation functions have been proposed [1], and it is hoped knowledge of relaxation processes holds the key to the structural and molecular nature of various materials. However, there are many difficulties in the interpretation of relaxation functions. For example, there is considerable uncertainty in the measurement of dielectric constants of electrolytes of various concentrations [2]. In some instances, inconsistencies have been observed between established theories and quantitative measurements. There are also problems presented by electrode polarization and contact impedance for aqueous dielectrics [1,3].

This paper outlines a simple technique for estimating the volume fractions of ballotini glass and nickel particles co-existing in suspension. The technique can be extended to other types of particulate dispersions provided that at least one of the dispersed phases has a frequency dependent conductivity. Analysis is based on the superposition of the conductivity spectra of the respective dispersed phases.

2. CONDUCTIVITY - CONCENTRATION RELATIONS

For suspensions of solids in aqueous solution, many electrical measuring techniques relate volume fractions of solids to the ratio of the suspension's conductivity to that of the medium of dispersion. The Maxwell-Wagner [4,5,6,7] theory for a dispersion of spherical particles in an aqueous medium is expressed as:

$$\kappa^* = \kappa_m^* \frac{2\kappa_m^* + \kappa_p^* - 2\Phi\,(\kappa_m^* - \kappa_p^*)}{2\kappa_m^* + \kappa_p^* + \Phi\,(\kappa_m^* - \kappa_p^*)} \qquad (1)$$

where κ^* = complex conductivity of the suspension (mS/cm);
κ_m^* = complex conductivity of the medium;
κ_p^* = complex conductivity of the disperse phase; and
Φ = volume fraction of the dispersed phase.

While making use of real conductivities, eqn. (1) can be simplified to:

$$\Phi = \frac{2(1-K/K_m)}{2+K/K_m} \quad \text{for } \kappa_p \ll \kappa_m , \qquad (2)$$

and

$$\Phi = \frac{K/K_m - 1}{K/K_m + 2} \quad \text{for } \kappa_p \gg \kappa_m . \qquad (3)$$

Eqns. (2) and (3) are believed to be accurate for sparse dispersions ($\Phi \ll 1$) [4].

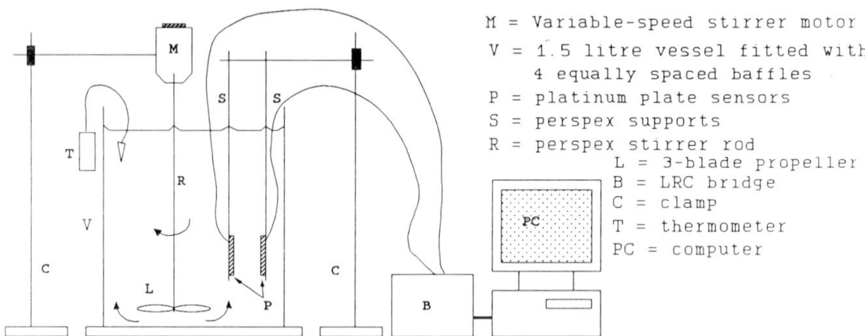

Fig. 1 Experimental set-up.

3. EXPERIMENTAL METHODS

Experiments were carried out to validate eqns. (2) & (3). The mediums of dispersions were 50 mM and 20 mM KCl solutions made in distilled water. The dispersed phases were spherical particles of ballotini glass and nickel obtained from commercial sources. Glass has very low conductivity, whereas nickel is highly conductive. The glass particles (50 μm mean size) were soaked in 2M HCl for about 24 hours with occasional stirring. They were later washed clean with distilled water and left to dry. The nickel particles are finer than 15μm. They were carefully handled to minimize the formation of oxides at the surface. Fig. 1 shows the experimental set-up. The electrodes are made of platinum foil. This choice is influenced by the fact that platinum is stable and non-polarizable. For this work, the platinum electrodes used were not treated with platinum-black [8]. The electrode plate area was 6.25 cm^2, and plate separation was 2 cm giving a cell constant of 3.125 cm.

The stirrer arrangement allowed for agitation of the particles. The stirred vessel was fitted with 4 equally spaced baffles. The temperature of the mixture was monitored from a thermometer and observed to be in the range 22°C to 24°C. The LRC bridge used was a Wayne Kerr 6425 with a frequency range of 20 Hz to 300 kHz. The LRC bridge was controlled by a personal computer which was also used for data storage and processing.

4. RESULTS AND ANALYSIS

Figs. 2 and 3 show the conductivity spectra where the glass and nickel particles were separately dispersed in 20 mM KCl electrolyte. The conductivity of the electrolyte is nearly constant above 10 kHz, but decreases exponentially below 10 kHz due to polarization and contact impedance at the interface of the electrodes and electrolyte. Observe that glass does not significantly alter the conductivity spectrum of the electrolyte other than create a depression in the conductivity in proportion to it's volume fraction. On the other hand, the effect of nickel on the conductivity of the electrolyte is strongly frequency dependent. Interestingly, the spectra (Fig. 3) show a conductivity depression below a frequency, f_c, and a conductivity increase beyond f_c. At f_c the conductivity of the suspension is equal to the conductivity of the electrolyte (medium of dispersion) irrespective of the volume fraction of nickel. A plausible explanation for this conductivity trend is that each nickel particle can be imagined as having a conducting core bounded by a non-conducting shell phase [9]. The impedance (capacitive) of the shell phase varies with frequency. At higher frequencies, it is short-circuited allowing the true conductivity of the nickel to manifest. From observation, the frequency of convergence f_c increases with increase in the concentration of the electrolyte.

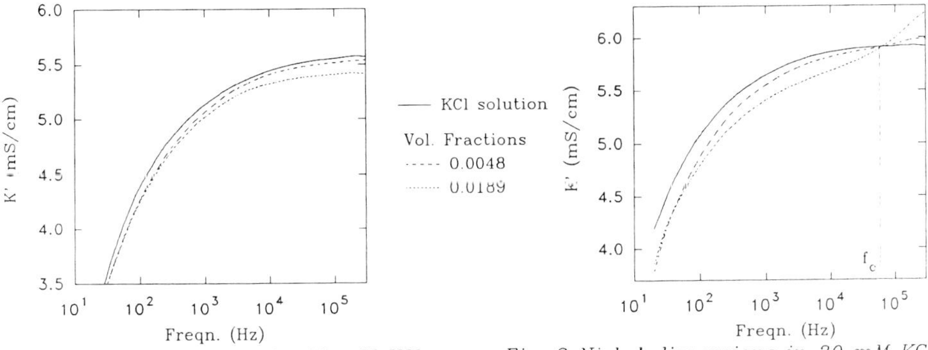

Fig. 2 Glass dispersions in 20 mM KCl

Fig. 3 Nickel dispersions in 20 mM KCl

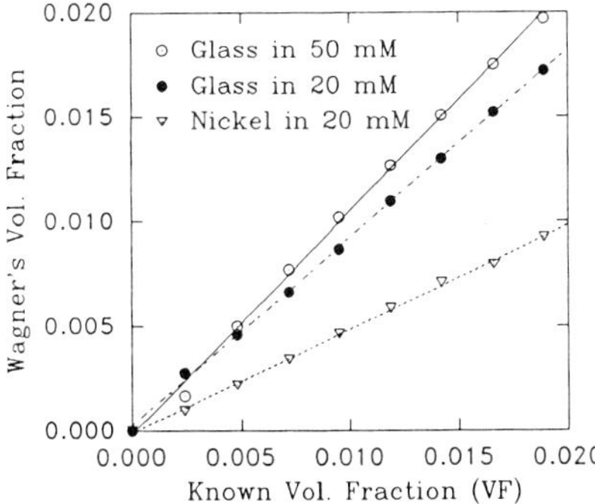

Fig. 4 Glass and Nickel particles separately dispersed

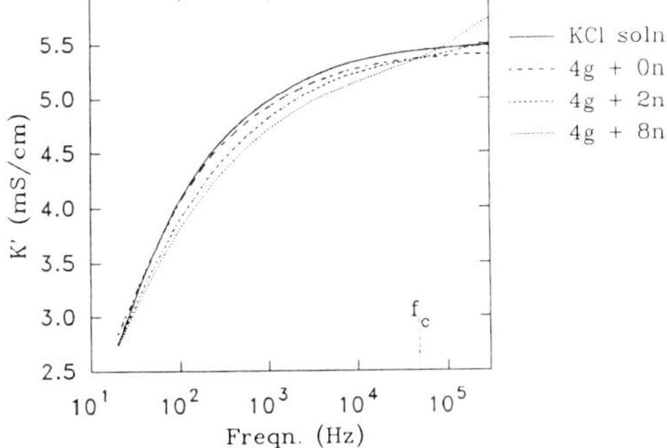

g = n = 1.685ml of glass or nickel added to 700ml of 20 mM KCl solution.

Fig. 5 Adding Nickel to Glass suspension

Fig. 4 is the result of a direct application of eqns. (2) and (3). It shows two plots for glass: one for dispersions in 50 mM KCl, and the other for dispersions in 20 mM KCl. The slope of the line tends to 1 for increasing electrolyte concentration (directly related to conductivity). The plot for nickel is also linear and the slope of the line tends to 1 for a decreasing electrolyte concentration. The computation of eqn. (3) for nickel was at 150 kHz at which the effects due to the shell-phase are virtually eliminated. For glass eqn. (2) was computed at 60 kHz, and it really does not matter if it is computed at 150 kHz since the spectra (Fig. 2) is almost flat beyond 60 kHz.

Fig. 5 shows the conductivity spectra for the case where nickel particles are added to a suspension of a fixed volume of glass. The convergence frequency f_c remained at 60 kHz, thus suggesting that there is no measurable inter-particle type interaction. Consequently, the resultant suspension can be modelled as two lossy capacitors in series. Given a mixture of nickel and glass particles dispersed in 20 mM KCl electrolyte, the respective volume fractions can be estimated by applying the following equations:-

The volume fraction of glass is:
$$\Phi_g = \frac{2(1-\kappa_1/\kappa_{m1})}{2+\kappa_1/\kappa_{m1}} C_g \qquad (4)$$

The volume fraction of nickel is:
$$\Phi_n = \frac{(\kappa_{m1}-\kappa_1+\kappa_2)/\kappa_{m2}-1}{(\kappa_{m1}-\kappa_1+\kappa_2)/\kappa_{m2}+2} C_n \qquad (5)$$

where κ_1 = conductivity of the suspension at 60 kHz;
κ_2 = conductivity of the suspension at 150 kHz;
κ_{m1} = conductivity of the medium at 60 kHz;
κ_{m2} = conductivity of the medium at 150 kHz;
$\kappa_{m1} - \kappa_1 = \kappa_d$ = depression in conductivity at 60 kHz.

For glass, the correction factor is:

$C_g = 152.02 - 53.25\kappa_{m1} + 4.6944\kappa_{m1}^2$ (6) for $5 \le \kappa_{m1} \le 5.7$ mS/cm, and $C_g = 1$ (7) for $\kappa_{m1} > 5.7$ mS/cm.

For nickel, the correction factor is:
$$C_n = -70.666 + 25.921\kappa_{m1} - 2.3093\kappa_{m1}^2 \qquad (8)$$
for $5 \le \kappa_{m1} \le 5.7$ mS/cm.

The eqns. (4) and (5) are a result of modifying eqns. (2) and (3), while exploiting the convergence conductivity behaviour of nickel. The correction factors were empirically derived from the slope variations observed in Fig. 4. The correction factors give a good tolerance in preparing 20 mM KCl solution. They can also be employed in correcting the effect of temperature variations on the conductivity of the electrolyte. Figs. 6 and 7 are plots based on the use of the corrected Wagner's eqns. (4) and (5) for estimating the respective volume fractions of glass and nickel. The accuracy of the estimated results is better than $\pm 10\%$ where the volume fraction of each particle type is less than 2%.

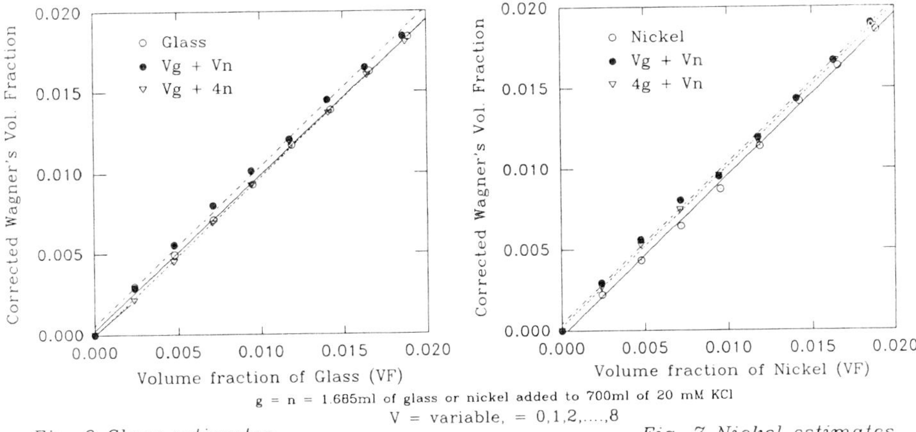

g = n = 1.685ml of glass or nickel added to 700ml of 20 mM KCl
V = variable, = 0,1,2,....,8

Fig. 6 Glass estimates

Fig. 7 Nickel estimates

5. HIGH VOLUME FRACTIONS

High volume fractions (up to 17.5%) of nickel suspensions were also investigated. Fig. 8 shows the resulting conductivity spectra. There is no significant change in the critical frequency, f_c. Application of eqn. (5) did however produce some non-linearities (Fig. 9) for volume fractions greater than 8%. This is to be expected as Maxwell-Wagner's equations were derived for very sparse dispersions. Hanai's equations [4,5] may be able to rectify these non-linearities, and could be investigated in future. It is worth mentioning that there exists a strong linear correlation between the percentage change in conductivities (from 60 to 150 kHz) and the volume fraction of nickel as can be seen in Fig. 10. A further study of this may yield the easiest method for high volume fraction characterization.

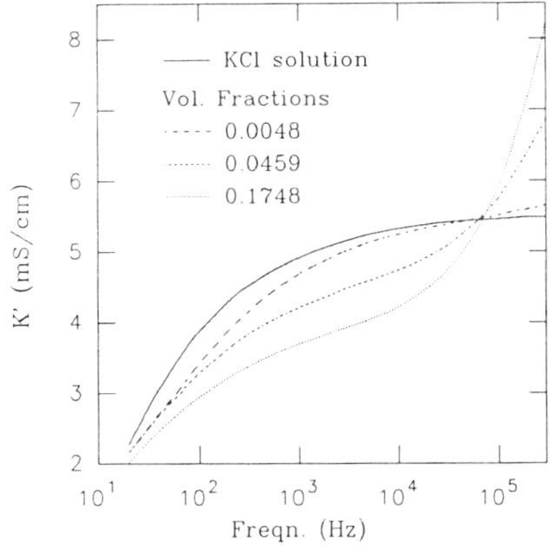

Fig. 8 Nickel dispersions – high vols.

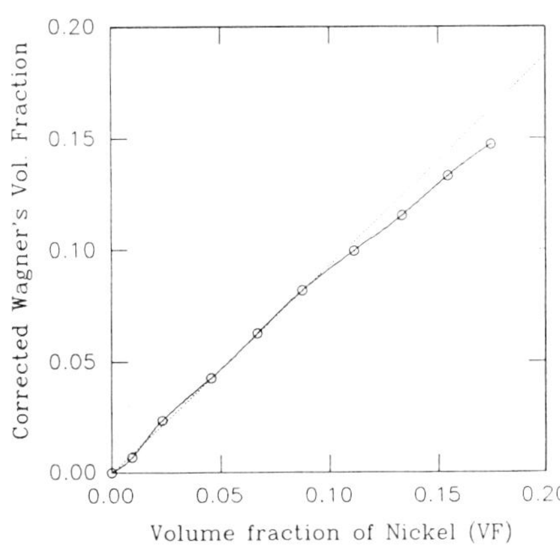

Fig. 9 Nickel estimates – high vols.

Beginning with 17.5% volume fraction of nickel in suspension, glass particles were added until both particle types were each 15%, making a total particle concentration of 30%. Eqn. (4) was employed to estimate how much glass was in suspension. The resulting graph is Fig. 11. A good accuracy can be observed for glass estimates where the glass content is greater than ½ the content of nickel.

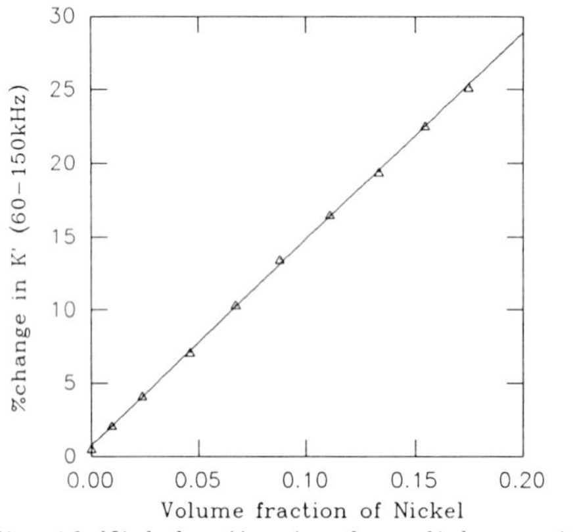

Fig. 10 Nickel estimates from %change in K'

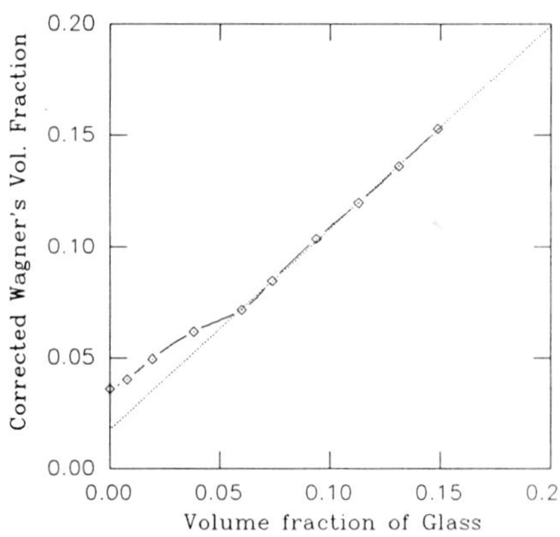

Fig. 11 Glass estimates – high vols.

6. CONCLUSION

The technique reported here allows an estimation of the volume fractions of two dispersed phases co-existing in suspension. It can however be extended to account for dispersions where the number of dispersed phases is greater than two provided that each phase exhibits a distinct conductivity spectra. A good accuracy has been recorded for volume fractions less than 2%, and there is a good prospect of extending the technique well beyond 2% particle concentration. The technique will be further developed to enable tomographic imaging [10] of particulate dispersions.

REFERENCES

1. Birks, J. B., and Hart, *Progress in Dielectrics Vol. 3,* London Heywood & Co. Ltd., London, (1961)
2. Smith, C. P., *Dielectric Behavior and Structure,* McGraw-Hill Book Co., Inc., USA, (1955)
3. Kuhn, A. T., *Techniques in Electrochem., Corrosion and Metal Finishing,* John Wiley & Sons Ltd., New York, (1987)
4. T. Hanai, *Kolloid-Zeitschrift,* **171**, 23 (1960)
5. T. Hanai, *Kolloid-Zeitschrift,* **175**, 61 (1961)
6. Maxwell, J. C., *A Treatise on Electricity and Magnetism Vol. 1,* Dover Publications, Inc., New York, (1954)
7. Wagner, K. W., *Arch. für Electrotechnik,* **2**, 371 (1914)
8. Mayer, D. F. and Saville, D. A. *Journal of Colloid and Interface Science,* **131,** No. 2, 448 (1989)
9. Asami, K. and Irimajiri, A., *Bull. Inst. Chem. Res. Kyoto Univ., Japan,* **63**, No. 2, 259 (1985)
10. Plaskowski, A., Beck, M. S., Thorn, R., Dyakowski, T., *Imaging Industrial Flows,* Submitted to IOP press, U.K., (1993)

Investigation of the effects of flow parameters on electrostatic field distribution in electrical capacitive tomography sensors for flow imaging

S H Khan F Abdullah

Measurement and Instrumentation Centre
Department of Electrical, Electronic and Information Engineering
City University, Northampton Square
London EC1V 0HB, UK

ABSTRACT - This paper investigates the effects of flow parameters, mainly the permittivities of flow components ε_1, ε_2 and flow concentration β on electric field distribution in electrical capacitive tomography (ECT) sensor systems for imaging two-phase flows in real time. The quantitative evaluation of the nonlinear effects of the above parameters is necessary in order to develop and adopt effective image reconstruction algorithms and improve image quality. The investigation is carried out by finite element (FE) modelling of 2D electrostatic fields in a 12-electrode ECT system designed for imaging two-phase flows of different permittivities. Various results produced in the form of field plots and sensor sensitivity distribution between electrodes are compared and analyzed.

1. INTRODUCTION

The electrical capacitive tomography (ECT) is a novel, fast and noninvasive flow imaging technique which relies on the change in capacitances between capacitive sensor electrodes due to the change in permittivities, concentration and spatial distribution of flow components. The cross section of such a 12-electrode ECT system is shown in Figure 1 which consists of 12 capacitive electrodes mounted symmetrically on the surface of the special insulating section of the flow pipeline. The earthed interelectrode radial screens reduce large capacitances between adjacent electrodes and increases measurement accuracy. The outer screen is also earthed and acts as a shield against stray fields. The idea of using such a system for imaging multiphase flows was first proposed by Beck *et al* [1] and the practical implementation of the idea was made in an 8-electrode ECT system by Huang *et al* [2] at UMIST. Today, the area of ECT flow imaging has gained considerable momentum and the recent successful testing of the industrial prototype of such a 12-electrode system [3] has firmly established the potential of such systems in diversified industrial applications [4-6].

The data acquisition in the above system is carried out by imposing a constant potential to one of the electrodes (called the 'active electrode') and measuring the capacitances between this and rest of the electrodes (the 'detecting electrodes', kept at zero potential) by discharging the active electrode and measuring the discharging current (proportional to the unknown capacitance) from the detecting electrodes [7]. For a given pair of electrodes a narrow region of positive sensitivity can be found within which an unit dielectric increase leads to an increase in their capacitance measurements. These sensitivity data (known as the 'permittivity map') for all possible combination of electrodes are obtained by repeated finite element (FE) modelling of electrostatic fields in the electrode system (forward problem) and used together with the above capacitance measurement data to solve the inverse problem of reconstructing the cross sectional image of the flow components [8]. Usually, for a given ECT system this permittivity mapping is done once and the same data are used to reconstruct the images of various flows. In this, it is assumed that the electric field distribution between electrodes does not depend on such flow parameters as the flow regime /pattern (e.g. core, annular, stratified etc.), flow concentration (β) and dielectric properties of flow components (permittivities ε_1 and ε_2). Here, the flow concentration β (in %) is defined as $\beta=(S_2/S_1)100\%$ where, S_2 and S_1 are cross sectional areas of the higher permittivity flow component (ε_2) and flow pipe ($S_1=\pi R_1^2$) respectively. Simulation results show that the electric field distribution in ECT systems depend not only on their various geometric (number of electrodes N, θ, δ_1, δ_2, scgp, scth in Figure 1) and material (ε_{pw}, ε_{fil}) parameters [9, 10, 8] but also on the above flow parameters [11]. In the following sections the effects of flow concentration β and permittivities of flow components ε_1 and ε_2 are investigated by the FE modelling of electric fields in the 12-electrode ECT system shown in Figure 1. For a given ECT system the study of this nonlinear dependence of electric fields on flow parameters is vitally important to justify and develop the appropriate iterative reconstruction algorithms for satisfactory image reconstruction [12, 13], especially in those cases when the nonlinear effects are clearly evident.

2. FINITE ELEMENT MODELLING OF ELECTROSTATIC FILEDS

The interaction of the electrostatic field generated as a result of the potential difference between active and detecting electrodes with the dielectric flow components distributed in the sensing (imaging) area Ω of the electrode system is given by the following Laplace's equation:

$$\nabla.[\,\varepsilon(x, y)\,\nabla\Phi(x, y)\,] = 0 \qquad \text{in } \Omega \qquad (x, y) \in \Omega \tag{1}$$

For a given permittivity distribution of flow components $\varepsilon=\varepsilon(x, y)$ the above eqaution is numerically solved by the finite element method (FEM) [14, 15] which gives the unknown potential distribution $\Phi=\Phi(x, y)$ in the 2D region Ω. In this it is assumed that (a) there is no fringing field effects due to the finite length of electrodes, (b) within the electrode length the flow component distribution does not change spatially along the axial direction of the pipeline, (c) permittivities of flow components remain constant and do not depend on the field, (d) the dielectric medium in Ω is piece-wise homogeneous and isotropic. Figure 2 shows the typical FE model of the electrode system shown in Figure 1. For radially symmetrical flows like core and annular flows only a half model of the electrode system is used for modelling purposes (Figure 3) as in those cases the electric field distribution between electrodes are symmetrical (Figure 4). FE models shown in Figures 2 and 3 are used to simulate various two-phase flow regimes shown in Figure 5. Using pre-processing facilities of commercial FE packages, for example PE2D [16] various model definition operations can be rationalized to obtain different models of

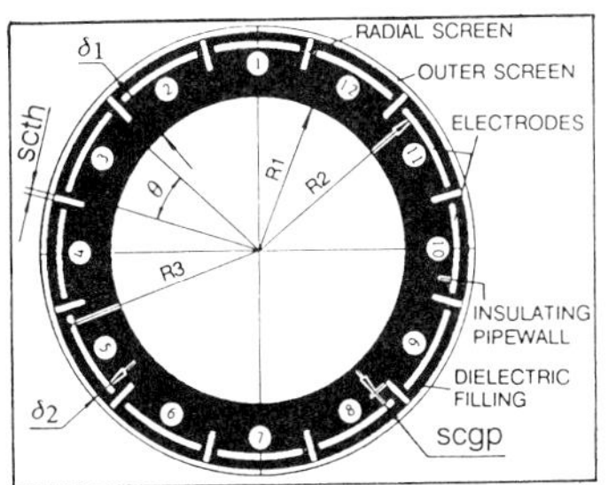

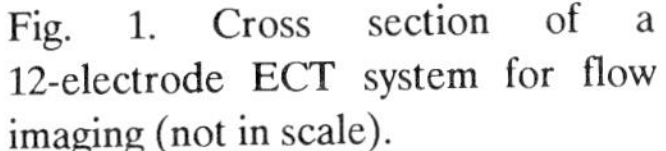
Fig. 1. Cross section of a 12-electrode ECT system for flow imaging (not in scale).

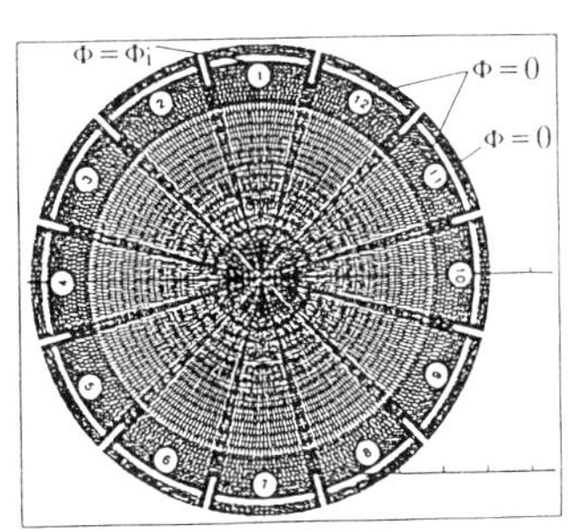

Fig. 2. Finite element model of the 12-electrode ECT system (full model).

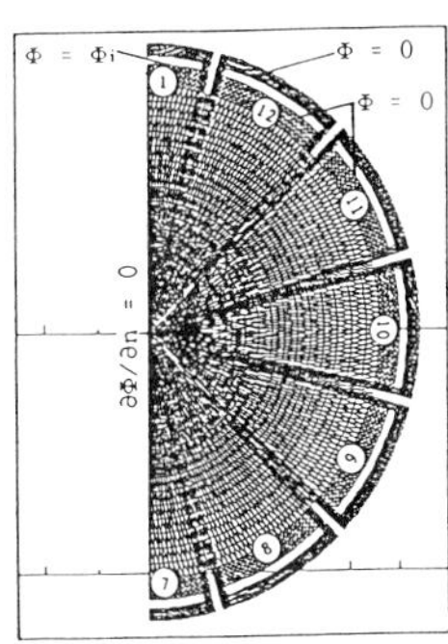

Fig. 3. Finite element model of the 12-electrode ECT system (half model).

the electrode system easily and quickly [17]. Knowing the potential distribution $\Phi=\Phi(x, y)$ in the sensing area Ω field vectors **E**, **D** and capacitances C between electrode pairs are calculated using $C=Q/V$ where, V is the potential difference and Q is the total charge distributed on the detecting electrode. The total charge Q is calculated using Gauss's law [18]: $Q=\int_s \mathbf{D}.\mathbf{ds}$. If the surface of the detecting electrode or a surface very close to it is chosen as the integration surface s then $C=(\int_s D\,ds)/V$ where, D is the magnitude of vector **D**.

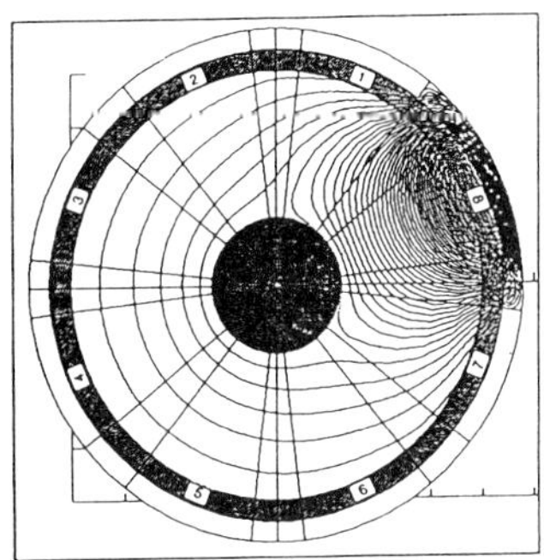

Fig. 4. Equipotential contours in an 8-electrode ECT system showing symmetry in their distribution (core flow).

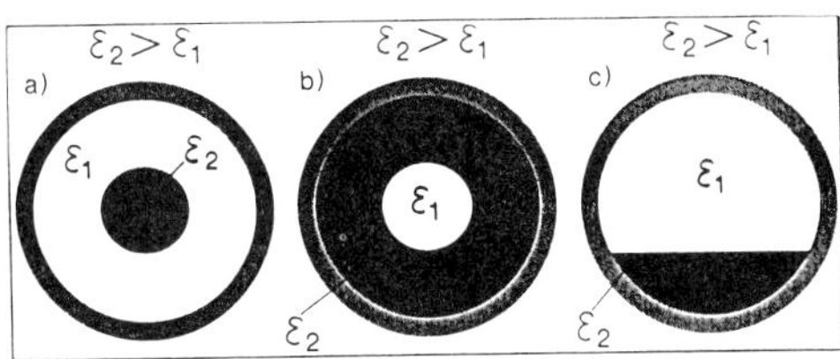

Fig. 5. Various two-phase flow regimes likely to be imaged by ECT systems: a) core flow; b) annular flow; c) stratified flow.

3. EFFECTS OF FLOW PARAMETERS ON ELECTROSTATIC FIELD DISTRIBUTION

Some of the modelling results showing the effects of flow parameters like ε_1, ε_2 and flow concentration β are presented in Figures 6-9. Figure 6 shows the variations in electric field distribution in the electrode system with permittivity ε_2 for 25% annular flow regime. As can be seen from Figures 6a through 6e for a given flow concentration β electric field at the central region of the flow pipe becomes weaker as ε_2 increases resulting in the concentration of field in the pipe wall (Figures 6d, e). Irrespective of the flow regime (core or annular) this 'deflecting' (shielding) effect of the higher permittivity flow component becomes more evident as the flow concentration β increases (Figures

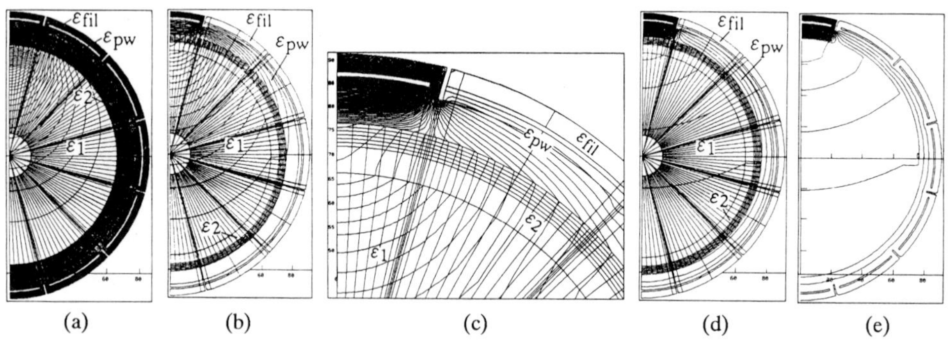

Fig. 6. Effects of permittivity ε_2 on electric field distribution (equipotential contours) in the 12-electrode ECT system (annular flow): (a) $\varepsilon_2=3$, $\delta_1=5$ mm, scgp = 1 mm; (b) $\varepsilon_2=80$; (c) $\varepsilon_2=80$, the area in the vicinity of the active electrode is zoomed to show, in detail the distribution of equipotential contours in the pipe wall and near the electrode; (d) $\varepsilon_2=1000$; (e) $\varepsilon_2=1000$, as in (d) but only the equipotential contours are shown. Other parameters: $R_1=76.2$ mm, $\theta=25.8^{\circ}$, scgp = 1 mm, $\delta_2=5$ mm, $\varepsilon_{pw}=5.9$, $\varepsilon_{fil}=4$, $\varepsilon_1=1$, $\beta=25\%$; for (b)-(e) $\delta_1=10$ mm, scgp = 2 mm (material regions are not shaded to show clearly the potential contours).

7a-e). In the case of high β and ε_2 annular flow the rarity of equipotential contours at the centre of the flow pipe (Figures 7a, b) indicates virtually no interaction of the field with the low permittivity (ε_1) flow component resulting in the uncertainty in its detection and imaging. These 'deflection' and distortion of the electric field distribution with flow concentration β can also been seen for core flows (Figures 8a-e). However, some differences in the nature of these changes in core and annular flow regimes of various concentrations can be revealed by careful comparison of Figures 6 and 8. Quantitatively,

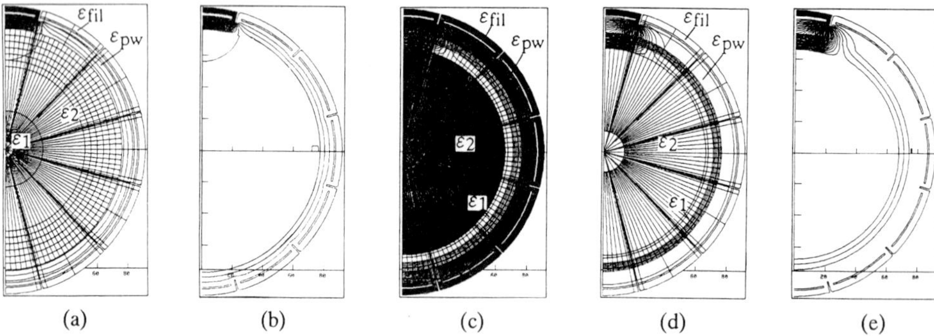

Fig. 7. Effects of high flow concentration β and permittivity ε_2 on electric field distribution (equipotential contours) in the 12-electrode ECT system: (a) annular flow, $\varepsilon_2=1000$, $\beta=90\%$ (material regions are not shaded); (b) as in (a) but only the equipotential contours are shown; (c) core flow, $\varepsilon_2=1000$, $\beta=75\%$; (d) as in (c) (material regions are not shaded); (d) as in (c) but only the equipotential contours are shown. Others parameters: $\delta_1=10$ mm, scgp = 2 mm, $\varepsilon_1=1$; rest of the parameters are the same as in Figure 6.

some of the above effects are presented in Figures 9a and 9b which show the variation of sensitivity distribution between sensor electrodes with flow concentration β and permittivity ε_2. Here, the sensitivity is defined in terms of normalized capacitances C^n_{ij} between electrode pairs i-j: $C^n_{ij}=(C^{\beta}_{ij}-C_{0ij})/(C^f_{ij}-C_{0ij})$ where, C_{0ij}, C^f_{ij} and C^{β}_{ij} are respectively capacitances between i-j when the flow pipe is empty ($\beta=0$, $\varepsilon=\varepsilon_1=1$), full of material $\varepsilon=\varepsilon_2$ ($\beta=100\%$) and has a two-phase flow of given concentration $\beta\neq 0$ ($\varepsilon_1\geq 1$,

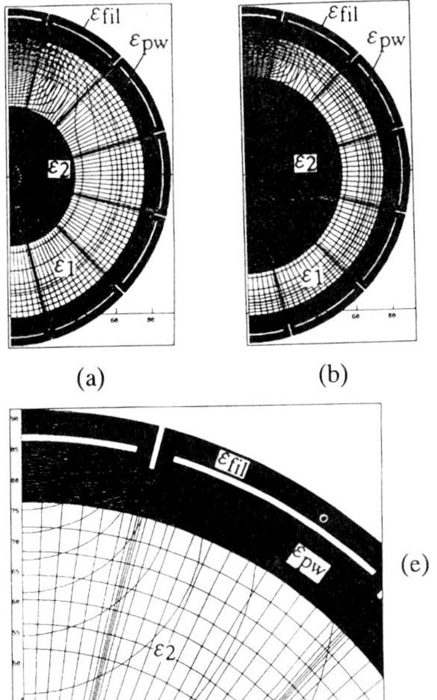

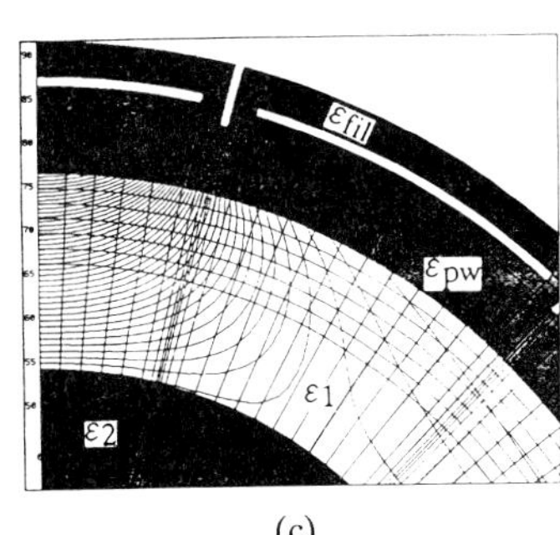

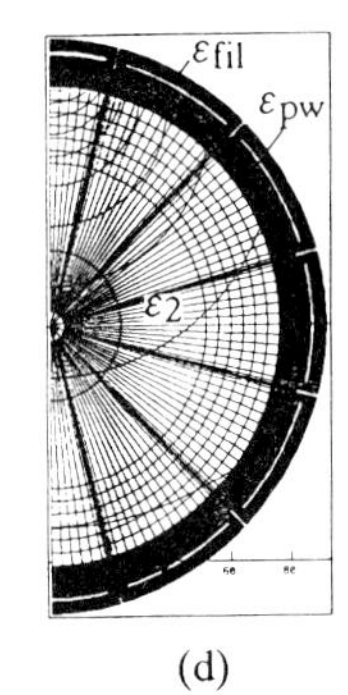

Fig. 8. Effects of flow concentration β on electric field distribution (equipotential contours) in the 12-electrode ECT system (core flow): (a) $\beta=25\%$; (b) $\beta=50\%$; (c) $\beta=50\%$, the area in the vicinity of the active electrode is zoomed to show, in detail the distribution of equipotential contours; (d) (e) $\beta=100\%$ (full pipe); regions with permittivity ε_2 are not shaded. Other parameters: $\delta_1=10$ mm, scgp=2 mm, $\varepsilon_1=1$, $\varepsilon_2=80$; rest of the parameters are the same as in Figure 6.

$\varepsilon_2>1$). It has been found that in order to ensure satisfactory image quality it is necessary to limit C^n_{ij} so that $0\le C^n_{ij}\le 1$ [8]. Considering this, the negative effect of increasing the permittivity ε_2 on C^n_{ij} is evident from Figures 9a and 9b. Here annular flow regime is chosen as an example as it usually gives higher normalized capacitances in comparison with the core flow [11, 19].

(a)

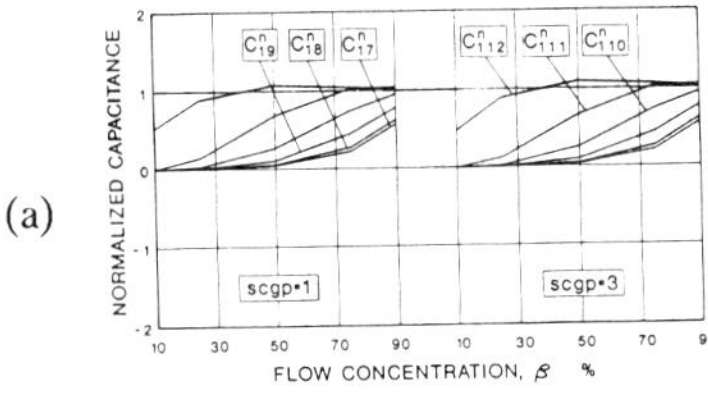

(b)

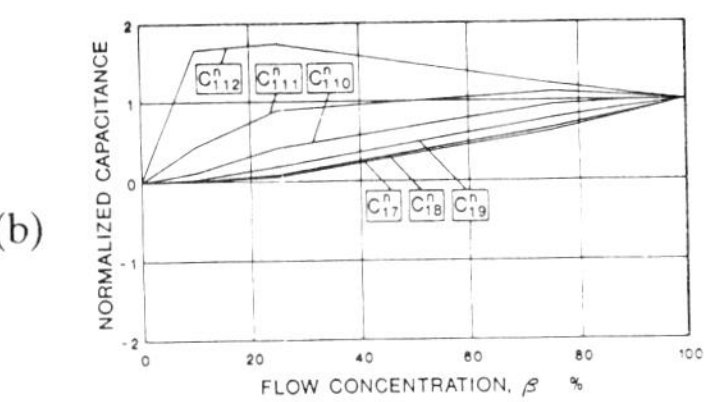

Fig. 9. Effects of permittivity ε_2 and flow concentration β on normalized capacitances (sensitivity distributions) in the 12-electrode ECT system (annular flow): (a) $\varepsilon_2=3$; (b) $\varepsilon_2=80$. Other parameters: $\delta_1=5$ mm, $\varepsilon_1=1$; for (a) $R_1=76.2$ mm, $\theta=25.8^{o}$ and for (b) $R_1=68$ mm, $\theta=24^{o}$, scgp=2 mm.

4. CONCLUSIONS

It has been shown that various flow parameters like the flow concentration β and permittivities of flow components have considerable effect on the electric field distribution in ECT systems. This, in turn affects the sensitivity (normalized capacitances) between sensor electrodes which ultimately determines the image quality. High permittivity and high concentration core and annular flows tend to 'deflect' the field away from the central region of the flow pipe. This further reduces the sensitivity of the inherently low sensitivity central region and for annular flows this would make the lower permittivity flow component at the centre of the flow pipe difficult to detect (image).

5. ACKNOWLEDGEMENTS

The authors would like to thank SERC for financing the work under an SERC Grant, Professor M. S. Beck and Dr. C. G. Xie from UMIST and Dr. S. M. Huang from Schlumberger Cambridge Research Limited, UK for their long standing collaboration.

6. REFERENCES

[1] Beck M. S., Plaskowski A. and Green R. G.: 'Imaging for measurement of two-phase flows', in VERET, C. (Ed.): 'Flow Visualization IV', Proceedings of the 4th International Symposium on Flow Visualization', 26-28 August, 1986, (Hemisphere Publishing Corporation), pp. 585-588.

[2] Huang S. M., Plaskowski A. B., Xie C. G. and Beck M. S.: 'Tomographic imaging of two-component flow using capacitance sensors', J. Phys. E: Sci. Instrum., 1989, **22**, pp. 173-177..

[3] Beck M. S., Hoyle B. S. and Lenn C. P.: 'Process tomography in pipelines', SERC Bulletin, 1992, vol. 4, no. 10, pp. 24-25.

[4] Dickin F. J., Hoyle B. S., Hunt A., Huang S. M., Ilyas O., Lenn C., Waterfall R. C., Williams R. A., Xie C. G. and Beck M. S.: 'Tomographic imaging of industrial process equipment: Techniques and applications', IEE Proceedings-G, February 1992, vol. 139, no.1, pp. 72-82.

[5] Plaskowski A., Bukalski P., Habdas T. and Skolimowski J.: 'Tomographic imaging of process equipment - application to pneumatic transport of solid material', *in* 'Sensors: Technology, systems and applications', Ed. K. T. V. Grattan (Bristol: Adam Hilger, 1991), pp. 221-226.

[6] Hammer E. A. and Nordtvedt J. E.: 'The application of a venturi meter to multiphase flow measurement', *in* 'Sensors: Technology, systems and applications', Ed. K. T. V. Grattan (Bristol: Adam Hilger, 1991), pp. 233-238.

[7] Huang S. M., Xie C. G., Thorn R., Snowden D. and Beck M. S.: 'Design of sensor electronics for electrical capacitance tomography', IEE Proceedings-G, 1992, **139**, (1), pp. 83-88.

[8] Xie C. G., Huang S. M., Hoyle B. S., Thorn R., Lenn C., Snowden D. and Beck M. S.: 'Electrical capacitance tomography for flow imaging: system model for development of image reconstruction algorithms and design of primary sensors', IEE Proceedings-G, 1992, **139**, (1), pp. 89-98.

[9] Khan S. H. and Abdullah F.: 'Computer aided design of process tomography capacitance electrode systems for flow imaging', *in* 'Sensors: Technology, systems and applications', Ed. K. T. V. Grattan (Bristol: Adam Hilger, 1991), pp. 209-214.

[10] Khan S. H. and Abdullah F.: 'Validation of finite element modelling of multielectrode capacitive system for process tomography flow imaging', *in* 'Tomographic Techniques for Process Design and Operation', edited by M. S. Beck *et al*, CEC Brite Euram Computational Mechanics Publications (1993).

[11] Khan S. H. and Abdullah F.: 'Finite element modelling of electrostatic fields in process tomography capacitive electrode systems for flow response evaluation', Presented at the IEEE International Magnetics Conference, INTERMAG 93, Stockholm, 13-16 April, 1993.

[12] Chen Q., Hoyle B. S. and Strangeways H. J.: 'Electric field interaction and an enhanced reconstruction algorithm in capacitance process tomography', *in* 'Tomographic Techniques for Process Design and Operation', edited by M. S. Beck *et al*, CEC Brite Euram Computational Mechanics Publications (1993).

[13] Abdullah M. Z., Quick O. C. and Dickin F. J.: 'Quantitative algorithm and computer architecture for real-time image reconstruction', *in* 'Tomographic Techniques for Process Design and Operation', edited by M. S. Beck *et al*, CEC Brite Euram Computational Mechanics Publications (1993).

[14] Silvester P. P. and Ferrari R. L.: 'Finite elements for electrical engineers' (Cambridge: Cambridge University Press, 1990) 2nd Edition.

[15] Zienkiewicz O. C. and Morgan K.: 'Finite elements and approximation' (New York: John Wiley & Sons Inc., 1983).

[16] 'The PE2D reference manual', Vector Fields Limited (Oxford: 1991).

[17] Khan S. H. and Abdullah F.: 'Finite element modelling of multielectrode capacitive systems for flow imaging', IEE Proceedings-G, 1993 (to be published).

[18] Ramo S., Whinnery J. R. and Duzer T. V.: 'Fields and waves in communication electronics', (New York: John Wiley & Sons, Inc., 1965), pp. 754.

[19] Khan S. H., Xie C. G. and Abdullah F.: 'Computer modelling of process tomography sensors and systems', *in* 'Process Tomography: Principles, Techniques and Applications' (to be published).

Development of capacitance measurements towards tomographic imaging of flames

R.He[+], C.M.Beck*, R.C.Waterfall[+], M.S.Beck[+]
+ Department of Electrical Engineering and Electronics, UMIST, PO Box 88, Manchester, M60, 1QD, U.K.
* Shell Research Ltd, Thornton Research Centre, Chester, CH1, 3SH, U.K.

ABSTRACT: A feasibility study using capacitance measurements towards tomographic imaging of flames in an internal combustion engine is presented. The experimental results have shown that flame position can be located and flame size measured, and absence of flame can be detected. The tests indicate that electrical capacitance measurements of a combustion zone could be used to characterise the physical and chemical process in an internal combustion engine. It is envisaged that Capacitance Tomography may provide a means of non-intrusive imaging of the combustion processes.

1. INTRODUCTION

In the last 15 years, a wide variety of new experimental diagnostic techniques have been developed to facilitate in-depth investigation of engine combustion phenomena. In particular, the development of laser scattering techniques has facilitated non-intrusive measurements with high spatial resolution in both premixed [1] and diffusion [2] flames. Doping of fuel-oxidant mixture with specific photosensitive molecules has been used for imaging of combustion and flow processes [3][4]. There can be no doubt that laser-based tomographic diagnostics have a place in engine combustion research. However, laser-based techniques can only be used in an optically accessible research engine and the instrumentation is usually expensive to install and operate. In particular, with the advent of 4-valve cylinders, optical access through the cylinder head has become difficult.

In this paper, a novel method of using a capacitance tomographic technique to characterise the combustion phenomena in an internal combustion engine is proposed. Experimental results show that the capacitance sensor is a convenient, reliable and inexpensive method for measuring both flame position and flame size, as well as identifying no-flame in an internal combustion engine. The spatial resolution is a function of electrode geometry, and the temporal resolution is compatible with that required for many engine studies.

2. FLAME MEASUREMENT BASIS

Capacitance tomography is a well-established technique for non-invasively imaging the cross-section of process pipelines and vessels which contain multiple components with different dielectric constants [5][6]. The technique is based on different components having different permittivities, enabling a direct measurement of component volumetric concentration to be made.

In an internal combustion engine, ions are created by the process of chemi-ionisation during the combustion of hydrocarbon fuels. This process has been shown by many previous workers including Clements and Smy [7]. Ionisation also results as a by-product of the combustion process. Typically one pair of ions is produced for every million carbon atoms burned [8]. The combustion products are therefore only weakly ionised, with the highest ion concentration (10^{18}-10^{19} ions/m^3) in the flame reaction zone. Outside the reaction zone, the ion concentration decays rapidly to values around 10^{14} ions/m^3 [7]. Therefore, the flame reaction zone and the no-flame zone will have different relative permittivities and this will give a proportional change in capacitance.

3. EXPERIMENTAL DETAILS

The experimental flame measurement system used is shown in figure 1. and figure 2. The sensor electronics, based on the earlier work by W.Q.Yang *et al* [9] and S.M.Huang [10] is applied to the measurements of flames within an idealised bounded internal combustion engine model. The combustion cylinder is of a standard configuration, except that the simulated cylinder head has been machined to support the capacitance electrodes mounted at equal spacing around the circumference of the insulation wall, made of bakelite. The electrical field from the sensors is inhomogeneous but symmetrical around the cylinder axis. The sensitivity to variation in flame size would be highest close to the adjacent electrodes. A laboratory Bunsen burner was used in this study, where the fuel emerges from a 1-cm diameter metal screen within an annular air feed. The premixed or diffusion (depending on the air inlet port setting) flame has a luminous region extending from a height of about 3 cm to 10 cm above the top of the burner, depending on the gas tap setting. All the work described in this paper was performed using natural gas, predominantly methane, drawn from a single source. A methane flow meter was used to measure the gas flow. Particular attention has been paid to locate the flame position and to measure the relationship between the relative capacitance change with the flame size. Additional work has been undertaken to examine the effect of varied air/fuel ratio and whether the no-flame condition is detectable. The above will provide information on the potential spatial and temporal resolution of such measuring systems.

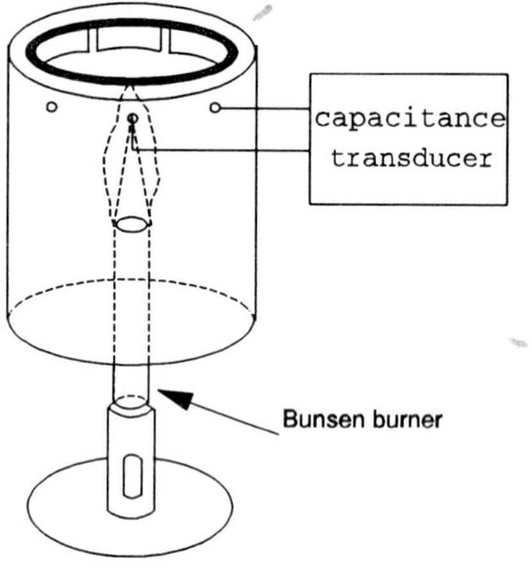

Figure 1. Side view of the flame measurement system

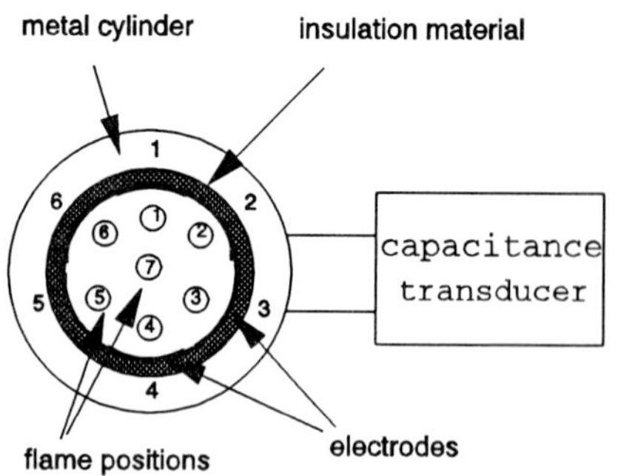

Figure 2. Top view of the flame measurement system

4. RESULTS AND DISCUSSION

Figure 3 shows the capacitance change between adjacent electrodes as a function of flame position when electrode 4 was supplied with a positive potential V_C (which is referred to as the source electrode), and the capacitance between it and electrode 5 (which is refered to as the detection electrode) was measured. Capacitance change is defined as that relative to the capacitance measured with no flame present. Electrodes 1, 2, 3 and 6 were earthed and have a guarding function. The flame size was about 4 cm high above the burner head. Figure 2 shows the flame positions within the chamber. It is apparent from the graph that there is a measurable increase in the capacitance between the electrodes system when the flame is present, even if the flame is not nearest the measuring electrodes.

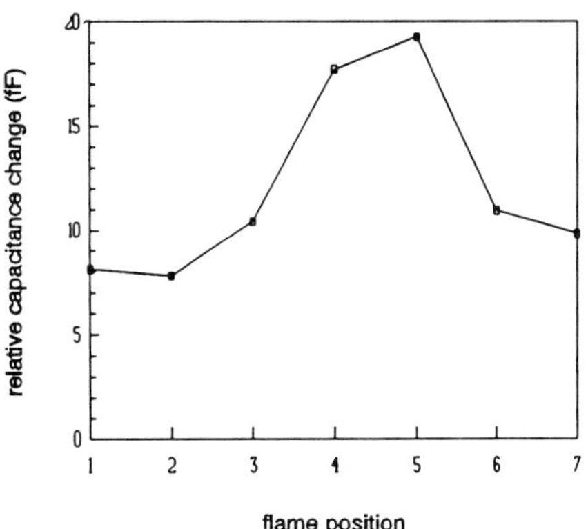

Figure 3. Capacitance change between adjacent electrodes related to flame position

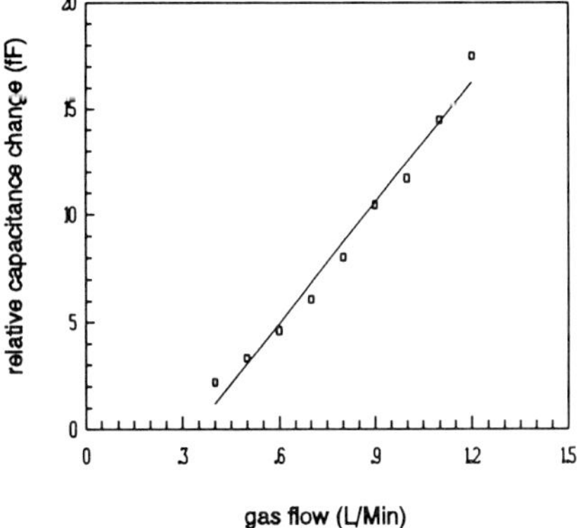

Figure 4. Capacitance change between adjacent electrodes with the change of gas flow

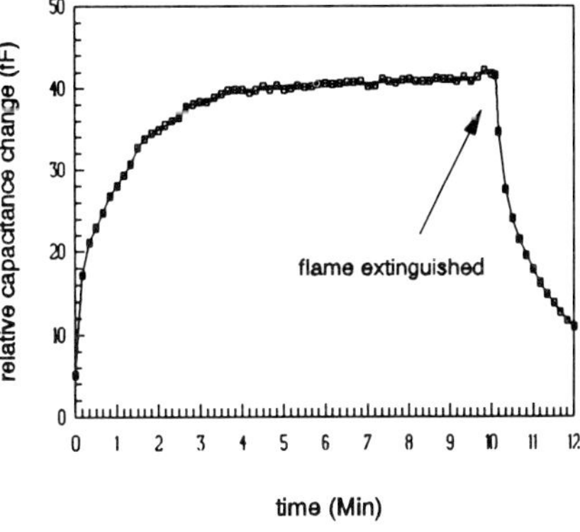

Figure 5. Capacitance change between adjacent electrodes as flame is introduced then extinguished

Figure 4 shows the capacitance change between adjacent electrodes with the increase of the gas flow. Clearly, the capacitance change is proportional to the gas flow and has a linear relationship. It can be seen from the graph that, the minimum detectable gas flow is about 0.4 L/Min. Below this flow, the burner flame cannot be sustained. The relative capacitance change could also be expected to vary with air/fuel ratio and the flame character. As the air inlet port of the Bunsen burner was closed, so the air/fuel ratio of the premixed flame was reduced and the flame colour changed from cone blue to soft blue. A decrease in the relative capacitance change was detected. With the air inlet port closed off, a yellow diffusion flame results, and compared with the premixed flame, a smaller relative capacitance change was detected. The explanation of the effect could be that flame temperature rises with an increased air/fuel ratio, thus raising the ionisation level in the reaction zone, resulting in an increase in capacitance.

Figure 5 shows the capacitance change against time after putting out the flame. It can be seen that the capacitance decreases abruptly as the flame goes out, indicating the no-flame condition. It should be pointed out that as the insulation wall was made of bakelite material, which cannot resist high temperature. When the gas flow-rate was increased above a certain level, the insulation material was charred and became conductive, yielding erroneous results. Subsequently, machinable ceramic material, which has an upper temperature limit of approximately 1000°C, has been used to replace the bakelite.

5. CONCLUSIONS

Our aim is to image the combustion phenomena inside an internal combustion engine every 1° of crank angle at up to 36,000 frames/s. The preliminary results show that the measurement is sensitive to flame position, and the flame properties within the combustion chamber can cause a change in this measurement, using the capacitance transducer designed and built by the Process Tomography Group at UMIST. These tests indicate the potential of such a technique when applied to combustion phenomena. In a logical extension to this feasibility study, the capacitance tomography technique might be able to provide a means of non-intrusive imaging of combustion processes and thereafter to monitor flame propagation in an internal combustion engine. However, the combustion phenomenon inside an internal combustion engine is a complicated dynamic process and many factors need to be carefully considered. For example, the temperature and chemi-ionisation concentration levels will change during the combustion process. Both the resistive and reactive components of the impedance need to be measured.

ACKNOWLEDGEMENT

This project is funded by Shell Research Ltd., Thornton Research Centre. Dr C.G.Xie and W.Q.Yang are acknowledged for their helpful discussions.

REFERENCES

1.Goix, P., Paranthoen, P., and Trinite, M., *Combustion and Flame*, **81**, pp 229-241 (1990)

2.Stepowski, D, Cabot, G, *Combustion and Flame*, **88**, pp 296-308 (1992)

3.Ziegler, G.F.W., Zettlitz, A., Meinhardt, P., Herweg, R., Maly, R. and Pfister, W., *SAE Tech. Pap.*, 881634 (1988)

4. Arnold,A., Becker, H., Suntz, R., Monkhouse, P. and Wolfrum, J., *Optics Letters,* **15**, No. 15, pp 831-833 (1990)

5. Xie, C.G., Huang, S.M., Hoyle, B.S., Thorn, R., Lenn,C., Snowden, D and Beck, M.S., *IEE Proceedings-G*, **139**, No., 1, pp 89-98, (1992)

6. Huang, S.M., Xie, C.G., Thorn, R., Snowden, D and Beck, M.S., *IEE Proceedings-G*, **139**, No., 1, pp 83-88, (1992)

7. Clements, R.M., Smy, P.R., *Journal of Applied Physics*, **47**, pp 505-509, (1976)

8.Lawton, J., Weinberg, F.J., "*Electrical Aspects of Combustion*", pp 221-225 Carendon Press (1969)

9. Yang, W.Q., Stott, A.L., *ECAPT*, pp 74-78, (1992)

10. Huang, S.M., PhD thesis, University of Manchester, (1986)

Design and fabrication of segmented capacitance level sensor for an oil separation tank

W.Q.Yang, H.X.Wang, C.G.Xie, M.R.Brant and M.S.Beck
Process Tomography Group, Department of Electrical Engineering and Electronics
UMIST P O Box 88, Manchester M60 1QD, UK

ABSTRACT: A segmented capacitance sensor has been designed and fabricated for measuring multi-interface levels in an oil separation tank. The sensor can detect not only the multiple interfaces between different materials, such as gas/oil, oil/water and water/sludge, but foam layer as well. This paper describes the design specifications and the fabrication methods.

1. OVERVIEW OF CAPACITANCE LEVEL SENSORS

Capacitance sensors are widely used in many laboratory and industrial measurements, including displacement, pressure, moisture, component concentration, etc. [1]. About 40 years ago, John Fielden and Georg Endress first utilised capacitance to measure level. Since then, more and more applications of capacitance level sensors have been found [2].

Compared with other types of level sensors, such as sight, force, pressure, ultrasonic, etc., the capacitance sensors have the following advantages [3]:

(1) no moving parts;
(2) simple and rugged;
(3) corrosion resistant and easily cleaned;
(4) appropriate temperature and pressure range, explosion-proof;
(5) satisfactory accuracy and resolution, minimum maintenance.

The main problem of a conventional capacitance sensor is that it needs temperature compensation because the dielectric constants of the measured materials are sensitive to temperature. For example, if temperature changes from 0°C to 100°C, the dielectric constant of water will change dramatically from 88 to 48 [3].

Usually a cylindrical capacitance sensor is used to measure the liquid level in a vessel. A capacitance probe is mounted in the centre of the vessel as one electrode and the wall of the metallic vessel acts as another electrode. The variation of the liquid level results in capacitance change. This is the simplest and most common version. However, this kind of capacitance sensor suffers from the following limitations:

(1) It requires a prior knowledge of the dielectric constant of the measured material which is often difficult to know.
(2) If there exist more than one interfaces in the vessel, i.e. more than two kinds of materials inside, the cylindrical sensor can not locate the levels at all.

In recent years, multi-electrode capacitance level sensors have been developed, e.g. the parallel segmented capacitance sensor of the Shell Company in Holland [4], the multi-electrode single-plate capacitance sensor of UMIST [5] and the multiplexed capacitance array sensor of the Manchester Metropolitan University [6].

Evidently, by using the multiple electrode capacitance sensors, it is easy to locate multiple interfaces. More importantly, only relative differences of the dielectric constants of the measured materials are needed and any absolute variations of the dielectric constants will not cause significant measurement error.

2. BACKGROUND AND REQUIREMENTS

The separation tank in an offshore oil field contains gas, oil, water and sludge. Usually there are foam layer between gas and oil, and emulsion layer between oil and water (see Fig.1). To make full use of the separation equipment, it is necessary to locate these interfaces of gas/foam, foam/oil, oil/emulsion, emulsion/water and water/sludge so that the separation process can be monitored and controlled.

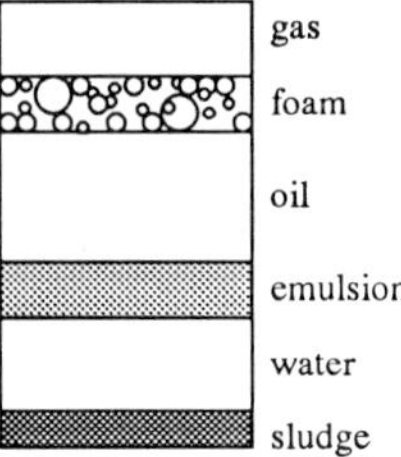

Fig. 1 *Materials in Separation Tank*

If a capacitance sensor with multiple electrodes is used, it is relatively easy to identify what material is between each pair of electrodes because different materials have different dielectric constants.

The resolution of a parallel multi-electrode sensor is at least the height of one electrode. In practice, if one level lies in some pair of electrodes (see Fig.2), the precise position of the level can be calculated by interpolation.

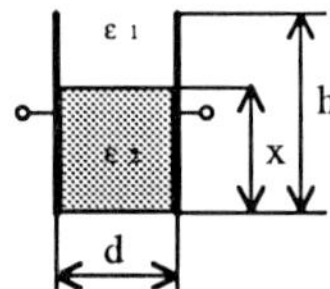

Fig.2 *One Pair of Electrodes*

$$C = \frac{\varepsilon_o wh}{d}[\varepsilon_1 + (\varepsilon_2 - \varepsilon_1)\frac{x}{h}] \quad (1)$$

where, C -- the measured capacitance, pF;

ε_o -- the permittivity of free space, 8.85 pF/m;

ε_1, ε_2 -- the dielectric constants of two different materials;

d -- the distance between two electrodes, m;

w, h -- the width and height of electrode respectively, m;

x -- the level position, m.

In an oil separation tank, foam often builds up and severe foaming can reduce the effective capacity of the tank, even cause shut-down of the separation equipment. Therefore, it is necessary to detect the foam layer so that the anti-foaming injection facilities can operate. Unfortunately, there is no any foam sensor available so far [7].

Since the structure of foam is inherently unstable, with bubbles continually forming, moving and collapsing, it is possible to detect the presence of foam by measuring the capacitance fluctuation from the electrodes immersed in foam layer if a capacitance sensor is used [5].

For oil separation tank application, the sensor size is often limited by the size of available installation holes, generally 2" in diameter. In addition, the level sensor should be corrosion-resistant (e.g. H_2S), heat-resistant (up to 80°C), pressure-resistant (up to 35 PSI) and intrinsically safe [8].

Summaring, the capacitance level sensor for an oil separation tank should have the following characteristics:

(1) locating multi-interface levels;
(2) detecting foam layer;
(3) fitting to 2" installation hole;
(4) corrosion, heat, pressure resistant and intrinsically safe.

3. SENSOR DESIGN AND FABRICATION

The proposed capacitance level sensor utilises parallel-plate structure as shown in Fig.3 because it has good linearity and resolution compared with single-plate structure [5]. The excitation plate has many segmented electrodes on it and the detection plate has only one electrode. The excitation signal will be applied on the excitation electrodes in sequence and the detection electrode will be connected to a charge amplifier. The capacitance between one excitation electrode and the detection electrode may be roughly considered as that between two electrodes of the same size as the excitation electrode.

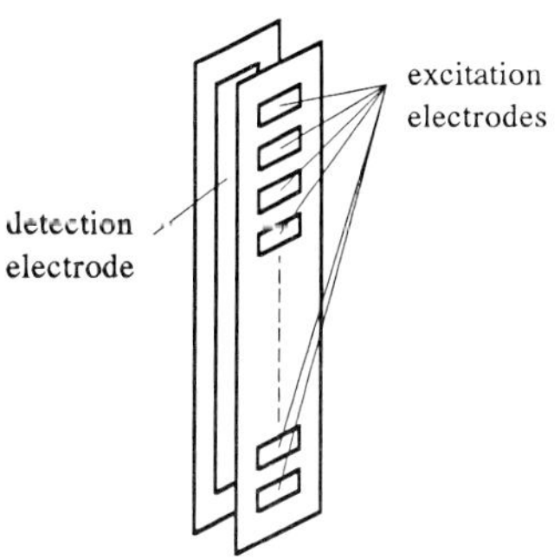

Fig.3 *Parallel-plate Structure*

Theoretically, if an oil separation tank contains 6 types of materials as shown in Fig.1, i.e. 5 interfaces need to be located, it may be possible to obtain a solution by using 5 excitation electrodes. However, this is often impractical because

(1) the prior knowledge of dielectric constants of all materials is still required,
(2) if two interfaces lie in a same excitation electrode, it is impossible to identify them,
(3) if the excitation electrodes are too big, the fluctuation signal representing foam information will be smoothed or filtered out and the capacitance sensor can not detect the foam layer.

Therefore, the number of excitation electrodes should be much more than the number of the measured materials and the excitation electrodes should be small enough to detect the signal fluctuation caused by foam.

Obviously, if the overall length of the sensor is fixed, the height of each excitation electrode will be reduced with an increase in the number of excitation electrodes. However, the more excitation electrodes, the smaller standing capacitance between each excitation electrode and the detection electrode and therefore a more sensitive capacitance measuring circuit is required. In addition, more excitation electrodes require more multiplexers to address each excitation electrode. So there is a trade-off between the number of excitation electrodes and the complexity of the capacitance measuring circuit.

The width of the excitation electrodes and the detection electrode, and the spacing between them are limited by the diameter of the installation hole (2").

Based on above considerations, a prototype sensor with 64 excitation electrodes has been designed and fabricated. Its main dimensions are listed in Table 1.

***Table 1** Dimensions of Sensor*

width of excitation and detection electrodes, (w)	20 mm
height of each excitation electrode, (h)	10 mm
gap between two adjacent excitation electrodes	2 mm
height of detection electrode	808 mm
spacing between excitation electrodes and detection electrode, (d)	22 mm

Ignoring the fringe field, the standing capacitance between each excitation electrode and the detection electrode for air can be calculated by equation (1), 0.08 pF. Now the Process Tomography Group in UMIST has developed a stray-immune high frequency capacitance measuring circuit with a sensitivity of 2 V/pF and a resolution of 0.035 fF [9]. So the capacitances of the sensor can be easily measured.

Since the measured capacitances are quite small, the stray capacitances have to be seriously considered. A five-layer PCB has been built for the excitation plate. The function of each layer is listed in Table 2.

***Table 2** Five-layer PCB for Excitation Electrodes*

layer	**function**
1	segmented excitation electrodes
2	earthed screen
3	analogue signal wires
4	earthed screen
5	digital signal and power supply wires

The analogue signal wires are sandwiched between the two layers of earthed screens. The stray-immune capacitance measuring circuit only measures the capacitances between the excitation electrodes and the detection electrode and the stray capacitances have no effect on the measurement [9].

To address 64 excitation electrodes, four multiplexers are mounted on the five-layer PCB, with each multiplexer addressing 16 channels. The circuit of the sensor is shown in Fig.4.

The two PCBs, with excitation and detection electrodes, are housed in a stainless steel tube (1.87" outer diameter). Epoxy resin is used to encapsulate these PCBs and to make the stainless steel tube and the PCBs into an integral element. Epoxy resin will prevent the PCBs from corrosion caused by sea water, oil and some chemicals.

The housing tube has the following features:

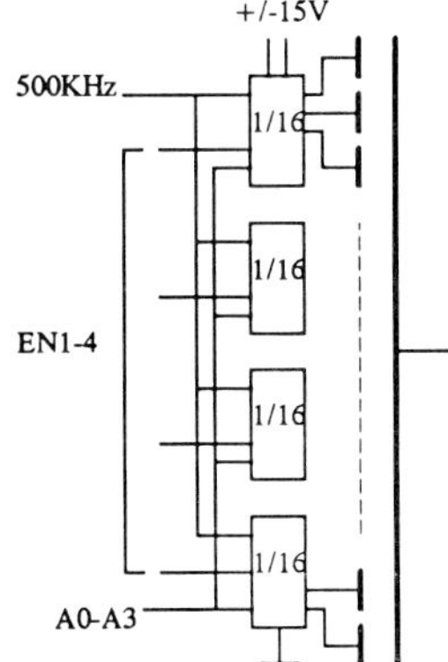

Fig.4 Circuit of Sensor

(1) sturdy support body to fix the two parallel plates and keep the distance between the two plates constant;
(2) compact construction to fit 2" installation hole;
(3) streamline outline to reduce the impulsive force of flow and fouling by some viscous materials;
(4) earthed metal shield to screen external electromagnetic interference.

A flange will be mounted on the top of the sensor in order to fit the sensor to a practical separator. A pressure-resistant connector will be fixed on the flange because the pressure inside the separator is often much higher than atmospheric pressure.

4. RESULTS AND DISCUSSION

Some experiments have been carried out in the laboratory. The sensor has been put into a container with oil and water for four months. No leakage into the PCBs has been found.

Although the stray capacitances between the excitation electrodes and the earthed screens are quite large (about 200 pF), it has no effect on the measurement of the capacitances between the excitation electrodes and the detection electrode (0.08 pF for air), showing the effectiveness of the five-layer PCB and the stray-immune circuit design.

Some quantitative results have been obtained. Fig.5 shows one of the results where the interfaces of air/oil and oil/water lie in electrode 33 and 44, respectively. Evidently it is quite easy to identify what material each excitation electrode is facing in terms of the measured capacitances [10].

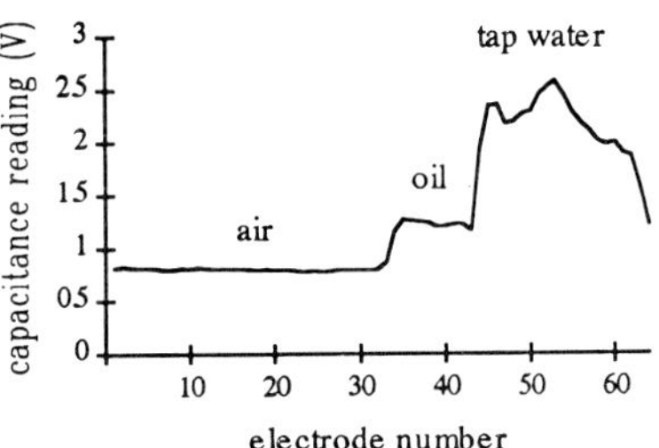

Fig.5 Capacitance Reading

For intrinsic safety, a revised sensor circuit (refer to Fig.4) may be necessary. Three kinds of isolation will be employed [8]:

(1) Since the frequency of the analogue signals is too high (500 KHz) to use usual barriers, only transformers can be used for them.
(2) Opto-electronic isolators are suitable for digital signal isolation.
(3) Zener barriers can be used for the power supply of the multiplexers inside the sensor.

5. CONCLUSION

The segmented capacitance level sensor is suitable for multi-interface level measurements for oil separation tank. It can not only locate the interface of gas/oil, oil/water or water/sludge in terms of their different dielectric constants but also detect foam layer by analysing the capacitance fluctuation. The fabrication methods employed have been experimentally proven, e.g. five-layer PCB, epoxy resin encapsulation, steel tube housing, etc. For the more stringent environment in an offshore separator, ceramic insulation for the capacitance level sensor is being developed.

ACKNOWLEDGEMENT

The authors are grateful to the Marinetech North West for supporting this work.

REFERENCES

[1] S.M.Huang, A.L.Stott, R.G.Green and M.S.Beck, Electronic Transducers for Industrial Measurement of Low Value Capacitances, *J. Phys. E: Sci. Instrum.* **21** 1988, pp 242-250

[2] N.Watmough, Capacitance for Level Measurement, *Measurement + Control,* **21** 1988, pp 15-19

[3] C.H.Cho, Measurement and Control of Liquid Level, *Publishers Creative Services, Inc.* 1982

[5] User Manual, Level Gauging System for Service Stations, B.V.ENRAF-NONIUS DELFT, Holland, 1983

[6] T.M.Shi, C.G.Xie, S.M.Huang, R.A.Williams and M.S.Beck, Capacitance-based Instrumentation for Multi-interface Level Measurement, *Meas. Sci. Technol.* **2** 1991, pp 923-933

[7] P.W.Southern and R.J.Deloughry, Imaging of Oil/gas/water/sand Interface Levels in An Oil Separation Vessel, *Proc. of ECAPT'93,* Karlsruhe, Germany, 25-27 March 1993, pp 77-80

[9] I.Bennett, Limitations of Existing Instruments, *Proc. of The Conference on Development in Production Separation System,* London, 4-5 March 1993

[10] M.R.Brant, W.Q.Yang and M.S.Beck, Safety Aspects in Instrument Design -- Explosion Hazards, *Proc. of ECAPT'93,* Karlsruhe, Germany, 25-27 March 1993, pp 197-200

[11] W.Q.Yang and A.L.Stott, Low Value Capacitance Measurements for Process Tomography, *Proc. of ECAPT'92,* Manchester, 26-29 March 1992, pp 74-81

[12] W.Q.Yang, H.X.Wang, M.R.Brant, C.G.Xie, Y.W.Chen, R.C.Waterfall and M.S.Beck, Multi-interface Level Measurement System Using Segmented Capacitance Sensor, *Proc. of ECAPT'93,* Karlsruhe, Germany, 25-27 March 1993, pp 81-84

Electrodynamic sensors for process tomography

A R Bidin[1], R G Green[1], M E Shackleton[1], A L Stott[2] and R W Taylor[3].

1. Sheffield Hallam University, Sheffield.
2. Process Tomography Group, UMIST, Manchester.
3. Electronic Engineering Department, York University, York.

ABSTRACT: This paper presents the initial work done to investigate the use of electrodynamic sensor arrays for producing tomographic images of flow conditions at a cross-section of a pneumatic conveyor. Single particle models have been investigated which predict the induced charge due to: 1) a static particle and 2) a particle moving at constant velocity past a sensor. An array of eight sensors have been built and used to test the models. Results for the static model are presented and a simple reconstruction method for single particles presented based on them.

1 INTRODUCTION.

This paper investigates the application of electrodynamic sensors to the imaging of solids being conveyed either pneumatically or by gravity. These images are required for two reasons: 1) as a method of investigating how the solids are behaving within the conveyor and 2) as a method which can determine the solids mass flow rate. Many materials become electrostatically charged during transport, primarily by friction of fine particles amongst themselves and abrasion on the walls of the conveyor [Cross,87]. Sensors mounted in the wall of the conveyor can detect the presence of this charge. After appropriate signal conditioning it is possible to reconstruct an image of the transported material within the conveyor.

Many applications of light phase conveying exist and the potentially high signal bandwidth of the electrodynamic sensor, 10kHz, makes it suitable for particles travelling at speeds up to 100m/s.

This paper is organised into three sections. The sensor electronics is presented in the first section. The second section develops models to predict the effect of 1) a single, static, charged particle and 2) a single, moving, charged particle on a small sensor placed off the axis of the charge. Results are obtained using the sensors and compared with the predictions of the model. Finally an initial approach to image reconstruction is described.

2 THE ELECTRODYNAMIC MEASUREMENT SYSTEM.

The generation of electrical charge in the pneumatic transport of materials is a problem in many industries. The magnitude of the charge is dependent on various factors such as the type of material, whether conducting or non conducting, the quantity involved, moisture level, air flow rate and physical dimensions of the conveyor.

The term electrodynamics is used to describe the measurement of a moving charge. One of the earliest works using electrical charge measurements or electrodynamics to detect flowing materials is by King [King, 1973]. He developed a method of relating the electrodynamic noise level, solids mass flow rate and particle-electrode collision rate to theoretical ideas of intensity of turbulence within a horizontal two phase pipe flow.

In an industrial application, yarn velocity in the textile process was measured [Featherstone, Green and Shackleton, 82]. The yarn, which generates an electrical charge as it rubs on machine spindles, is passed through two metal guides which act as sensors. The sensor output voltages are cross-correlated to obtain the transit time for the charge to travel between the two points.

Electrodynamic process tomography system consists of three main components: the sensors, the transducer array and data acquisition system and the image reconstruction and display system (figure 1).

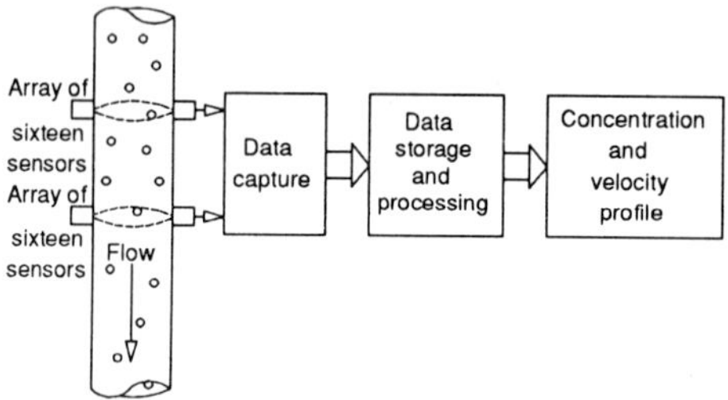

Figure 1. An electrodynamic process tomography system.

2.1 THE 8-CHANNEL ELECTRODYNAMIC TRANSDUCER.

The transducer system is passive (the field is due to the conveying material) and consists of two basic components: a small electrode as the sensing device and the signal processing electronics.

Eight electrodes are used and the sampling has to be fast to ensure the sensors detect the same charge during the flight of the particle through the sensing volume.

The range of particle sizes in this work is 1-300 microns. The maximum amount of charge that particles of this size range accumulate is estimated to be in the range of 10^{-5} to 10^{-1} volts [Shackleton,81]. The overall block diagram of the eight-channel transducer is shown in figure 2.

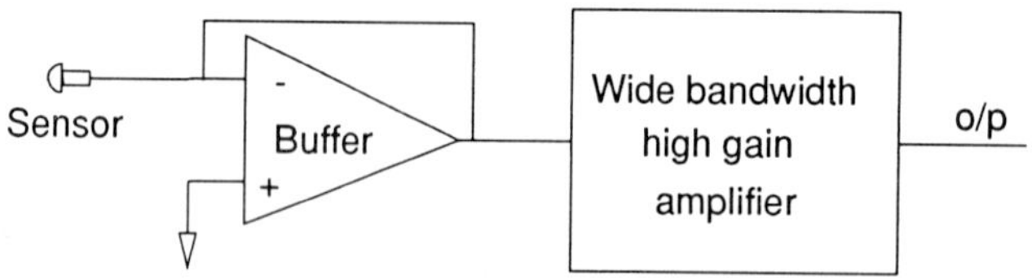

Figure 2. Transducer block diagram.

The input to each channel of the transducer is from a guarded buffer receiving outputs from the sensor. In this way the effects due to capacitance from the wires in the active field is minimised and the sensitivity of the output measurement is ensured.

3 A THEORETICAL MODEL OF BOUNDARY VOLTAGE PROFILES IN THE FLOW CONVEYOR.

The model is for a single particle and is formed in two parts. The first determines the voltage which is induced into a sensor by a single static charge, q. This is then expanded to the response on the sensor of a moving charge.

For a single charged particle, assumed to be a point charge of value q, it can be shown that for a given sensor, radius r_e, with surface area πr_e^2 normal to the flux, that the charge induced in the i'th sensor will be proportional to q.

$$Q_{induced} = \frac{k r_e^2 q}{4 r_i^2} = \frac{k_1 q}{r_i^2}$$

Hence,

$$V_{induced} = k_3 \frac{q}{r_i^2}$$

The distance from the charge to the sensor, figure 3, is given by the geometrical relation

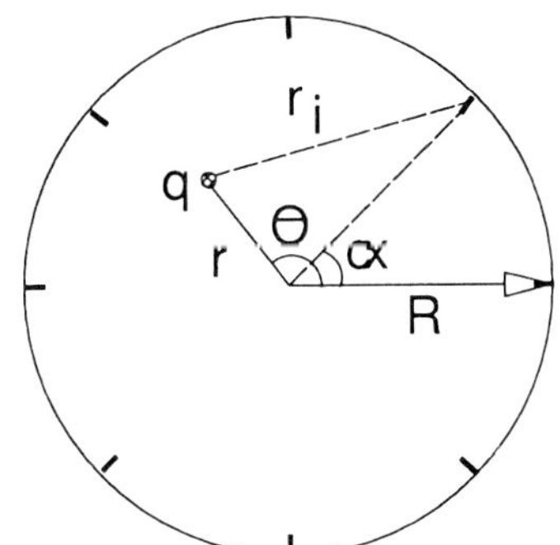

Figure 3. Geometrical relationship between sensor and particle.

$$D_i = \sqrt{r_i^2 + R^2 - 2Rr_i \cos(\theta - \alpha)}$$

The angle θ is the radial position of the charge while α is the sensor location from a reference axis. Note that the notation D is used to describe a generic distance between sensor and charge.

Because q and k are unknowns in the above equation, the voltages are normalised over the maximum voltage to cancel out any constant terms in the model due to sensor or pipe geometry and size of charge.

From this static model, an estimation of the profile of the boundary voltages is produced. It is possible to simulate different positions of the charge by changing the values of r and θ. However, changing θ will only give an angular offset to the profiles.

The response to a moving particle is determined as follows. Assume a sensor at the origin, and a particle (charge q) moving at constant velocity v along the line x=D, z=0 (figure 4).

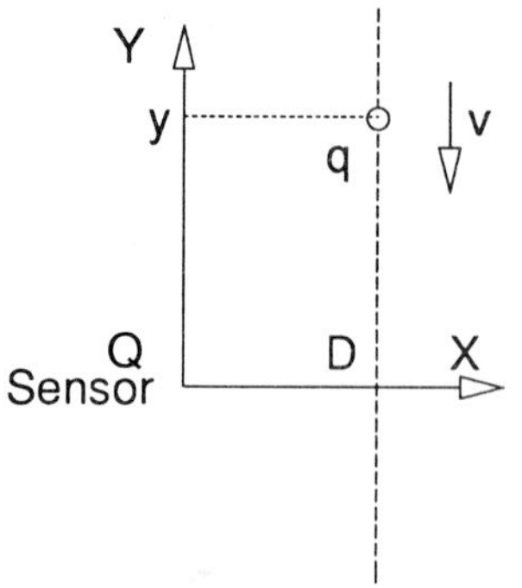

Figure 4.

Assuming the inverse square law relationship holds, then the induced charge Q is given by:

$$Q = -kq\frac{1}{D^2 + y^2}$$

where y=vt and t=0 when the particle is directly opposite the electrode. The equation then becomes

$$Q = -kq\frac{1}{D^2 + v^2t^2}$$

This charge Q is varying with time and generates a current $\frac{dQ}{dt}$ which results in the sensor output voltage. The current, i, is given by

$$\frac{dQ}{dt} = 2kq\frac{v^2t}{(D^2 + v^2t^2)^2}$$

Typical results for the induced charge Q and resulting voltage V are shown graphically in figure 5. Measured voltages for four sensors are shown in figure 6.

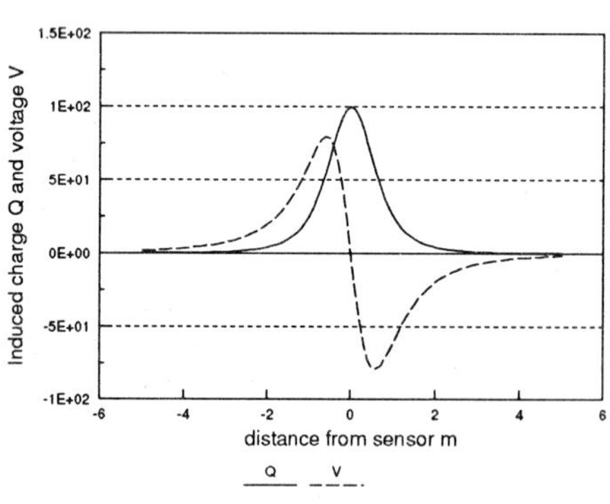

Figure 5. Results of modelling for a moving charge.

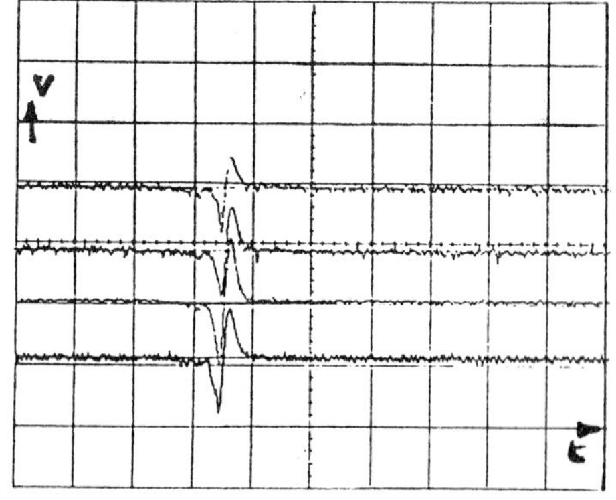

Figure 6. Measured voltages for a moving charge.

4 MODEL VERIFICATION.

An experimental rig was designed to verify the model. A solid particle, made from plastic material, is electrostatically charged and dropped through a narrow, 10mm external diameter, non-conducting tube to specify its path through a PVC pipe of 100mm diameter, mounted with 8 electrodes located at equally spaced positions around the pipe wall. Figure 7 shows a schematic diagram of the experimental set up.

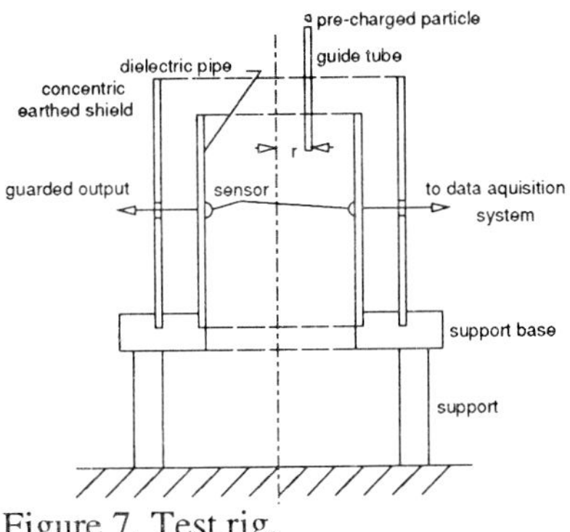

Figure 7. Test rig.

The sensors are connected to two 4-channel 20 Mhz oscilloscopes to capture the measurements which are then transferred onto printed graphs. The responses from each sensor in the time domain compare favourably with the predicted values.

The results were plotted on special graphs where the theoretical profiles were already plotted. The results are plotted in such a way that the sensor with the maximum value is fixed as the reference sensor at $\theta = 0$ and the readings from the other sensors plotted on the graph in a cyclic manner. Typical results are shown in figures 8 and 9.

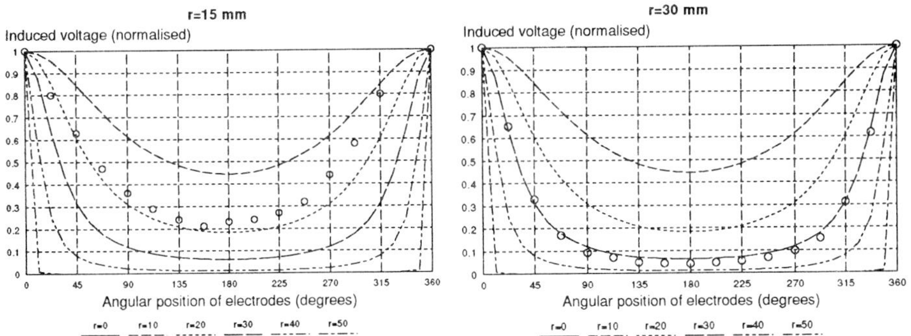

Figure 8. Typical result for r=15mm. Figure 9. Typical result for r=30mm.

The position of a single charged particle can be reconstructed from the measurements as follows. From the voltage profile around the boundary of the pipe wall the position where the voltage is maximum indicates which sensor is closest to the charge inside the pipe.

The radial position of the charge is determined from another graph. The maximum and minimum range values are calculated for a range of radial charge positions, which are then plotted to correlate the range and radial positions as shown in figure 10. Then, knowing the range value from the voltage profile, obtained by the method described above, the radial position of the charge can be determined. In reconstructing the image, the task of determining the values r and θ is therefore achieved.

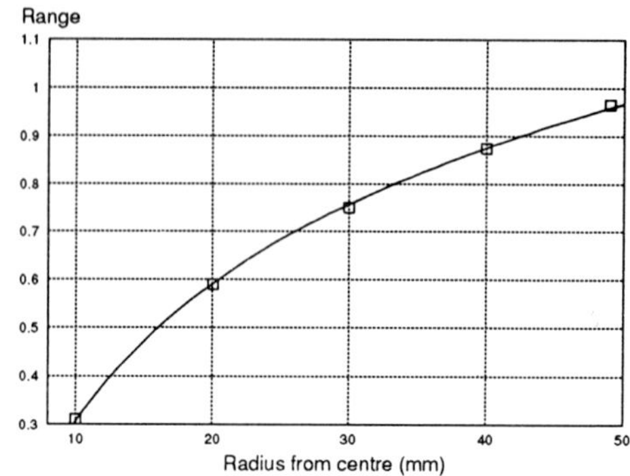

Figure 10. Relationship between max/min ratio and radius.

The results from the experiment show a reasonable agreement with the induction model. The results are close to the estimated profiles for all charge locations. The dynamic model needs further development because it predicts that the induced charge will increase without bounds as the charge approaches the sensor.

5 REFERENCES.

Cross,J "Electrostatics; Principles, problems & Applications". Adam Hilger, 1987.

Featherstone,AM; Green,RG; Shackleton,ME "Yarn velocity measurement". J Phys E: Sci Instrum, vol 16, 1983.

Gregory,I "Shot velocity measurement using electrodynamic transducers". PhD thesis, UMIST, 1987.

King,PW "Mass flow measurement of conveyed solids by monitoring of intrinsic electrostatic noise levels". Pneumotransport 2, 5-7 Sept 1973, Surrey.

Shackleton,ME "Electrodynamic sensors for process measurement". MPhil thesis, University of Bradford, 1981

A distributed pressure sensor which utilises electrical impedance tomography

I Basarab-Horwath*, P Lacey
School of Engineering Information Technology
Sheffield Hallam University
Sheffield S1 1WB

Abstract

The development of a pressure sensor to measure distributed loads is described. The sensor uses a conductive planar element which can be deformed by an applied load. The deformations are measured as changes in resistance using electrical impedance tomography. The results reported here show that a relatively simple analysis of the collected data set can measure the magnitude and position of the applied load.

The main objective of the work described in this paper is to develop a sensor to measure distributed pressures. The applied pressures produce a graded response in a deformable conductive sensor element. The space-distributed changes in dimension due to the applied load result in local changes in resistance; these can be measured using electrical impedance tomography. Such a sensor would be used to obtain tactile information.

Tactile sensing provides information about the state of contact between a gripper and object and the magnitude and distribution of that contact. Tactile sensing normally refers to skin-like properties: that is, properties with which areas of force sensitive surfaces are capable of reporting graded signals and parallel patterns of touching (1. 2. 3). Tactile sensors are based on direct measurement force transducers - these force transducers are arranged in a two-dimensional area and measure the mechanical deformation produced by the applied force acting on each transducer. Tactile sensors have been constructed using a wide variety of technologies, eg resistive, capacitive, optical and inductive. These are well documented in the literature (1, 2, 3). There are advantages and disadvantages with each technology. However, the basic concept has been to produce an array of tactels and to interrogate each tactel to determine the applied load. In the sensor described in this paper, there are no individual tactels; instead the whole 'array' is interrogated.

The sensor element in a conductive elastomer. The elastomeric materials used in this reported experimental work were obtained as samples from an industrial company manufacturing rubber products. They are designed to be used in sensor applications; force/resistance curves are shown in Figure 1. It can be seen that there are two distinct types of material; one type gives a graded response while the other type has a switching characteristic.

The main problem encountered in the construction of the transduction element was the need for good, reliable electrical connections to the elastomer sheet. These connections had to be such that the elastomer sheet could be removed and replace with other elastomer sheets. It has been found from previous work that gluing electrodes to the polymer surface does not produce a reliable connection. Care has to be taken to limit any electrode pressure so as not to damage the surface of the material. It is difficult to obtain the same contact pressure for each electrode.

* All correspondence should be addressed to Dr I Basarab.

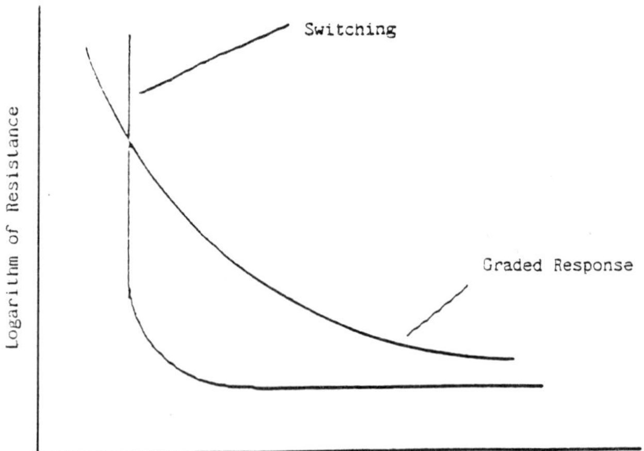

Figure 1. Force-resistance curves for the elastomer

The electrode configuration for the planar element is shown in figure 2. This shows 16 electrodes spaced around the periphery of the sensor element. For the data and results described in this paper, these electrodes are part of an adjacent-electrode tomographic system.

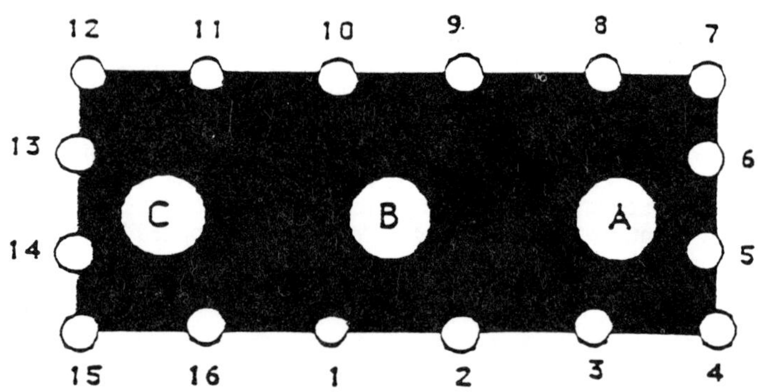

Figure 2. Diagram of the electrode configuration for the elastomer sheet

The block diagram for the system is shown in Figure 3. A variable-frequency current source is used to drive a constant current between adjacent pairs of electrodes; each electrode pair is selected using two multiplexers.

Differential boundary voltages are measured between adjacent electrodes by using a pair of multiplexers and an instrumentation amplifier. The output of the instrumentation amplifier is converted to a dc level using an rms to dc converter before being processed by the ADC. The software to control the multiplexers and ADC is written in C; the data acquisition rate is limited by the ADC.

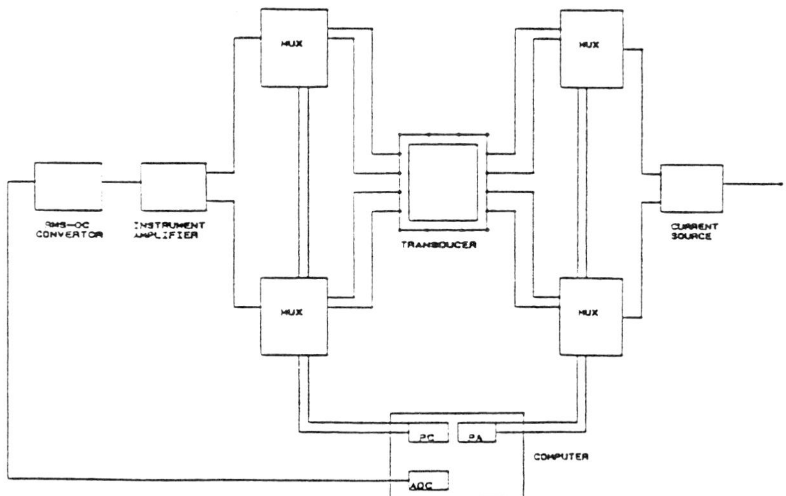

Figure 3. Block diagram of tomographic data collection system

Variations between electrical connections to the elastomer sheet will, to some extent, be minimised by using a constant current drive and differential voltage measurement. However, in order to minimise differences between each testing of a new material and also to overcome problems due to drift and noise, a set of readings was taken before each load was applied and also during the application of the load. The resulting difference between these two sets of readings shows a clear relationship to the applied load. Each set of readings forms a matrix. Typical sets of readings or matrices (to 1 d.p.) are presented in figures 4, 5 and 6. Figure 4 is a set of results for no applied load. Reassuringly, the matrix shows virtually no change from the previous null reference reading. This can be contrasted with figure 5, which is the matrix for a large pressure exerted near to electrodes 5 and 6. Figure 6, which shows the matrix for a smaller pressure exerted at the other 'end' of the element, near to electrodes 13 and 14, can also be compared with figure 5, to show the difference in the matrix due to the position as well as the magnitude of the applied load.

Each of these matrices could be used as the input data to a reconstruction algorithm so as to produce an image of the applied load. Image reconstruction is typically done using either a 'back-projection' method or a modified Newton-Raphson technique in order to solve the large number of simultaneous (sparse) non-linear equations. These methods are computationally very intensive and require, where speed is important, hardware specifically designed for numerical work. A very much simpler method of data analysis is presented here.

-0.0	-0.0	-0.0	-0.0	0.0	0.0	0.0	0.0	0.0	0.0	-0.0	-0.0	-0.0
-0.1	-0.1	-0.1	-0.1	-0.0	-0.0	-0.1	-0.1	-0.1	-0.0	-0.0	-0.0	-0.0
0.0	0.0	0.0	0.0	0.0	-0.0	-0.0	0.0	0.1	0.1	0.1	0.1	0.1
0.1	0.1	0.1	0.1	0.1	0.1	0.1	0.0	0.0	0.0	0.0	0.0	0.0
0.0	0.0	0.0	0.0	0.0	-0.0	-0.0	-0.1	-0.1	-0.0	-0.0	-0.0	-0.0
-0.0	-0.0	-0.1	-0.2	-0.2	-0.1	0.0	0.1	0.1	0.1	0.1	0.1	0.1
0.1	0.1	0.1	0.1	0.1	0.1	0.1	0.1	0.1	0.1	0.0	0.0	0.0
0.0	0.0	-0.0	-0.0	-0.0	-0.1	-0.1	-0.0	-0.0	-0.0	-0.0	-0.0	-0.0
-0.0	-0.0	0.0	0.0	0.0	0.0	0.0	0.0	0.0	0.0	0.0	0.0	0.0
0.0	0.0	0.0	0.1	0.1	0.1	0.1	0.1	0.0	0.0	0.0	0.0	0.0
0.0	0.0	0.1	0.1	0.1	0.1	0.0	0.0	0.0	0.0	0.0	0.0	0.0
0.1	0.1	0.1	0.1	0.1	0.1	0.1	0.0	0.0	0.0	0.0	0.0	0.1
0.1	0.1	0.1	0.1	0.1	0.1	0.1	0.1	0.1	0.0	0.0	0.0	-0.0
-0.0	-0.1	-0.1	-0.1	-0.1	-0.1	-0.1	-0.1	-0.1	-0.0	0.0	0.1	0.1
0.0	-0.0	-0.0	-0.0	-0.0	-0.0	-0.0	-0.0	-0.0	-0.0	0.0	0.0	0.0
0.0	0.0	0.0	0.0	0.0	0.0	0.0	0.0	0.1	0.1	0.1	0.2	0.2

Figure 4. Matrix for no applied load

The matrix of readings shown in Figures 4, 5 and 6 is the set of voltage readings taken from the boundary of the transducer element using adjacent current drive electrodes. There are 16 electrodes and thus there are 13 readings for each projection, giving a 16 x 13 matrix.

4.8	4.8	4.8	4.8	4.8	4.6	3.6	2.9	2.3	1.8	1.4	1.2	1.5
2.1	2.4	2.5	2.3	2.1	1.8	1.4	1.2	0.9	0.7	0.7	0.8	1.0
1.4	1.9	2.1	1.9	1.6	1.3	1.0	0.8	0.7	0.6	0.4	0.7	1.1
1.7	2.1	1.9	1.6	1.3	1.1	0.9	0.7	0.4	0.3	0.5	1.0	1.5
2.0	1.8	1.5	1.3	1.0	0.8	0.7	0.3	0.2	0.4	0.9	1.5	2.0
1.8	1.6	1.3	1.1	0.9	0.6	0.4	0.2	0.5	0.9	1.5	1.9	1.7
1.5	1.2	1.0	0.8	0.7	0.6	0.9	1.0	1.3	1.8	2.0	1.8	1.5
1.2	1.0	0.8	0.7	0.7	1.0	1.5	1.9	2.0	1.9	1.7	1.5	1.2
1.0	0.8	0.6	0.5	0.8	1.4	1.8	1.9	1.9	1.7	1.4	1.2	1.0
0.8	0.6	0.6	0.8	1.4	1.8	1.9	1.8	1.7	1.4	1.2	1.0	0.8
0.6	0.6	1.1	1.8	2.3	2.4	2.3	2.1	1.8	1.5	1.2	0.9	0.7
0.5	0.4	0.7	0.9	1.1	1.1	1.0	0.8	0.7	0.6	0.5	0.4	0.4
0.3	0.3	0.3	0.3	0.2	0.2	0.2	0.2	0.2	0.1	0.1	0.1	0.0
0.0	-0.0	-0.1	-0.1	-0.1	-0.1	-0.1	-0.1	-0.1	-0.1	-0.1	-0.1	-0.2
-0.2	-0.2	-0.1	-0.1	-0.1	-0.1	-0.1	-0.0	-0.0	0.0	0.0	0.1	0.0
0.0	0.1	0.2	0.3	0.3	0.2	0.2	0.2	0.2	0.2	0.1	0.0	0.0

Figure 5. Matrix for a large load applied near to electrodes 5 and 6

A column matrix or vector is constructed from each matrix of readings by summing each row (1-16) and then summing each column (1-13), producing a vector with 29 elements.

Figures 7, 8 and 9 show the value of each element, v_j, plotted against its position j in the vector.

Figure 7 shows the results for a continuous graded response material with loads of 4 kg, 7 kg and 10 kg applied at position A. The no-load vector is shown for comparison purposes.

0.1	0.1	0.1	0.2	0.3	0.5	0.5	0.4	0.5	0.4	0.4	0.3	0.2
0.1	0.1	0.1	0.2	0.8	0.9	0.8	0.6	0.5	0.4	0.4	0.3	0.2
0.2	0.1	0.3	0.7	0.9	0.6	0.4	0.3	0.3	0.3	0.2	0.2	0.1
0.1	0.3	0.9	1.1	0.9	0.7	0.6	0.5	0.4	0.3	0.3	0.2	0.2
0.2	0.1	-0.2	-0.5	-0.7	-0.7	-0.6	-0.5	-0.4	-0.3	-0.2	-0.2	-0.1
-0.1	-0.2	-0.6	-0.8	-0.7	-0.7	-0.6	-0.5	-0.4	-0.3	-0.2	-0.2	-0.1
-0.0	-0.0	-0.1	-0.3	-0.8	-1.3	-1.7	-1.9	-1.8	-1.7	-1.4	-1.2	-1.0
-0.7	-0.6	-0.6	-0.8	-1.2	-1.4	-1.5	-1.4	-1.2	-1.1	-0.9	-0.7	-0.6
-0.4	-0.3	-0.3	-0.4	-0.4	-0.3	-0.3	-0.3	-0.2	-0.2	-0.1	-0.1	-0.1
0.5	0.7	0.7	0.6	0.6	0.4	0.4	0.3	0.2	0.2	0.2	0.4	0.8
1.0	0.9	0.8	0.6	0.5	0.4	0.3	0.2	0.2	0.3	0.4	0.8	1.0
1.0	1.0	0.9	0.9	0.7	0.6	0.5	0.4	0.4	0.5	1.0	1.2	1.1
0.9	0.8	0.7	0.6	0.5	0.4	0.4	0.4	0.5	1.0	1.1	0.8	0.6
0.4	0.2	0.1	-0.0	-0.1	-0.1	-0.1	0.2	0.7	0.9	0.7	0.5	0.4
0.2	0.1	0.0	-0.0	-0.0	-0.0	0.2	0.3	1.0	0.8	0.7	0.5	0.4
0.3	0.2	0.2	0.1	0.1	0.4	0.9	-1.1	0.9	0.7	0.6	0.4	0.3

Figure 6. Matrix for a small load applied near to electrodes 13 and 14

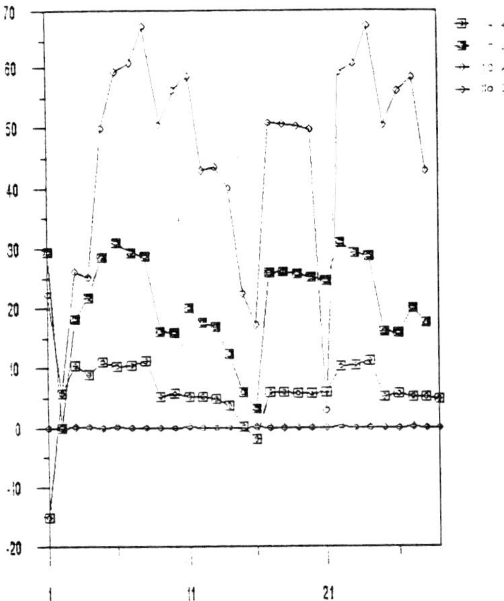

Figure 7. Continuous graded response material; Magnitude of element plotted against element index number for various loads at a fixed position

Figure 8 shows the results for the on/off or switch response material with 2, 6 and 10 kg applied at position C; again, the no-load vector is shown. These results can be compared with those shown in Figure 7. Both figures show that both materials exhibit a graded response to applied load. It can be seen that not only is the overall response different for the different materials but that the position of the element with the minimum value has moved, corresponding to a change in the position of the applied load.

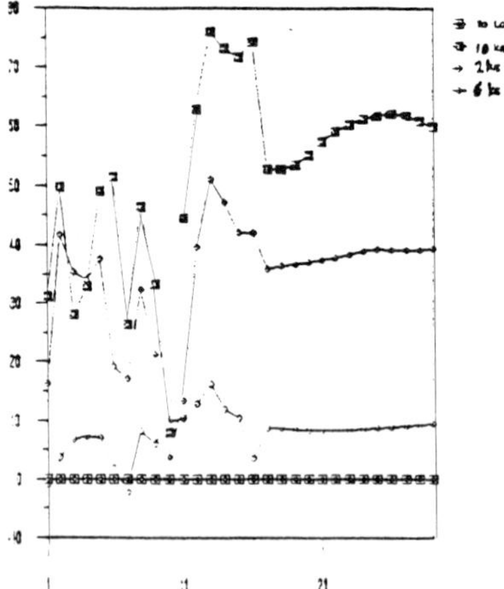

Figure 8. Switch response material; Magnitude of vector element plotted against element index numbers for various loads at a fixed position

This is illustrated more clearly in Figure 9 which shows the results for a 6 kg load applied at the three different positions shown in Figure 2. The minimum value for each set of readings moves as a result of the change in the position of the applied load. From this diagram, it can also be seen that the 'waveforms' have different sizes and shapes for the different positions of applied load.

These presented results show clearly that it is possible to use a conductive elastomer as a transduction element in a tomographic system and that simple, fast analysis of the collected data can be used to measure the magnitude and location of applied loads.

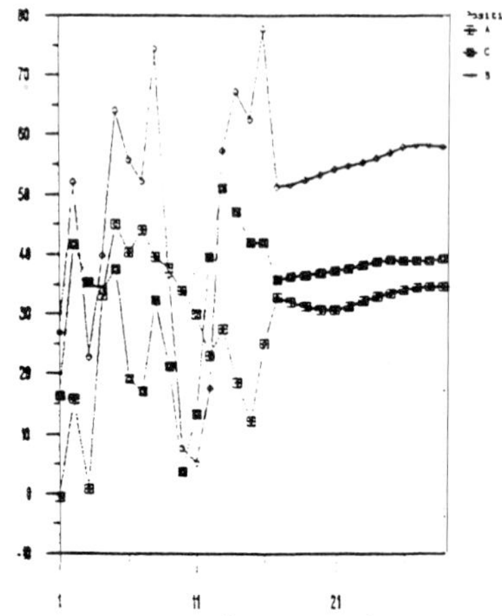

Figure 9. Switch response material; Magnitude of vector element plotted against element index number for a 6 kg load at 3 different positions.

References

1 Harman L D, Automated Tactile Sensors. Int J Robotics Res, 1, 2, 2-32, 1982
2 Ruocco S R, Robot Sensors and Transducers, Halstead Press and OUP
3 Webster J G, Tactile Sensors for Robotics and Medicine, John Wiley and Sons, 1988

Acknowledgements I would like to thank Miss Marie Broxholme for typing this paper.

Section F

BIOMEDICAL SENSORS

Measurement of light scattering in the eye using ellipsometric techniques

M H Miran Baygi and P A Payne
Department of Instrumentation and Analytical Science, UMIST, Manchester M60 1QD

ABSTRACT: The aqueous humour fills the volume between the cornea and the lens of the eye and is normally a very clear liquid. It is renewed continually by a process of ultrafiltration and this process can be compromised as a result of damage or clinical interference. Should this happen, then blood products (proteins, cell debris, etc) compromise the near-perfect optical properties causing light scattering and a reduction in the efficiency of light reaching the retina. Measurement of the effects of such interference are crucial in order to study the effects of disease processes and clinical treatment.

1. INTRODUCTION

Unwanted light scattering in the aqueous humour (Figure 1) may arise from the breakdown of the blood-aqueous barrier whose function otherwise is to maintain the normal aqueous humour [1]. The breakdown of the barrier may, for example, arise from inflammation of the anterior chamber. Large particles and/or increased levels of proteins may find their way into this region and scatter light thus reducing the amount of light available to the retina. Clinically, slit-lamp microscopy has been used to measure and grade the concentration of these particles in the eye and more recently and with greater success use has been made of the flare-cell meter [2]. It is of clinical importance to obtain other information such as size, shape and composition of these scatterers.

Ellipsometry is an optical technique which is based on exploiting the polarisation transformation that occurs as a beam of polarised light interacts with a reflector, transmitter or scatterer. We have used this technique to measure light scattering in vitro from standard turbidity suspensions of formazin. We have shown that ellipsometry is a viable technique which could be developed for light-scattering measurement in the eye.

2. ELLIPSOMETRY

The theory of ellipsometry has been discussed in detail elsewhere [3]. Briefly, there are three types, namely reflection, transmission and scattering ellipsometry. Scattering ellipsometry or polarimetry has been widely used in studying light scattering by particles of biological or non-biological nature [4,5,6,7]. Ellipsometric measurement are taken in order to obtain information about the particles such as their size, size distribution, shape, refractive index, optical activity, and other parameters. It is evident that scattering ellipsometry is particularly suited to the study of light scattering in the aqueous humour.

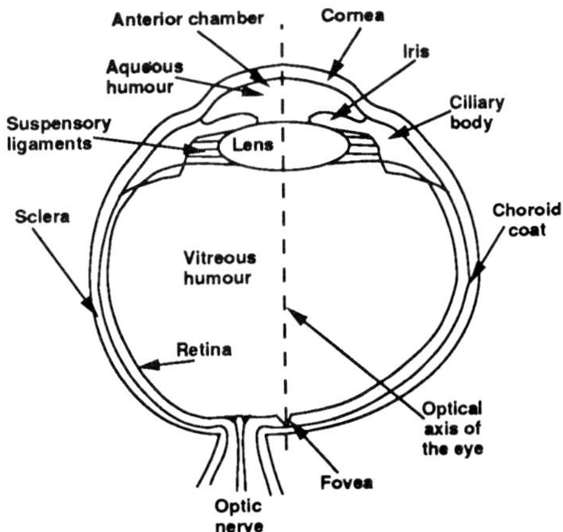

Figure 1 *Cross-sectional diagram of the human eye*

In scattering ellipsometry, a beam of light is mathematically represented by a 4 x 1 column vector known as the Stokes vector. Taken together, the parameters of this vector completely describe the polarisation state of the light beam. This is changed as the light beam interacts with polarisation-altering media (eg, polariser, retarder, reflector, scatterer etc). Each of these optical components may be represented by a 4 x 4 matrix known as a Mueller matrix. If the incident-beam Stokes vector is denoted by $\mathbf{S}_i$ and the Mueller matrix of the interacting medium is denoted by **M** then the Stokes vector of the outgoing beam $\mathbf{S}_o$ is obtained by:

$$\mathbf{S_o} = \mathbf{MS_i} \tag{1}$$

or, in full, as:

$$\begin{bmatrix} I_o \\ Q_o \\ U_o \\ V_o \end{bmatrix} = \begin{bmatrix} M_{11} & M_{12} & M_{13} & M_{14} \\ M_{21} & M_{22} & M_{23} & M_{24} \\ M_{31} & M_{32} & M_{33} & M_{34} \\ M_{41} & M_{42} & M_{43} & M_{44} \end{bmatrix} \times \begin{bmatrix} I_i \\ Q_i \\ U_i \\ V_i \end{bmatrix} \tag{2}$$

where I, Q, U and V are the four Stokes parameters.

For a series of polarisation-altering media 1, 2, 3, ... , the combined effect of all these elements is determined by multiplying their associated Mueller matrices M_1, M_2, M_3, ... in a sequential order:

$$\mathbf{S_o} = \ldots\ldots\ldots\ \mathbf{M_3M_2M_1S_1}. \tag{3}$$

The Mueller matrix representing the scattering of light is termed the scattering matrix. The scattering matrix, in general, contains 16 non-zero and independent elements. For a randomly oriented and homogenous particulate medium with particle sizes of about that of the wavelength and not optically active, the scattering matrix can be reduced to the matrix shown below:

$$\begin{bmatrix} S_{11} & S_{12} & 0 & 0 \\ S_{12} & S_{22} & 0 & 0 \\ 0 & 0 & S_{33} & S_{34} \\ 0 & 0 & -S_{34} & S_{44} \end{bmatrix} \tag{4}$$

The scattering matrix for a collection of isotropic spherical particles has the greatest degree of symmetry and can be reduced to:

$$\begin{bmatrix} S_{11} & S_{12} & 0 & 0 \\ S_{12} & S_{11} & 0 & 0 \\ 0 & 0 & S_{33} & S_{34} \\ 0 & 0 & -S_{34} & S_{33} \end{bmatrix} \tag{5}$$

ie, $S_{11} = S_{22}$ and $S_{33} = S_{44}$.

For particles having sizes much smaller than the working wavelength ($<\lambda/10$), $S_{34} = 0$ and therefore the scattering matrix reduces further to the familiar Rayleigh scattering matrix given below:

$$\begin{bmatrix} S_{11} & S_{12} & 0 & 0 \\ S_{12} & S_{11} & 0 & 0 \\ 0 & 0 & S_{33} & 0 \\ 0 & 0 & 0 & S_{33} \end{bmatrix} \tag{6}$$

3. SIMULATION

The elements of the scattering matrix are angle-dependent functions of wavelength, particle size, shape and composition. They are usually normalised to S_{11} to suppress intensity fluctuation between measurements. An exact solution for the Mie-scattering matrix exists thus providing a useful tool in predicting approximately the outcome of an ellipsometric measurement. We have adopted the simulation programme developed by Bohren and Huffman [8] and modified it to include our own criteria and conventions. Using this programme the predicted light scattering patterns for three different sizes of spherical particles were calculated and are shown in Figures 2 to 4. The particles are assumed to have a refractive index of 1.59 (latex) suspended in an aqueous solution with a refractive index of 1.33. The working wavelength has been assumed to be 670 nm.

For small particles (Rayleigh particles), the degree of polarisation ($-S_{12}/S_{11}$) (Figure 2) is positive over all scattering angles and shows a single maximum at 90°. As the particle size increases the number of maxima and minima increases and the maximum peak moves towards higher scattering angles. For these larger particles the degree of polarisation may take negative values as shown.

Similar features are apparent when examining the normalised element S_{33}/S_{11} (Figure 3). The response for Rayleigh particles is symmetrical either side of the 90° scattering angle, crossing $S_{33}/S_{11}=0$ line at this angle. As the size increases the crossing point shifts towards larger angles. The difference between Mie-scattering and Rayleigh scattering matrices is that the normalised element S_{34}/S_{11} is zero for Rayleigh particles and therefore this element is expected to be very sensitive to size of the particles. As shown in Figure 4 the response is zero for the Rayleigh particles and as the size increases large responses are obtained.

There are both similarities and differences between scattering by spherical and non-spherical particles. The degree of polarisation ($-S_{12}/S_{11}$) by some non-spherical particles is positive over a wide range of scattering angles. $S_{22} = S_{11}$ and hence $S_{22}/S_{11} = 1$ for spherical Mie particles. However, S_{22}/S_{11} deviates from unity for non-spherical particles quite markedly. For spherical Mie particles $S_{33}/S_{11} = S_{44}/S_{11}$, however the equality becomes an inequality for non-spherical particles. Therefore, either examination of how closely the equalities $S_{22}/S_{11} = 1$ or $S_{33}/S_{11} = S_{44}/S_{11}$ are satisfied can be an indicator of the shape of the particles.

4. EXPERIMENTAL PROCEDURE

A photograph of the experimental set up namely, an optical-bench-mounted ellipsometer is shown in Figure 5. The light source is a laser diode (670 nm), the output of which is altered to a desired state using a polariser and a retarder (a quarter-wave plate) in the incident beam. The

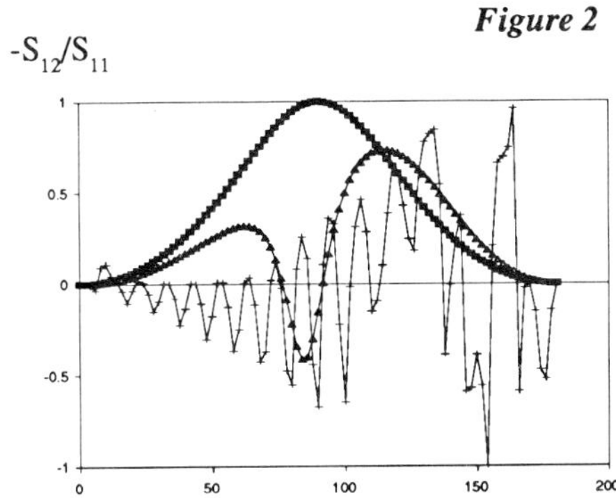

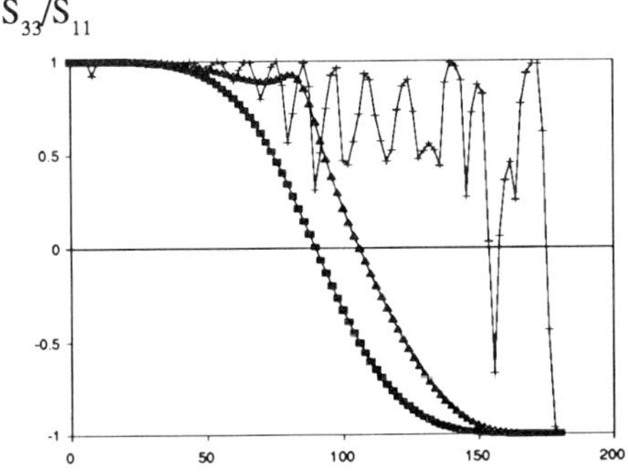

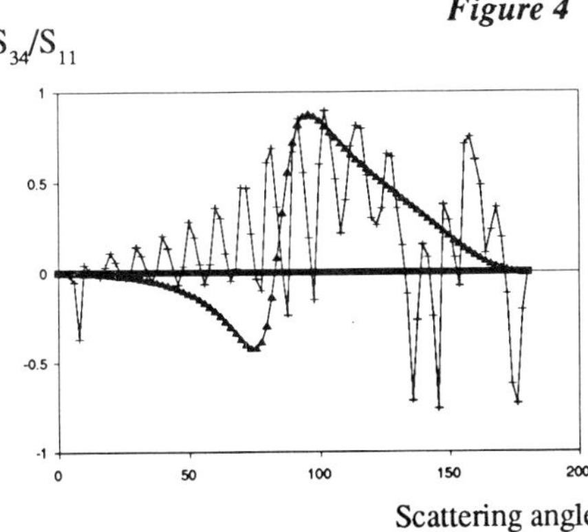

Key: □ Particle radius 0.0345 μm
Δ Particle radius 0.240 μm
\+ Particle radius 1.505 μm

Figures 2 to 4
Computer simulation of scattering patterns for three particle sizes

light scattered from the sample was then analysed by another polariser (analyser) followed by the detector, a photo-multiplier tube. To suppress stray light the input to the laser diode was modulated and the output of the detector was connected to a lock-in amplifier.

A series of latex particles of radii a = 0.0345 μm, a = 0.24 μm and a = 1.505 μm were used to calibrate the ellipsometer. The sizes were consistent with the sizes used to create the simulation results.

We also prepared a standard turbidity suspension employing formazin which is used in the measurement of the quality of water [9]. The optical properties of the resulting scatterers suspended in the aqueous solution was then determined using the ellipsometer.

Figure 5 *The bench-mounted ellipsometer*

Measurements were taken over an angular scattering range of 50 to 130°. Although with this limitation some valuable information might be lost, the practical situation where scattering in the eye is concerned would not allow a wider range to be covered. The azimuthal angles of the polariser (P), retarder (C) and the analyser (A) used in our measurements together with the resulting detector response D as a function of measured elements are listed as follows:

P = 0,	no retarder,	A = 0	$\rightarrow D_1 = S_{11} + 2S_{12} + S_{22}$
P = 0,	no retarder,	A = 90°	$\rightarrow D_2 = S_{11} - S_{22}$
P = 90°,	no retarder,	A = 0	$\rightarrow D_3 = S_{11} - S_{22}$
P = 90°,	no retarder,	A = 90°	$\rightarrow D_4 = S_{11} - 2S_{12} + S_{22}$
P = 45°,	no retarder,	A = 45°	$\rightarrow D_5 = S_{11} + S_{13} + S_{31} + S_{33}$
P = 45°,	no retarder,	A = -45°	$\rightarrow D_6 = S_{11} + S_{13} - S_{31} - S_{33}$
P = -45°,	no retarder,	A = 45°	$\rightarrow D_7 = S_{11} - S_{13} + S_{31} - S_{33}$
P = -45°,	no retarder,	A = -45°	$\rightarrow D_8 = S_{11} - S_{13} - S_{31} + S_{33}$
P = 45°,	C = 0,	A = 45°	$\rightarrow D_9 = S_{11} - S_{14} + S_{31} - S_{34}$
P = 45°,	C = 0,	A = -45°	$\rightarrow D_{10} = S_{11} - S_{14} - S_{31} + S_{34}$
P = -45°,	C = 0,	A = 45°	$\rightarrow D_{11} = S_{11} + S_{14} + S_{31} + S_{34}$
P = -45°,	C = 0,	A = -45°	$\rightarrow D_{12} = S_{11} + S_{14} - S_{31} - S_{34}$

Hence the normalised elements $-S_{12}/S_{11}$, S_{22}/S_{11}, S_{33}/S_{11}, S_{34}/S_{11}, S_{13}/S_{11}, S_{31}/S_{11} and S_{14}/S_{11} can be calculated using the above equations.

5. RESULTS AND CONCLUSIONS

The results of the ellipsometric measurements for formazin are show in Figures 6 to 12. The degree of linear polarisation ($-S_{12}/S_{11}$) reaches its peak at 90°. The peak value is slightly less than 1. The first conclusion is that the particles are small compared with the wavelength (Rayleigh type) but a similar pattern has been predicted for some large non-spherical particles [10] though with considerably lower peak value. The scattering pattern for S_{33}/S_{11} results in the same conclusions. For this pattern some valuable information is missing in the range 130 to 180°. Examining the scattering pattern for S_{22}/S_{11} and S_{34}/S_{11}, however, confirms our findings about the formazin particles, ie, that these particles are Rayleigh type. The reason is that S_{22}/S_{11} which is an indicator of non-sphericity is 1 within our instrumentation accuracy and S_{34}/S_{11} which is an indicator of size is zero over all the measured scattering angles. Note also that the normalised elements

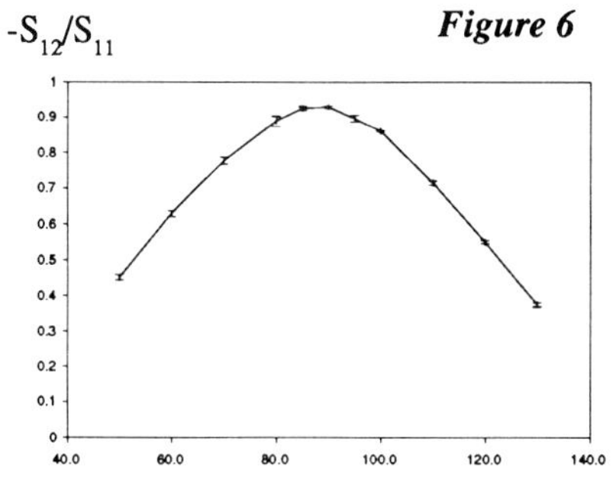

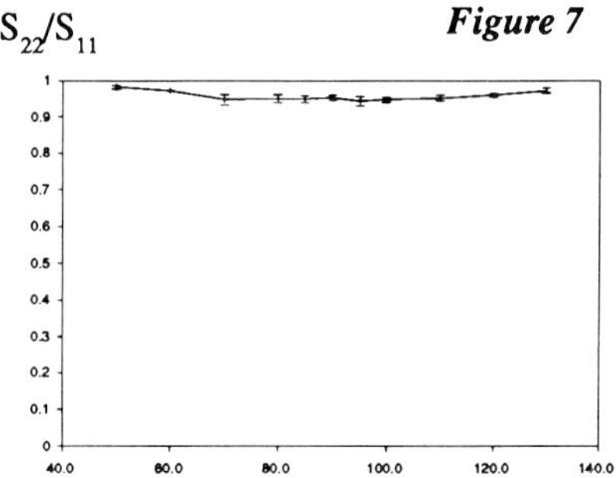

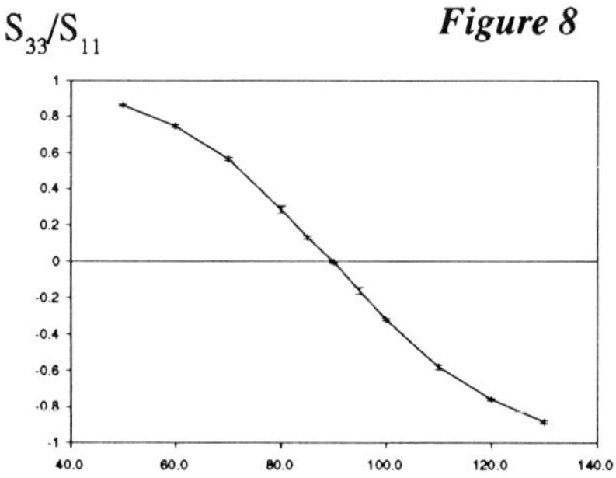

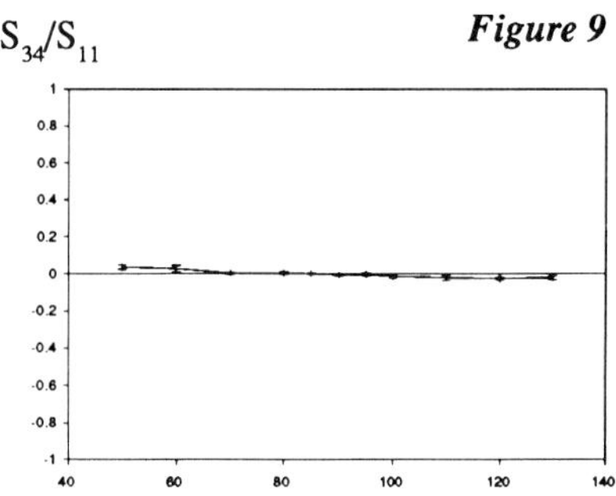

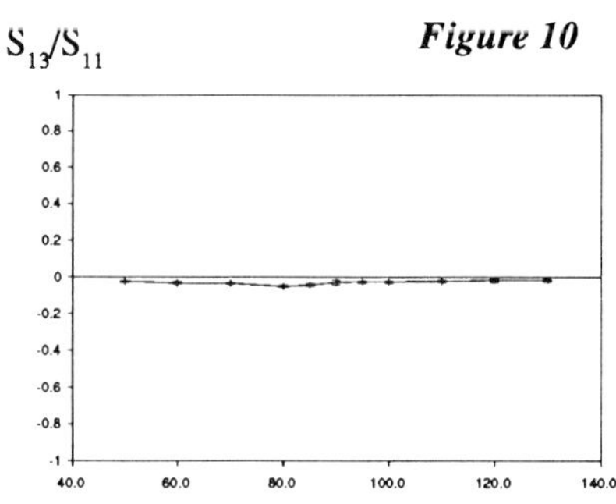

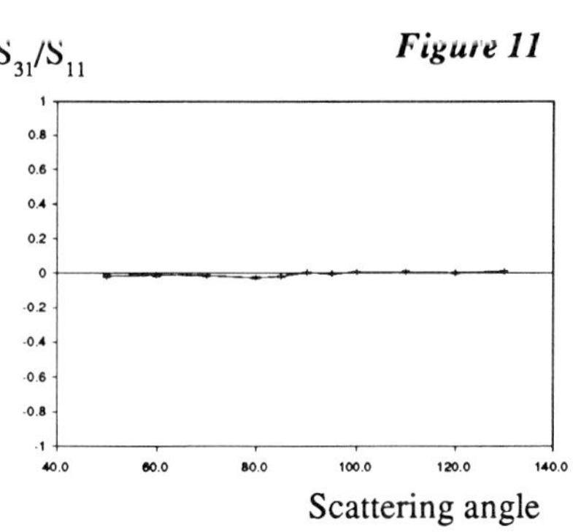

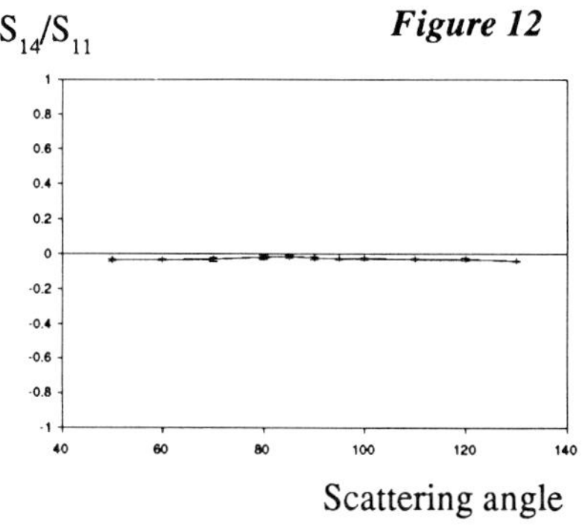

Figures 6 to 12
Measured results using formazin, plotted as average plus or minus one standard deviation (number of experiments = 6)

S_{13}/S_{11}, S_{31}/S_{11} and S_{14}/S_{11} are zero and therefore the use of the reduced scattering matrix is justified.

Figure 13 is a photograph of the formazin particles taken by a scanning electron microscope. From the photograph the average size of the particles is about 0.1 μm though there appears to be some smaller particles in the background that have not come into focus. This further confirms our finding about the size of the particles.

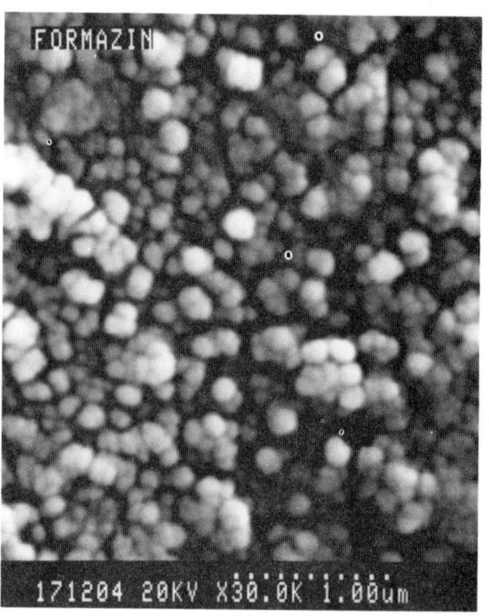

Figure 13 *Scanning electron microscope image of formazin on a glass slide*

6. FUTURE WORK

The ellipsometer shown in Figure 5 needs to be made small and portable for clinical use. It can be made fully automatic with or without moving parts and under computer control. There are several published articles [11-13] that report on the polarisation properties of the various layers in the eye. We need to consider and measure these in order to know by how much the ellipsometric results may be affected when conducting measurements on the real eye. It may be wise to construct an accurate model eye.

REFERENCES

1. Sawa, M., Tsurimaki, Y., Tsuru, T. and Shimizu, H. (1988) New quantitative method to determine protein concentration and cell number in aqueous in vivo, *Jpn. J. Ophthalmol.*, **32**: 132-142
2. Sawa, M. (1990) Clinical application of laser flare-cell meter, *Jpn. J. Ophthalmol.*, **34**: 346-363
3. Azzam, R.M.A. and Bashara, N.M. (1987) *Ellipsometry and polarised light,* North-Holland, Amesterdam
4. Bickel, W.S., Davidson, J.F., Huffman, D.R. and Kilkson, R. (1976) Application of polarisation effects in light scattering: A new biophysical tool, *Proc. Nat. Acad. Sci. USA*, **73**(2): 486-490
5. van de Merwe, W.P., Huffman, D.R. and Bronk, B.V. (1989) Reproducibility and sensitivity of polarised light scattering for identifying bacterial suspensions, *Applied Optics*, **28**(23): 5052-5057
6. Perry, R.J., Hunt, A.J. and Huffman, D.R. (1978) Experimental determination of Mueller scattering matrices for nonspherical particles, *Applied Optics*, **17**(17): 2700-2710
7. Kuik, F., Stammes, P. and Hovenier, J.W. (1991) Experimental determination of scattering matrices of water droplets and quartz particles, *Applied Optics*, **30**(33): 4872-4881
8. Bohren, C.F. and Huffman, D.R. (1983) *Absorption and scattering of light by small particles*, Wiley, New York
9. Ives, K.J., Atkin, J.R. and Thompson, R.P. (1968) Measurement of turbidity, *Effl. Wat. Treat. J.*, **8**: 342-348 and 406-409
10. Hofer, M. and Glatter, O. (1989) Mueller matrix calculations for randomly oriented rotationally symmetric objects with low contrast, *Applied Optics*, **28**(12): 2389-2400
11. van Blokland, G.J. (1985) Ellipsometry of the human retina in vivo: preservation of polarisation, *J. Opt. Soc. Am. A*, **2**(1): 72-75
12. van Blokland, G.J. and Verhelst, S.C. (1987) Corneal polarisation in the living human eye explained with a biaxial model, *J. Opt. Soc. Am. A*, **4**(1): 82-90
13. klein Brink, H.B. (1991) Birefringence of the human crystalline lens in vivo, *J. Opt. Soc. Am. A*, **8**(11): 1788-1793

Time-resolved phosphorescence anisotropy measurements of an antibody–antigen complex

Liqun Yang, D.McStay, A.J Rogers[2] and P.J Quinn[3]

School of Applied Sciences, The Robert Gordon University, St Andrews St, Aberdeen, U.K
[2]Department of Electronic & Electrical Engineering, King's College London, Strand, London, U.K.
[3]Biochemistry Section, Division of Life Sciences, King's College London, London, U.K.

Abstract
The rotational mobilities of antigens and antigen-antibody complexes labelled with the triplet probe eosin have been examined by measurement of the time-resolved phosphorescence depolarization of the probe, excited by a laser pulse from a frequency-doubled Nd:YAG laser and the time-resolved anisotropy calculated using a Marquardt curve fitting procedure. Enhancement of the anisotropy change by conjugating one of the species to a relatively large polymer bead is reported. The experimental results indicate that the time-resolved phosphorescence depolarization could be exploited to study the kinetics of antigen and antibody interaction.

Introduction

Radioimmunoassays have been the most widely used immunological assays due to their sensitivity and accuracy over a wide range of antigen concentration. However, these assays have several major drawbacks: possible health hazards of the radioisotopes, special requirements for handling reagents, training of staff, special storage and disposal of wastes, limited useful lifetimes of the isotope probes, expensive instrumentation, and the slow counting rate of the radioactivity. As a result, many different analytical methods have been developed as alternatives, including enzyme immunoassays, spin-label, metalloi- ,and optical (chemiluminescence, fluorescence, and phosphorescence) immunoassays. Among these alternatives, optical immunoassay techniques, especially fluorescence techniques are the most important ones, due to their distinct advantages in potential sensitivity, wide dynamic range, variety of probes and selective techniques available, fast and simple assay procedures etc. Using fluorescence techniques, single molecule detection has been claimed.[2] Many fluorescence probes with different emission lifetimes and different spectroscopic properties have been successfully developed.[3,5] Photo-selective principles that have been exploited, include intensity, lifetime, steady-state-polarization and fluorescence anisotropy[6,7,8]. Among these, the fluorescence depolarization immunoassay enjoys the greatest utility but it is not without certain limitations, such as the short duration of the fluorescence emission (i.e. short emission lifetime) and transient interference of the scattering of light. These disadvantages restrict application of the technique to the studies of small biomolecules, and exclude its use to study macromolecules, such as cell surface antigens or large viruses. As a result, two novel techniques have been developed: (a) using rare-earth chelate labels and (b) using phosphorescence labels. Both labels have two distinct advantages, namely long emission lifetime and large Stoke's shifts. The selectivity employed in both techniques is currently based on the intensities (or lifetimes) of photoemissions, which are virtually independent of the size of the molecules under study.

In this paper, a phosphorescence anisotropy selective technique, which may provide a more direct format for a phosphorescence immunoassay, has been developed to identify antigens, antibodies and their complexes via their relative sizes

Theory

It is well known that when a spherical rigid macromolecule, which is undergoing Brownian motion, is excited by a pulse of linearly-polarized light, the decay of the luminescence anisotropy (r(t)) takes the form of a single exponential decay law which can be expressed as:

$$r(t) = r(0)\exp(-t / \Phi(t)) \tag{1}$$

and can be calculated from the resulting orthogonally-polarized phosphorescence components directly:

$$r(t) = \frac{(I_{//}(t) - I_{\perp}(t))}{(I_{//}(t) + 2I_{\perp}(t))} = \frac{y(t)}{d(t)} \tag{2}$$

where $I_{//}(t)$ and $I_{\perp}(t)$ are the emission components with their polarization directions parallel and perpendicular to that of the excitation light respectively, y(t) and d(t) are the difference functions of the two orthogonally polarized emission components and the total emission intensity, both functions also obey an exponential decay law; r(0) is the initial anisotropy, which is basically a function of the wavelength of the excitation and emission and the internal molecular conformation and $\Phi(t)$ is the rotational correlation time, which is related to the size (volume) or the rotational diffusion coefficient of the molecule. The variation of rotational correlation time of BSA and a complex of BSA-AntiBSA with solution viscosity is shown in figure 1. For a heterogeneous system, consisting of N homogeneous spherical rotating species, the linearity property of the anisotropy r(t) yields:

$$r(C_1 \ldots C_N\, \lambda,t) = \sum_{i=1}^{N} F(C_i).r_i(\lambda,t) \tag{3}$$

where index i enumerates the N species present, C_i is the population of i-th species, $r_i(\lambda,t)$ and $F(C_i)$ are the anisotropy of i-th homogeneous rotating species and contribution of this anisotropy to the whole averaged signal r.

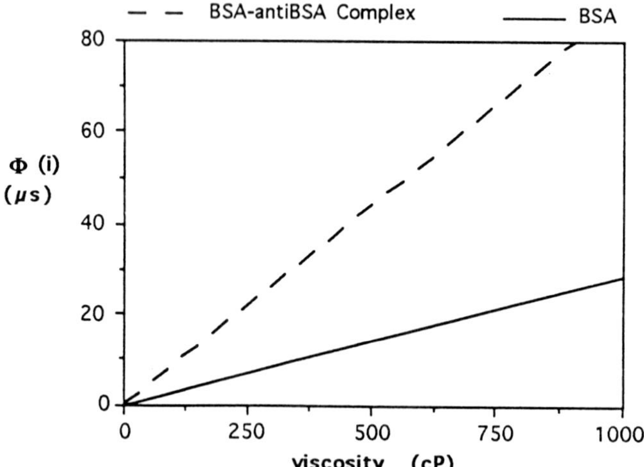

Figure 1: The variation of rotational correlation time of BSA and a complex of BSA-AntiBSA with solution viscosity

Using equation.(1) and (2), equation (3) can be rewritten as

$$\frac{(I_{//}(t) - I_{\perp}(t))}{(I_{//}(t) + 2I_{\perp}(t))} = \frac{\sum_{i=1}^{N} y(C_i, \lambda, t)}{\sum_{i=1}^{N} d(C_i, \lambda, t)} \tag{4}$$

as indicated above $F(C_i)$ represents the contribution of the i-th species to the total signal. Provided all species present are of the same quantum yields, i.e. the all the species are labelled with same probe and the antigenic reaction does not quench probe emissions, $F(C_i)$ will then simply be proportional to its population C_i according to the linear property of $r(t)$, and may be denoted as:

$$F(C_i) = C_i / \sum_{i=1}^{N} C_i \tag{5}$$

$r_i(\lambda,0)$ is the initial anisotropy of i-th species and is a function of wavelength and molecular internal structure. By fixing the wavelengths of the excitation and the measurement, the initial anisotropy will be a constant $r(0)$, determined by the molecular conformation. The anisotropy measured (r), can then be described as a function of the normalized concentration of the species

$$C_i / \sum_{i=1}^{N} C_i \tag{6}$$

and the moment (t) at which the measurement was performed. In this way, the rotating species present in the sample can be identified by analysis of the rotational correlation times contained in the time-resolved anisotropy decay, and the population of each species estimated.

Experiment

To demonstrate the utility of the phosphorescence anisotropy selective technique, bovine serum albumin (BSA) (Sigma, defatted, Fraction V), molecular weight of 67 kDa, and mouse monoclonal anti-bovine serum albumin antibody (Sigma Chemical Co., mouse IgG 2a isotope) have been employed as antigen and antibody respectively to create rotating species of different sizes. The triplet probe, eosin-5'-isothiocyanate (Molecular Probes Inc.) has been covalently attached, using standard techniques, to the antigen to provide the phosphorescence emission signals required for the measurement. Actual concentrations of labelled protein in the initial sample collection were measured using the method of Lowry et al[9] with unlabelled bovine serum albumin as standard, and was found to be 10.7 mg/ml (1.55×10^{-4} M). The concentration of eosin molecules bound to the BSA was determined from the spectral absorption at 529nm, using $\varepsilon = 8.3 \times 10^4$ mol^{-1} cm^{-1} and was found to be 2.064×10^{-4} M. The stoichiometry of the protein to eosin was therefore 1: 1.33. Samples were prepared with different ratios of the antigen to the monoclonal antibody in order to create different populations of labelled antigen-antibody complex and labelled antigen species respectively. The reaction equation for creating the three species labelled in this reaction can be expressed as:

$$[BSA^*]_E + [AntiBSA] \Longleftrightarrow [2BSA^*\text{-}AntiBSA]_E + [BSA^*\text{-}AntiBSA]_E + [BSA^*]_E \tag{8}$$

where the subscript E denotes the species labelled with eosin, "*" indicates the molecules to which the eosin label was covalently attached. The [AntiBSA] and [BSA-AntiBSA] represent the species of free anti-BSA and the complex of BSA-anti-BSA respectively. The species of BSA-antiBSA complex, in fact, consisted of two species namely [BSA-antiBSA] and [2BSA-antiBSA], due to the bivalent binding feature of the IgG antibody. However, the rotational correlation times of the two species are expected to be similar as there is only a small difference of their overall size (about 25%). Because of the small difference in size of the two complex species our analysis has been performed assuming an average rotational correlation.

Instrumentation

Figure 2 shows the block diagram of the phosphorimeter developed for this application. The instrument employs a pulsed, frequency doubled Nd:YAG laser (pulse width, FWHM =5ns, λ=532 nm) as an excitation source. This laser has an adjustable output energy of up to 22 mJ, and a repetition rate of up to 100 Hz. During the measurement, horizontally polarized output pulses from the laser are turned to a vertically polarized exciting beam by passing through two Glan-Taylor prisms which are positioned such that the axes of polarization of the first and second prisms are kept at angles of 45^0 and 90^0, with respect, to that of the laser output polarization. This arrangement also reduces the output power of the pumping light to about 25% of its original power. The resulting orthogonally polarized components of the phosphorescence emission are selected, by passing through two sheet polarisers and two long-pass optical filters (cut off 665nm), and simultaneously collected by two photomultiplier tubes. All of these elements are oppositely arranged with the sample in the middle, forming a "T" format measurement geometry with the incident laser beam. The collected signals are digitized by a 8 bit digital storage adapter (Thurlby, DSA524), where the difference between the two components is also calculated and stored. The output from the adapter is subsequently transferred to a microcomputer, which operates a specially developed interactive program. Prior to the phosphorescence measurement, the two measurement channels of the phosphorimeter were balanced using the phosphorescence emission components of a free eosin sample in water which was excited by the laser pulse.

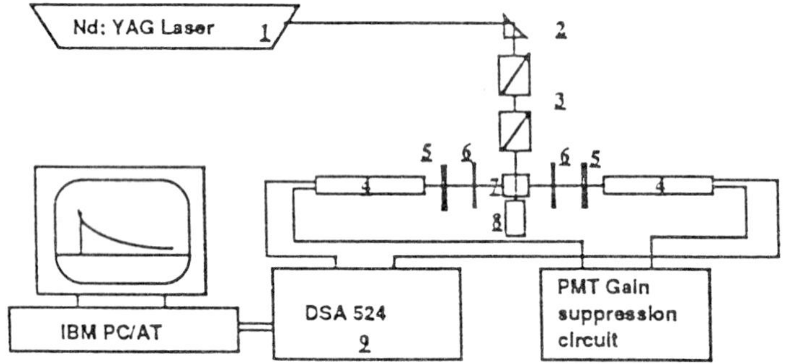

1) Nd:YAG Laser (λ = 532nm, FWHM = 5ns); 2)Prism; 3) Polarisors; 4) Photomultipliers; 5) Optical Filters; 6) Sheet polarisers; 7) Sample; 8) Beam dump; 9) Digital storage adaptor.

Figure 2: The phosphorescence depolarisation measurement system

Experiments were performed on sample preparations with population ratios of the anti-BSA to the BSA up to 0.7. The samples were placed in 1x0.5cm rectangular quartz cells fitted with a stopcock. Prior to the phosphorescence measurement, the samples were deoxygenated by gently blowing argon gas into the sample cuvette for 25min and re-incubated at 30°C for a further 30min. to eliminate the possible disturbance of the temperature turbulence to the antigen-antibody reaction during the deoxygenation procedure. The anisotropy measurements were then performed and the time-resolved phosphorescence anisotropy (r(t)) calculated according to the equation (4), and the rotational correlation times and their associated weighting function calculated via a Marquardt curve fitting procedure, and therefore, the rotating species in the samples and their population estimated.

Results

The time-resolved phosphorescence anisotropies obtained from several combinations of antibody and antigen are presented in Figure 3. The lower traces are the signals of the components with their polarisation direction parallel to the excitation source, and the upper traces the signals of the perpindicular components Figure 3 (a) shows the phosphorescence emissions of a homogeneous population of eosin-labelled BSA, and Figure 3(b) and (c) show the time-resolved phosphorescence from the samples of heterogeneous populations of BSA and complex of BSA-AntiBSA species. The concentration ratio of anti BSA to BSA in these samples were 0.1 and 0.245 respectively. The difference between the two components in the photograph is proportional to the total phosphorescence anisotropy measured according to the equation (2) and (4). From Figure 3 it can be seen that phosphorescence anisotropy increases as the relative concentration of antibody-antigen complex increases.

In order to distinguish the bound and unbound antigen more clearly and to provide a better signal to noise ratio, antibody was immobilized on a large surface by coupling the antibody to active beads. The enhanced phosphorescence anisotropy signals obtained from labelled antigen bound to immobilized antibody is shown in Figure 4 which shows two orthogonally polarized emission components from a sample homogeneous antigen (BSA) labelled with eosin-5'-isothiocyanate, while Figure 4 (b) and (c) show the same signals obtained from samples of heterogeneous populations of antigen and beads conjugated antibody. The relative concentration of the beads conjugated antibody to antigen in the latter samples were 0.67 and 0.8 respectively. Figure 4 shows essentially consistent observations as recorded in figure3. The overall anisotropy of the sample increases following the increase of the relative concentration of the antibody to antigen, or in other words, the relative population of the antigen-antibody complex to free antigen.

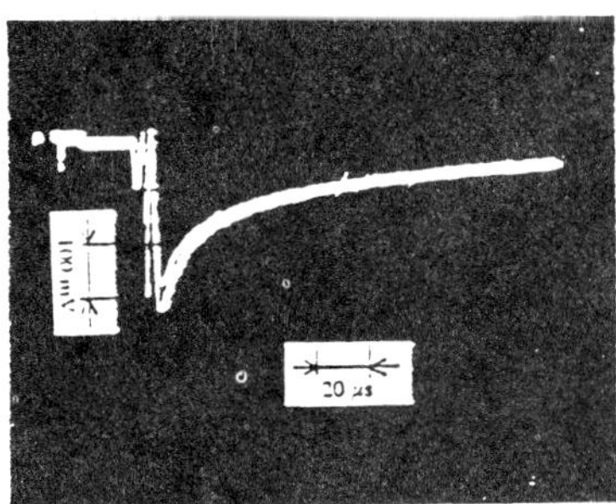

a

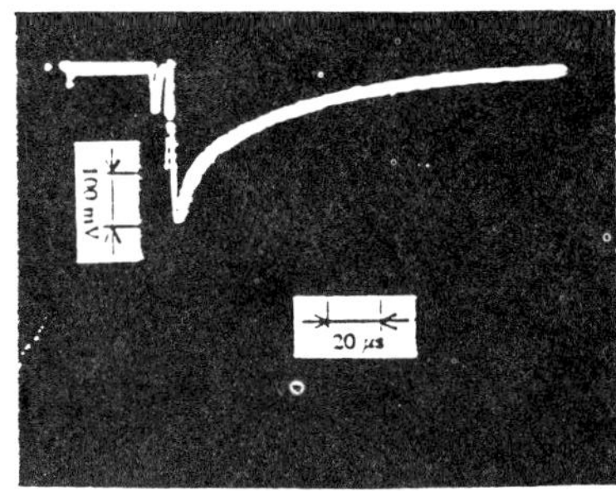

a

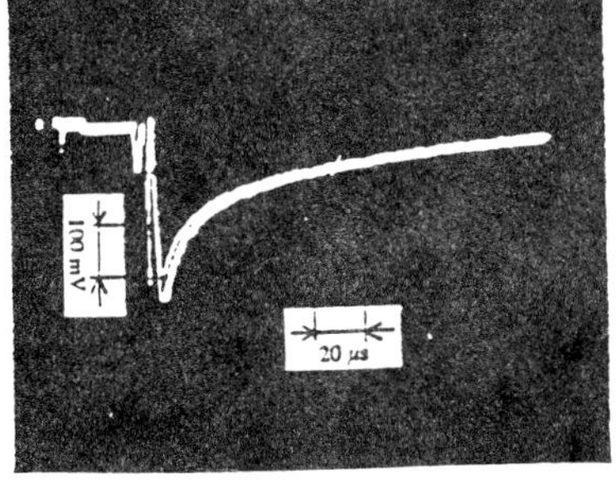

b

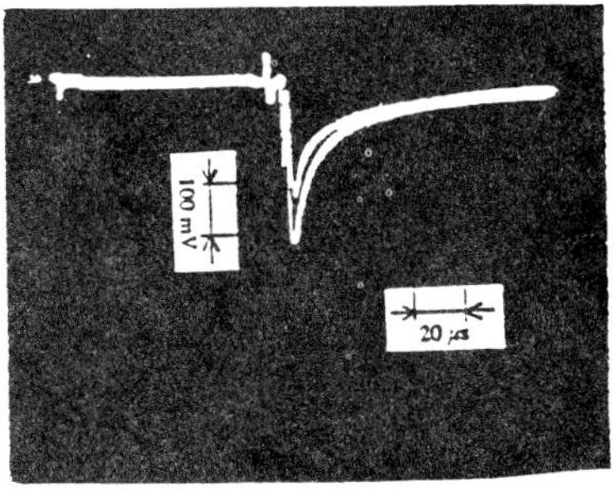

b

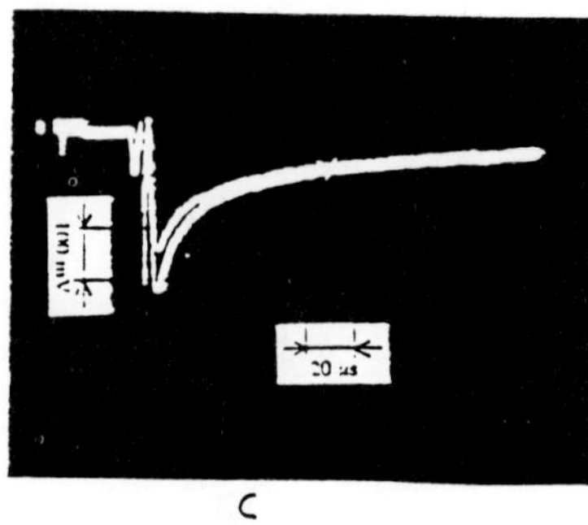

Figure 3: The time resolved phosphorescence of the BSA-AntiBSA mixture. The upper trace is the perpindicular and the lower the parallel polarised component for samples of: (a) homogeneous eosin-labelled BSA, (b) hetrogeneous population of antibody to antigen (concentration of antibody to antigen = 0.1), (c) concentration of antibody to antigen = 0.245

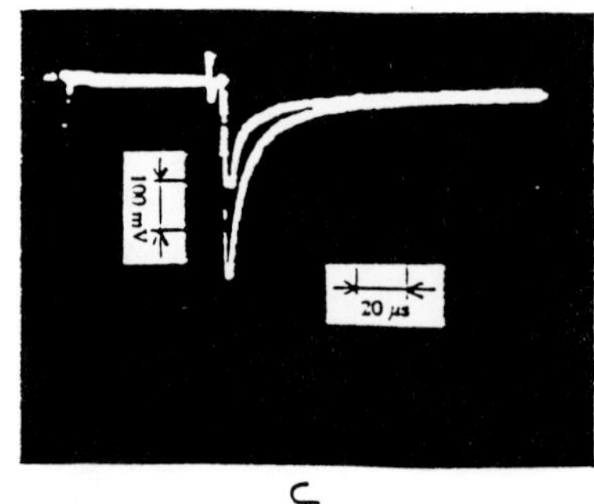

Figure 4: The enhanced phosphorescence anisotropy signals from immobilised antibody interacting with a free antigen The upper trace is the perpindicular and the lower the parallel polarised component for samples of: (a) homogeneous eosin-labelled BSA; (b) hetrogeneous population of antibody to antigen (concentration of antibody to antigen = 0.67), (c) concentration of antibody to antigen = 0.8

Conclusions

As discussed above, the primary experimental results indicate that (a) the total phosphorescence anisotropy signals from a sample consisting of species of different sizes increases as the relative proportion of the lager species increases; (b) the rotational correlation times and weighting functions observed in such systems are likely to be species-dependent and population-dependent respectively, therefore, the rotational correlation times could be used to identify the different species. Obviously to derive actual concentration of the species detailed data analysis via curve fitting is required. Initial analysis indicates that the rotational correlation times and the corresponding weighting functions are likely to be species-dependent and population dependent respectively which is consistent with a change in the rotational dynamics of the labelled molecules due to the multiple populations. The results indicate that the technique may be applicable to immunoassays.

Acknowledgements

Support of the K.C.Wong foundation and The Chinese Education Commission to L.Yang is acknowledged. The research was aided by grants from the KCL Research Strategy Fund.

References

1. Devices, D.R. and Padlan, E.A.; *Annu. Rev. Biochem.*, 59, 439-473, (1990)
2. Yalow, R.S. and Berson, S.A.; *J. Clin. Invest.*, 39, 1157, (1960)..
3. T.Olsson and A.Thore, in *Immunoassays for the 80's*, A.Voller, A.Bartlett, D.Bidwell, Eds., in-Press, UK., ppl 13-125, (1981).
4 P.R.Hangland; *Handbook of Fluorescent Probes and Research Chemicals*, Molecular Probes Inc: Eugene, (1989-1991).
5. A.A.Amkraut; *Immunochemistry, 1*, 231-235, (1964).
6. F.V.Bright and L.B.McGown, *Talanta*, 32, 15, (1985).
7. F.V.Bright; *Applied Spectroscopy*, 42(7), 1245-1250, (1988).
8. M.D. Barkey, A.A. Kowalcryk, and L. Brand, *J. Chemi. Phys.*, 75(1), 3581-3593, (1981).
9. O.H. Lowry, N.J. Rosebough, A.L.Farr, and R.J.Randall, *J.Biol.Chem.*,193, 256-275, (1951).
10. R.J. Cherry, *Methods in Enzymology*, 54, 47-61, (1979).

The development of a fibre optic respiratory plethysmograph (FORP)

A.T. Augousti and A. Raza

School of Applied Physics, Kingston University, Kingston, Surrey KT1 2EE

Introduction

Respiratory tidal volumes are usually measured using a spirometer [1]. However these devices are invasive, in that a nose-clamp is placed on a subject forcing them to breathe through a tube gripped in the mouth. As well as being uncomfortable, spirometers are bulky and are not convenient to use under certain circumstances, for instance respiratory monitoring during an NMR scan, or they may cause distress in respiratory measurement of neonates in intensive care. Non-invasive devices which do not require connection to the airway have been developed such as those based on the use of magnetometers for measurement of antero-posterior diameters of rib cage and abdomen [2], and more recently the respiratory inductive plethysmograph (RIP) [3,4], which is based on direct detection of changes of cross sectional areas. However, the transduction principles involved in both of these devices renders them useless for operation in harsh electromagnetic environments, such as the interior of an NMR scanner.

This paper details the construction of a prototype of an optical analogue of the RIP which will be capable of operation in a wider range of environments than hitherto.

Optomechanical Design

This sensor consisted of a red light emitter and detector unit, two sensor bands and a dedicated P.C. to display results. The experimental arrangement is shown in figure 1 with a detailed enlargement of the sensor bands shown in figure 2.

The light emitter and detector unit consisted of two separate channels and was based upon two red light emitting diodes centred on 660 nm wavelength [5] and two corresponding photodiode

[5]. One pair of emitters and detectors was used for each sensor band. Light was guided from each LED to its sensor band and then back to the photodiode via a single 1 mm internal diameter polymer optical fibre. The fibre was contained within the sensor band in a series of loops. Numerous methods of wrapping the fibre around the body and attaching it to the elastic bands were investigated and the optimum arrangement has been shown in figure 2. Each sensor band was made by sandwiching the fibre between two strips of elastic bandage (Tubigrip). The fibre was looped repeatedly and secured along the first elastic bandage by sewing a second thinner strip down over these loops. Elasticised thread was used to allow the bands to stretch freely. The two ends of the fibre that protruded out of the band were taped down to stop the fibre being retracting when the band expanded.

The bands were designed to wrap around the torso (chest and abdomen) and enabled the fibre loops within the band to diminish in radius as the torso cross-section increased. The more the torso expanded, the greater the reduction in loop radius that occurred and therefore the greater the attenuation of light propagating along the fibre. The sensor is therefore essentially an intensity based device utilising the principle of bending losses in a fibre. The sensitivity of each channel increases as the loop radius decreases, but the resultant intensity diminishes. The unstretched loop radius was chosen so as to optimize the sensitivity of each channel whilst ensuring that the signal strength remained at an acceptably high level. The sensitivity was further increased by increasing the number of loops within a sensing band.

The light was detected using a standard photodiode, amplified and input directly into a PC using the internal A/D converter. This performed a degree of signal processing and displayed the results graphically.

For testing purposes, each band was sewn together at the ends to form a tube with a given band radius. One tube was placed around the upper torso (chest) and one around the lower (abdomen). Following calibration of each sensing band, it was now possible to crudely determine the volume enclosed between the tubes by using a short computer program. The program assumed a simple linear extrapolation between the sensing bands and thereby calculated the volume of a normal conic section. The program also included a primitive warning system, alerting the operator when there was no volume change or if the volume change was below a pre-set value.

Since this was a proof of principle investigation, the sensor bands were made much shorter and a rubber balloon was used as a test object in place of a human torso. This was inflated and deflated by a hand pump, and allowed a straightforward calibration between signal strength and band radius. Note that the latter quantity is distinct from the radius of individual loops within the band.

Results

Figure 3 shows a calibration graph of the transmitted intensity within a channel versus the band radius. This graph is relatively linear and also reproducible upon cycling (to within 5-10%). In order to determine accurately the volume changes in the torso, both tubes were calibrated independently.

During operation of the sensor, the computer displayed a graph of volume change against time and also recorded minimum and maximum volume intakes. An example of a typical display from the computer screen can be seen in figure 4.

Discussion

A crude proof-of-principle fibre optic respiratory plethysmograph (FORP) has been constructed. It operates using a differential fibre bending loss mechanism, and consists of two independent channels. The device is interfaced directly to a PC. Following calibration of each channel, a simple algorithm calculates the volume enclosed by the bands, and displays volume changes in real time. A certain degree of operator control is available through a menu system, for instance minimum warning settings may be chosen (for instance depth of tidal volume, frequency of respiration etc.). It is envisaged that the device will eventually be suitable for use in an NMR environment.

References

1. **Cameron JR and Skofronick JG** *Medical Physics* John Wiley and Sons London 1978 130-2

2. **Grimby G, Bunn J, and Mead J** Relative contribution of rib cage and abdomen to ventilation during exercise **J. Appl. Physiol.** 1968 **24** 159-66

3. **Cohn MA, Watson H, Weisshaut R, Stott F, and Sackner MA** A transducer for non-invasive monitoring of respiration *ISAM Proceedings of the 2nd International Symposium on Ambulatory monitoring,* London, Academic Press 1975 119-28

4. **Sackner JD, Nixon AJ, Davis B, Atkins N and Sackner MA** Non-invasive measurement of ventilation during exercise using a respiratory inductive plethysmograph I. *Amer. Rev. of Resp. Dis.* **122** 1980 867-71

5. Farnell Electronic Components April 1993 p545

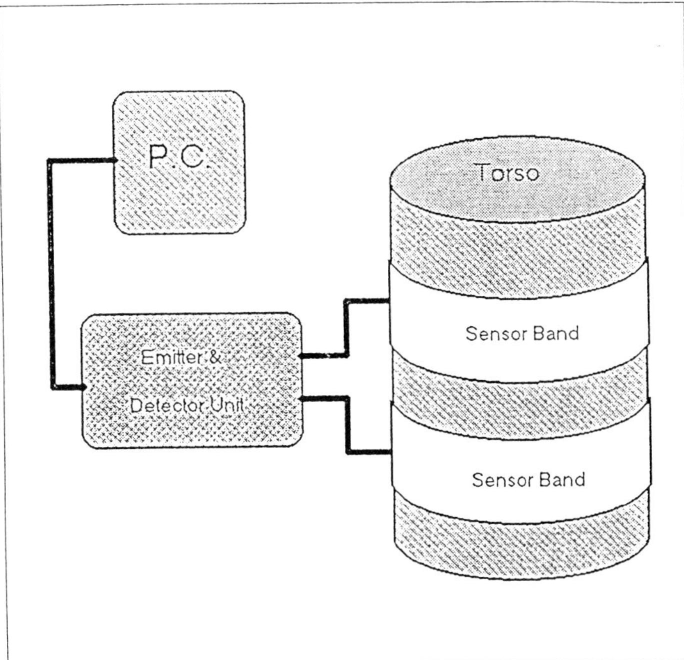

Figure 1 Experimental Arrangement.

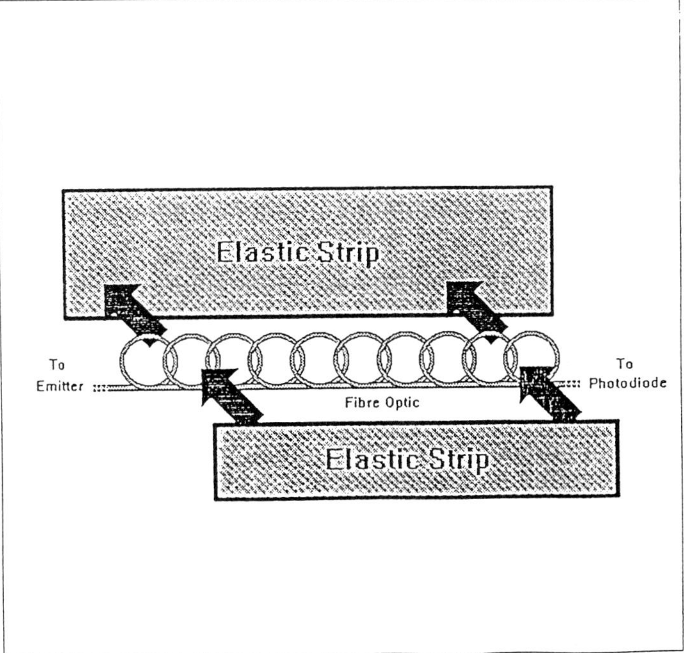

Figure 2 Construction of Sensor Band

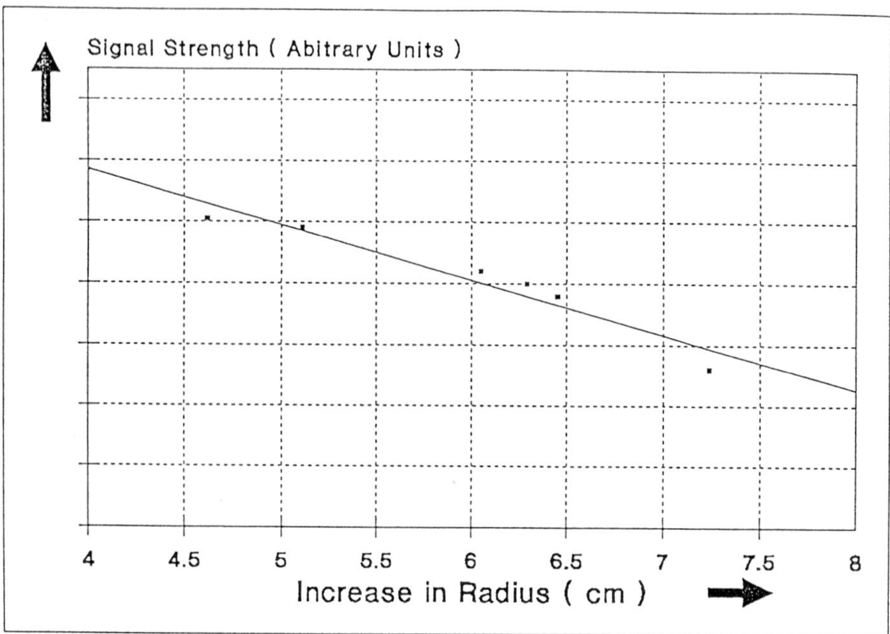

Figure 3 A graph of Signal Strength verses change of Sensor Band Tube Radius

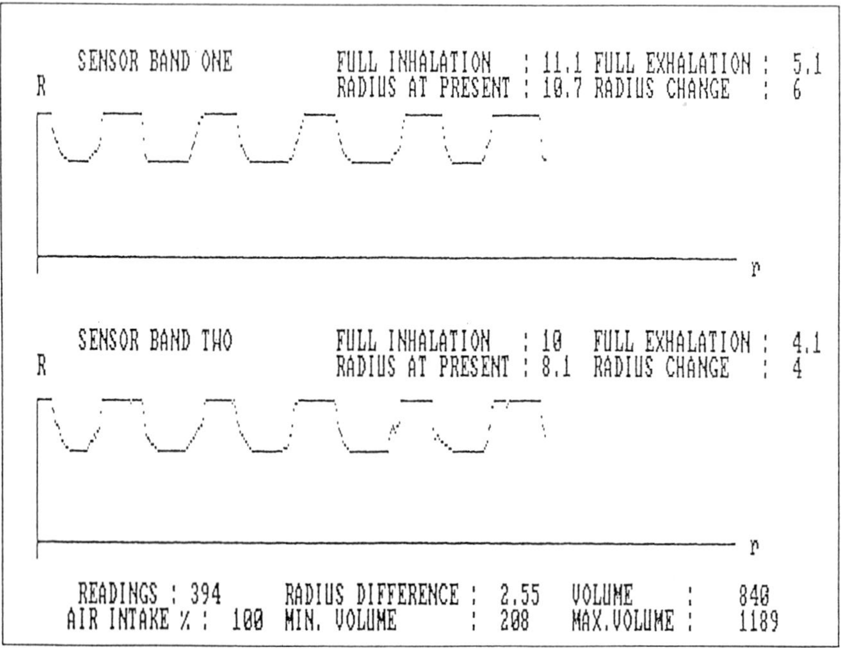

Figure 4 Examlple of typical screen output using two sensor bands on a test balloon.

Design and fabrication of a new piezoelectric bone strain gauge

V.R. Singh and Sanjay Yadav
National Physical Laboratory, New Delhi-110 012, INDIA

ABSTRACT: Bone is now a well established piezoelectric material. A piezoelectric bone strain gauge is developed by using a thin element segment cut out from a solid bone. This element is of rectangular shape and has two lead wires attached to its two ends. Both the ends of the bone element are coated with a conductive silver paste to obtain electrical connections. The lead wires are then soldered on to both the ends. The gauge is excited by an a.c. oscillator and is bonded on to a cantilever made of a thin bronze sheet. The free end is bent downwards with the application of mechanical stress. The resultant electric output is displayed on an oscilloscope or measured with an electrovoltmeter. Non-electrical parameters are measured with the calibrated electrical output of the gauge. Micro-displacements, jaw forces, vibrations etc. are measured with the help of such a gauge for various biomedical and scientific applications. Design and fabrication aspects of such a new piezoelectric strain gauge are given in this paper in detail.

INTRODUCTION

Strain gauges are extensively used these days for monitoring strain of various structures [1-4]. In the medical field, semi-conductor and conventional resistance wire gauges are used for the measurement of blood pressure and other bio-physical movements. Mainly, gauges are of two types, the resistance type and semi conductor type, both in bonded and un-bonded forms. Piezoelectric strain gauges are, however, used in dynamic measurements. Generally, quartz and PZT (Lead Zirconium Titanate) materials are used in these gauges. Recently, the authors have developed a new piezoelectric material from natural bone [2]. A new type of piezoelectric strain gauge by using the bone as the piezoelectric material has been developed in the present work. Design and fabrication aspects of such gauges are given in detail, in this paper.

DESIGN AND FABRICATION

Bone samples were procured from a local butchers shop and were cleaned first manually and then chemically. Elements were cut out from the mid-diaphysis of the solid wall of the bone in the form of a rectangular strip (size 5 mm wide, 30 mm long and 1 mm thick). A low speed hexa blade was used for cutting purposes in the stream of water, used as a coolant. This cutting method was chosen to avoid any pathological changes in the bone matrix. The electrical contact was provided with firing on silver using air dry colloidal silver paint on either side of the bone element. Measurements of piezoelectric properties were taken on these samples and found to be of use. Electrical leads were attached to two ends of the thin gauge element.

MEASUREMENTS AND DISCUSSION

One end of the bone strain gauge element bonded on a thin bronze sheet was kept fixed in a cantilever fashion and the other end was kept free to be attached to the measurement position. A special light-weight platform, made of "cork" rubber, was used to put known weights (in grams) at the free end. The gauge was excited with an oscillator. Electrical output with the application of the known weight stress was monitored as linear up to a certain range. Also, the strain developed in a structure under study was measured in terms of micron displacement. The gauge factor viz. sensitivity, defined as the ratio of incremental change in resistance (ΔR) in gauge resistance (R) and strain (ratio of incremental length Δι in length 1) was found to be between 10 to 40. The value of the gauge factor in the case of semiconductor strain gauges is about 100 and in the case of resistance wire gauges about 2. Thus, bone piezoelectric gauges are better than the ordinary resistance gauges. The stress response of the gauge is referred to frequency response relating to a uniform stress applied to the gauge to its output voltage. This response is in a particular direction. The resonance was found to occur at around 30 KHz but it was better over higher frequencies up to 50 KHz, in this case in comparison to that of piezoelectric ceramic type elements at the frequency range of 200 Hz to 15 KHz. The size of the sensing element is directly proportional to the wavelength of sound. Dynamic vibrations can be measured very successfully with the present gauge in various scientific and biomedical applications.

CONCLUSION

A new strain gauge based on bone as piezoelectric sensing element is developed.

REFERENCES

1. Singh, V.R., Ahmed, A and Yadav, S., Electrical and mechanical properties of bone , J. Instn. Engrs Ind. 68, IDP-I, 1988, 43-47.
2. Yadav, S. and Singh, V.R., Development of a piezoelectric bone diagnostic probe J. Meas. Sci. Tech. (UK) 2, 1992, 1155-1158.
3. Yadav, S. and Singh, V.R., Development of a bone piezoelectric microphone pick-up for vibration measurements Innov. Tech. Biolog. Med. 11 (I), 1990 ,89-95.
4. Singh, V.R., Yadav, 5. and Ahmed, A. A piezoelectric bone hydrophone for medical ultrasound application , Proc. 10th Int. IEEE-EMBS Conference, New Orleans (USA), Nov. 4-7, 1988, pp 755-756.

Section G

SENSOR APPLICATIONS

Sensors in manufacturing systems

D.A. Bradley
Engineering Department, Lancaster University, Lancaster LA1 4YR

ABSTRACT: The move towards flexible production units means that a modern manufacturing system cannot be considered solely in terms of the manufacturing processes involved but as an integration of engineering design with production technology and support services such as warehousing. Each of these aspects places specific and particular demands on sensor requirements and sensor performance in providing the information necessary to maintaining the required levels of system performance. The paper examines each of the above aspects of a manufacturing system in turn and considers the associated sensing requirements both in terms of current technology and with reference to areas of future development.

1. Introduction

Modern industry increasingly demands manufacturing systems and production facilities which can react quickly to changes in market conditions, a requirement which often involves rapid change-overs from one product to another in order to be able to offer a wide product range, an approach which generally involves batch sizes which are too small for more conventional production systems. The result has been the introduction of flexible production units or flexible manufacturing systems based around a combination of computer numerically controlled machine (CNC) tools, robots and automated materials handling together with automated transport systems.

This change in the structure of the production unit has been accompanied with a change in the organisation of the production process to accommodate techniques such as just-in-time parts delivery to reduce stock levels and, more recently, concurrent or simultaneous engineering to integrate the design and manufacturing processes and reduce the time required for product development. Additionally, the demand for increased quality and the introduction of quality standards such as BS5750 and ISO9000 have placed increasing demands on the need for improved control of all aspects of product design and manufacture.

The result is that manufacturing can no longer be considered purely in terms of the sequence of processes that are involved in the production of a single component or of a product containing a large number of assemblies, sub-assemblies and components but must be considered as a complete system covering all aspects of the design, development, production, marketing and support of the product[1,2].

In the context of such a system, sensors and sensing systems play a variety of roles ranging from the provision of support for the design and development process, through the control and operation of the production processes and the associated support services such as stock control and component delivery to the in-service monitoring of the performance of the finished product.

The boundaries between these areas are generally vague and ill-defined within any individual manufacturing system. For instance, in a flexible manufacturing system such as that suggested by figure 1, the support facilities in the form of the tool store and delivery system may be considered to be just as much an integral part of the of the production process as the CNC machine tools and robots making up the local 'islands of automation'. Similarly, the design and development of the product cannot be properly carried out without an understanding and knowledge of the performance of the production system on which the product is to be manufactured.

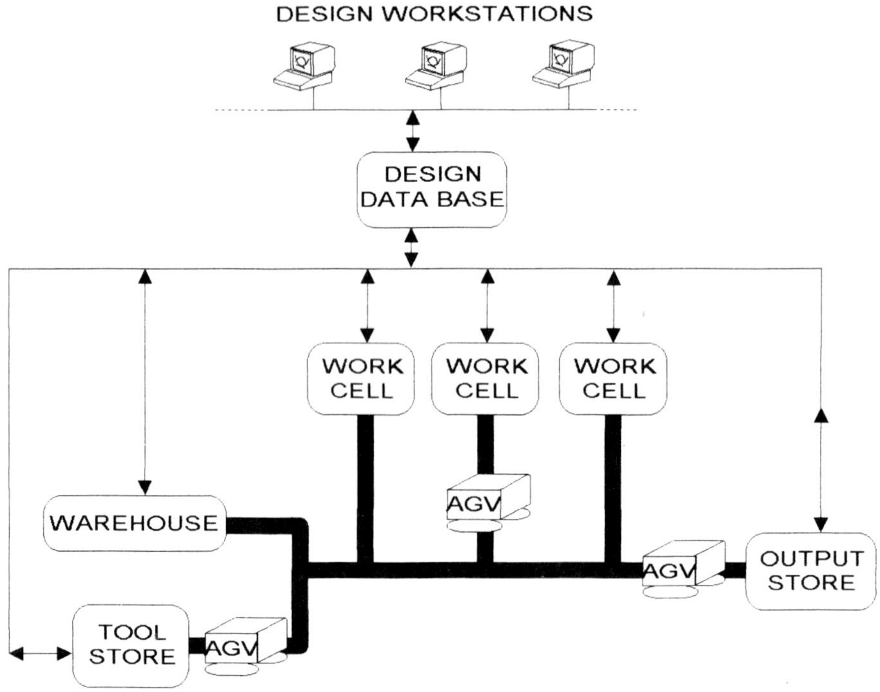

Figure 1. A manufacturing system

For the purpose of the paper the sensing requirements are considered in the context of three particular elements of the overall manufacturing process as follows:

Production - This covers all aspects of the processes involved in the production of the finished product and includes the operation of items of equipment such as CNC machine tools and robots together with quality matters.

Support - Support systems are taken to be those elements of the manufacturing system that facilitate production through the storage and delivery of parts and components. Items such as automatic guided vehicles (AGVs) and automated warehousing are therefore in this category.

Design - This covers all aspects of the engineering design process together with the development phase leading to the final design. It encompasses prototype testing as well as the definition and selection of assemblies and sub-assemblies.

The paper therefore considers current and likely future approaches to sensing within the overall context of manufacturing systems and makes suggestions for a new and improved generation of sensors to enhance the overall performance of such systems.

2. Sensors in the Production Process

Within a production environment sensors are required to monitor a wide range of physical parameters and to provide information on the dimensional and physical properties of the output from the production process as suggested by table 1 while table 2 sets out some of the criteria associated with the operation and classification of sensors[3,4]. In addition, items of equipment such as CNC tools and robots integral to the production process and whose operation often dictates the quality of the finished product are in turn dependent for their operation on the performance of their internal sensors.

The production process itself can be further broken down into the following major areas:

Process Monitoring
Condition Monitoring
Machine Control
& Materials Handling

each of which presents particular problems and requirements in terms of sensor performance.

Length	Breadth	Depth
Profile	Contour	Angle
Position	Orientation	Roundness
Flatness	Surface finish	Temperature
Weight	Colour	Acceleration
Force	Torque	Velocity
Structural integrity	Thickness	Density
Pressure	Time	Volume

Table 1. Possible measured properties

Accuracy	Tolerances	Repeatability
Linearity	Resolution	Sensitivity
Hysteresis	Bandwidth	Natural frequency
Monotonicity	Frequency response	Impulse response

Table 2. Factors affecting the quality of a measurement

2.1 Process monitoring

With increasing levels of automation, the influence of the sensing structure on the operation of the production process has correspondingly increased as responsibility for maintaining acceptable levels of performance has moved away from the operators to the systems themselves. For instance, the growth in the use of automated inspection methods both off-line, in-line and on-line has led to the development of highly sophisticated co-ordinate measuring machines using both touch and optical probes such as those in figure 2 to check component dimensions[5,6,7].

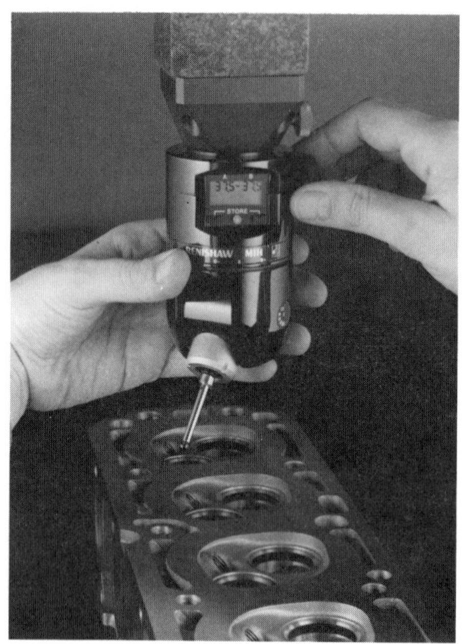

Manual indexable head

OP5M laser probe

Figure 2. Touch and optical probes
(Courtesy of Renishaw)

In a conventional monitoring system individual items are sent to the checking station in turn as suggested by figure 3 and the results monitored using the techniques of statistical process control.

The resulting feedback process does not normally allow for the immediate correction of the manufacturing process. For this reason, attention is being given to the development of the techniques of in-process gauging in which the dimensional accuracy of the workpiece is monitored during manufacture and appropriate compensation introduced directly.

With in-process gauging, each operation carried out by a machining centre would, for instance, be continuously monitored and the results immediately fed-back to control operation, as suggested by figure 4. The close linkage that can be achieved between the product and the manufacturing process by this means has a number of likely and significant benefits in terms of quality and, if linked to the known characteristics of the machine could reduce error rates in order to 'get it right first time'.

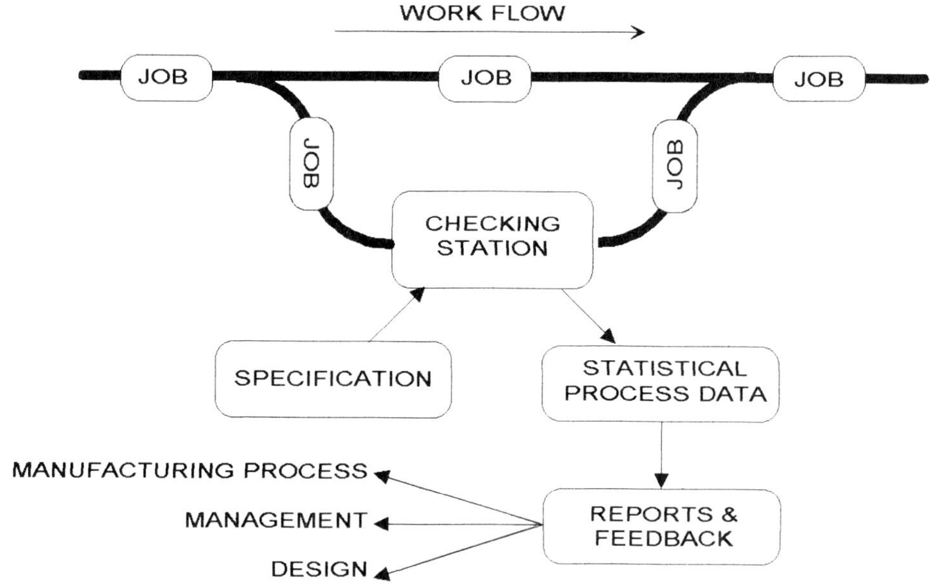

Figure 3. Process monitoring using a checking station

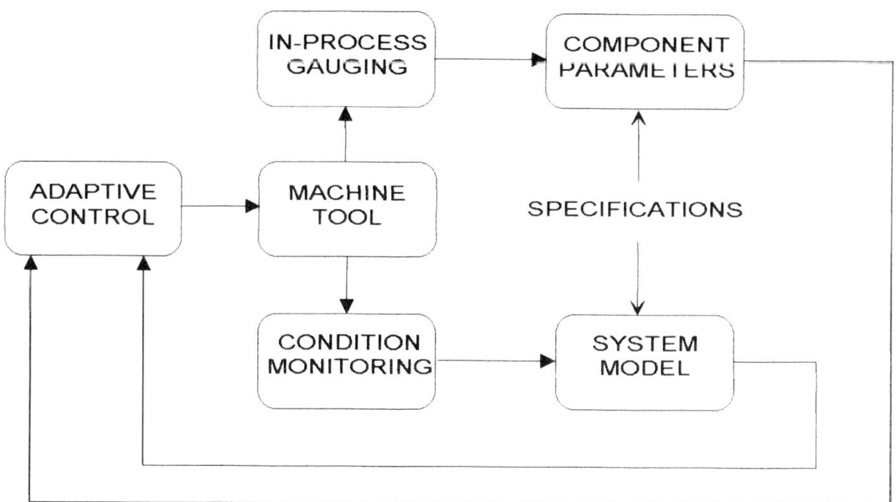

Figure 4. In-process gauging combined with direct feedback and machine adaptive control

The increasingly widespread adoption of in-process gauging by a range of industries will require the provision of high performance, fast response sensors, largely based on the use of non-contact techniques, with high levels of reliability and incorporating a range of self-test features together with enhanced signal processing.

Other areas of manufacturing have the same problems of process control and quality as those associated with machining processes but present different range of sensing and sensor problems by virtue of the different properties of the process. For instance, the spinning and weaving industries are faced with the need to measure the thickness[9], and indeed the 'hairiness', of yarn and the weight per unit area of the woven material at production rates of the order of several metres per second. Optical techniques, such as the Barco-Dextralog 'Barcoprofile' system which uses a shadowing technique to measure yarn diameters to a resolution of 0.01mm, have been developed for these and other purposes and are currently incorporated into on-line process control systems.

2.2 Condition monitoring

Condition monitoring can be seen as an extension of the concept of in-process gauging to encompass the performance of the machine, to monitor factors such as tool wear and provide information on maintenance requirements[10,11] and give advanced warning of possible system failures, particularly where operation may continue to be essential normal up to the instant of failure. Table 3 sets out some of the possible techniques that may be used as part of condition monitoring.

On-line		
Vibration analysis	Shock pulse monitoring	Sound signature analysis
Stress measurements	Torque measurements	Position sensing
Visual imaging	Flow measurement	Trace element sensing
Temperature measurement	Flow measurement	
Off-line		
Wear particle analysis	Lubricant condition	Filter debris analysis
Remote imaging	Scanning electron microscopy	

Table 3. Condition monitoring techniques

A machine tool is usually a noisy and often dirty environment subject to high mechanical stresses and vibrations and presents a serious challenge to the designer of any instrumentation system, particularly with regard to condition monitoring. What is needed in this environment are sensors and sensor systems which are capable of providing the design and production engineer with a clear insight into and information on the status of the manufacturing process in a form which can be used to control that process.

2.3 Machine control

The control of machine tools and industrial robots depends upon the provision of a range of internally generated sensor data covering all aspects of their operation. However, in order to ensure effective operation under all conditions the measurement of basic position data may not be sufficient. Consider the case of an industrial robot. Current robots are be designed to be structurally stiff to minimise problems of error introduced as a result of both static and dynamic deflections. This in turn leads to their being relatively massive in relation to their payload. Similar arguments also apply to many machine tools designs.

A particular problem in relation to the deployment of robotic technologies is the need to define the operation of the end effector or gripper, the operation of which is usually dependent upon the provision of appropriate sensing within the gripper itself. Requirements for effective gripping may include factors such as object detection and ranging, measurement of gripping forces and torques[12,13], detection of slippage and for unknown object types some form of shape recognition[14]. In many instances effective operation will require the incorporation of sensors within the end effector 'fingers' themselves.

The incorporation of integral sensing[15] within the structure of a robot or machine tool together with the use of advanced materials in their construction and the embedding of appropriate actuators and micro-actuators when taken together with the adoption of advance control and operating strategies as suggested by figure 4, offers the possibility of developing a new range of production machinery which is more capable than the current generation but which are lighter, more reliable and consume less energy.

2.4 Materials handling

Effective materials handling plays an important role in the efficient operation of many manufacturing systems by ensuring that parts and materials are presented to the process in the right order and at the right time. Many current systems rely either on the pre-orientation of parts prior to delivery or the use of vision, acoustic and other techniques to detect the presence and orientation of the required component.

There are however certain types and classes of materials, such as limp materials and including cloth, which present particular sensing and gripping problems, especially with regard to the ability of the system to locate and grip a single layer of fabric or material from a pile of such material. This not only requires that the gripper correctly locates the material but also that it incorporates appropriate sensing to detect the separation of a single layer[16].

The materials handling system also has a major role to play in the sorting of, and rejection where necessary, of components prior to their entry into the production process. For instance, there is often the need to ensure that items are properly presented to the appropriate machine tool or robot. This requires the use of sensing which will determine the presence and the orientation of the object or objects prior to manipulation to change the orientation if necessary. As an example of a sorting problem, consider the need in the food industry to identify items which are not of the required quality prior to processing. Illustrative of the type of techniques that have been evolved, a system has been developed for the sorting of peas that uses a combination of a helium/neon laser to check colour and an argon laser to measure texture. The output of the measurement is used to control 64 compressed air jets to eliminate unacceptable peas. The system claims an accuracy of 99% at a rate of 8 tonnes of peas an hour[17]. Similar systems have been developed for the sorting of chipped potatoes and diamonds!

3. Sensors in Support of Manufacturing

The support services associated with the manufacturing processes include the full range of warehousing and goods delivery systems, including automatic guided vehicles, but can also be taken to encompass areas such as environmental monitoring and safety systems which provide the framework within which the automated systems operate.

3.1 Warehousing and stock control

The concept of 'just-in-time' manufacturing envisages minimum stock levels for the majority of system components and sub-assemblies. This means that the warehousing and stock control systems must be capable of dealing with a large, and often highly variable, stock list. The use of electronic tagging of components for rapid identification together with highly automated pick-and-place functions are essential elements of a high-volume, high-throughput system.

Less critical in terms of the rate of throughput but just as essential to the operation of any flexible manufacturing system is the operation of the tool store. Here, the operation must take account not only of the particular tooling demands but also of the need to refurbish tools being returned and their re-entry to stock.

3.2 Transport systems

Automatic guided vehicles (AGVs) using guidance techniques based on the use of buried cables and floor markings are now well established within manufacturing environments and are used for the transport and delivery of parts, components and tools[18]. The next stage of development has been the introduction of free-ranging AGVs (FRAGVs) using systems of beacons and reference markers and capable of a limited degree of self-navigation. Further flexibility of operation will be achieved by the combination of improved sensing, particularly vision based, and enhanced signal processing techniques[19,20].

3.3 Safety systems

Machine tools, and in particular robots, must be provided with a safe operating environment within which they can carry out their function without risk to personnel, to themselves or to other equipment. This requires the use of a sensor system capable of mapping the local environment and of detecting the presence of both static and mobile obstacles within that environment. Current systems based on the use of combinations of technologies such as vision, lasers and ultrasound are not as yet capable of providing the level of integrity required with adequate response times[21,22].

4. Sensors in Engineering Design

CAD and ECAD systems are now well established tools in engineering design and enable not only the visualisation of products but in some cases the processes involved in their manufacture. However, on considering current design support systems in more detail, a number of problems become apparent in relation to their operation and use. In particular, while processing times continue to reduce, the rate at which an operator can input data to the computer system remains essentially fixed at around 10 characters per second for whatever form of input is being used. Also, even with the use of high resolution solid modellers, the two dimensional on-screen image does not allow the design team the full multi-degree-of-freedom visualisation that would often be desirable.

A prime requirement in the development of advanced, computer based design support tools lies with the development of systems which facilitate communication between team members in a manner more akin to the ways in which they would do when seated round a table. With regard to visualisation, the requirement is for a means by which the designer can 'enter into' the design environment.

4.1 Virtual Reality

High performance Virtual Reality (VR) systems[23] using head mounted displays have to date been largely developed for applications such as flight simulators where the high costs, typically several million pounds, associated with the performance can be justified[24]. In the design environment associated with a manufacturing system, Virtual Reality represents a possible means by which the level of interaction with the CAD system can be enhanced, for instance by permitting an almost literal 'walk through' of the design of the product or process or allowing the consideration of a production process from the point-of-view of a particular part or component and to directly manipulate and adjust the operation of items such as robots in virtual space[25].

To be effective, Virtual Reality systems need to deploy sophisticated sensors to track head and hand movements and to monitor other physical motions such as the flexing of the users fingers. A particular problem is that of providing an effective tactile feedback to enable the gripping of objects in virtual space. Techniques based on head mounted displays and data gloves are the basic tools for current virtual reality systems but are often bulky and uncomfortable for the user.

For the techniques of Virtual Reality to become more generally applicable and to be fully integrated into the design process there is the need to develop lighter and more sensitive motion sensors capable of accommodating the data rates required and of handling the large number of degrees-of-freedom associated with, for example, the motion of the human arm, and all at an acceptable price.

5. The Future

As manufacturing systems develop in performance and intelligence sensors will be required to provide increasingly complex information at higher data rates and at enhanced performance levels. This will require the introduction not only of new sensor technologies, perhaps with an emphasis on new and novel forms of non-contact sensing, incorporating features of self-calibration and self-test integrated with enhanced signal conditioning and processing techniques, including neural networks[26,27,28,29,30].

Strategic level	User specified production goals
Tactical level	Production goals are analysed to specify robot function
Task level	Define tasks to be completed in relation to the specified goal
Action level	Decompose individual tasks into an appropriate sequence of actions
Servo level	Decompose actions into the drive commands for individual joints

Table 4. Task planning for an intelligent robot

Consider the advanced robot system suggested by figure 5, the operation of which can be considered in relation to various levels of operation as set out in table 4 the associate sensor hierarchy for which is defined in table 5.

For a system such as that of figure 5 in which sensors are required to provide information about system and world conditions there is a need not only for the development of the appropriate sensors but of the associated systems necessary to aid and support their choice, definition and application across a wide range of functions such as those set out in table 6.

Perception level	Analysis of task model to infer the state of the world, consequences of actions and the possibility of collisions
Model level	High level geometric and world model based control in a multi-sensor environment. High level sensor integration and data fusion.
Measurement level	Physical system model. Conversion of sensors signals to values representing physical quantities. Detection of dangerous situations.
Physical level	Individual sensors and transducers with local signal processing and signal conditioning. Interface electronics and hardware.
	After McKerrow, 'Introduction to Robotics', p548

Table 5. A sensor hierarchy for an intelligent robot system

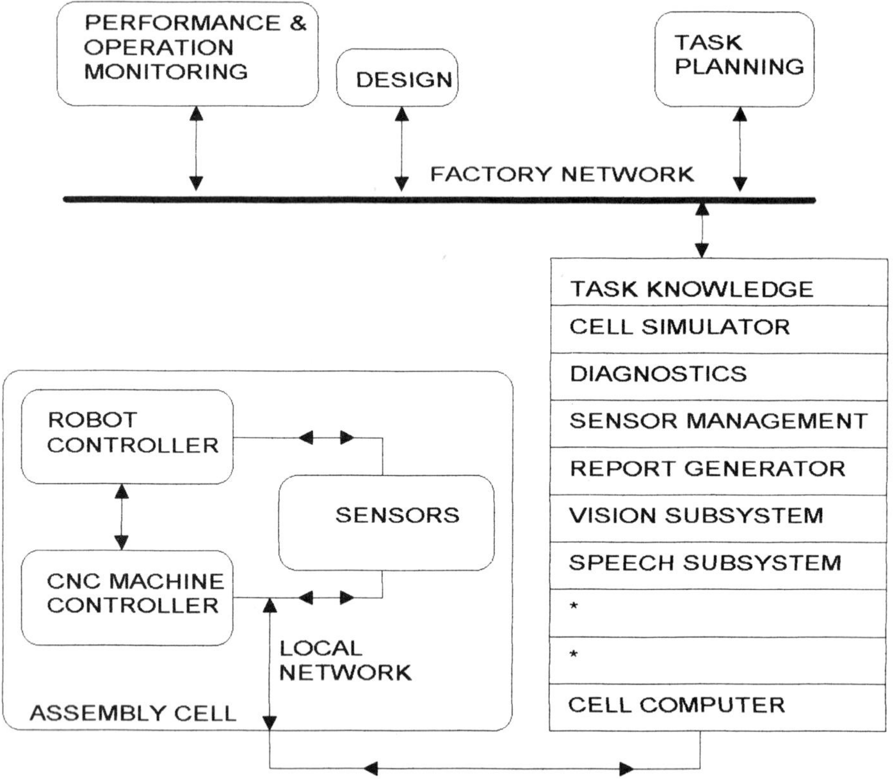

Figure 5. An intelligent robot system

6. Conclusions

Sensors form an integral part of manufacturing systems and indeed it is generally their performance that determines the overall performance of the system. Current sensor technologies can provide good to high levels of performance over a wide range of functions involved in current manufacturing technology. However, developments such as lightweight, dextrous robots, intelligent robots, advanced machine tools and virtual reality will require the introduction of a new range of integral and remote sensors capable of directly monitoring an increasing range of system and world parameters.

As the integrity and performance of the system increasingly depends on the operation of an integrated mix of sensors there is an accompanying need to ensure that the required levels of sensor performance is maintained at all times. The incorporation of features of self-calibration and built-in-test must therefore be a major feature of sensor development as will be the requirement to interface with complex distributed and embedded information processing structures.

Parameter measurement for control	Correction of errors in robot and world models
Location of objects	Condition monitoring
Detecting and avoiding failures	Detecting and avoiding collisions
Path planning	Monitoring interaction with environment
Monitoring changes in the environment	Inspection and quality control

Table 6. Sensor applications in intelligent manufacturing systems

References

1 Hartley, J.R., (1992), *Concurrent Engineering: Shortening lead times, raising quality and lowering Costs*, Productivity Press, Cambridge, Mass
2 Eversheim, W. & Gross, M., (1991), *Trends and experiences in applying simultaneous engineering*, Computer Integrated Manufacturing, 7th CIM-Europe Conference, Turin, Italy, pp 3-11
3 Jones, B.E., (1989), *Industrial metrology and sensors*, IEEE Conf. VLSI & Microelectronic Applications in Intelligent Peripherals & their Interconnection Networks, Paper 3, p 8
4 Heitor, M.V., (1991), *Advanced sensor systems for the application of CIME technologies in the process industry: A review*, Industrial Metrology, Vol 2, pp 1-31
5 Butler, C., (1991), *An investigation into the performance of probes on coordinate measuring machines*, Industrial Metrology, Vol 2, pp 59-70
6 Butler, C., (1992), *A smart sensor for precision position measurement*, Conf. on Mechatronics - The Integration of Engineering Design, Dundee, pp 87-90
7 Purves, W.K. & Wright, D.A., (1991), *Microprocessors on coordinate measuring machines*, Conf on Mechatronics - Designing Intelligent Machines, Cambridge, pp37-46
8 Katebi, M.R., McIntyre, J. & Grimble, M.J., (1992), *Integrated process and control design for fast coordinate measuring machine*, Conf. on Mechatronics - The Integration of Engineering Design, Dundee, pp 109-116
9 Cassidy, T., (1991), *The relationship between woollen card web density and light obscuration characteristics*, Wool Science Review, No 67, paper 3
10 Nicholas, J.R., (1991), *Instruments of predictive maintenance*, Advances in Instrumentation and Control, Proc ISA91, Vol 1, pp 489-498
11 IEE Colloquium on Condition Monitoring for Fault Diagnosis, October 1991

12 Monkman, G.J., (1992), *A rapid response thermal tactile sensor*, Conf. on Mechatronics - The Integration of Engineering Design, Dundee, pp 81-86

13 Heilala, J. & Ropponen, T., (1991), *Mechatronic design concept for intelligent robot grippers*, Conf on Mechatronics - Designing Intelligent Machines, Cambridge, pp23-30

14 Regtien, P.P.L., (1992), Tactile imaging, Sensor and Actuators A, Vol 31, pp 83-89

15 Huijsing, J.H., (1992), *Integrated smart sensors*, Sensors and Actuators, Vol 30, pp 167-174

16 Taylor, P.M., Taylor, G.E., Wilkinson, A.J. & Gibson, I., (1991), *Mechatronics in automated apparel manufacture*, Conf on Mechatronics - Designing Intelligent Machines, Cambridge, pp1-4

17 Sensor Review, Vol 12, No 3, 1992, p 37

18 Bouguechal, N-E, Bradley, D.A. & Chaplin, R.V., (1988), *The use of proximity sensors in the navigation of AGVs*, SENSOR88, Nuremberg, pp 401-416

19 Stephens, P., Robbins, M. & Roberts, M., (1983), *Truck location using retroreflective strips and triangulation with laser equipment*, 2nd European Conf. on Automated Manufacturing, Birmingham, pp 271-282

20 Reid, I. & Brady, M., (1992), *Model-based recognition and range ,imaging for a guided vehicle*, Image and Vision Computing, Vol 10, No 3, pp197-207

21 Zhang, LiLi, Bradley, D.A. & Seward, D.W., (1993), *Safety critical systems for robot excavation*, 10th International Symposium on Automation and Robotics for Construction, Houston

22 Derby, S., Graham, J. & Meagher, J., (1985), *A Robot Safety and Collision Avoidance Controller*, Robot Safety, International Trends in Manufacturing Technology Vol 4, Springer-Verlag, pp 237-246

23 Ellis, S.R., (1991), *Nature and origins of virtual environments: A bibliographical essay*, Computing Systems in Engineering, Vol 2, No 4, pp 321-347

24 Vince, J., (1992), *VR - Promises, promises!*, Intelligent Tutoring Media, Vol 3, No 1, pp 29-31

25 Takahashi, T. & Sakai, T., (1991), *Teaching robot movement in Virtual Reality*, IEEE/RSJ Workshop on Intelligent Robots and Systems IROS'91, Osaka, Japan, pp 1583-1587

26 Rogers, A.J. & Handerek, V.A., (1991), *Novel methods for distributed optical-fibre sensing*, SPIE Vol 1586 Distributed and Multiplexed Fibre Optic Sensors, pp 2-12

27 Grattan, K.T.V. & Palmer, A.W., (1992), *Optical-fibre sensor technology: a challenge to microelectronic sensors?*, Sensors and Actuators A, Vol 30, pp 129-137

28 Kleinschmidt, P. & Hanrieder, W., (1992), *The future of sensors, materials science or software engineering*, Sensors and Actuators, Vol 33, pp 5-17

29 Pedrycz, W., (1990), *Fuzzy sets in pattern recognition: methodology and methods*, Pattern Recognition, Vol 23, No 1/2, pp 121-146

30 Villalobos, L. & Gruber, S., (1991), *Measurement of surface roughness parameter using a neural network and laser scattering*, Industrial Metrology, No 2, pp 33-44

Measurement and control with serial digital networks

J.R. Jordan

University of Edinburgh, Dept. of Electrical Engineering,
Mayfield Road, Edinburgh, EH9 3JL.

ABSTRACT: Fieldbus systems have been designed for each of the major measurement and control market areas. This paper introduces the IEC International Fieldbus, the Home Automation Bus and the Medical Information Bus as examples of this activity. The impact of serial bus systems on sensor development is discussed in the context of typical fieldbus instrumentation.

1. INTRODUCTION

Serial data highways for measurement and control applications were discussed very soon after the first digital computer was used to automate a process control plant. Since then a large number of serial bus systems have been produced or are under development. For example: MIL-STD-1553B (widely used military bus), IEEE 1118 (based on the Intel Bitbus), ARINC 629 (civil aviation bus), CAnet (vehicles), CEbus (USA home automation), medical information bus (IEEE P1073), IEC international fieldbus and LON (the Echelon proprietary network which is attempting to become a de facto international standard). The last four of those listed are based on the ISO open system interconnect communication model. Over the last ten years the ISO communication model has strongly influenced standards developed for information technology.

Two of the listed serial bus systems are proprietary (the Echelon LON and the CAnet). The others, apart from the military bus and IEEE 1118, are examples of pre-product standard development where the standard eventually defines the basic features and performance of a product. A wide range of application areas are covered by the listed standards, nevertheless the stated aim of the Echelon bus system is to become the de facto standard for a significant number of these applications.

This paper introduces three of the above listed standards (the international fieldbus, the home automation bus and the medical information bus). The aim being to highlight their similarity and the large duplication of work carried out in the serial bus area. The impact of serial bus systems on sensor development is then discussed in the context of typical fieldbus instrumentation.

2. PROCESS CONTROL AND FACTORY AUTOMATION

A number of proprietary serial highways have been developed (and continue to be developed) but although such systems are often technically successful they restrict users to purchasing equipment from a limited number of vendors. An international standard is being developed by the International Electrotechnical Commission (IEC) to specify fieldbus equipment that will allow multi-vendor systems to be constructed offering a large measure of instrumentation interchangeability. Strictly the term fieldbus should be reserved for serial data highways defined by the international standard. However by common usage a fieldbus is any serial data highway interconnecting field instrumentation for measurement and control purposes.

The international fieldbus (IFB) standard will specify physical media, messaging structure and protocols to be used by a digital, serial, communication link between field devices located in a manufacturing (process and factory) area and higher level systems located in a supervisory (control) area. The development of a fieldbus with intelligent nodes will undoubtedly change the function of supervisory areas (they will become less complex) since, for example, control algorithms and, maintenance related data processing and collection will be carried out by the field devices in addition to the supervisory system.

Commonly quoted advantages of fieldbus systems include reduced material and installation costs, reduced weight of wiring, systems easy to change, improved accuracy of signal transmission and increased information flow. However it is the higher level of automation that will arise from the use of the fieldbus that will have the most impact on users. A fieldbus system interconnecting intelligent instrumentation in a manufacturing organisation will provide more information about the process and about the performance of plant equipment. This may allow process safety to be enhanced and, pollution emission and levels reduced to meet new legislation. In addition a fieldbus system will facilitate the construction of cost efficient smaller and more flexible plants to meet the start-up and shut-down requirements of the product and demand changes required by the market place, enable improved control and instrumentation to be implemented to ensure compliance with quality control standards which demand traceable manufacture and maintenance procedures, and lead to the incorporation, at low cost, of improved fault finding and self diagnostics features. It is clear that users will want the IFB to offer a multivendor capability, enable faulty devices to be quickly replaced (in less than 30 minutes) and enable IS certification to be maintained if devices are replaced, place no position restrictions on fieldbus devices, be usable by current staff with no requirement for a skill increase, and use tree connection topology rather than a loop. The maintenance and fault detection (i.e. added value) possibilities of the fieldbus which lead to a reduction in down time and greater information transfer are seen to be more important to potential users than the expected reduction in installation costs.

Like the majority of groups working in this area the IEC committee developing the fieldbus standard have adopted the ISO seven layer architecture to define the communication link. A consequence of this is that, although a reduced number of layers is used, a long and complicated message structure is required with the result that the IFB may not be suitable for, so called, hard real time control. The international fieldbus activity started in the mid-1980's. Draft Physical Layer and Data Link Layer documents have been published. A complete review of the fieldbus was presented by the competing and collaborating groups at the recent Institute of Measurement and Control Conference on the fieldbus (IMC, 92).

A layered communication architecture facilitates a controlled approach to message construction. Each layer performs a specified operation on the data (information) to be passed over the bus with each layer only being able to interact with the layer immediately above (or below). The ISO seven layer communication model was devised to allow very complex, large, networks to operate reliably. The fieldbus will involve relatively small networks with data transfer operations usually involving short messages. Hence the IFB will not require the full seven layer model and will use three layers, namely; the Physical Layer, the Data Link Layer and the Application Layer. Since some of the functionality of the removed layers is required for the IFB, its Data Link Layer and Application Layer will not be exactly as defined by the ISO. In addition an eighth layer (the Users Layer) will be required to provide a high level, standard, natural language interface for users.

Several groups are supporting the IEC fieldbus standard committee. In the USA the Instrumentation Society of America (ISA) provides a strong input through its SP5O committee. In the UK the AMT7 committee of the BSI coordinates the fieldbus work. Strong inputs are obtained from France and Germany who already have national fieldbus standards: in France the FIP project and in Germany the PROFIBUS project. Both of these groups have integrated circuits and software to support their bus and with a significant user base they command a major market position. Fortunately all of these groups appear to be supporting the IEC fieldbus but with such a large number of technical aspects to consider the committee is making slow progress towards the production of draft documents.

The physical layer part of the international fieldbus standard specifies a screened twisted pair with low (31.2Kbps) and high (1/2.5Mbps) speed voltage modes, and a high (1Mbps) speed current mode. A clip-on transformer connection is specified for the current mode bus with a maximum cable length of at least 750 metres. The current mode bus uses a 20KHz current fed sine wave to supply power to remote devices and the 1MHz digital data signal is added to the sine waveform. A radio frequency (wireless) option is at an advanced stage of preparation but a fibre optic link option has, as yet, received little attention.

The groups working on the data-link layer specification have experienced great difficulty in arriving at a document compactly expressing the consensus view. One group supported the use of centralised media access which ensures that sampling periods are maintained while another group supported the use of token-passing media access control. Token-passing allows changes to be easily made (i.e. adding and removing devices) but sample rate jitter can be experienced at high traffic loads. Timed tokens methods can be envisaged with bus idle times sufficient to ensure that fixed scan times are achieved. A draft compromise document has been produced which is complicated with many implementation options.

The instrumentation industry has worked hard to establish the international fieldbus standard. It will be successful if it allows multivendor systems to be easily constructed. Unfortunately conformance to a standard will not guarantee interoperability and interchangeability of equipment. It remains to be seen if the Echelon Corporations is successful in establishing a de facto serial highway standard based on its neuron chip and local operating network software.

The Echelon system supports a multi-media approach (including twisted pair, powerline and radio) to the interconnection of field devices with data rates from 4.8Kbps to 1.25Mbps. The protocol used follows the OSI communication model and provides services at all of its seven layers. The silicon implementation of their protocol with their neuron chip simplifies implementation. The neuron chip uses three microcomputers: one implements layers 1 and 2 (the data link layer implements the CSMA/CD protocol) one implements layers 3-6 and one is dedicated to the application layer.

The availability of a working integrated circuit implementation supported by development and user software could be a powerful incentive for instrumentation vendors to adopt the Echelon system.

3. HOME AUTOMATION

The communication requirements of home automation are similar to that of industrial process control and factory automation. Several groups are working to produce a standard defining the serial interconnection of domestic equipment. The Electronic Industries Association (USA) formed its Consumer Electronics Bus (CEBus) standard committee in 1984 (Hanover, 1989). A Japanese EIA group is working on a similar standard specification for what has been called the Home Automation Bus. In Europe several large companies (including Electrolux/Zanusi, GEC, Philips, Siemens, Thomson and Thorn-EMI) formed a group to work on a home automation communication standard. An ISO/IEC joint committee is working to unify this activity and produce an international standard defining the serial interconnection of domestic equipment.

The diversity of equipment used in Home Systems and the need to support both the retro-fit and the new house situation has resulted in a multi-media approach to in-house communication with powerline, twisted pair, co-axial cable, infra-red and radio frequency links currently specified. All of the home bus systems have used or proposed the use of the OSI communication model. Unlike the previously described industrial fieldbus the home bus will require the network layer (layer three) of the OSI seven-layer model to be included in the standard specification. The router, a special purpose device (implementing the lower three layers of the communication model), is used to link different media networks together. A sparse stack version of the OSI model is proposed with stacks 1,2,3, and 7 defined. For user convenience it is desirable that a natural language interface to the Application Layer (seven) of the standard is developed. The success of the home bus system will be critically dependent on the widespread acceptance of a standard user interface.

Many product opportunities will arise once the bus standard is established and accepted. Interactive communication within the home will facilitate: management of security and safety, and energy and water consumption; control of heating, lighting, air conditioning, appliances and entertainment systems. External links (e.g. use the telephone network) will facilitate interactive communication with outside bodies (e.g. banking and information sources) and working from home will become a practical proportion even for manual and measurement related work. Telemedicine (health monitoring using remote hospital resources), telemetering (remote monitoring by gas, electricity and water utilities) and telecontrol (remote operation of domestic systems - cookers, heating, video recording and surveillance) will offer significant opportunities to the manufacturers of domestic equipment.

4. MEDICAL INSTRUMENTATION

It is surprising that a standard is not available to define the interconnection of medical devices to a patient care computer system. Acute patients care will benefit most from a standard interface that would enable medical devices to operate with a host computer in a compatible, vendor-independent fashion. This application area requires frequent addition and removal of equipment by medical staff and therefore provision for the automatic identification of a device whenever a connection is established is a prime requirement. A non-technical user interface must be provided for medical staff.

The IEEE have sponsored a standard defining the interconnection and control of medical devices. It has been published as a draft standard, (Franklin, 1989), called the Medical Information Bus standard, IEEE P1073. The MIB uses the familiar stem and fan topology. A Bedside Communication controller links with devices located at the end of fan elements. A multidrop bus links the bedside into a supervisory host systems. The OSI seven layer model is used with the data link layer specified as a subset of the international standard HDLC protocol and for the user applications interface a medical device data language (MDDL) has been devised. This has many of the properties of a natural language interface. The physical layer is specified by E1A-485 operating at 375Kbps. A shielded six wire cable (organised as three twisted pairs) is used with one pair providing half-duplex data transmission, one pair supplying 12VDC current limited to 250mA and one pair providing a timing signal. The important design objective for this bus was the minimisation of user effort. An MIB network should automatically detect device connections, establish the connection and identify devices - the user merely plugs in a data cable and switches on the device.

5. FIELDBUS INSTRUMENTATION

Despite the attractions of the fieldbus approach to measurement and control, potential users are indicating that they are not expecting to pay significantly more for fieldbus instrumentation. In addition, for the extra cost, they are expecting added features that will be facilitated by the digital communications link, e.g. fault detection algorithms in the field devices and an automatically up-dated maintenance distributed data base. Clearly extensive use will be made of silicon circuit technology and, indeed, the complexity of fieldbus standard specifications will be largely hidden from users by being encapsulated in dedicated silicon systems.

For users the most attractive feature of the international (IEC) fieldbus standard will be its use to specify systems that allow multi-vendor equipment to be easily constructed. The key feature here will be an ability to interchange devices on the bus without changing the operating system software. It is desirable that all field devices should be linked by the same type of fieldbus. Hence in addition to the basic measurement and control instrumentation field devices used for condition monitoring and safety systems should be connected by a fieldbus. This will require additional signal processing functions to be implemented in field devices and the use of, say, dual redundant buses to improve reliability. Users will require a system reliability at least equivalent to that achievable by current field instrumentation. Users will ultimately control the success of fieldbus by their buying policy but the long term success of fieldbus will only be assured by users working with vendors to create the standards.

Looking at the typical electronic circuit requirements of field instrumentation it can be seen that the additional component cost of a fieldbus device could be quite small. Typical major elements (e.g. for a process control pressure transmitter) could include a transceiver, protocol circuit, microcomputer and sensor interface circuit. In addition signal processing circuit may be used to give added functionality. An internal serial bus (e.g. I^2C) could be used to reduce wiring complexity on the circuit board. For simple sensors (e.g. limit switches or frequency output devices) it would be better if the protocol circuit could directly accept a small number of inputs, say one analogue and a few digital inputs. It would be attractive to have the protocol circuit available as a "silicon layout" so that analogue and digital interface components could be combined with it to form a dedicated sensor single chip interface to the fieldbus. (Jordan, 1992).

Remote devices will often require power to be supplied via the bus. Power over the bus is a feature of the IEC fieldbus standard. For example a current driven 1Mbps bus is specified with a lower frequency, 20KHz, signal added to provide a power supply signal. Transformer coupling is used so galvanic isolation can be achieved. If intrinsic safety is required a small number of devices can be supplied from a barrier located at the supply end of the bus. Alternatively the bus cable can be flame proofed and intrinsically safe barriers installed at each appropriate take-off point.

Fibre optic has been specified as an option in the IFB standard but it has not yet been defined. In this case of course a separate power supply cable or a battery is required. It is interesting to note that, the fibre optic link can be designed to operate at very low power (100 μw at a 100Kbps data rate). Even at 1Mbps a fibre optic 1553 system operated successfully with a power budget of approximately 30mW (Jordan, 1992). The adoption of fieldbus techniques will increase the need for low power sensor systems.

From the fieldbus point of view a sensor should be considered to be a remotely located electronic instrumentation system. Many sensors already include complex electronic systems for signal conditioning and calibration. Increasing the electronic circuit complexity does not necessarily mean a large increase in component or manufacturing cost. Users will be looking for added value and this can be succintly stated as being more information without a wiring or software penalty.

The full benefits of the fieldbus approach to system automation will be realised when sensors (actuators) implement error detection and control algorithms. This will enable control via the bus and lead to a simplification of the conventional control room.

6. CONCLUSIONS

A uniform information technology approach to system automation, from management level down to field devices, will provide benefits to vendors and users. Vendors restricted to a particular market are clearly duplicating work as indicated by the serial bus examples presented in this paper. A single integrated protocol circuit covering the majority of applications is attractive. This would require a very large concensus to the achieved which may be facilitated by the low cost of a device developed for the large potential market. The Echelon system may be able to build-up a sufficiently large user base to satisfy these requirements by establishing a de facto standard.

Access to communication over the bus will open-up considerable opportunities for sensor manufacturers to add value to their products. The additional measurement of ancillary variables and the measurment of high frequency noise for fault detection purposes can be noted as two possibilities. In addition calibration control via the bus will be possible with minimum intervention by maintenance personnel. A close involvement with integrated circuit technology will be required.

7. ACKNOWLEDGEMENTS

The support of the Science and Engineering Research Council is gratefully acknowledged.

8. REFERENCES

Franklin (1989), D.F. and D.V. Ostler. The P1073 Medical Information Bus. IEEE Micro, October, pp. 52-60.

Jordan (1992), J.R., S. Lytollis and D.W. Kent. A fibre optically extended fieldbus. Measurement Science and Technology, 3(9), pp. 902-908.

Inst. of Meas. and Control (IMC, 92). Proceedings of FIELDBUS '92, London, November.

Hanover, G. (1989). Networking the intelligent home. IEEE Spectrum, October pp. 48-49.

A new cylindrical structure load cell with integral resonators

Cheshmehdoost A, and Jones B E
The Brunel Centre for Manufacturing Metrology, Brunel University, Uxbridge, Middlesex UB8 3PH, UK

ABSTRACT: The measurement of force in many engineering environments is mainly achieved using strain-gauge based load cells. The novel developed cylindrical structure load cell described has multiple *Double-Ended Tuning Fork* (DETF) resonators whose resonant frequencies are a function of the applied force. The load cell has four integral DETFs and can be used in tension or compression. The design methodology, drive and pick-up methods, and performance of the device are explained. Characteristics features of the new load cell are high resolution, good linearity, repeatability and stability.

1. INTRODUCTION

The accurate measurement of physical quantities such as mass, force, and strain is of prime importance for many industries. The force measurement devices are commonly known as *load cells* and normally utilize the strain-gauge technology. The performance of such a load cell is limited due to associated problems such as creep. Furthermore, most strain-gauge based load cells have an undefined zero-load output when they are operated between tension and compression regions. These problems have increased the need to develop load cells based on other technologies. Good sensing devices for the measurement of force are mechanical resonators of various structures whose natural frequencies are a function of applied force.

Mechanical resonators have excellent stability and potentially low hysteresis. These factors together with the fact of outputs in frequency form have made them the subject of considerable practical interest. Frequency domain transducers have many advantages over conventional analogue sensors in that their output can be measured accurately with inexpensive frequency counters, they have high reliability, low error rates, low susceptibility to degradation of transmitted signals by electrical interference, and low dependence on change in electrical characteristics with time [Ref. 1].

The double-ended tuning fork (DETF) basic structure consists of twin beams (tines) which are coupled at their roots (figure 1). When one of the tines in this symmetrical structure is excited at its fundamental frequency, the opposite tine resonates in sympathy. Due to symmetrical vibration of the DETF, the vibrational counter forces R_1 , R_2 and bending moments M_1 , M_2 cancel at the roots, so that at each end of the resonator beyond the roots no vibration occurs [Refs. 2, 3]. This unique property of the DETFs allow them to be clamped beyond their vibrational roots without significant loss of vibrational energy or reduction in their mechanical quality factor Q.

The DETF resonators can be characterised in an open-loop configuration using a spectrum analyser. A piezoelectric chip is used to energize one of the tines, and a second piezoelectric chip attached to the opposite tine is used to detect the natural frequency of the DETF. To maintain the high Q DETF in continuous resonance an electronic circuit is used to provide feedback and phase control around a closed-loop. Because of the high Q of the resonator the perturbing effects of cross-talk between the piezoelectric chips and electronic amplification circuit are minimal. Therefore, the performance of the DETF resonator may be assumed to be entirely mechanical.

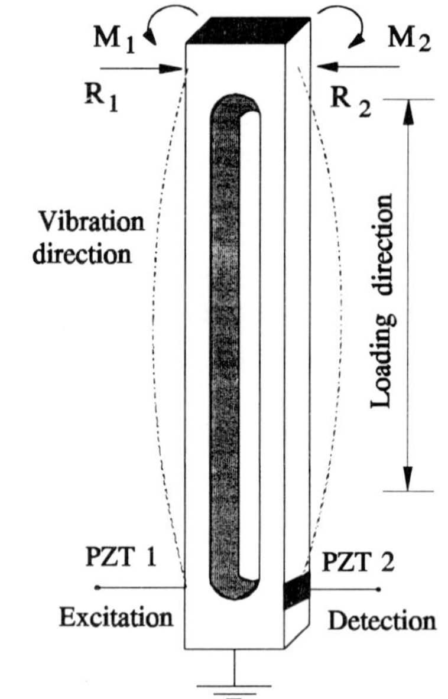

Figure 1 Configuration of a DETF resonator.

2. THEORY

The fundamental frequency of a beam is a function of its parental material property, and dimensions. This frequency increases with increasing applied axial force, and the equation governing the vibration of a beam under tension T is given by the following expression;

$$T \frac{d^2Y}{dX^2} - E\,I \frac{d^4Y}{dX^4} = \rho\, S \frac{d^2Y}{dt^2} \tag{1}$$

where

I = second moment of area = (W.H)/12
E = Young's modulus of the material
ρ = density of the material
S = cross-sectional area of the beam (W.H)
t = time, and

W and H are thickness and width of the beam, respectively [Ref. 4]. Assuming a simple harmonic motion and applying the boundary conditions, the natural frequency f_{01} when T=0 (axial force T = 0) is given by the expression below;

$$f_{01} = \frac{a^2}{2\,\pi\,l^2}\sqrt{\frac{E\,I}{\rho S}} \tag{2}$$

where

a = value determined by mode number n of vibration $\approx$ (1+2n) $\pi/2$, and
l = length of the beam [Ref. 5].

However, when T $\neq$ 0, the natural frequency of the beam under axial force f_{T1} is given by;

$$f_{T1} = f_{01}\sqrt{1 + \frac{1}{a^3}\tanh\frac{a}{2}\left(a\tanh\frac{a}{2} - 2\right)\frac{l^2\,T}{2\,E\,I}} \tag{3}$$

These equations define the motion of a DETF under stress and they may be used to design a DETF to operate at a particular frequency; it is also possible to determine the total frequency shift of a DETF under full-scale load. However, in practice other important parameters should be considered in order to realize a DETF-based load cell [Ref. 3].

3. DESIGN METHODOLOGY

The load bearing capability of a DETF is determined by its material's yield strength and cross-sectional area [Ref. 6]. In designing the high capacity DETF-based load cell a simple approach of integrating the DETFs and their support was chosen (i.e. DETFs and supports are fabricated from one piece of material). This integration approach reduces the potential hysteresis due to the mismatch of materials of the DETFs and the support. Four equi-spaced DETFs were fabricated in the walls of a hollow stainless steel cylinder at 90° relative to each other as shown in figure 2.

The cylinder's four identical DETFs have tines with 1 mm x 4 mm cross-sectional area and a length of 20 mm. The load cell is designed for operation either in tension or compression. The load bearing capability of this type of load cell may be easily varied by changing the dimensions of the supports. Finite element analysis has been used to establish an optimum static loading model for the device. The 20 kN load cell is quite insensitive to off-axial loading which may occur in real applications. This insensitivity is achieved by addition of the frequency excursions of all four DETFs around the cylinder.

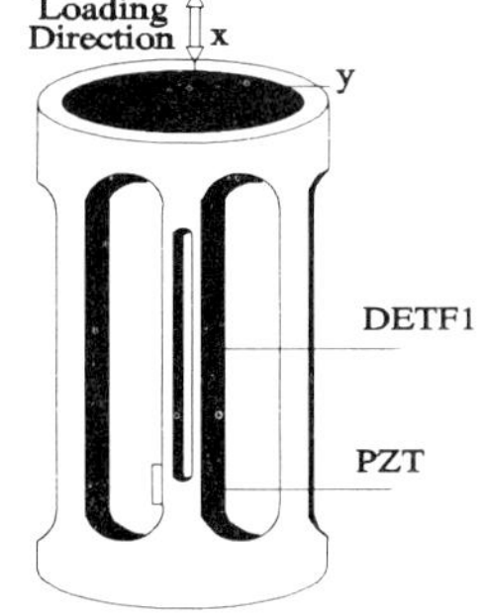

Figure 2 Cylindrical load cell.

4. RESULTS

The cylindrical DETF based load cell was tested under open-loop condition. The open-loop test involves the use of a spectrum analyser to obtain the frequency response of the DETFs. The natural frequency and mechanical quality factor of each DETF was measured under zero load conditions. The natural frequencies of all four DETFs were found to be around 6000 Hz (less than 5% variation between them). A typical frequency response curve of one of the DETFs is shown in figure 3. This DETF has a natural frequency of 6029 Hz and a Q greater than 1500.

The open-loop characterisation of the DETFs showed that the lower the drive voltage the higher the Q. The high level of drive voltage causes loss of energy from the system by means of sound and/or heat. Frequency response curves also showed that the value of the Q is almost independent of the applied load (i.e. Qs of DETFs were >1500 for various loads).

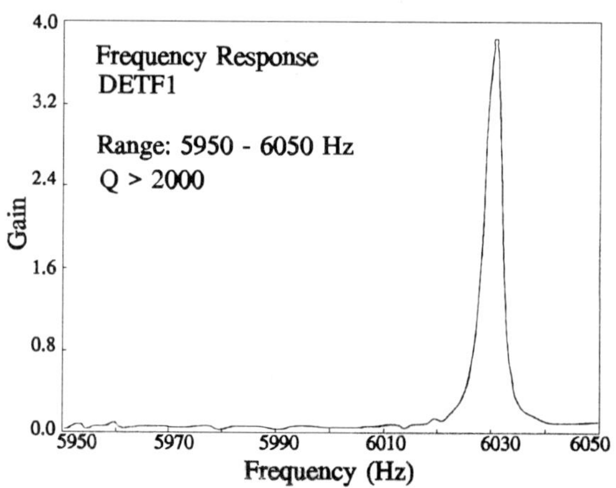

Figure 3 Frequency response of the DETF1.

An electronic closed-loop circuit tracks the resonant frequency (output of each DETF) and incorporates a digitally generated 90° phase-shift to bring input and output in phase. This closed-loop arrangement is shown in figure 4. This arrangement keeps each device in continuous oscillation, and as the structure is loaded each closed-loop circuit tracks the resonant frequency and maintains respective oscillation. The output frequency of each DETF is monitored using a frequency counter. All characterisation of the load cell was carried out using this arrangement.

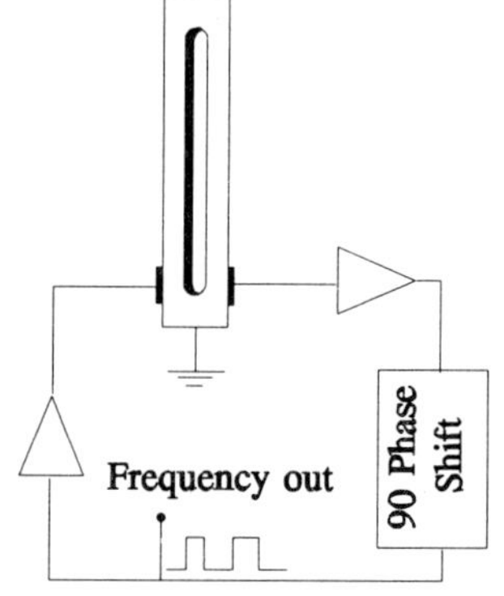

Figure 4 DETF closed-loop arrangement.

Figure 5 shows a typical continuous loading curve for the load cell monitored from one of the DETFs for a load cycle from 10 kN tension to 10 kN compression. The graph shows a total frequency excursion of approximately 500 Hz corresponding to 20 kN. The total frequency excursion or span is approximately an 8.5% shift in the natural frequency of the DETF. The characteristic loading curve also shows that the load cell output is linear. The region of transition from tension to compression shows no discontinuity.

Characteristics of the load cell are indicated in table 1. The load cell was tested under different conditions and the fundamental frequencies of all the DETFs were monitored. Because of the compensation method (described in section 3) no significant off-axial loading was noted and the results shown are typical of one of the DETFs. The load cell has a safety margin of 100% .

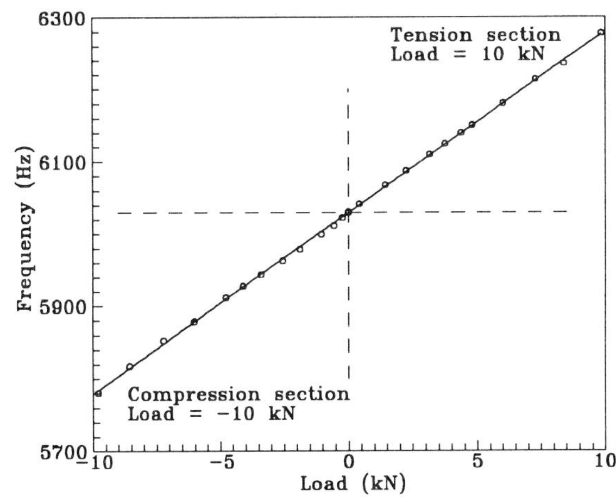

Figure 5 Typical loading curve of the DETF load cell.

Table 1 Typical characteristics of the DETF based load cell.

Parameter	Value (Hz)	% of span
Nat. Freq.	6030	-
Freq. shift (20 kN)	~500	100
Repeatability	0.2	0.04
*Stability (0 k)	0.2	0.04
" (2 kN)	0.5	0.10
" (5 kN)	0.5	0.10
Linearity	5.0	≈1.00
Sensitivity	(1 Hz/40 N)	
Resolution	(better than 1 part in 2000)	
Stiffness	(less than 100μm displacement for FS)	

* Stabilities are measured over 4 hour tests.

5. CONCLUSION

The performance of the new load cell based on the resonating DETF structure has shown that this technology has the potential to be utilized in fields of mass, pressure, and force measurement. The benefits of this technology may be summarised as follows;

i) the DETF-based load cell has low power consumption (less than 2 mW), and a frequency output which makes it ideal for interfacing with digital systems;
ii) the load cell characteristics indicate good repeatability, linearity, stability, and high resolution;
iii) the load cell requires only simple electronics, thus reducing the manufacturing cost.

6. ACKNOWLEDGMENTS

The authors wish to acknowledge the support provided by SERC, DTI and MECMESIN Ltd.

7. REFERENCES

1. Langdon R M, *Resonator Sensors - A Review,* J. Phy. E:Sci. Inst., Vol. 18, 1985, 103-112.
2. Ueda T, Kohsaka F, and Ogita E, *Precision Force Transducers Using Mechanical Resonators,* Measurement, Vol. 3, No. 2, April-June 1985, 89-94.
3. Cheshmehdoost A, Halliwell M J, and Jones B E, *Design of a Miniature High Q Double-Ended Tuning Fork Resonator Employing Non-contact Drive and Pick-up Arrangements for Force Measurement,* BSSM, Int. Weighing Conf., July 1989, Paper 16.
4. Morse P M, *Vibration and Sounds,* McGraw-Hill, N.Y., 1948.
5. Harada K, Ikeda K, and Ueda T, *Precision Transducers Using Mechanical Resonators,* Yokogawa Technical Report No. 1, 1984, 1-9.
6. Cheshmehdoost A, O'Connor B, Jones B E, *Characteristics of Force Transducer Incorporating a Mechanical DETF Resonator,* Sensors & Actuators A, 25-27 1991, 307-312.

A miniature dew-point hygrometer based on capacitance measurement

P D Harris, M K Andrews

Industrial Research Ltd, P O Box 31-310, Lower Hutt, New Zealand

ABSTRACT: Results from a dew-point hygrometer intended for long term operation in near-condensing environments around 0°C are described. The sensor comprises an interdigitated capacitor and a diode temperature sensor on a silicon chip which is cooled by a small Peltier unit, a platinum resistance ambient sensor, and a software-controlled servo loop for establishing and maintaining a preset dew load. Satisfactory operation between 12% and 100% RH is confirmed, with a response time to step changes of about 0.5 sec. The effect of contamination is evaluated, and possible signatures distinguishing dew and frost are noted.

1. INTRODUCTION

It is the perception of experienced users of humidity probes that polymer-based capacitor sensors lack long term stability and reliability in near condensing atmospheres of high humidity such as found in cool stores and freezers. Such environments are typically in the range of -20°C to +5°C, with humidity in the vicinity of 90%. In some cases, it is desirable to control humidity at as high a level as possible to minimise product weight loss, without the possibility of frost formation. The measurement of RH by dew point in such an environment is attractive because it offers a fundamental measurement which is repeatable, using a probe which if necessary may be self-cleaned by condensing cycles.

Measurement of dew point by optical scattering has been the subject of extensive investigation in past years, and lead to studies of the physics of nucleation on real surfaces and an understanding of the effects of surface contamination (e.g. [1], [2], [3]). Regtien [4] has surveyed solid state humidity sensors and presented some measurements of capacitance versus dew load for several versions of an interdigitated capacitor on a silicon chip. Capacitance sensing of dew load is attractive because it eliminates optical sources and ambiguities which can arise in dynamic situations when a light scattering system is presented with a flooded mirror. Implementation on a chip gives surfaces which are chemically stable, and the possibility of integrating a temperature sensor in the surface of the silicon, a material of high thermal conductivity, immediately adjacent to the capacitor.

In this work we report results of a dew-point device based on a Peltier-cooled silicon chip which contains an interdigitated capacitor and diode temperature sensor. The result is a sensor head which measures high humidity reliably, and which, because of its small size, requires only 1 watt to achieve 30° of cooling and has a time constant of less than 0.5s when responding to step changes of humidity. The system runs under software control, which may be adapted to provide an indication of the existence of contamination on the chip and possibly to distinguish between frost and water on the chip.

2. SENSOR DETAILS

Figure 1 shows a sketch of the system used. The chip is 2x5mm to suit the size of the smallest readily available Peltier cooler, but could be considerably smaller. The capacitor comprises interdigitated polysilicon fingers 0.5μm high, 5μm wide and 5μm apart. These are lightly oxidised, and coated with 0.1μm of LPCVD silicon nitride, a material with a high resistance to hydration. This is exposed directly to the atmosphere through openings in the two glass layers on the chip, vapox and overglass. SEM photographs show that the polysilicon edges are slightly undercut, a feature which may impact on the nucleation of dew. The silicon nitride surface itself is very smooth. The dry capacitance is approximately 1 pf, and rises a further 2 pf when the surface has heavy condensation. The chip is bonded to a 1 watt Peltier cooler and mounted in a 25mm tube equipped with a viewing port, and aspirated by a small in-line fan. The airflow was set to be turbulent in the tube (velocities above 1 m/s) to maximise transfer of vapour to the chip, but no significant difference in response speed has been detected for airflow variations in the range 1-4 m/s. The diode temperature sensor is fed at a constant current of 15μA and chip temperature is calculated from its forward voltage drop. The diode is calibrated against the upstream ambient temperature sensor, a PT100 sensor. A linear approximation gives temperature accurate to 0.1°C. However, for dew point, it is the depression which is important, and this more reliably obtained from the diode voltage change than from the difference in two absolute measurements of temperature. This is particularly true for accurate measurements at high humidity. The Peltier cooler is fed by a digitally controlled power source in response to capacitance measurements, which are made by a synchronous detection of the capacitor admittance at 100kHz. It was found necessary to measure capacitance with a resolution of 1fF and have 16-bit resolution of the power source to the Peltier to implement a stable response because of the sensitivity of the capacitor to dew load. The control loop runs under software control at a speed of 100 measurements/second.

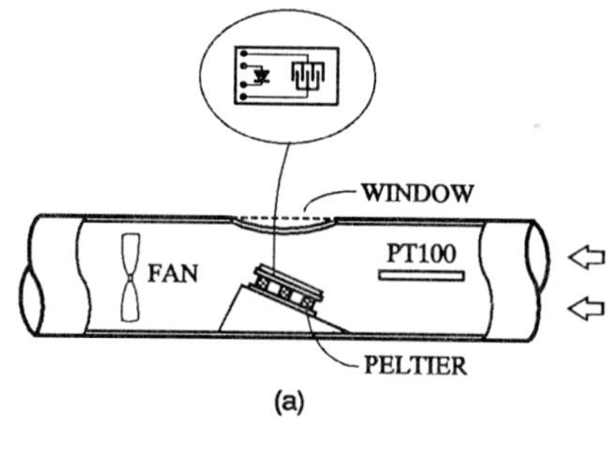

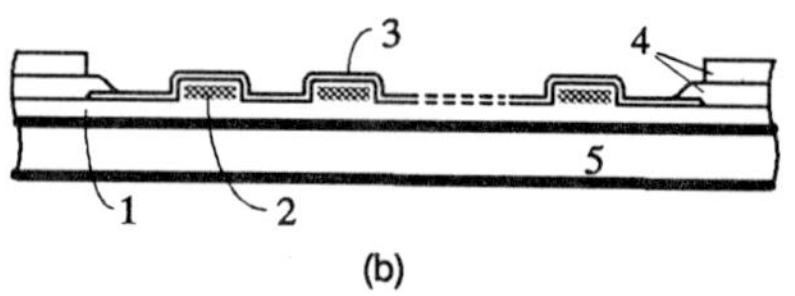

Figure 1(a). Schematic of sensor head. (b). Cross section through capacitor. Layers are 1, thermal oxide; 2, polysilicon; 3, silicon nitride; 4, reflowed vapox glass and overglass

3. CONDENSATION REGIME

Regtien [4] reported the sensitivity of the capacitance of several microelectronic interdigitated structures to dew coverage. By comparing dew formation on the nitride covered capacitor with that on the planar, silicon dioxide covered region, we conclude that the fingers have a major influence not only on the nucleation of dew, but its subsequent behaviour. Viewing under a 100x microscope shows that condensation occurs first at the edges of the fingers, and structure cleanliness modifies subsequent development. In a clean structure, condensation runs along the corner of the channels between fingers; it may accumulate at the ends of the channels. Condensate then spreads across the channels to flood them. On evaporation, these sheets disappear suddenly with a fair degree of uniformity. On the tops of the polysilicon fingers, and in the areas away from the capacitor, isolated droplets remain the predominant form of deposit. If the capacitor is not clean, ragged edges of water grow outwards from the edges of the fingers into the channels, which flood locally. The water then seems to lack the ability to provide a uniform distribution by flow along the channels.

Figure 2 shows the approach to dew point for a contaminated capacitor, and the same surface after cleaning. The change in equilibrium capacitance from its dry value is plotted against temperature depression. While capacitance changes are seen when water is forming at the corners of the channels, the greatest sensitivity is found when a layer of water extends between adjacent fingers. With the clean surface, the difference in equilibrium temperature for dew loads corresponding to 0.8 and 1.8 pf is .02° which is close to the temperature resolution; the subsequent drift down with time as the set point is cycled (points 1-5) of .05° could be ambient condition changes. The indicated dew point is practically independent of load, the best indication that the true dew point has been found [2]. The situation is quite different for the dirty capacitor. Significant capacitance is seen more than 0.5° above the dew point indicated by the clean sensor, and there is a small (.05°C) hysteresis in temperature between loading and unloading dew. However, the capacitance curve is steepening as the dew point is approached. If the capacitance set point is above 1.5 pf, the difference between "clean" and "dirty" is little more than 0.1°, leading to an error of 1-2% in the calculation of RH.

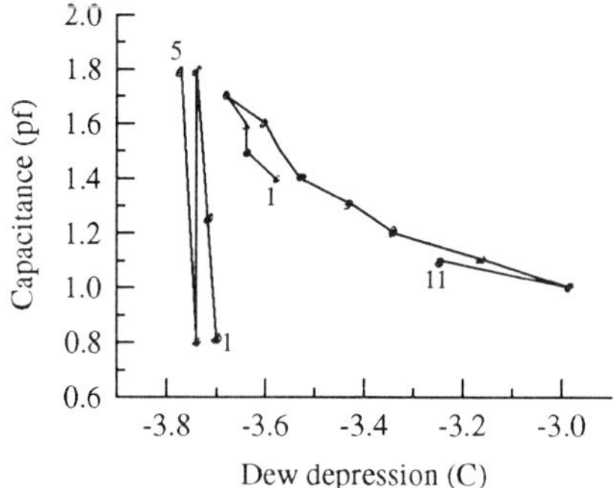

Figure 2. Equilibrium Capacitance versus temperature for a clean sensor cycled near the dew point (steep curve points 1-5) and a contaminated sensor (points 1-11).

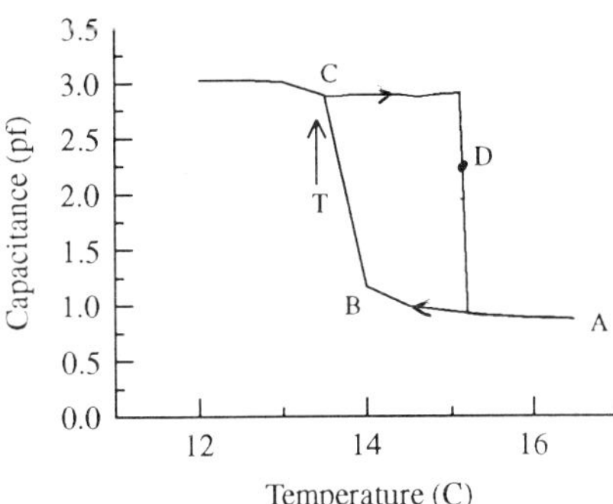

Figure 3. Hysteresis curve of sensor used to learn and establish the dew point as described in the text.

Because of the potentially rapid changes in equilibrium capacitance, the sensor approaches the dew point via a learning cycle based on the "hysteresis curve" obtained when the chip is cooled below its dew point and then slowly heated (Figure 3). The chip is not in equilibrium with its dew load during the section A B C. Cooling ceases at C when a particular dew load is reached at temperature T which is below the dew point. The servo loop then begins in a mode which primarily sets the rate of change of capacitance to zero (since this defines dew point). The final control is to bring the constant value of capacitance to the set point D, on the steep portion of the curve.

In practice, on reaching point C, the sensor is dried by an abrupt 1°C temperature increment, and then abruptly recooled to temperature T. This is to establish a pattern of dew load which minimises the water drawn along fingers to the channel ends, an area of low capacitance sensitivity. However as described later, this phase may enable frost and dew to be distinguished.

The effects of the control algorithm are seen in the response to step changes in humidity. Capacitance control ($\dot{C}=0$) is gained after a fraction of a second, but the value will be at a different static value of capacitance (and therefore more or less close to the dew point, depending on the slope of the C-T curve). The capacitance is slowly brought back to its preset value over several seconds.

HUMIDITY PERFORMANCE

Initial tests of the device have been made using saturated salt solution in a controlled temperature chamber at 20°C at 12, 33, 85 and 95% RH, and comparisons made with a high quality polymer capacitance reference probe located in the same airstream. RH was calculated from the dew point and ambients temperature using the expressions given in [5]. At 95% RH, the reference probe showed small instabilities and an increased time constant. Static humidity values from the dew point sensor were within 1% of those expected for the salts. Successful tests were conducted with KCl solutions between +20°C and -5°C (RH values 83 - 86% corresponding to dew points from about 8° to -6°C). It is probable that all these dew points, which range down to -8° for LiCl at 20° ambient, represent supercooled water rather than frost; observations with a microscope at -7°C certainly indicate that the condensate is water. However, the part of the learn cycle where the temperature is incremented 1° to lose water shows systematic behaviour with dew point which may indicate a transition from dew to frost. Figure 4 shows the time taken for the temperature increment to shed the load. This increases slowly as dew point falls almost independently of ambient temperature until -10°, after which there is a marked increase in time which is caused by the time taken to begin to lose significant amounts of water. The figure also indicates a discontinuity in the maximum rate of change of capacitance (ie rate of water loss) at the same temperature.

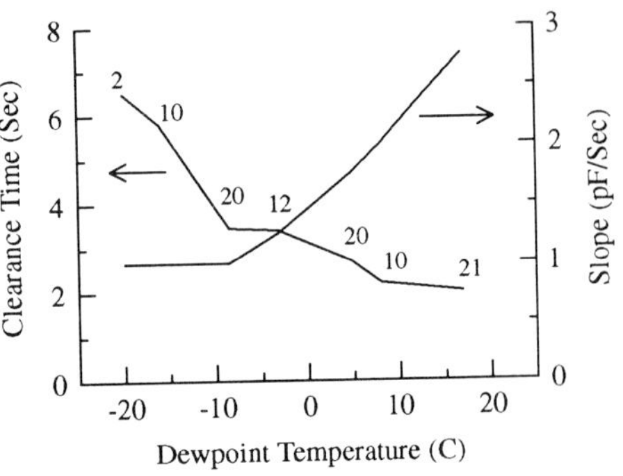

Figure 4. Time for 1° temperature increment near dew point to clear the dew load, and the maximum rate of change of capacitance encountered. Both curves show a discontinuity near -10° which may indicate a transition from dew to frost. Numbers beside points indicate ambient temperature.

The small size and high temperature slew rates available mean that the device can track rapid humidity excursions. An example is shown in Figure 5 in which the airstream is switched to room ambient. The time constant for the change is 0.5 seconds. The capacitor achieves stable control ($\dot{C}=0$) in a fraction of a second, but the value of capacitance may be higher than its previous value. An error is introduced into the calculated RH while the chip temperature is modified to drive the equilibrium load back to its previous value. Such errors are usually a fraction of 1% unless conditions are changing rapidly. To illustrate the dynamic response and immunity to saturated conditions the sensor was installed in a breathing tube to monitor the RH of inhaled and exhaled air. The results are shown in Figure 6. The values exceeding 100% represent the situation described previously where the chip has capacitance control but is over-compensating temperature to adjust the static dew load. Such a humidity sensor may have medical applications in monitoring the humidification of oxygen and anaesthetics where it is necessary to supply a highly humidified gas stream to the patient.

CONCLUSION

Preliminary results from a dew point sensor which utilises an interdigitated capacitor structure have been described. The area of application envisaged is in near condensing atmosphere at low temperature. The sensor can be operated such that "dirty surface" dew point results are within 0.2° of the clean (true) results but the surface cleanliness could be estimated from capacitance measurements. The programmed learn cycle enables the rates of change of precipitation to be monitored sensitively near the dew point; systematic behaviour is seen and it is believed that signatures for dew and frost can be distinguished. The fast dynamic response may lead to applications other than intended.

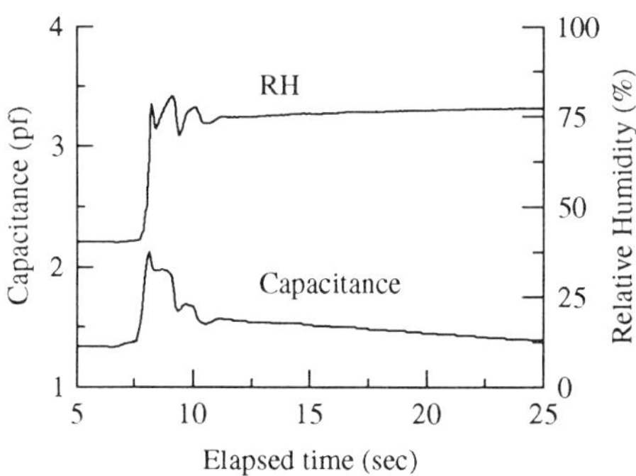

Figure 5. Response of sensor to abrupt change of humidity from suaturated salt solution to room ambient.

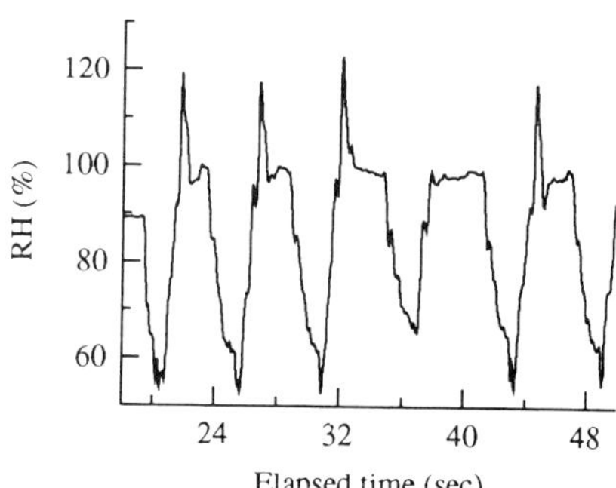

Figure 6. Humidity of air inhaled and exhaled directly through the sensor. The points above 100% RH are explained in the text.

REFERENCES

[1] Wylie R G, "On the hysteresis of adsorption on solid surfaces", Aust. J. Sci. Res. A5, 1952, p288.

[2] Wylie R G, Davies D K and Caw W A, "The basic process of the dewpoint hygrometer", Principles and Methods of Measuring Humidity in Gases, Vol I, Ed. A Wexler, Reinhold, NY 1965, p125.

[3] Brewer A W, "The dew- or frost-point hygrometer", Principles and Methods of Measuring Humidity in Gases, Vol I, Ed. A Wexler, Reinhold, NY 1965, p135.

[4] Regtien P P L, "Solid state humidity sensors", Sensors and Actuators, 2, 1981/82, p85.

[5] IHVE Guide, Section C1, Inst. of Heating and Ventilation Engineers, London 1970.

Design issues for medium resolution metal resonant force transducers

A. Kennedy, W. O'Connor
Mechanical Engineering Department, University College Dublin, Belfield, Dublin 4, Ireland.

ABSTRACT: When a beam is axially loaded its natural frequency of transverse vibration changes. If the beam can be made to resonate and that frequency change measured, the applied load can be determined. Transducers based on this principle have been successfully developed for high resolution measurements (better than 10 ppm). This paper examines the equations which define their theoretical performance. It looks at their advantages over conventional strain-gauged load cells and at their potential use in medium resolution measurement applications (1000 ppm to 100 ppm).

1. INTRODUCTION

Conventional load cells, in which the deflection of an elastic spring element under an applied load is measured using strain gauges, dominate the medium resolution force transducer market. Resonant force transducers have two major inherent advantages over these;

1. Digital output: Strain-gauged load cells require expensive A-to-D conversion when interfacing with modern digital measurement and control systems. The frequency output of a resonant transducer can be measured easily and inexpensively with simple digital electronic components. Furthermore electrical noise problems are avoided.
2. High sensitivity: Because of the small strains allowed in a metal before it yields, conventional devices are inherently insensitive. Their sensitivity is maximised by operating at stresses approaching the metals fatigue strength. At that level of stress, however, creep and hysteresis in the receptor material limits their resolution to approximately 100 ppm. Vibrating beam resonant transducers on the other hand offer high sensitivity even at low strains and can therefore achieve higher resolution measurements.

This paper shows that resonant force transducers are only suited to a relatively small range of rated loads. Sample calculations are included for a 500N 'double-ended tuning fork' transducer design.

2: TRANSDUCER CHARACTERISTIC

The natural frequency of transverse vibration of an unloaded fixed-fixed beam of constant rectangular cross-section is (see appendix for meaning of terms)

$$f_0 = \frac{a^2}{2\pi l^2}\sqrt{\frac{EI}{\rho A}} \text{ Hz.}$$

If there is external damping on the beam by a surrounding fluid, such as air the natural frequency is altered according to the equation

$$f_{\rho} = \frac{f_0}{\sqrt{1+\frac{B\rho_0 b}{\rho d}}} \text{ Hz.}$$

When a positive axial load, *F*, is applied to the beam the frequency, *f*, increases

$$f = f_{\rho}\sqrt{1+KF} \text{ Hz,}$$

where $K = \frac{cl^2}{EI}$ N^{-1}.

Maximum sensitivity is achieved when the beam vibrates in its fundamental mode. Then a = 4.73004 and c = 0.0245775. The sensitivity of resonant frequency to applied force is given by

$$S_F = \frac{1}{f}\frac{df}{dF} = \frac{1}{2[F+1/K]} \text{ N}^{-1}.$$

3: TEMPERATURE SENSITIVITY

The temperature sensitivity of a vibrating beam is given by the equation

$$S_T = \frac{1}{f}\frac{df}{dT} = \frac{\frac{1}{2}C}{\frac{RT^2}{P}+TC} + \frac{\frac{1}{2}[e+\alpha]}{[1+eT+\alpha T]} - \frac{\frac{1}{2}KF[e+2\alpha]}{1+KF} \ {}^{\circ}\text{C}^{-1}.$$

where $C = \frac{Bb}{\rho d}$ $Kg^{-1}m^3$. The first component is due to the change in surrounding air density with temperature. The other two are due to thermal expansion and the change in the metals elastic modulus with temperature. Unless a material with a very low thermoelastic coefficient (such as Elinvar) is used, the first component is relatively insignificant. Therefore

$$S_T \approx \tfrac{1}{2}[e+\alpha] - \frac{\frac{1}{2}KF[e+2\alpha]}{1+KF} \ {}^{\circ}\text{C}^{-1}.$$

If there is no temperature compensation used the change in output of the transducer with temperature is

$$\frac{dF}{dT} = \frac{S_T}{S_F} = \frac{1+KF}{K}[e+\alpha] - F[e+2\alpha] \ \text{N}^{\circ}\text{C}^{-1}.$$

The maximum error per unit temperature change is then

$$E_{TEMP} = \frac{1}{KF_{RANGE}}[e+\alpha].$$

The frequency shift at rated load , F_{RANGE}, is

$$\Delta f_{RANGE} = \sqrt{1+KF_{RANGE}} - 1.$$

So to minimise E_{TEMP} *K* should be maximised, which also maximises ΔF_{RANGE}.

4: STRAIN SENSITIVITY

It can be shown that the strain sensitivity of a resonant transducer output is approximately given by

$$S_{\varepsilon} = \frac{1}{f}\frac{df}{d\varepsilon} = -2 - \nu + \frac{KF[2+4\nu]}{1+KF}.$$

The maximum strain sensitivity is more than twice that of a conventional deflection-measuring load cell. If operated at similar strain levels, therefore, creep and hysteresis would limit the performance even more than with a conventional device. As mentioned above, however, resonant transducers can operate at considerably lower strain levels. The maximum stress encountered is simply

$$\sigma_{RANGE} = \frac{F_{RANGE}}{A} \text{ Nm}^{-2}.$$

5: LOAD RANGE

It can be shown from the above equations that

$$F_{RANGE} \propto \frac{\Delta f_{RANGE}\, Ebd^3}{l^2}.$$

For a practical transducer Δf_{RANGE}, E, b, and l are relatively fixed so that

$$F_{RANGE} \approx K_1 d^3 \text{ N}.$$

In order to measure low loads the thickness of the vibrating beams must be very small. The lower limit on practical load range is set, therefore, by the ability to manufacture suitably fine resonators.

The stress in the metal beam at rated load is

$$\sigma_{RANGE} = \frac{F_{RANGE}\, KEd^2}{12cl^2} \text{ Nm}^{-2}.$$

Therefore

$$\sigma_{RANGE} \approx K_2 d^2 \text{ Nm}^{-2}$$

or

$$\sigma_{RANGE} \approx K_3 F_{RANGE}^{2/3} \text{ Nm}^{-2}.$$

The maximum load range is therefore set by the strength of the material used. It can be seen that vibrating beam resonant force transducers can be designed for only a relatively narrow range of rated loads.

6: FREQUENCY MEASUREMENT

The period of an electric signal (and hence the frequency) can be measured inexpensively using digital circuitry. Pulses from a high frequency reference clock are counted over the duration of the input signal period. The frequency is then calculated using

$$f_{CALC} = \frac{f_{CLOCK}}{N} \text{ Hz}$$

where N is the count and f_{CLOCK} is the reference frequency. Without the added complexity of measuring fractions of a cycle there is an inherent ±1 error in this count so that the error in calculated frequency is

$$\Delta f_{CALC} = \frac{2}{N} + E_{CLOCK}$$

where E_{CLOCK} is the error in the standard. The error in calculated force due to this error is

$$E_{CALC} = \frac{\Delta f_{RANGE} + 1}{\Delta f_{RANGE}} \cdot \Delta f_{CALC} \cdot$$

If E_{CLOCK} is insignificant the time taken to obtain a given measurement accuracy is therefore

$$T = \frac{2}{\Delta f_{CALC} \cdot f_{CLOCK}} \text{ s.}$$

The reference clock frequency is limited to about 100MHz if readily available and inexpensive electronic components are to be used. The error E_{CLOCK} is about 10 ppm/°C for a quartz crystal based clock. This can be reduced to 0.2 ppm by sufficiently precise temperature compensation or regulation.

7: EXAMPLE

A first step in designing a force transducer resonator is to choose a suitable material with known properties. This example looks at a 17-4 PH Stainless Steel double-ended tuning fork arrangement as shown in fig. 7.1 which allows the mounting and points of application of the forces to be at vibrational nodes. The resonator has two beams so that the load on each is half the input load. In this example the load range is 500N and the required resolution is 100 ppm (i.e. to within 0.5N). The properties of the steel are E = 188GPA, σ_y = 1200MPa, $\alpha = 11\times10^{-6}$/°C, $e = -450\times10^{-6}$/°C.

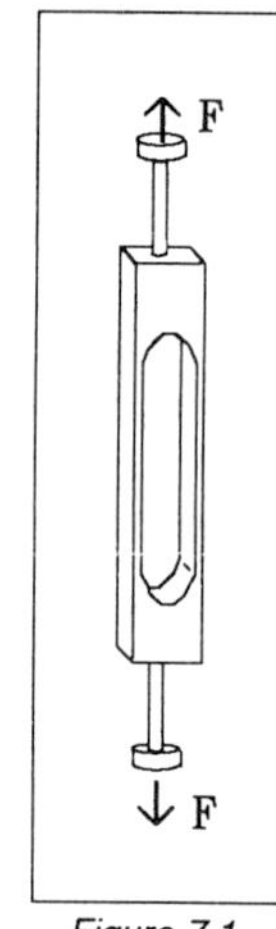

Figure 7.1.

Since the thermoelastic coefficient is high, temperature compensation will be necessary. If the temperature probe used has an accuracy of 0.1°C and the overall temperature error must be less than 50 ppm then $E_{TEMP} = 500\,\text{ppm}/°\text{C} = \frac{1}{KF_{RANGE}}[e + \alpha]$. Since F_{RANGE} is 250N then K must be greater than 3.5×10^{-3}.
Suitable dimensions for the resonator beams are therefore l = 25mm, b = 5mm, d = 0.35mm. These values give $K = 4.6\times10^{-3}$.

The maximum stress in the beams would be $\sigma_{RANGE} = \frac{F_{RANGE}}{A} = 140\,MPa$. This is sufficiently small to insure that the error due to creep and hysteresis would be significantly less than 100 ppm. The resonant frequency of the device would vary from approximately 2790Hz to 4090Hz over the input load range ($\Delta f_{RANGE} = 0.466$).

The error in the calculated force after frequency measurement is then $E_{CALC} = \frac{1.466}{0.466} \cdot \Delta f_{CALC}$, so that the reference clock error must be in the region of 1 ppm/°C.

8: CONCLUSION

In order to attain high resolution measurements with resonant force transducers, the metals used must be very carefully chosen, very precise temperature compensation is required, and highly stable reference frequency sources are necessary. For these reasons the devices are relatively expensive to manufacture and calibrate. Transducers for lower resolution applications could be manufactured far less expensively. They could be made only for load ranges up to about 1KN however.

REFERENCES

[1] T. Ueda, F. Kohsaka, and E. Ogita, "Precision force transducers using mechanical resonators", *Proc. of the 10th Conf. of IMECO TC-3 on Measurement of Force and Mass*, pps. 17-22, 1984.
[2] "Current Advances in Sensors", *IOP Publishing Ltd.*, 1987.
[3] H. Zulliger, "Precise measurement of small forces", *Sensors and Actuators*, vol.4, pps. 483-495, 1983.
[4] W. Weinstein, "Microperformance of Metals", *Machine Design*, Dec. 11, pps. 174-181, 1969.
[5] M. Frerking, "Methods of Temperature Compensation", *36th Annual Frequency Control Symposium*, pps. 564-570, 1983.

APPENDIX: Notation

a, c	constants for a given mode of vibration.
l, b, d	length, width, and thickness of the vibrating beam.
EI	flexural stiffness of the beam.
E	elastic modulus of beam metal.
ρ	density of the beam metal.
A	cross-sectional area of the beam.
ρ_0	density of the surrounding gas.
K	the variable $\frac{cl^2}{EI}$ N^{-1}.
R	gas constant.
T	ambient temperature.
P	surrounding gas pressure.
B	constant for a given resonator geometry (most easily calculated empirically).
e	thermoelastic coefficient of beam metal.
α	coefficient of thermal expansion of beam metal.
ν	Poissons' ratio of the beam metal.

A resistive sensor for the high microwave pulsed power density measurement in a free space

M.Dagys, Ž.Kancleris, R.Simniškis

Semiconductor Physics Institute, Goštauto 11, Vilnius 2600, Lithuania

M.Bäkström, U.Thibblin and B.Wahlgren

Saab-Scania AB, Saab Military Aircraft, Linköping S-581 88, Sweden

ABSTRACT: The resistive sensor together with horn antenna was used to detect high microwave pulsed power density in a free space. Experiments were performed in S (2.75 GHz) and X (9.2 GHz) frequency bands. It was established that microwave pulsed power density in a free space up to a few hundreds of kW per m^2 can be detected by such a unit in a both frequency bands.

1. INTRODUCTION

Since radiolocation and telecommunication systems are continuously growing up in the world high frequency strong electromagnetic field becomes inalienable part of the environment. On the other hand, high power microwave (mw) pulses are proposed to be used as a directed energy weapon to damage the electronic system of flying objects. Thus in order to increase reliability of the electronic equipment used in the aircraft and satellites, testing of this equipment for microwave radiation has to be performed. For this purpose short high power mw pulses are usually used. Therefore, for mw power control the problem of great importance is to develop sensors which will be able to measure high power mw pulses in a free space.

One of the most perspective devices for such type measurements are resistive heads where a resistive sensor made from n-type Si [1] is used as a sensitive element. The main goal of such sensor in comparison with Shotkey or point contact diode, which is also used for mw pulse power measurement, lies in following: (i) possibility to measure high mw pulse power directly in a transmission line without using direct couplers or attenuators, (ii) high value of the output signal is available (up to 10 V), (iii) high reliability and overload resistance.

In the present paper it is shown that high power mw pulses in a free space can be measured making use of the resistive head connected to the horn antenna.

Experiments were carried out in S and X frequency bands.

2. THE PERFORMANCE OF THE RESISTIVE SENSOR

The operation of the resistive sensor is based on electron heating effect in a semiconductor [2]. Thus, it is a bar shaped piece of a semiconductor with ohmic contacts on the ends, which is mounted into the transmission line. The sketch of the sensor placed inside the rectangular waveguide is shown in Figure 1. The resistive sensor is also connected into DC circuit together with a current source. When the electric field of mw pulse heats electrons in the semiconductor their mobility decreases [2], the resistance of the sensor increases and voltage pulse U_s appears in the DC circuit. The resistance change is proportional to the mw electric field strength in the sensor's volume. The latter, in turn, is proportional to the mw power P. Thus, measuring U_s mw pulse power in the transmission line can be determined.

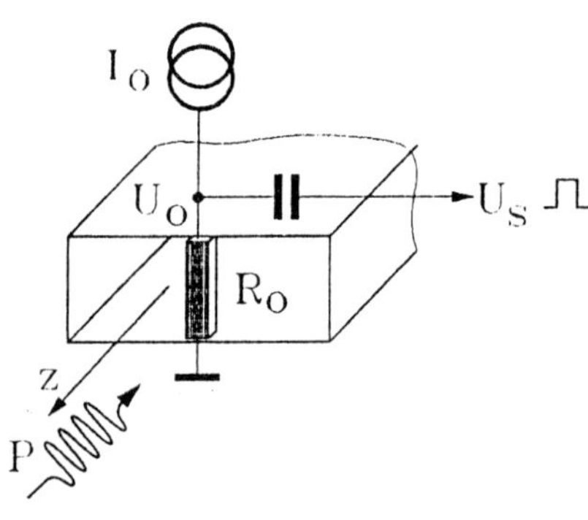

Figure 1. Bar shaped resistive sensor in the rectangular waveguide.

3. EXPERIMENTAL SETUP

To detect mw pulse power density in a free space two resistive power heads being able to detect maximum mw pulse power up to 100 kW and 5 kW in S (2.75 GHz) and X (9.2 GHz) frequency bands were designed and produced. The resistive sensor which length is less than the narrow wall of the waveguide was used in these heads. It was mounted between the special diaphragm and the wide wall of the waveguide. Such construction allows to decrease the voltage standing wave ratio. It was less than 1.1 for both power heads. DC voltage pulse was amplified 5-6 times by properly shielded preamplifier. It was positioned in close proximity to the resistive sensor. Maximum output signal after the preamplifier was 10 V.

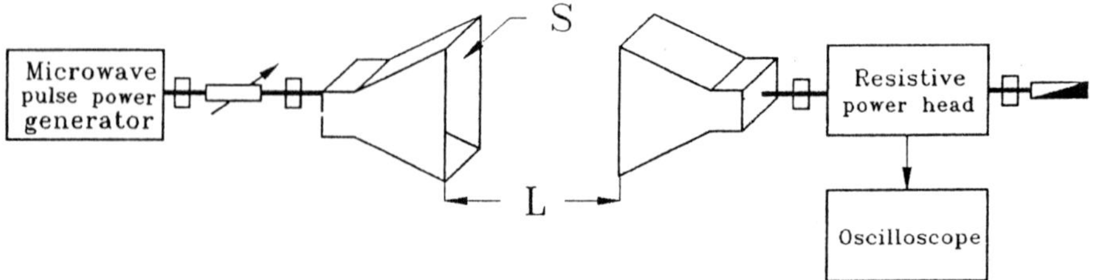

Figure 2. Experimental setup for mw pulse power density measurement in a free space.

Making use of the developed power heads mw pulse power density measurement in a free space has been performed. A block diagram of the experimental setup shown in Figure 2 was used. It is seen that to generate the electromagnetic field in free space a horn antenna is employed. The analogous horn antenna connected to the power head together with a matched load serves as mw pulse power density detector. Mw pulse duration was 3 μs, repetition rate 12.5 Hz and maximum available mw pulse power was 100 kW and 50 kW in S and X frequency band, respectively. The horns with cross-sectional area 0.106 m^2 and 0.0121 m^2 were employed.

4. EXPERIMENTAL RESULTS

If a longitudinal axis of the horn antenna is supposed to be adjusted parallel to mw power flux, mw power P getting into waveguide can be obtained from simple geometrical consideration

$$P = W_p \eta S, \tag{1}$$

where W_p is a mw power flow density (Pounting's vector) in a free space, η is the efficiency and S is a cross-sectional area of the horn antenna. We have measured mw pulse power P getting into waveguide when the receiving horn antenna is placed at some distance L from the transmitting one. Then assuming that the efficiency of the horn antenna $\eta=1$ and making use of expression (1) the mw power density in a free space has been determined.

Experimental results for S and X bands are shown in Figures 3 and 4, respectively. It is seen that at a distance $L<10\lambda$, where λ is a wave length in a free space, the

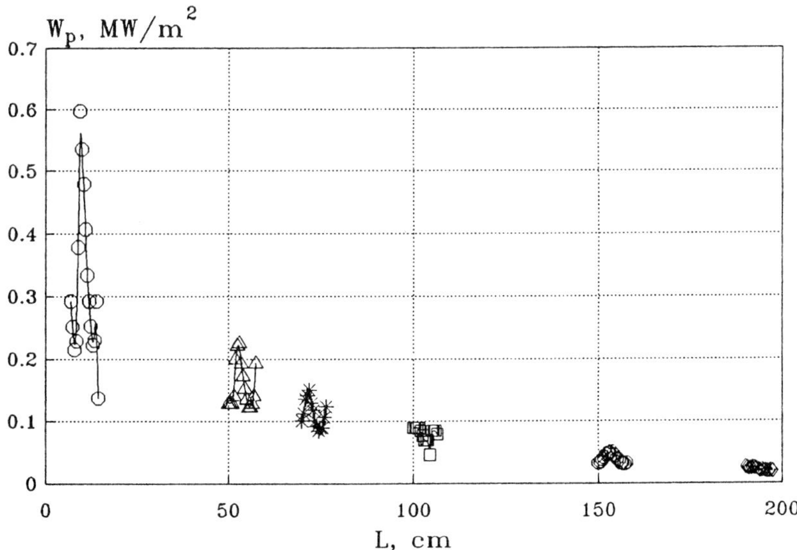

Figure 3. Mw pulse power density dependence on the distance between horns in S band at a frequency 2.75 GHz, $\lambda=10.9$ cm.

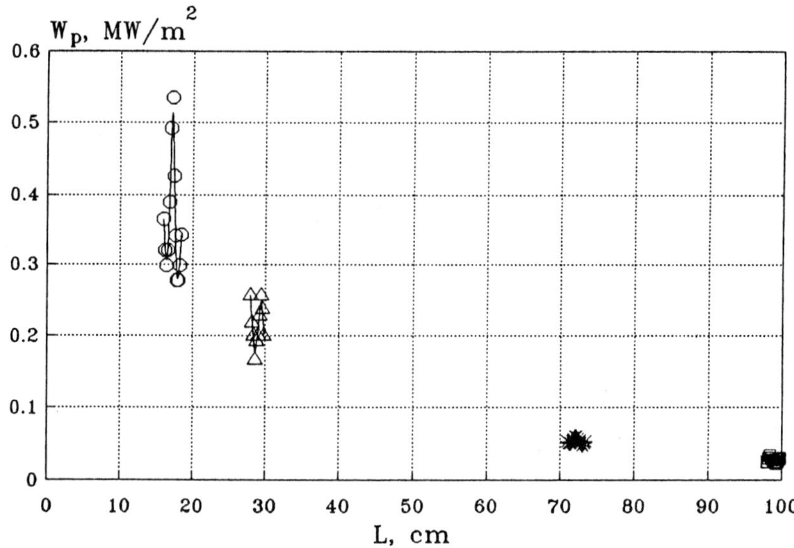

Figure 4. Mw pulse power density dependence on the distance between horns in X band at a frequency 9.2 GHz, λ=3.26 cm.

interaction between horns occurs and the standing wave is observed. It also means that efficiency of horn antenna is lower than unity and this fact has to be taken into account if precise measurement of W_p is attempted. As one could expect, at a greater distance pulse power density decreases as L^2. Switching off DC supply of the resistive sensor the value of the parasitic signal induced by the external electromagnetic field in the electronic circuit can be determined. We got that in the worst case the parasitic signal ranges less than 7% of the useful one.

5. CONCLUSION

It was shown that high level mw pulse power density can be measured in a free space with the help of the resistive power head connected to the horn antenna. Such a unit was able to detect pulse power density up to a few hundreds of kW per square meter. The value of mw electric field detected by our unit is approximately three times grater than the peak electric field value proposed by JAA for certification of a high intensity radiated field environment in a both frequency bands.

REFERENCES

1. Baltušis R., Dagys M., Simniškis R. Proceedings of 22nd European Microwave Conference, p.169, 1992.
2. Conwell E.M. High field transport in semiconductors. New York and London: Academic Press, 1967.

Magneto resistive (MR) thin film sensor

O.R.Khan, Member, IEEE
Department of Physics, Kenyatta University, P.O.Box 43844, Nairobi, Kenya

ABSTRACT: Magneto resistive (MR) Thin film sensors, can detect magnetic fields in a wide range of sharp resolution and high density applications. These devices are suitable for an integrated approach to fabricate magnetic recording replay thin film Heads, which are fabricated mainly with the applications of sputtering, Photo Masking, Chemical etching, wire bonding and encapsulating techniques in a clean room environments.
These techniques have a considerable impact on production and improving performance, miniaturization and cost.
This paper highlights the design optimisation methods and fabrication techniques of MR thin film read / write Heads.

1. INTRODUCTION

The seed of information technology (IT) revolution was sown in the Bell laboratories in 1947 with the invention of transistors by Brattain, Baidin and Shocklay.That has now acquired a very strong root of a tree in form of very large scale integration (VLSI) technology and its application to fabricate micro electronics devices.

Information storage engineering (ISE) is an offshoot of IT revolution and the fabrication of thin film magneto resistive (MR) read/write heads is a very significant application of ISE. These sensors can detect and measure magnetic flux or information in a range of high density with sharp resolution. Miniaturisation has been a very important virtue of ISE and IC. Therefore the ideal sensor would be cheap , efficient and reliable in addition to offer a simple interface to its functional and controlling electronics.

This paper highlights the technical details of designing methods and fabrication techniques of MR sensor within a magnetic recording and replay thin film Heads.

Iwasaki [1] has realised the full potential of perpendicular recording and replay process by employing a single pole head with the two layer medium. This application improved the high density and sharp transition resolution in magnetic recording and replay process. MR heads are suitable primarily for discs memory applications as they have no media contact problems.
There have been many improvements and modifications to the very basic design of a MR head [2]. The problems of linearisations and resolutions were tackled in the most significant designs of Phillips [3] and IBM [4] and [5].

The use of MR material, its geometry, general theory have been described to exploit the recording and replay characteristics of such devices in the paper [6].
Feasibility aspects of recording and replay theory and performance of thin film MR head were presented earlier [7] in an investigation of thin film perpendicular magnetic recording / replay devices.

In this paper the design features and fabrication steps of MR read / write Heads are described, using computer aided design (CAD) methods and employing GAELIC language format supported by Scientific and Engineering Research council (SERC) [8] facility. The Photo masking fabrication techniques allow for very small size and cheap production of these devices. The use of CAD and the application of electron beam lithography facility (EBLF) provided by SERC made it possible for extreme miniaturisation and economically viable, despite of the sophistication and diversity of the production process.

2. DESIGN

2.1 Design features using CAD

The fabrication masks of the thin film MR heads are designed using the SERC based GAELIC language format [8]. Each thin film shape is represented by a language statement indicating the type of shape, that is, Rectangle, Polygon, Track or Text with the origin and the mask number. A group or the sum of groups of shapes can be produced at any specified location in a variety of possible orientations.
Language statements can also be written for the text on the masks in a required scaling factor. A language file can thus be developed specifying the masks for an integrated design to make a micro structure of an electronic or a magnetic circuit.

With a language file of any design, the user then has the option, either of describing these shapes in GAELIC language compiler or entering the shapes graphically for conversion into GAELIC, at a Textronics or Sigmex terminal using interacting editor. The edited files can be processed to plot the design for the whole device or for a component of the device at the Rutherford Appleton Laboratory by sending a print request along with the file on the network "JNET".

The plots files can further be edited for any small or drastic modification up to a satisfactory design incase the prints show any design discrepancy but certainly before ordering the masks.

2.2 Fabrication Masks for MR heads

The final compiled file is processed to plot the shapes on either large scale using an interactive plotter or in micro shapes on the masks using EBLF facility [8] at the Rutherford Laboratories as part of Interactive computing facility.

Such a GAELIC suite of programs (MRH1) [9] were used to define various shapes in steps using nine masks. These devices have three different MR thin film designs to achieve maximum economy during the fabrication.These are shown in figure 1. Some of the parameters of the material of the sensor used, are reported earlier [9] and [10].

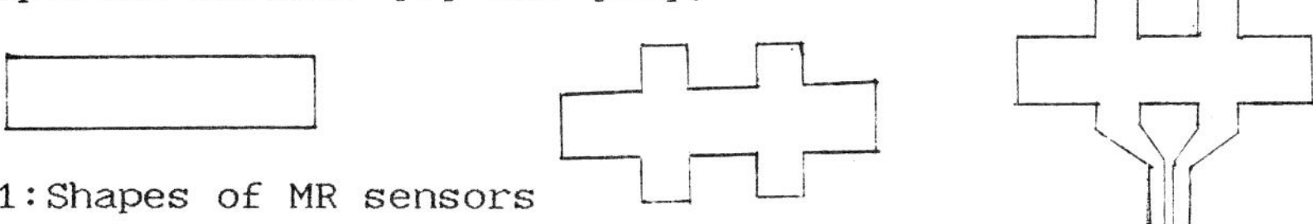

Fig. 1:Shapes of MR sensors

A magnified copy of one of the devices on the substrate using all the masks as a whole is shown in figure 2 and a diagram of the design of the coil winding and the pole is shown in figure 3. It is planned to have 416 devices on the substrate with three different shapes of sensors. Each device has two heads with its own mark number.

1. Upper flux guide
2. Device Numbers
3. Alignments Marks
4. Connection Boxes

Fig. 2:Thin film MR head

1. Flux guide hole
2. Sensor (Fin Type)
3. Sacrificial
4. Recording coil
5. Lower Flux guide
6. Thin Film Pole
7. MR Connections
8. Coil Pad

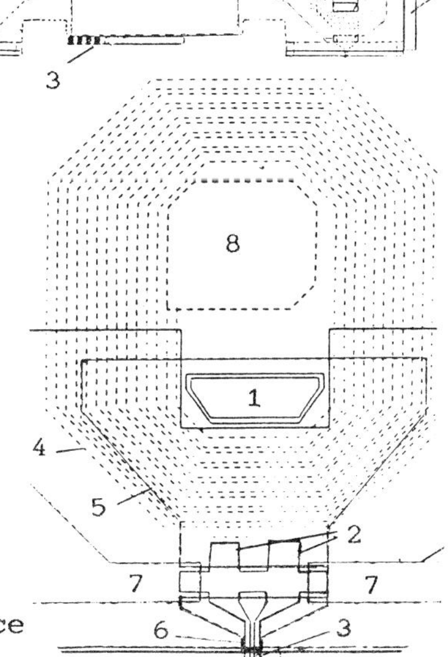

Fig. 3:Coil winding and MR device

The overlapping features is strongly discouraged in any design description, as this will have a deleterious effect on line width which could lead to the failure of the device.
Three standard industrial mask plate sizes are available at the EBLF, in addition to two standard silicon wafer substrate sizes for non mask making applications.
A plate size of 100 x 100 mm(4"x 4") with a thickness of 1.5 mm (0.06") is used of standard low expansion Borosilicate glass.Positive acting electron beam resists are used as the positive process.This produces lithography with better resolution and edge acuity than the negative process.

3. FABRICATION

3.1 Fabrication of thin film heads.

Miniaturisation has been an abiding virtue of IC, therefore its production techniques applied in the fabrications, that is, the connections overlap the sensors edges and gold layers have been designed for all connections.
Layers of thin films of magnetic and electronics conductive materials are deposited usually in high vacuum on a suitable substrate. Microcomposition and thickness of metallic films are important for good performance and durability in information storage and retrieval technology. By maintaining high vacuum and controlling critical parameters, higher quality films are deposited. All depositions are planned to be done using the Nordico sputtering equipment.

For the fabrication of the devices the masks are to be used in the steps sequence during Photo masking process at an industry-standard clean room at the department of Electrical and Electronic Engineering, South West University, Devon.

3.2 Steps in fabrication Scheme.

3.2.1 First Pump down

A) Deposition of nickel Iron (Ni-Fe) film of 0.5 µm thickness.
B) Use of mask 1 to achieve the upper flux guide within the device as shown shaded in the figure 4.

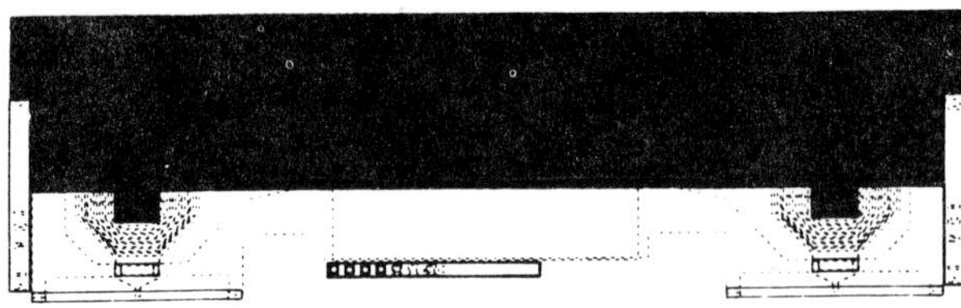

Fig. 4

3.2.2 Second Pump down

A) Deposition of SiO_2 (Insulation) layer of 0.3 µm thickness.
B) Removal of SiO_2 layer for the areas of flux guide holes, sensors and sacrificial using mask 2, and mask 3 for device number and alignment marker boxes as shown shaded in figure 5.

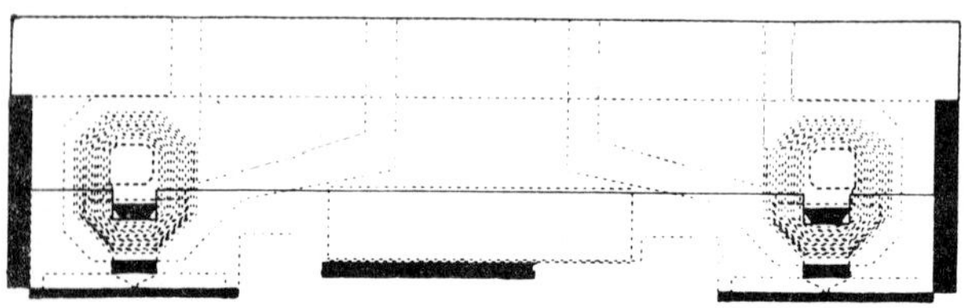

Fig. 5

3.2.3 Third pump down

A) Deposition of Ni-Fe film for sensors, Flux guide holes and sacrificial of 50 nm thickness on to the substrate.
B) Use of mask four to retain Ni-Fe film for the shapes (shaded) for flux guide hole sensors and sacrificial of 50 nm thick, as shown in figure 6.

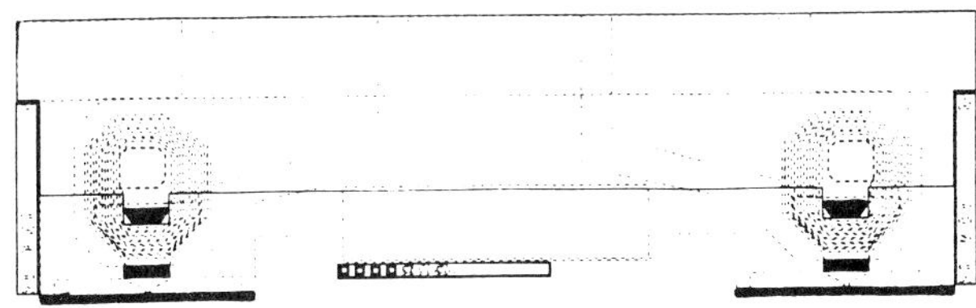

Fig. 6

3.2.4 Fourth Pump down

A) Deposition of SiO2 layer of 0.3 µm thickness to protect sensors and sacrificial.
B) Removal of SiO2 layer from the areas for flux guide holes boxes for device Number and alignment Markers using mask 5.
C) Use of mask 6 also to remove SiO2 layers from the areas for connections of Sensors and sacrificial. All shaded areas as shown in figure 7 can be achieved using steps B and C.

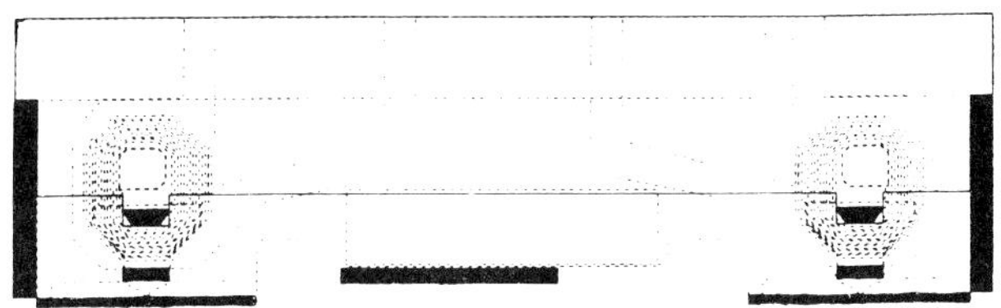

Fig. 7

3.2.5 Fifth Pump down.

A) Deposition of gold of 1 um thickness for coil pads, coils their connections, all linked connections for Sensors and sacrificial.
B) Use of mask 7 for all the shapes of coil pads, coils, their connections, all connections and boxes for mask Number and Alignment markers.
C) Use of mask 5 (repeat) to remove gold layer from the areas for flux guide holes, boxes for device number and Alignment Markers. Shaded areas in figure 8 show the results of steps B and C.

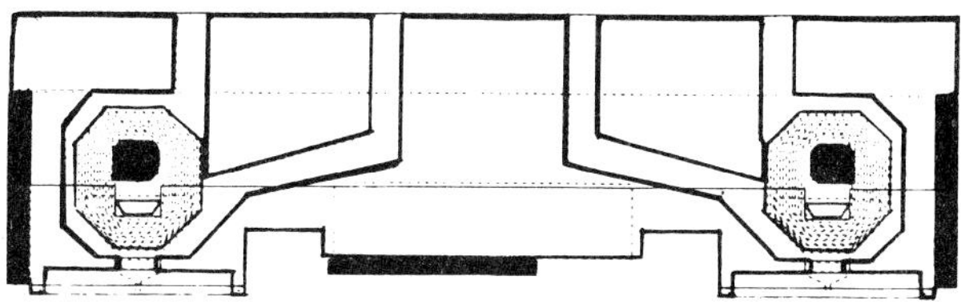

Fig. 8

3.2.6 Sixth Pump down

A) Deposition of SiO2 layer of 0.3 μm thickness to protect coils and parts of some connections.
B) Use of mask 8 to remove SiO2 layer from the areas of coil pads and connection boxes.
C) Use of mask 5 (Repeat 2nd time) to remove SiO2 layer from the areas for flux guide hole and boxes for device number and alignment markers. Steps B and C provide the shapes shown in figure 9.

Fig. 9

3.2.7 Seventh Pump down.

A) Deposition of Ni-Fe film of 0.5 μm thickness for the areas for lower flux guide, pole and device numbers.
B) Use of mask 9 to achieve the shapes of flux guide (Lower), pole and device numbers as shown in shaded areas in figure 10.

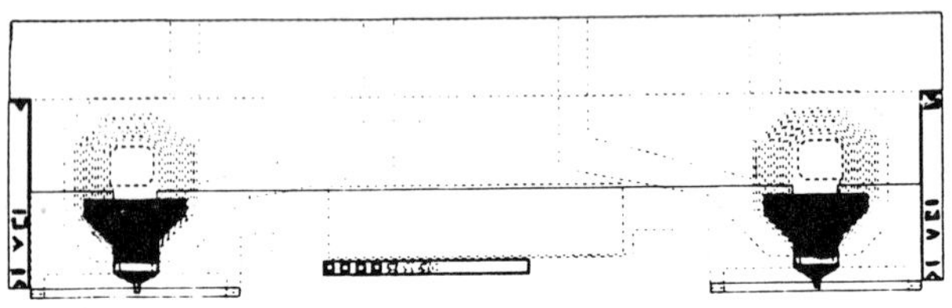

Fig. 10

3.2.8 Eighth Pump down.

A) Deposition of SiO2 layer of 0.3 μm thickness to protect the lower flux guide and pole.
B) Use of mask 8 to remove the SiO2 layer for the areas of the coil pads and connection boxes.
C) Use of mask 3 (repeat) insures the removal of SiO2 from boxes for device number and alignment marks to show the numbers and markers.All shaded areas in figure 11 would be attainable as a result of steps B and C.

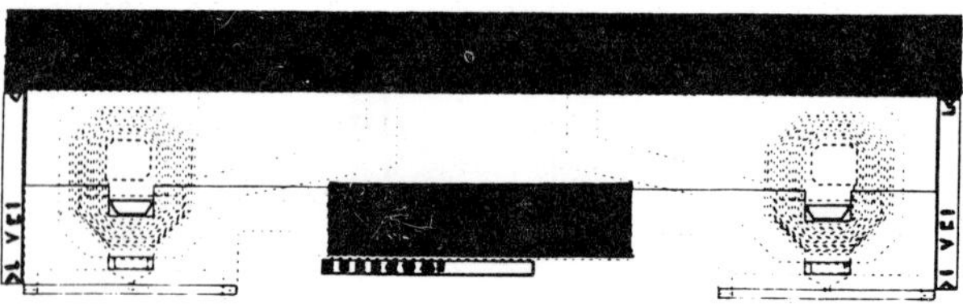

Fig. 11

4. RESULTS AND CONCLUSION

Some of the essential technology of ISE is described with a contribution to use CAD methods and computer interactive EBLF application for the fabrication of MR thin film read/write Heads. A feasibility scheme to fabricate these devices is written to realise the logical steps during the production.

5. ACKNOWLEDGMENT

The author would like to thanks the colleagues at Kenyatta University,Nairobi for the moral support. Financial assistance was provided by SERC, UK,during the project period.

6. REFERENCES

[1] Iwasaki s. 1981," Wave length response of perpendicular Magnetic recording",IEEE Trans.on Magnetics,17(6), p 2535.

[2] Hunt R.P , 1971, " A magneto resistive readout transducer" IEEE Transactions in Magnetics Vol. Mag. 7 (1) p.150.

[3] Kuizk et al., 1975," The Barberpole, a linear magnetoresistive head." IEEE Transactions in Magnetics Vol. Mag. 11 (5).

[4] Thompson D.A., 1974," Magneto resistive transducers in high density magnetic recording". A.I.P. conf. Proc. Magnetism and Magnetic Materials No. 20, pp 528 533.

[5] Bajorek et al., 1974, 2 An integrated magneto resistive read, inductive write high density recording head" A.I.P Conf. Proc. Magnetism and Magnetic Materials No. 20, pp 548-549.

[6] Jeffers F. and Karsh H.,1984,"Unshielded Magneto resistive Heads in High density recording", IEEE Trans. on Magnetics, Vol. Mag. 20(5), pp703-708.

[7] Khan O.R., 1990,"Thin film Magnetic read/Write devices" , Eighth International conf. IEE 319,p 34, 1990.

[8] Notes "Form EBLF/1 Rutherford Appleton Lab.,18 Feb. 1981.

[9] Khan O.R., 1988,"Thin film MR read/write devices for computer memory application Technology", Appendix 7, SERC Report Dept.,Elect. & Elec. Engineering,Plymouth Polytechnic, Devon, UK.

[10]Mapps D.J., et al., " A double Bifilar Magneto resistor for earth's field detection"BH01", Digests of the Intermag. Con.,Tokyo, Japan, 1987.

Author Index